AF355987

Analysis of Dynamics of Railway Vehicles

Analysis of Dynamics of Railway Vehicles

Collection Editor

Suchao Xie

Basel • Beijing • Wuhan • Barcelona • Belgrade • Novi Sad • Cluj • Manchester

Collection Editor
Suchao Xie
Central South University
Changsha
China

Editorial Office
MDPI AG
Grosspeteranlage 5
4052 Basel, Switzerland

This is a reprint of the Topical Collection, published open access by the journal *Applied Sciences* (ISSN 2076-3417), freely accessible at: https://www.mdpi.com/journal/applsci/topical_collections/ Dynamics_Railway_Vehicles.

For citation purposes, cite each article independently as indicated on the article page online and as indicated below:

Lastname, A.A.; Lastname, B.B. Article Title. *Journal Name* **Year**, *Volume Number*, Page Range.

ISBN 978-3-7258-2829-6 (Hbk)
ISBN 978-3-7258-2830-2 (PDF)
https://doi.org/10.3390/books978-3-7258-2830-2

Contents

About the Editor

Suchao Xie

Suchao Xie is a Full Professor and Head of the Equipment Department at the School of Transportation Engineering, Central South University. He has been selected as an Outstanding Young Scholar in Hunan Province and a Leading Talent in Scientific and Technological Innovation in Hunan Province. He was also chosen as one of the top 2% scientists globally for the year 2023. He serves as an Editorial Board Member and Guest Editor for *Applied Sciences*, an Editorial Board Member for *Innovation Discovery*, and a Young Editorial Board Member for journals such as the *Journal of Central South University* and the *Journal of Railway Science and Engineering*. His research primarily focuses on high-speed railway operation safety, vehicle structure analysis and optimization, and acoustic metamaterials. He has published over 120 academic papers in reputable domestic and international journals and has applied for 57 public invention patents, of which 36 have been granted. As the primary national internal auditor, he has successfully included 15 testing projects in the China National Accreditation Service for Conformity Assessment (CNAS) and 23 testing projects in the China Metrology Accreditation (CMA).

Preface

The dynamics of railway vehicles play a crucial role in ensuring the safety, stability, and efficiency of rail transportation systems. As the demand for faster, more reliable, and environmentally friendly rail travel continues to grow, it is imperative to develop a deep understanding of the complex interactions between vehicles, tracks, and their surrounding environment. This reprint, *Analysis of Dynamics of Railway Vehicles*, brings together cutting-edge research and practical insights from leading experts in the field to provide a comprehensive overview of the latest methodologies and applications in railway vehicle dynamics.

The reprint covers a wide range of topics, including the modeling and simulation of vehicle–track interactions, stability analysis, noise and vibration control, and the optimization of suspension systems. It explores the use of advanced computational techniques, such as multibody dynamics, finite element analysis, and machine learning, to tackle the challenges associated with the design and operation of railway vehicles. The authors also address the importance of experimental validation and the integration of measurement data with numerical models to enhance the accuracy and reliability of the analysis.

By bringing together the latest research findings and industry best practices, *Analysis of Dynamics of Railway Vehicles* serves as an invaluable resource for researchers, engineers, and decision-makers in the railway sector. It provides a solid foundation for the development of innovative solutions that can enhance the safety, reliability, and sustainability of rail transportation, thereby contributing to the growth and competitiveness of the industry in the face of evolving global challenges.

Suchao Xie
Collection Editor

 *applied sciences*

Editorial

Special Issue on Dynamics of Railway Vehicles

Suchao Xie [1,2,3]

1 Key Laboratory of Traffic Safety on Track, Ministry of Education, School of Traffic & Transportation Engineering, Central South University, Changsha 410075, China; xsc0407@csu.edu.cn
2 Joint International Research Laboratory of Key Technology for Rail Traffic Safety, Changsha 410075, China
3 National & Local Joint Engineering Research Center of Safety Technology for Rail Vehicle, Changsha 410075, China

1. Introduction

High-speed Railway Vehicle systems have become integral to modern transportation infrastructure, offering a rapid, efficient, and environmentally friendly travel option. The development of High-speed Railway Vehicles involves a complex interplay of engineering disciplines, including safety engineering, computational algorithms, noise control technologies, material science, and passenger comfort optimization. Continuous innovation in these areas is essential to meet the increasing demands for speed, reliability, and sustainability. Enhancing the design and performance of high-speed trains not only improves the passenger experience, but also contributes significantly to economic growth and environmental conservation.

2. New Theories and Technological Progress

Safety is paramount in high-speed rail operations. The integration of advanced algorithms and real-time monitoring systems has significantly improved safety assessments and risk mitigation strategies. Cai et al. [1] investigated the influence of wheel–rail contact algorithms on the running safety assessment of trains under earthquake conditions. Their study emphasizes the critical role of accurate modeling in predicting train behavior during seismic events, which is essential for ensuring passenger safety and infrastructure integrity. Advanced monitoring technologies, such as fiber optic sensing and Internet of Things (IoT) devices, are being employed for the real-time structural health monitoring of railway infrastructure [2]. These systems enable the early detection of potential faults, allowing for timely maintenance and preventing accidents. Moreover, machine learning algorithms are being utilized to predict and prevent derailments. Li et al. [3] developed a predictive model using machine learning techniques to assess derailment risks based on various operational parameters.

Understanding and controlling the dynamic behavior of high-speed trains is crucial for passenger safety and comfort. Yu et al. [4] developed a flow-induced vibration hybrid modeling method to analyze the dynamic characteristics of a U-section rubber outer windshield system. This research aids in predicting and mitigating vibration issues caused by aerodynamic forces, enhancing structural integrity and passenger comfort. Additionally, computational algorithms are being used to optimize suspension systems. Chen et al. [5] applied genetic algorithms to optimize the suspension parameters of high-speed trains, resulting in improved stability and ride comfort.

Noise pollution is a significant challenge in high-speed rail operations, affecting both environmental compatibility and passenger comfort. Yan et al. [6] conducted a comprehensive review of recent research into the causes and control of noise during high-speed train movement. Strategies such as aerodynamic optimization, sound-absorbing materials, and noise barriers are critical in mitigating noise levels. Innovative materials like acoustic metamaterials are being explored for their exceptional sound absorption

Citation: Xie, S. Special Issue on Dynamics of Railway Vehicles. *Appl. Sci.* **2024**, *14*, 11062. https://doi.org/10.3390/app142311062

Received: 15 November 2024
Accepted: 26 November 2024
Published: 28 November 2024

properties [7]. These materials can be integrated into train components and infrastructure to reduce noise transmission effectively. Furthermore, advancements in wheel and rail design have contributed to noise reduction. Thompson et al. [8] studied the impact of wheel and rail roughness on rolling noise and proposed design modifications to minimize noise generation.

Enhancing passenger comfort is a key objective in high-speed rail development. Bao et al. [9] introduced a mobile device-based train ride comfort-measuring system, enabling the real-time assessment of ride quality and allowing operators to make adjustments to improve the passenger experience. Advanced HVAC systems are being designed for optimal thermal comfort while minimizing energy consumption [10]. Moreover, ergonomic seat designs and cabin layouts are being developed to improve passenger well-being during long journeys [11].

Efficient maintenance strategies are crucial for the reliability and safety of HSR systems. Predictive maintenance, powered by AI and big data analytics, allows for the anticipation of equipment failures before they occur, significantly reducing downtime and maintenance costs [12–14]. Chen et al. [15] developed a data-driven fault diagnosis system for high-speed trains using deep learning techniques. Their model can detect anomalies in real time, enhancing the safety and reliability of train operations.

Emerging technologies such as artificial intelligence [16,17], IoT [18–20], and quantum computing are poised to revolutionize high-speed rail systems. AI algorithms are being developed for autonomous train operations, enhancing trains' safety and efficiency. Hyperloop technology will require newer super materials to meet the demands of this new technology, promising unprecedented speeds and energy efficiency [21,22]. Moreover, the development of superconducting materials may enable the creation of the next generation of maglev trains, offering frictionless travel at ultra-high speeds [23].

3. Conclusions

The integration of safety engineering, advanced algorithms, noise reduction technologies, energy efficiency measures, and passenger comfort enhancements is driving significant advancements in high-speed rail systems. The studies discussed herein not only improve the operational efficiency and safety of high-speed trains, but also enhance their environmental sustainability and the passenger experience. Future research and development will continue to leverage emerging technologies to further advance high-speed rail systems, addressing challenges and meeting the growing demands of modern transportation.

Funding: This research was undertaken at Key Laboratory of Traffic Safety on Track (Central South University), Ministry of Education, China. The authors gratefully acknowledge the support from the Key Project of Scientific Research of the Hunan Provincial Department of Education (Grant No. 23A0017). This paper was also supported by the National Natural Science Foundation of China (Grant No. 52202455) and the Science and Technology Innovation Program of Hunan Province (Project No. 2024RC1019).

Conflicts of Interest: The authors declare no conflict of interest.

References

1. Cai, G.; Zhu, Z.; Gong, W.; Zhou, G.; Jiang, L.; Ye, B. Influence of Wheel-Rail Contact Algorithms on Running Safety Assessment of Trains under Earthquakes. *Appl. Sci.* **2023**, *13*, 5230. [CrossRef]
2. Adeagbo, M.O.; Wang, S.M.; Ni, Y.Q. Revamping Structural Health Monitoring of Advanced Rail Transit Systems: A Paradigmatic Shift from Digital Shadows to Digital Twins. *Adv. Eng. Inf.* **2024**, *61*, 102450. [CrossRef]
3. Li, Y.; Ding, Y.; Zhao, H.; Sun, Z. Data-Driven Structural Condition Assessment for High-Speed Railway Bridges Using Multi-Band FIR Filtering and Clustering. *Structures* **2022**, *41*, 1546–1558. [CrossRef]
4. Yu, Y.; Lv, P.; Liu, X.; Liu, X. Flow-Induced Vibration Hybrid Modeling Method and Dynamic Characteristics of U-Section Rubber Outer Windshield System of High-Speed Trains. *Appl. Sci.* **2023**, *13*, 5813. [CrossRef]
5. Chen, X.; Yao, Y.; Shen, L.; Zhang, X. Multi-Objective Optimization of High-Speed Train Suspension Parameters for Improving Hunting Stability. *Int. J. Rail Transp.* **2022**, *10*, 159–176. [CrossRef]

6. Yan, H.; Xie, S.; Jing, K.; Feng, Z. A Review of Recent Research into the Causes and Control of Noise during High-Speed Train Movement. *Appl. Sci.* **2022**, *12*, 7508. [CrossRef]
7. Yan, H.; Xie, S.; Zhang, F.; Jing, K.; He, L. Semi-Self-Similar Fractal Cellular Structures with Broadband Sound Absorption. *Appl. Acoust.* **2024**, *217*, 109864. [CrossRef]
8. Thompson, D.J. Wheel–Rail Interaction Noise Prediction and Its Control. In *Handbook of Noise and Vibration Control*; John Wiley & Sons, Ltd.: Hoboken, NJ, USA, 2007; pp. 1138–1146. ISBN 978-0-470-20970-7.
9. Yuxue, B.; Bingchen, G.; Jianjie, C.; Wenzhe, C.; Hang, Z.; Chen, C. Sitting Comfort Analysis and Prediction for High-Speed Rail Passengers Based on Statistical Analysis and Machine Learning. *Build. Environ.* **2022**, *225*, 109589. [CrossRef]
10. Yuan, Y.; Li, S.; Yang, L.; Gao, Z. Real-Time Optimization of Train Regulation and Passenger Flow Control for Urban Rail Transit Network under Frequent Disturbances. *Transp. Res. Part E Logist. Transp. Rev.* **2022**, *168*, 102942. [CrossRef]
11. Jung, E.S.; Han, S.H.; Jung, M.; Choe, J. Coach Design for the Korean High-Speed Train: A Systematic Approach to Passenger Seat Design and Layout. *Appl. Ergon.* **1998**, *29*, 507–519. [CrossRef]
12. Bustos, A.; Rubio, H.; Soriano-Heras, E.; Castejon, C. Methodology for the Integration of a High-Speed Train in Maintenance 4.0. *J. Comput. Des. Eng.* **2021**, *8*, 1605–1621. [CrossRef]
13. Nagy, R.; Horvát, F.; Fischer, S. Innovative Approaches in Railway Management: Leveraging Big Data and Artificial Intelligence for Predictive Maintenance of Track Geometry. *Teh. Vjesn.* **2024**, *31*, 1245–1259. [CrossRef]
14. Tang, R.; De Donato, L.; Besinović, N.; Flammini, F.; Goverde, R.M.; Lin, Z.; Liu, R.; Tang, T.; Vittorini, V.; Wang, Z. A Literature Review of Artificial Intelligence Applications in Railway Systems. *Transp. Res. Part C Emerg. Technol.* **2022**, *140*, 103679. [CrossRef]
15. Chen, H.; Jiang, B.; Ding, S.X.; Huang, B. Data-Driven Fault Diagnosis for Traction Systems in High-Speed Trains: A Survey, Challenges, and Perspectives. *IEEE Trans. Intell. Transp. Syst.* **2022**, *23*, 1700–1716. [CrossRef]
16. Ramos, A.; Castanheira-Pinto, A.; Colaço, A.; Fernández-Ruiz, J.; Alves Costa, P. Predicting Critical Speed of Railway Tracks Using Artificial Intelligence Algorithms. *Vibration* **2023**, *6*, 895–916. [CrossRef]
17. Yin, J.; Zhao, W. Fault Diagnosis Network Design for Vehicle On-Board Equipments of High-Speed Railway: A Deep Learning Approach. *Eng. Appl. Artif. Intell.* **2016**, *56*, 250–259. [CrossRef]
18. Fraga-Lamas, P.; Fernández-Caramés, T.M.; Castedo, L. Towards the Internet of Smart Trains: A Review on Industrial IoT-Connected Railways. *Sensors* **2017**, *17*, 1457. [CrossRef]
19. Zhong, G.; Xiong, K.; Zhong, Z.; Ai, B. Internet of Things for High-Speed Railways. *Intell. Converg. Netw.* **2021**, *2*, 115–132. [CrossRef]
20. Ai, B.; Cheng, X.; Kürner, T.; Zhong, Z.-D.; Guan, K.; He, R.-S.; Xiong, L.; Matolak, D.W.; Michelson, D.G.; Briso-Rodriguez, C. Challenges Toward Wireless Communications for High-Speed Railway. *IEEE Trans. Intell. Transp. Syst.* **2014**, *15*, 2143–2158. [CrossRef]
21. Belingardi, G.; Cavatorta, M.P.; Duella, R. Material Characterization of a Composite–Foam Sandwich for the Front Structure of a High Speed Train. *Compos. Struct.* **2003**, *61*, 13–25. [CrossRef]
22. Li, Z.; Li, J.; Li, C.; Xie, X.; Yang, Z. Study on the Mechanism of Mechanical Properties Deterioration of Brake Disc Materials of High-Speed Train. *Eng. Fail. Anal.* **2024**, *156*, 107816. [CrossRef]
23. Li, X.; Zhang, J.; Huang, K.; Song, X.; Fang, J. Electromagnetic Design of High-Temperature Superconducting Traction Transformer for High-Speed Railway Train. *IEEE Trans. Appl. Supercond.* **2019**, *29*, 5501905. [CrossRef]

Review

A Review of Fault Diagnosis Methods for Key Systems of the High-Speed Train

Suchao Xie [1,2,3,*], Hongchuang Tan [1,2,3], Chengxing Yang [1,2,3] and Hongyu Yan [1,2,3]

1   Key Laboratory of Traffic Safety on Track, Central South University, Changsha 410075, China; 204201035@csu.edu.cn (H.T.); chengxing_yang_hn@csu.edu.cn (C.Y.); 204201034@csu.edu.cn (H.Y.)
2   School of Traffic & Transportation Engineering, Central South University, Changsha 410075, China
3   Joint International Research Laboratory of Key Technology for Rail Traffic Safety, Changsha 410075, China
*   Correspondence: xsc0407@csu.edu.cn

Abstract: High-speed train is a large-scale electromechanical coupling equipment with a complex structure, where the coupling is interlaced between various system components, and the excitation sources are complex and diverse. Therefore, reliability has become the top priority for the safe operation of high-speed trains. As the operating mileage of high-speed trains increases, various key systems experience various degrees of performance degradation and damage failures. Moreover, it is accompanied by the influence of external environmental high interference noise and weak early fault information. Thus, those factors are serious challenges for the condition monitoring and fault diagnosis of high-speed trains. Therefore, this paper summarizes the research progress and theoretical results of the fault detection, fault isolation, and fault diagnosis methods of the key systems of high-speed trains. Finally, the paper summarizes the applicability of the main methods, discusses the challenges and opportunities of condition monitoring and fault diagnosis of high-speed trains, and looks forward to improving its diagnosis level.

Keywords: high-speed train; key systems; fault diagnosis; feature extraction; pattern recognition

1. Introduction

With the advantages of high speed, large carrying capacity, safety, and all-weather operation, high-speed trains have become one of the most popular modern means of transportation. With the development of high-speed trains toward high speed and electrification, as well as the long operating mileage, higher requirements have been proposed for their safety and stability. Due to the long-running in various bad environments and with uncertain factors, some system components of the high-speed train will inevitably break down, even suffering serious failure [1]. If these faults are not discovered and handled in time, they eventually lead to disastrous consequences. A series of train accidents in recent years show that effective condition monitoring and fault diagnosis technology can provide a reliable solution to prevent or reduce the occurrence of similar incidents in the future. Therefore, it is very urgent and necessary to study the condition monitoring and fault diagnosis of high-speed trains to improve the safety of the rail transit system [2,3].

In recent years, good research results have been achieved in fault condition monitoring and diagnosis technology for mechanical systems [4]. Therefore, drawing on existing technologies can provide certain theoretical support for condition monitoring and fault diagnosis of high-speed trains, which play an important role in maintaining the safety of railroad transportation systems. However, there are still some difficulties in fault diagnosis of high-speed trains. Firstly, the diverse service environment of high-speed trains, with large speed fluctuations and unclear shock loads, leads to signals with strong nonlinearity [5,6]. Secondly, the early fault information of each key system is weak, and the fault signal is easily disturbed by the background noise, leading to greater difficulty in early fault feature extraction. In addition, the relationship between the system components

Citation: Xie, S.; Tan, H.; Yang, C.; Yan, H. A Review of Fault Diagnosis Methods for Key Systems of the High-Speed Train. *Appl. Sci.* **2023**, *13*, 4790. https://doi.org/10.3390/app13084790

Academic Editor: Ki-Yong Oh

Received: 13 March 2023
Revised: 30 March 2023
Accepted: 7 April 2023
Published: 11 April 2023

is extremely complex. Hence, the collected signals are not only independent but also have a certain correlation and often contain more interference components. To this end, this thesis summarizes the fault diagnosis techniques of key systems (bogie system, traction/braking system, and electrical/information control system) of high-speed trains in recent years, as shown in Figure 1. From Figure 1, signal processing, machine learning, and deep learning techniques are effective methods to achieve condition monitoring and fault diagnosis of each key system. Finally, the development of high-speed train fault diagnosis technology is forecasted. The main contributions of this study are as follows:

(1) The fault diagnosis methods of the bogie system of the high-speed train are presented.
(2) The fault diagnosis methods of the traction system and brake system of the high-speed train are summarized.
(3) The fault diagnosis methods of electric systems and information control systems of the high-speed train are overview.
(4) The applicability of main fault diagnosis methods for high-speed trains is discussed.

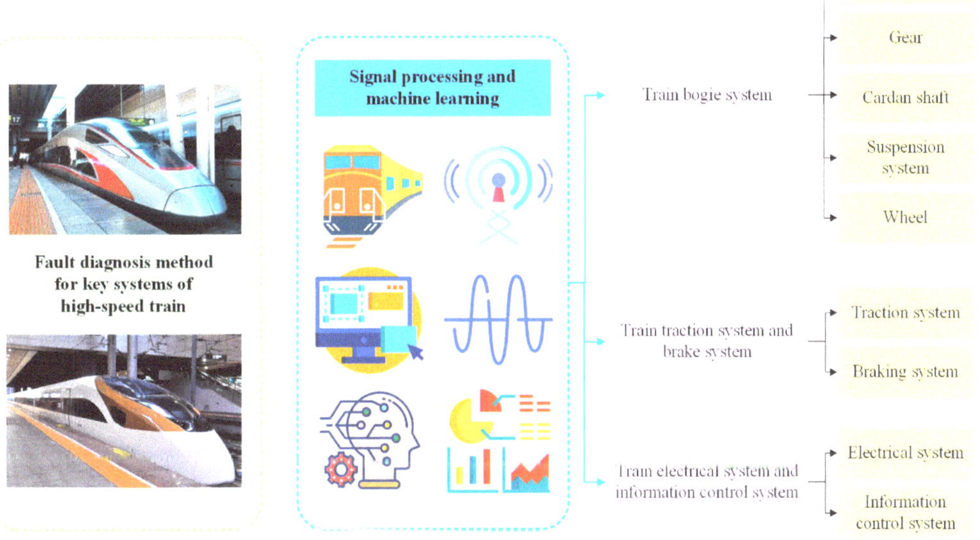

Figure 1. Fault diagnosis for key systems of high-speed trains.

2. Fault Diagnosis of High-Speed Train Bogie System

2.1. Fault Diagnosis of Train Bearing

Bearings are widely used in mechanical systems and are essential parts of high-speed trains. The bearings of trains work under heavy loads and alternating impact stresses for a long time and are highly susceptible to pitting, cracking, wear, and other failures. Their main diagnostic techniques are shown in Figure 2.

To address the challenge of the large data volume of high-speed trains, Liu et al. [7] proposed a method based on mapping approximate principal component analysis (PCA), and validated it with actual train operation data. To effectively analyze the vibration signals of bearings, Cao et al. [8] presented a method based on empirical wavelet transform (EWT) to effectively detect outer ring faults, rolling body faults, and composite faults of bearings. Due to the relative motion between the train and the detection system, the collected acoustic signal is distorted by the Doppler effect, so Zhang et al. [9] introduced the method of multiscale chirp path pursuit (MSCPP) and variable digital filter (VDF) to effectively suppress the Doppler effect and extract the bearing fault features. Ding et al. [10] proposed a detection method based on shock response convolutional sparse coding, thus improving the accuracy of bearing fault detection. To extract bearing fault features effectively, Li et al. [11] proposed an acoustic detection technique based on a sparse decomposition of resonant signals and singular value decomposition, which is superior to traditional methods. To improve the diagnostic accuracy of train bearing faults, Li et al. [12] advanced a multiscale morphological filter method based on feature selection using maximum correlation based on more than 30 feature indicators of vibration signals. Through the technique of time domain interpolation resampling, Zhang et al. [13] proposed an enhanced spline kernelized wavelet transform-based acoustic detection method for roadside acoustics, which effectively overcomes the distortion caused by the Doppler effect. Ding et al. [14] introduced a sparse representation method based on cyclic structure dictionary learning and applied it to the extraction of bearing fault features of trains, and the effectiveness of the proposed method was verified by simulated signals and measured signals. Zhao et al. [15] presented an improved harmonic product spectrum (IHPS) to identify multiple modulation sources hidden in the vibration signal, which not only effectively identifies the fault features but also eliminates the influence of non-fault modulation on the fault features. Huang et al. [16] optimized the key parameters of variational mode decomposition (VMD) and proposed an improved scale-space VMD method, which can automatically decompose the resonance band of bearing signals and further reveal the fault mode. Through a time-frequency data fusion strategy, Zhang et al. [17] advanced a method of time-frequency distribution (TFD) that can effectively separate the acoustic signals in the time-frequency domain and achieve an accurate diagnosis of bearing faults. Minimum entropy deconvolution (MED) enhances the impulse component of the signal, and Cheng et al. [18] designed a particle swarm algorithm optimized MED method, and the experimental results showed that the method could effectively diagnose bearing faults with a low signal-to-noise ratio and outperforms other methods. Through the singular value decomposition package (SVDP) and reconstruction of the Hankel matrix, Huang et al. [19] proposed a method of extended SVDP (ESVDP) and validated it with the failure data of train bearings. References [20,21] improved the conventional signal processing method and were able to successfully identify train-bearing faults. In addition, the results of references [22–24] show that fault identification of train bearings can be achieved based on morphological component analysis, data-driven analysis, and improved SVDP. Furthermore, the time-frequency features of the empirical modal decomposition of the signal are input to a support vector machine, which can achieve fault diagnosis of train bearings [25–27]. To improve the accuracy of fault diagnosis across domains, Shen et al. [28] proposed an adaptive convolutional ResNet-50 network model and validated it with train bearings. Zhang et al. [29] proposed an effective adaptive Gabor sub-dictionary approach that can extract fault features from the vibration signals of trains. With the autoencoder as the feature learning model, bearing fault types can be accurately identified through the weighted voting strategy and generative adversarial network (GAN) [30,31]. Xu et al. [32] introduced a time-frequency domain feature extraction method based on sound signals and vibration signals and applied it to condition monitoring and service life prediction of train bearings.

Therefore, the bearing fault diagnosis method for high-speed trains is mainly achieved by fault feature extraction and fault pattern recognition. In feature extraction, the main methods are EWT, singular value decomposition, VDF, empirical modal decomposition, MED, wavelet transform, VMD, TFD, etc. In pattern recognition, the main methods used are principal component analysis, support vector machines, dictionary learning, self-encoders, generative adversarial networks, etc.

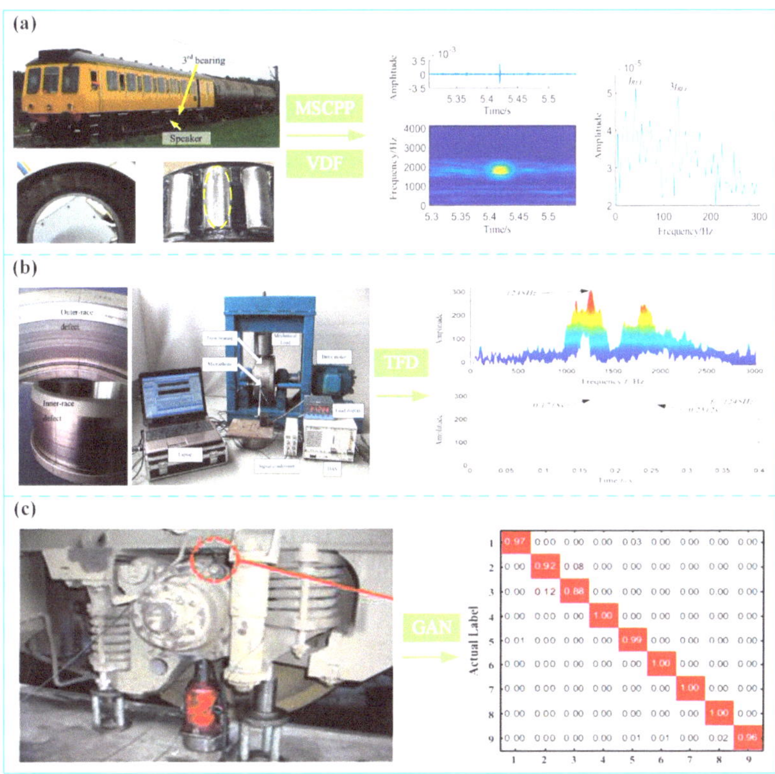

Figure 2. Fault diagnosis method of bearing: (**a**) multiscale chirp path pursuit (MSCPP) and variable digital filter (VDF) [9]; (**b**) time-frequency distribution (TFD) [17]; (**c**) generative adversarial network (GAN) [30].

2.2. Fault Diagnosis of Train Gear

Gears are an important part of high-speed train transmission systems, which usually work under harsh environmental conditions (such as sudden shock loads, unbalanced shaft conditions, heavy loads, large speed fluctuations, etc.), resulting in gears inevitably cracking, pitting, wearing, spalling, suffering tooth surface damage, etc. Their main diagnostic techniques are shown in Figure 3.

It is well known that gear failure is the main source of vibration and noise in high-speed trains. To effectively extract the fault features of train gears, Chen et al. [33] designed an improved complementary ensemble empirical mode decomposition (CEEMD), which not only reduces the problem of modal aliasing but also reduces the residual noise in signal reconstruction and obtains valuable intrinsic mode functions (IMFs). In order to detect the early failure of cracks in gears, Zhang et al. [34] proposed a time-frequency statistical feature (TFSF) method based on finite element analysis combined with modal analysis, which not only obtained the causes of crack generation but also effectively identified the different degrees of crack failure. To address the problem of weak extraction of weak fault features under strong background noise, Zhu et al. [35] presented an improved maximum correlated kurtosis deconvolution (MCKD) method, which can effectively diagnose multiple faults in train gears and outperforms some conventional methods. To extract fault information from the vibration signals of train gears, Wang et al. [36] advanced an improved empirical wavelet transform method based on EWT using the kurtosis index and inverse transformation technology and applied it to gear fault identification. Combining the advantages of spectral kurtosis and tunable Q-factor wavelet transform (TQWT), Long et al. [37] introduced an improved TQWT, which was verified by simulation signals and experimental signals of train gears. Zhu et al. [38] proposed a method based on singular value decomposition (SVD) and multipoint optimal minimum entropy deconvolution adjusted (MOMEDA) and applied it to the fault diagnosis of train gears. In order to reduce the degree of modal aliasing in empirical mode decomposition (EMD), Peng et al. [39] proposed a method based on a soft sieving criterion to optimize EMD and applied it to the simulated signals and the measured signals of train gear faults. Li et al. [40] developed a method to optimize VMD with a discrete difference evolutionary (DDE) algorithm, combined with MED for noise reduction, and successfully identified multiple faults in train gears.

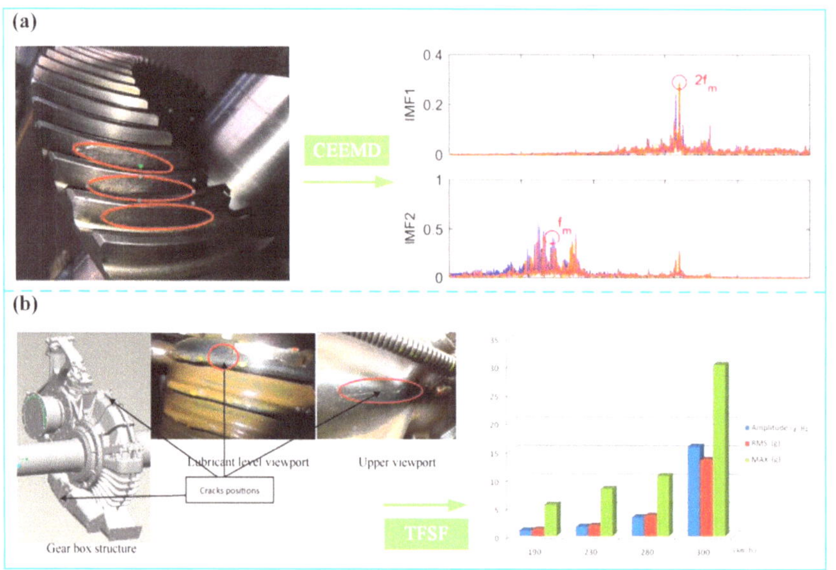

Figure 3. Fault diagnosis method of gearbox: (**a**) improved complementary ensemble empirical mode decomposition (CEEMD) [33]; (**b**) time-frequency statistical feature (TFSF) [34].

Therefore, gear fault diagnosis is mainly accomplished by fault feature extraction, and the main methods used include CEEMD, MCKD, wavelet transform, singular value decomposition, EMD, VMD, TFSF, etc.

2.3. Fault Diagnosis of Train Cardan Shaft

The cardan shaft is the key power transmission component of high-speed trains. It is connected to the drive motor and the gearbox through two universal joints, which transmit the traction force generated by the motor to the gearbox, and then drive the train forward. Additionally, it is a slender mechanical structure with little bending and stiffness, which makes it easy to appear to various degrees of deformation during the train's long run, causing abnormal vibration of other components, thus leading to damage. Since the cardan shaft not only needs to transmit traction power but also needs to coordinate complex motion relationships, its failure causes a major train accident. Their main diagnostic methods are displayed in Figure 4.

In order to quickly diagnose the misalignment fault of the cardan shaft of trains, Hu et al. [41] presented a method based on dual-tree complex wavelet packet transformation (DTCWPT), which greatly eliminated the interference of other mechanical noises in the vibration signal of the cardan shaft and was verified by train data. By using wavelet packet decomposition, Ding et al. [42] proposed a method based on double decomposition and double reconstruction (DDDR), which can effectively extract the fault features of the universal axis of the train and is better than wavelet decomposition and wavelet singular value decomposition. By using the Hamming window, Li et al. [43] designed a method based on an improved morphological filter that can effectively identify the characteristic frequencies and multiplicative information of train cardan shaft faults.

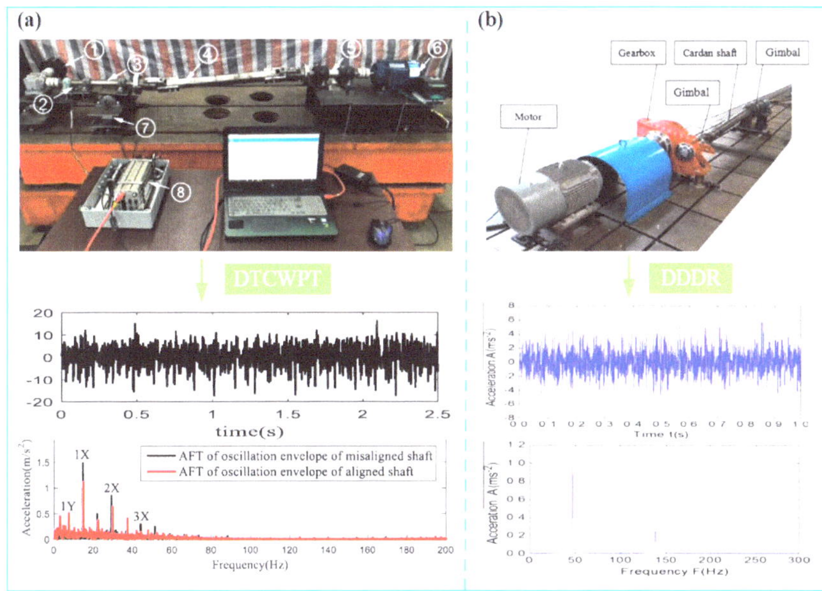

Figure 4. Fault diagnosis method of cardan shaft: (**a**) dual-tree complex wavelet packet transformation (DTCWPT) [41]; (**b**) double decomposition and double reconstruction (DDDR) [42].

Therefore, the fault diagnosis of the train cardan shaft is mainly made by fault feature extraction, and the main applied methods include wavelet packet transform, morphological filter, singular value decomposition, etc.

2.4. Fault Diagnosis of Train Suspension System

The suspension system is an important part of the high-speed train, which mainly consists of the frame, primary suspension, secondary suspension, drive unit, etc. The suspension system plays an important role in power transmission and load bearing, and its performance directly affects the train's running quality and safety. Their main diagnostic techniques are shown in Figure 5.

As the mileage of a high-speed train increases, its suspension system suffers from various types of fatigue damage and performance degradation, which increases the risk of train accidents. By taking advantage of the good extraction ability of deep learning, Zhang et al. [44] proposed a method based on a deep neural network (DNN) that can effectively identify multiple faults in a suspension system with an accuracy of 92.5%, which is better than traditional signal methods. To detect early faults in train suspension systems in a timely manner, Dumitriu et al. [45] calculated the correlation function of the vibration acceleration signals of a train under unbalanced forces and thus diagnosed the faults of the dampers. Due to the complex vibration signals and weak fault information of the train's suspension system, Gou et al. [46] presented a method based on VMD multiscale entropy combined with SVM and validated it with experiments. Based on the analysis of the vibration signal features of different components of the suspension system, Qin et al. [47] introduced a method based on the information entropy of the ensemble empirical modal decomposition to diagnose multiple fault types of the train suspension system effectively. Qin et al. [48] proposed a fault identification technique based on wavelet entropy features for typical faults in train suspension systems. Due to the strong noise and difficult feature extraction of the train suspension system, Guo et al. [49] proposed a method based on an ensemble deep belief network (EDBN) and combined it with SVM to achieve a diagnosis. In addition, the application of ensemble empirical mode decomposition (EEMD) entropy and wavelet packet energy with least squares support vector machines as classifiers can effectively identify faults in suspension systems [50,51]. To further improve the fault diagnosis accuracy of train suspension systems under small sample conditions, references [52,53] designed methods based on combined permutation entropy and EEMD permutation entropy, respectively. By using the time-frequency domain features of vibration signals and convolutional neural networks (CNN), Wu et al. [54] presented a multi-domain fusion (MDF)-based method to successfully diagnose faults such as air springs and transverse dampers of train suspension systems. Zhu et al. [55] advanced a feature extraction method based on energy entropy and singular entropy, and the effectiveness of the method was verified by simulated and measured signals of the train suspension system. Additionally, in order to improve the diagnostic accuracy of the suspension system, references [56,57] developed the dual-tree complex wavelet and singular value decomposition, respectively. Ye et al. [58] designed a method based on multiscale permutation entropy and linear local tangent space alignment (LLTSA), which was finally validated with tracking data of the suspension system of China's CRH3 high-speed trains. To detect early faults in high-speed train suspension systems as early as possible, Wu et al. [59] introduced a method based on improved ToMFIR, which was then validated with data from the Chinese CRH2 and was better than the conventional method. To address the problem of great difficulty in early fault diagnosis of the anti-yaw dampers of the suspension system of high-speed trains, Kaiser et al. [60] proposed a real-time monitoring method based on a nonlinear constraint model. To change the maintenance of train suspension systems from scheduled repair to condition repair, Zoljic et al. [61] proposed a method based on cubic Kalman filtering, which was verified experimentally. To improve the diagnostic accuracy and robustness of suspension systems for high-speed trains, Aravanis et al. [62] presented a method based on a functional model (FM), which was experimentally verified to have better performance compared to the traditional method.

Therefore, the fault diagnosis of the train's suspension system is mainly accomplished by feature extraction and pattern recognition. The main methods applied in feature extraction include permutation entropy, LLTSA, wavelet packet, VMD, MDF, FM, singular value decomposition, EEMD, etc. In pattern recognition, DNN, SVM, LSSVM, CNN, Kalman filters, etc., are mainly used.

2.5. Fault Diagnosis of Train Wheels

The wheels of high-speed trains are important components for carrying, guiding, traction, and braking. When the train is in normal service, due to the long-term interaction between the wheels and the track, it will have different degrees of wear and tear and even cracks, which affects the stability of the train. Therefore, in order to improve the safety and reliability of train operation, early failure of wheels (wear, cracks, etc.) should be detected as soon as possible, and maintenance should be carried out as soon as possible. Their main diagnostic techniques are shown in Figure 6.

To accurately identify defects in the wheels of high-speed trains, Skarlatos et al. [63] proposed a method based on fuzzy logic, which was experimentally shown to reduce the maintenance cost of trains. To be able to detect wheel defects in real time, Wang et al. [64] proposed a real-time detection method based on a Bayesian dynamic linear model, which can detect the degree of wheel damage and show the abnormal values. To obtain the effect of seasonal changes on wheels, Chi et al. [65] presented a method based on three Bayesian data-driven models with field monitoring data showing that the wear is more severe during the summer season than in other seasons. Due to the great difficulty of wheel fault detection in high-speed trains, Li et al. [66] advanced an algorithm of adaptive multiscale morphological filtering (AMMF) that can extract the wheel fault features in strong background noise. By establishing a wear model of the wheel, Zeng et al. [67] developed a physical data-driven approach to achieve accurate prediction of wheel performance degradation and effective assessment of the remaining service life. Chellaswamy et al. [68] designed a dynamic differential evolution algorithm to identify defects and compared it with the chaotic particle swarm algorithm and the genetic algorithm, and experimental results showed that the method is more effective. Liang et al. [69] introduced an adaptive noise canceling (ANC) method, and the experimental results showed that the method could effectively reduce noise interference and extract fault features from strong background noise. To accurately predict the early defects of wheels and reduce the maintenance time of railroad vehicles, Ward et al. [70] proposed a method to predict wheel wear. Lingamanaik et al. [71] collected signals and assessed wheel wear by installing various sensors (vibration acceleration, displacement, etc.) in the carriage, bogie, and wheelset. To identify fatigue cracks in high-speed train wheels more effectively, Zhou et al. [72] developed a method based on the energy entropy of EMD and the Elman neural network, which uses the energy entropy of EMD as a feature and the Elman as a classifier.

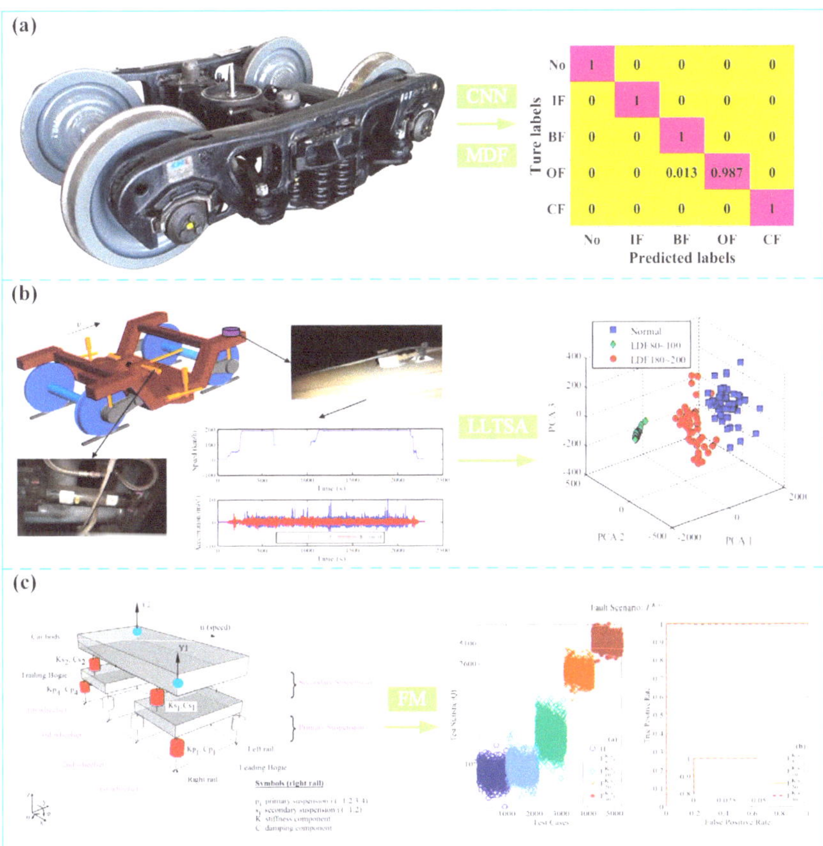

Figure 5. Fault diagnosis method of bogie suspension system: (**a**) multi-domain fusion (MDF)–convolutional neural network (CNN); (**b**) linear local tangent space alignment (LLTSA) [58]; (**c**) functional model (FM) [62].

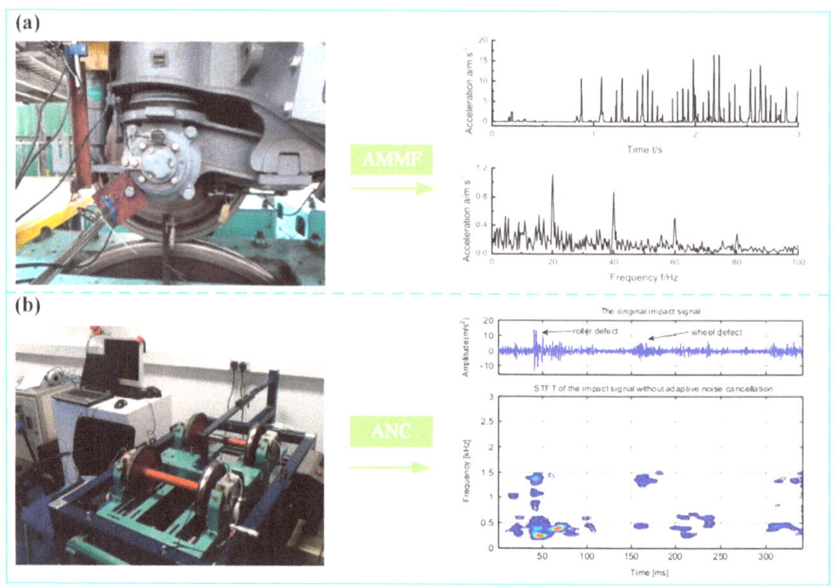

Figure 6. Fault diagnosis method of wheel-rail system: (**a**) adaptive multiscale morphological filtering (AMMF) [66]; (**b**) adaptive noise canceling (ANC) [69].

Therefore, the fault diagnosis of train wheels is mainly achieved by feature extraction and pattern recognition. In feature extraction, morphological filters, ANC, AMMF, EMD, and other methods are mainly applied. In pattern recognition, the main methods used include Bayesian and Elman neural networks, etc.

3. Fault Diagnosis of the High-Speed Train Traction System and Brake System

3.1. Fault Diagnosis of Train Traction System

As the heart of the whole high-speed train, the traction system of the train can generate traction power, which mainly consists of a traction transformer, traction power supply (rectifier, DC link, and inverter), and traction motor. Early failures of traction systems are often caused by unavoidable factors, such as the degradation of winding insulation and the aging of mechanical and electronic components, which in turn affect the safety of trains. Therefore, these early faults need to be detected in time, and their main diagnostic techniques are presented in Figure 7.

The traction system of high-speed trains will deteriorate under long-term operation, leading to various early failures. If these early failures are not successfully detected, they evolve into safety incidents. To detect early faults in the traction system, Chen et al. [73] introduced a method based on Kullback Leibler divergence and independent component analysis (ICA), and the experimental results showed that the method could diagnose three early faults in the train traction system with higher computational efficiency. To effectively diagnose early faults in train traction systems, Wang et al. [74] designed an improved relevant vector machine (RVM) approach based on a Bayesian framework and demonstrated the effectiveness of the method with practical cases. To improve the accuracy of fault diagnosis of train traction motors under external disturbances, Tian et al. [75] developed a method based on interference isolation with an unknown input observer. To identify early faults in the traction system of trains more accurately, Wu et al. [76] advanced a method based on improved total measurable fault information residual (ToMFIR) and validated the effectiveness of the proposed method by simulation models. Insulated-gate bipolar transistor (IGBT) is the main power supply component of traction systems in high-speed trains, and to improve the accuracy of fault diagnosis of IGBTs, Fei et al. [77]

presented a method based on wavelet transform and SVM. In order to diagnose the typical faults of traction systems effectively, both simulation-driven and data-driven methods are effective [78,79]. Zhang et al. [80] proposed a method based on statistical indicators and fuzzy clustering, which was verified by experimental data from the train traction control system. By constructing a nonlinear fault dynamics model for high-speed trains, Tao et al. [81] proposed a method for fault-tolerant control based on state observation.

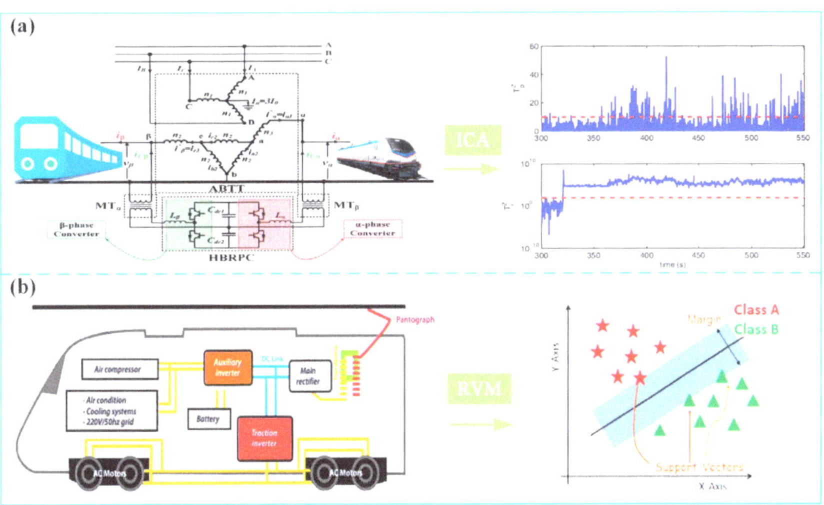

Figure 7. Fault diagnosis method of traction system: (**a**) independent component analysis (ICA) [73]; (**b**) relevant vector machine (RVM).

Therefore, the fault diagnosis of the train's traction system is mainly made by pattern recognition, and the main models used include ICA, relevant vector machine, ToMFIR, SVM, etc.

3.2. Fault Diagnosis of Train Braking System

The braking system of trains mainly consists of brake units, brake cylinders, and related brake components. The complex service environment of high-speed trains, coupled with long-distance driving, leads to various early failures that inevitably occur. If these early failures are not detected in time for maintenance, safety accidents may occur. Therefore, it is important to carry out fault diagnosis research on the train braking system, and its main fault diagnosis methods are shown in Figure 8.

Combining the concepts related to inter-variate variance (IVV) and the idea of four-stage division, Ji et al. [82] proposed a fault detection method based on refined IVV (R-IVV) and stage monitoring indicators, which can improve the accuracy of early fault diagnosis of the train's braking system. To address the problem of data imbalance in train brake system fault detection, Liu et al. [83] presented a diagnostic framework based on a support vector machine (SVM), which improved the computational efficiency from the aspects of feature brush selection, feature selection, and model construction, and verified that the method had better results through experiments. Sang et al. [84] advanced a method based on data domain description and a variable control limit, which can detect three kinds of early failures of high-speed train braking systems and effectively reduce the false alarm rate. Guo et al. [85] designed an improved convex packet vertex-based method to diagnose multiple faults in the train's braking system accurately.

Therefore, the fault diagnosis of train braking systems is mainly made by failure mode identification, and the main methods include IVV, feature selection, SVM, etc.

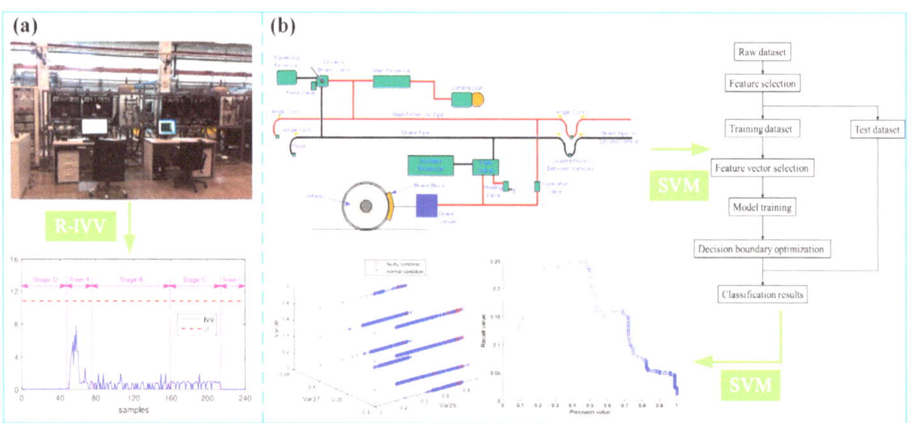

Figure 8. Fault diagnosis method of brake system: (**a**) refined inter-variate variance (R-IVV) [82]; (**b**) support vector machine (SVM) [83].

4. Fault Diagnosis of the High-Speed Train Electrical System and Information Control System

4.1. Fault Diagnosis of Train Electrical System

High-speed train electrical systems are extremely complex and prone to performance degradation and early failure of their internal components due to various uncertainties (overload, unexpected disturbances, and extreme environmental changes). If these early faults are not detected in time, it affects the safe operation of trains. Therefore, it is important to perform the diagnosis of early faults in the train electrical system, and the main diagnostic methods applied are presented in Figure 9.

The long-term operation of high-speed trains in various harsh environments makes it inevitable that the actuators and sensors in the electrical system will suffer performance degradation and even more serious failures. If not addressed in a timely manner, minor malfunctions can eventually turn into catastrophic consequences. To accurately diagnose early faults of asynchronous motors in high-speed trains, Wu et al. [86] constructed a kinetic model based on a squirrel-cage induction motor with a d-q coordinate system, which has high sensitivity and accuracy. To address the problem of serious difficulty in detecting compound faults in electrical systems under multi-source disturbances, Bai et al. [87] proposed a detection method that assigns different faults in coupled signals to separate subspaces, which can effectively improve diagnostic capability. To effectively detect early faults in the slanting support lines of the suspension chain network, Yang et al. [88] presented a deep learning detection method based on fast RCNN Resnet 101 and YOLO V2, which achieves the detection of faults and their localization with high accuracy and robustness. Yao et al. [89] proposed a deep learning foreign object detection method based on YOLO V3, which combines the advantages of migration learning and data augmentation and is not only easier to train but also improves the accuracy of foreign object detection in train electrical systems. To address the problems existing in the manual inspection of train electrical systems, Han et al. [90] advanced an automatic machine vision-based inspection method, which uses CNN to extract crack faults and then uses a fast algorithm to perform further identification of the region of interest. Combining the advantages of traditional hierarchical analysis, maximum absolute weighted residuals, and maximum entropy, Liu et al. [91] developed a comprehensive weighting algorithm (CWA), and the experiments proved that the proposed method has good practical application value. To detect early faults of capacitors in train electrical systems, Oukhellou et al. [92] designed a method based on Dempster–Shafer classification fusion, which improves the accuracy of fault detection. To increase the accuracy of diagnosing different fault states in train electrical systems, Liu et al. [93] proposed a discrete Hidden Markov method based on k-mean clustering, which uses Lloyd's algorithm to quantify the collected sample vector set and can effectively identify six different kinds of faults.

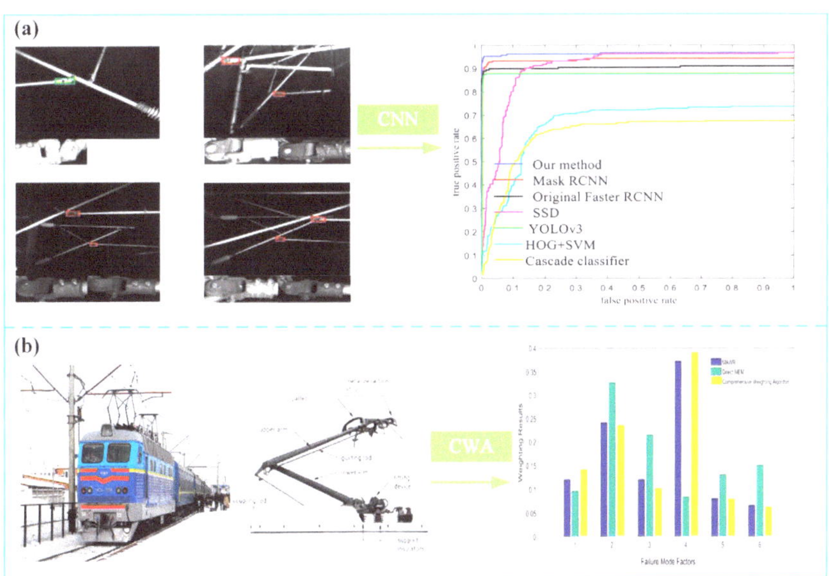

Figure 9. Fault diagnosis method of electrical system: (**a**) convolutional neural network (CNN) [90]; (**b**) comprehensive weighting algorithm (CWA) [91].

Therefore, the fault diagnosis of train electrical systems is mainly made by pattern recognition, mainly applying Resnet, CNN, CWA, YOLO, and discrete hidden Markov models.

4.2. Fault Diagnosis of the Train Information Control System

The information control system of high-speed trains mainly consists of a detection unit, a control unit, an input unit, an output unit, a communication unit, display equipment, and other auxiliary equipment. During the long service life of a train, the components of the information control system develop early failures, which affect the stability of the train. Therefore, it is important to carry out the fault diagnosis of the information control system of high-speed trains to ensure the safe operation of trains, and the main diagnostic methods applied are displayed in Figure 10.

To quickly detect faults in the information control system of trains, Xu et al. [94] introduced a big data-driven approach based on Hadoop, MySQL, and HDFS as the basic framework for realistic user-oriented fault diagnosis and fault prediction. To enable more accurate fault detection and isolation, Niu et al. [95] designed a method to improve the Petri net, which uses PCA to monitor the state of the vibration signal from the acquisition and can achieve the detection and isolation of information control system faults. To achieve a rapid diagnosis of early faults in information control systems, Cai et al. [96] developed an approach based on a train safety sensor network (TSSN), which is accomplished in concert with a multi-layer sensor, a train data center, and a ground data center. Huang et al. [97] advanced a method based on support vector regression (SVR) and Kalman filter (KF), which not only reduced the calculation time but also improved the accuracy of state prediction of the information control system. Aiming at the inaccuracy of fault diagnosis in the train information control system, Yin et al. [98] presented a method based on a vehicle-mounted deep belief network (DBN), which was verified by actual fault data of the Wuhan–Guangzhou high-speed railway and was better than the traditional method. In order to diagnose the faults of the information control system of high-speed trains more comprehensively, Liu et al. [99] proposed a method based on a deep fault tree and quantitative analysis, and the effectiveness of the proposed method was verified by experiments. Huang et al. [100] proposed a method based on fault trees and fuzzy D-S evidence inference, which was shown to be effective for fault prediction and diagnosis through experiments. To improve the accuracy of fault diagnosis in information control systems, Zhou et al. [101] designed a method based on a transferable belief model and multi-sensor data fusion, which was validated experimentally.

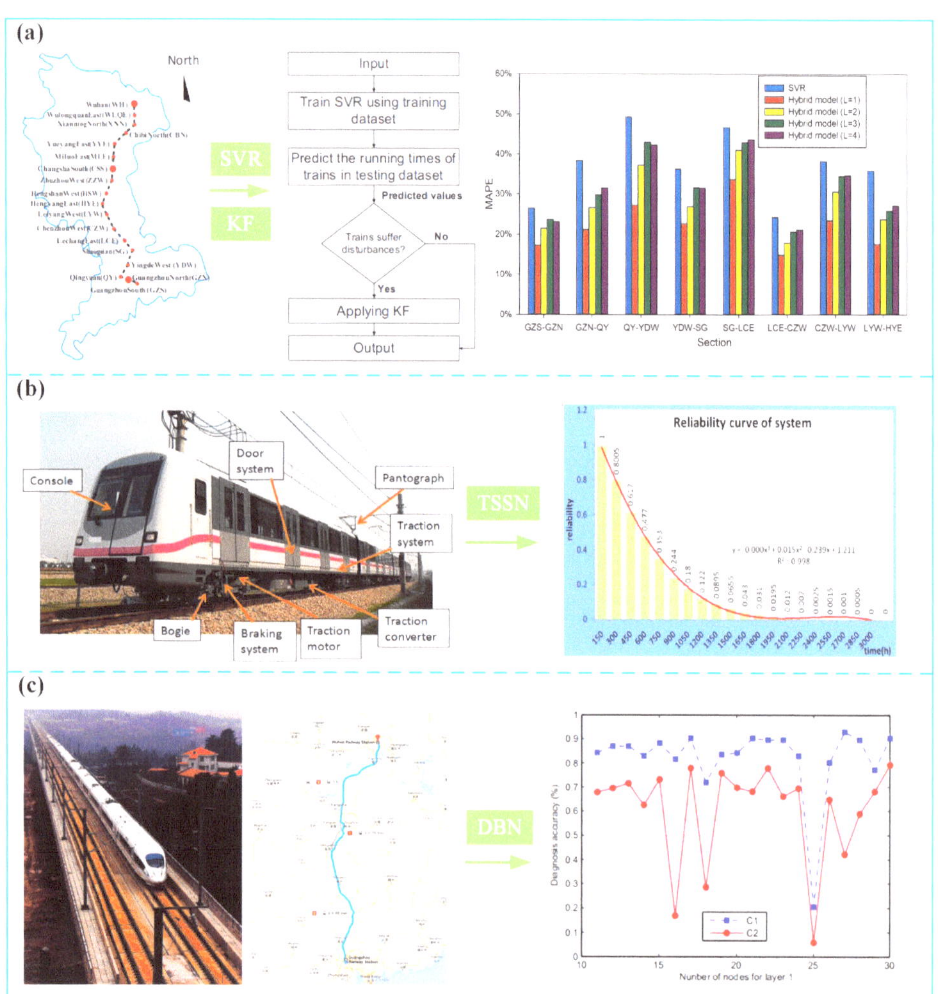

Figure 10. Fault diagnosis method of information control system: (**a**) support vector regression (SVR) and Kalman filter (KF) [97]; (**b**) train safety sensor network (TSSN) [96]; (**c**) deep belief network (DBN) [98].

Therefore, the fault diagnosis of an information control system is mainly made by pattern recognition, and the main methods used include PCA, the deep belief network, TSSN, the Kalman filter, the fault tree, D-S evidence, etc.

5. Applicability of Fault Diagnosis Methods for the High-Speed Train

In summary, the fault diagnosis of each key system of high-speed trains is mainly realized by fault feature extraction and fault pattern recognition, and the main methods used are shown in Figure 11. In addition, the applicability of the same method in different systems was categorized, as shown in Table 1. From Table 1, the Kalman filter can be applied to the fault diagnosis of the bogie system and the information control system at the same time. Secondly, support vector machines can be used not only in the bogie system (bearings, suspension system, and wheels) but also in the traction system and brake system fault diagnosis. Thirdly, the principal component analysis can enable not only the effective fault diagnosis of bogie systems (bearing and wheel) but also the fault diagnosis of information control systems. Additionally, morphological filters are mainly used in the fault diagnosis of bogie systems (bearing, cardan shaft, wheel) and information control systems. Furthermore, convolutional neural networks are mainly applied to the fault diagnosis of bogie systems and electrical systems. Therefore, the Kalman filter, support vector machine, principal component analysis, morphological filter, and convolutional neural network are widely used in train fault diagnosis and have good applicability and robustness in different systems. Moreover, the main methods used in the fault diagnosis of bogie systems are empirical modal decomposition, variational modal decomposition, wavelet transform, singular value decomposition, and support vector machines.

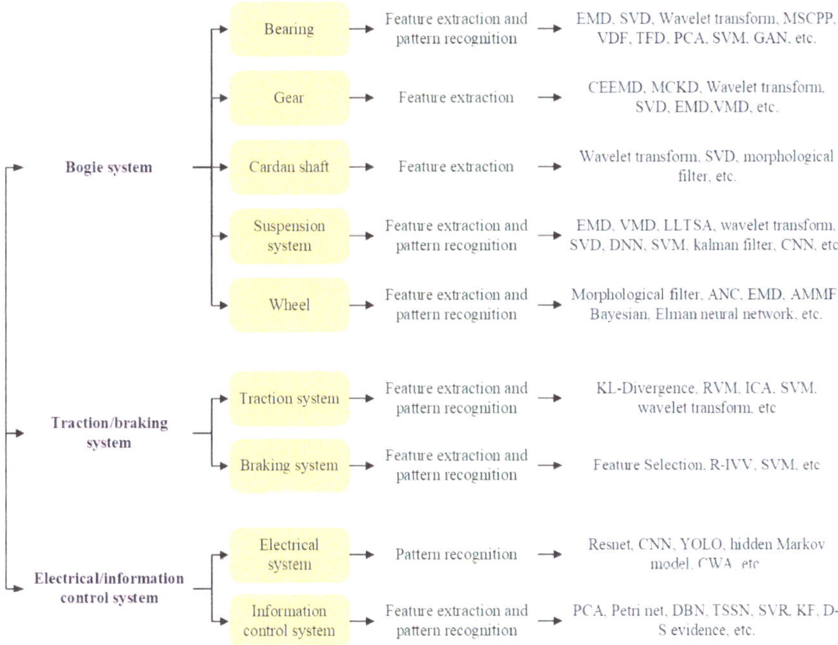

Figure 11. The main fault diagnosis methods used in the key systems of high-speed trains.

Table 1. Applicability of main fault diagnosis methods.

Methods	Key Systems
Empirical modal decomposition	Bogie system (bearing, gear, suspension system, wheel), etc.
Variational modal decomposition	Bogie system (bearing, gear, suspension system), etc.
Wavelet transform	Bogie system (bearing, gear, cardan shaft), traction system, etc.
Singular value decomposition	Bogie system (bearing, gear, cardan shaft, suspension system), etc.
Kalman filter	Bogie system (suspension system), information control system, etc.
Support vector machine	Bogie system (bearing, suspension system, wheel), traction system, braking system, etc.
Principal component analysis	Bogie system (bearing, wheel), information control system, etc.
Morphological filter	Bogie system (bearing, cardan shaft, wheel),
Convolutional neural network	information control system, etc.

6. Discussion and Conclusions

While enjoying the convenience and comfort of high-speed trains, increasing attention has been paid to their safety during service. In addition, as a piece of large and complex equipment, the performance degradation and early failure of various key systems inevitably occur during the service of high-speed trains, thus affecting the safe operation of trains. Therefore, condition monitoring and fault diagnosis of key train systems are of great value and are increasingly becoming a difficult challenge.

Through the previous review, it was known that the fault diagnosis of key systems of high-speed trains is realized by feature extraction and pattern recognition, but it is mainly carried out by the traditional manual inspection, test bench, and video monitoring, which still has many shortcomings. Firstly, existing fault diagnosis methods do not work online on a large scale. Secondly, the monitoring data collected is not comprehensive and objective enough, which restricts the in-depth research on the fault diagnosis of key train systems. Third, the practicality of the existing methods needs to be further improved to be applied to the fault diagnosis of high-speed trains. Therefore, this review has the following prospects:

(1) Developing online monitoring technology can ensure the reliability and safety of high-speed trains. This can not only quickly identify early failures of key systems but also predict performance degradation, thus establishing a long-term warning mechanism.

(2) By installing various sensors (voltage, current, vibration acceleration, displacement, and pressure) in the key system of the train, a large number of real train operation data and corresponding historical data can be obtained, which provides data support for train fault diagnosis based on the combination of big data and experience knowledge.

(3) With the advantages of machine learning and deep learning, a big data-driven condition monitoring and fault diagnosis platform for key systems of high-speed trains is being developed. This can shift from traditional planned repairs to condition repairs and forecast repairs, thereby reducing annual maintenance costs and preventing potential safety accidents.

Author Contributions: Supervision, S.X.; funding acquisition, S.X.; methodology, H.T.; writing—original draft, H.T.; visualization, H.T.; formal analysis, H.T.; validation, H.T.; resources, C.Y.; investigation, H.Y. All authors have read and agreed to the published version of the manuscript.

Funding: This research was undertaken at the Key Laboratory of Traffic Safety on Track (Central South University). The authors gratefully acknowledge the support from the National Key R&D Program of China (2022YFB4300300). This paper is also supported by the National Natural Science Foundation of China (52202455) and the Fundamental Research Funds for the Central Universities of Central South University (Grant No. 2021zzts0161).

Institutional Review Board Statement: Not applicable.

Informed Consent Statement: Not applicable.

Data Availability Statement: Not applicable.

Conflicts of Interest: The authors declare no conflict of interest.

References

1. Zhang, K.; Huang, W.; Hou, X.; Xu, J.; Su, R.; Xu, H. A Fault Diagnosis and Visualization Method for High-Speed Train Based on Edge and Cloud Collaboration. *Appl. Sci.* **2021**, *11*, 1251. [CrossRef]
2. Tan, H.; Xie, S.; Ma, W.; Yang, C.; Zheng, S. Correlation feature distribution matching for fault diagnosis of machines. *Reliab. Eng. Syst. Saf.* **2023**, *231*, 108981. [CrossRef]
3. Hamadache, M.; Dutta, S.; Olaby, O.; Ambur, R.; Stewart, E.; Dixon, R. On the Fault Detection and Diagnosis of Railway Switch and Crossing Systems: An Overview. *Appl. Sci.* **2019**, *9*, 5129. [CrossRef]
4. Zhao, M.; Fu, X.; Zhang, Y.; Meng, L.; Tang, B. Highly imbalanced fault diagnosis of mechanical systems based on wavelet packet distortion and convolutional neural networks. *Adv. Eng. Inform.* **2022**, *51*, 101535. [CrossRef]
5. Garramiola, F.; Poza, J.; Madina, P.; del Olmo, J.; Almandoz, G. A Review in Fault Diagnosis and Health Assessment for Railway Traction Drives. *Appl. Sci.* **2018**, *8*, 2475. [CrossRef]

6. Tan, H.; Xie, S.; Liu, R.; Ma, W. Bearing fault identification based on stacking modified composite multiscale dispersion entropy and optimised support vector machine. *Measurement* **2021**, *186*, 110180. [CrossRef]

7. Liu, Q.; Kong, D.; Qin, S.J.; Xu, Q. Map-Reduce Decentralized PCA for Big Data Monitoring and Diagnosis of Faults in High-Speed Train Bearings. *IFAC-PapersOnLine* **2018**, *51*, 144–149. [CrossRef]

8. Cao, H.; Fan, F.; Zhou, K.; He, Z. Wheel-bearing fault diagnosis of trains using empirical wavelet transform. *Measurement* **2016**, *82*, 439–449. [CrossRef]

9. Zhang, D.; Entezami, M.; Stewart, E.; Roberts, C.; Yu, D. A novel Doppler Effect reduction method for wayside acoustic train bearing fault detection systems. *Appl. Acoust.* **2019**, *145*, 112–124. [CrossRef]

10. Ding, J. Fault detection of a wheelset bearing in a high-speed train using the shock-response convolutional sparse-coding technique. *Measurement* **2018**, *117*, 108–124. [CrossRef]

11. Zhang, D.; Entezami, M.; Stewart, E.; Roberts, C.; Yu, D. Adaptive fault feature extraction from wayside acoustic signals from train bearings. *J. Sound Vibr.* **2018**, *425*, 221–238. [CrossRef]

12. Li, Y.; Liang, X.; Lin, J.; Chen, Y.; Liu, J. Train axle bearing fault detection using a feature selection scheme based multi-scale morphological filter. *Mech. Syst. Signal Proc.* **2018**, *101*, 435–448. [CrossRef]

13. Zhang, D.; Entezami, M.; Stewart, E.; Roberts, C.; Yu, D.; Lei, Y. Wayside acoustic detection of train bearings based on an enhanced spline-kernelled chirplet transform. *J. Sound Vibr.* **2020**, *480*, 115401. [CrossRef]

14. Ding, J.; Zhao, W.; Miao, B.; Lin, J. Adaptive sparse representation based on circular-structure dictionary learning and its application in wheelset-bearing fault detection. *Mech. Syst. Signal Proc.* **2018**, *111*, 399–422. [CrossRef]

15. Zhao, M.; Lin, J.; Miao, Y.; Xu, X. Detection and recovery of fault impulses via improved harmonic product spectrum and its application in defect size estimation of train bearings. *Measurement* **2016**, *91*, 421–439. [CrossRef]

16. Huang, Y.; Lin, J.; Liu, Z.; Wu, W. A modified scale-space guiding variational mode decomposition for high-speed railway bearing fault diagnosis. *J. Sound Vibr.* **2019**, *444*, 216–234. [CrossRef]

17. Zhang, H.; Zhang, S.; He, Q.; Kong, F. The Doppler Effect based acoustic source separation for a wayside train bearing monitoring system. *J. Sound Vibr.* **2016**, *361*, 307–329. [CrossRef]

18. Cheng, Y.; Zhou, N.; Zhang, W.; Wang, Z. Application of an improved minimum entropy deconvolution method for railway rolling element bearing fault diagnosis. *J. Sound Vibr.* **2018**, *425*, 53–69. [CrossRef]

19. Huang, Y.; Huang, C.; Ding, J.; Liu, Z. Fault diagnosis on railway vehicle bearing based on fast extended singular value decomposition packet. *Measurement* **2020**, *152*, 107277. [CrossRef]

20. Wang, T.; Zhang, B.; Sun, Q. Fault Diagnosis of High-speed Train Rolling Bearings Based on EWT-SVD Method. *Electr. Drive Locomot.* **2020**, *1*, 102–107.

21. Yang, H.; Wu, C.; He, L.; Long, Y. Application of an Impact Feature Extracting Method Based on WATV in Fault Diagnosis of High-speed Train Bearing. *Electr. Drive Locomot.* **2020**, *1*, 108–111.

22. Li, J.; Song, D.; Zhang, W.; Wang, Z.; Chen, B. Failure Diagnosis Method for Axle Box Bearing of High-speed Train Based on Morphological Component Analysis. *Railw. Locomot. Car* **2020**, *40*, 20–24.

23. Liu, Q.; Zhan, Z.; Wang, S.; Liu, Y.; Fang, T. Data-driven multimodal operation monitoring and fault diagnosis of high-speed train bearings. *Sci. China (Inf. Sci.)* **2020**, *50*, 527–539.

24. Huang, C.; Lin, J.; Yi, C.; Huang, Y.; Jin, H. Extended SVD packet and its application in wheelset bearing fault diagnosis of high-speed train. *J. Vib. Shock.* **2020**, *39*, 45–56.

25. Zhang, K.; Luo, Y.; Zou, Y.; Wang, C.; Song, X. Sample Correlation Improvement Based High Speed Train Fault Diagnosis Method. *China Mech. Eng.* **2018**, *29*, 151–157.

26. Jin, H.; Lin, J.; Wu, C.; Deng, T.; Huang, C. Diagnostic Method for High-Speed Train Bearing Fault Based on EEMD-TEO Entropy. *J. Southwest Jiaotong Univ.* **2018**, *53*, 359–366.

27. Feng, B. Fault Diagnosis of Rolling Bearing of High Speed Train Based on SVD-PE. *Modul. Mach. Tool Autom. Manuf. Tech.* **2018**, 108–110. [CrossRef]

28. Shen, C.; Wang, X.; Wang, D.; Que, H.; Shi, J.; Zhu, Z. Multi-scale convolution intra-class transfer learning for train bearing fault diagnosis. *J. Traffic Transp. Eng.* **2020**, *20*, 151–164.

29. Zhang, X.; Liu, Z.; Wang, L.; Zhang, J.; Han, W. Bearing fault diagnosis based on sparse representations using an improved OMP with adaptive Gabor sub-dictionaries. *ISA Trans.* **2020**, *106*, 355–366. [CrossRef]

30. Li, X.; Jiang, H.; Niu, M.; Wang, R. An enhanced selective ensemble deep learning method for rolling bearing fault diagnosis with beetle antennae search algorithm. *Mech. Syst. Signal Proc.* **2020**, *142*, 106752. [CrossRef]

31. Liu, S.; Jiang, H.; Wu, Z.; Li, X. Rolling bearing fault diagnosis using variational autoencoding generative adversarial networks with deep regret analysis. *Measurement* **2021**, *168*, 108371. [CrossRef]

32. Xu, G.; Hou, D.; Qi, H.; Bo, L. High-speed train wheel set bearing fault diagnosis and prognostics: A new prognostic model based on extendable useful life. *Mech. Syst. Signal Proc.* **2021**, *146*, 107050. [CrossRef]

33. Chen, D.; Lin, J.; Li, Y. Modified complementary ensemble empirical mode decomposition and intrinsic mode functions evaluation index for high-speed train gearbox fault diagnosis. *J. Sound Vibr.* **2018**, *424*, 192–207. [CrossRef]

34. Zhang, B.; Tan, A.C.C.; Lin, J. Gearbox fault diagnosis of high-speed railway train. *Eng. Fail. Anal.* **2016**, *66*, 407–420. [CrossRef]

35. Zhu, D.; Su, Y.; Sun, Q.; Long, Y. Application of BSO-MCKD in Incipient Fault Diagnosis of Gearbox Bearings of High-speed Train. *Railw. Locomot. Car* **2020**, *40*, 14–19.

36. Wang, T.; Zhang, B. Application of Improved Empirical Wavelet Transform in Fault Feature Extraction of Rolling Bearings. *Railw. Locomot. Car* **2019**, *39*, 53–58.
37. Long, Y.; Su, Y.; Gao, Y.; Li, Y.; He, L. Fault diagnosis of gearbox bearings of high-speed train applying adaptive TQWT. *China Meas. Test. Technol.* **2019**, *45*, 108–113.
38. Zhu, D.; Su, Y.; Yan, C. Fault Diagnosis of Gearbox Bearings of High-speed Train Based on the SVD-MOMEDA. *Electr. Drive Locomot.* **2020**, 144–148. [CrossRef]
39. Peng, D.; Liu, Z.; Jin, Y.; Qin, Y. Improved EMD with a Soft Sifting Stopping Criterion and Its Application to Fault Diagnosis of Rotating Machinery. *J. Mech. Eng.* **2019**, *55*, 122–132. [CrossRef]
40. Li, C.; Lin, J.; Hu, Y. Application of Optimization Parameters VMD and MED in Fault Diagnosis of Train Gearbox Rolling Bearings. *Electr. Drive Locomot.* **2020**, 142–147. [CrossRef]
41. Hu, Y.; Zhang, B.; Tan, A.C. Acceleration signal with DTCWPT and novel optimize SNR index for diagnosis of misaligned cardan shaft in high-speed train. *Mech. Syst. Signal Proc.* **2020**, *140*, 106723. [CrossRef]
42. Ding, J.; Lin, J.; Yu, S. Dynamic unbalance detection of Cardan shaft in high-speed train applying double decomposition and double reconstruction method. *Measurement* **2015**, *73*, 111–120. [CrossRef]
43. Li, Y.; Liu, W.; Lin, J. A morphology filter method for high speed train cardan shaft unbalance fault detection. *J. Vib. Eng.* **2018**, *31*, 176–182.
44. Zhang, Y.; Qin, N.; Huang, D.; Liang, K. Fault Diagnosis of High-speed Train Bogie Based on Deep Neural Network. *IFAC-PapersOnLine* **2019**, *52*, 135–139. [CrossRef]
45. Dumitriu, M.; Gheṭi, M.A. Cross-Correlation Analysis of the Vertical Accelerations of Railway Vehicle Bogie. *Procedia Manuf.* **2019**, *32*, 114–120. [CrossRef]
46. Gou, X.; Li, C.; Jin, W. Fault Diagnosis Method for High-Speed Train Lateral Damper Based on Variational Mode Decomposition and Multiscale Entropy. *J. Vib. Meas. Diagn.* **2019**, *39*, 292–297.
47. Qin, N.; Wang, K.; Jin, W.; Huang, J.; Sun, Y. Fault feature analysis of high-speed train bogie based on empirical mode decomposition entropy. *J. Traffic Transp. Eng.* **2014**, *14*, 57–64.
48. Qin, N.; Jin, W.; Huang, J.; Gou, X.; Jiang, P. Wavelet entropy used in feature analysis of high speed train bogie fault signal. *Appl. Res. Comput.* **2013**, *30*, 3657–3659.
49. Guo, C.; Yang, Y.; Jin, W. Fault Analysis of High Speed Train Based on EDBN-SVM. *Comput. Sci.* **2016**, *43*, 281–286.
50. Qin, N.; Jiang, P.; Sun, Y.; Jin, W. Fault Diagnosis of High Speed Train Bogie Based on EEMD and Permutation Entropy. *J. Vib. Meas. Diagn.* **2015**, *35*, 885–891.
51. He, D.; Chen, E.; Li, X.; Liu, Q. Research on fault diagnosis method of high-speed train running gear rolling bearing based on RS and LSSVM. *J. Guangxi Univ. (Nat. Sci. Ed.)* **2017**, *42*, 403–408.
52. Shi, G.; Li, X.; Jin, W.; Gou, X. Fault analysis of high-speed train bogie based on permutation entropy. *Appl. Res. Comput.* **2014**, *31*, 3625–3627.
53. Li, H.; Jin, W. Lateral damper fault diagnosis of high-speed train based on statistical characteristics of white noise and EEMD. *Appl. Res. Comput.* **2016**, *33*, 2648–2651.
54. Wu, Y.; Jin, W.; Huang, Y. Fault Diagnosis of High Speed Train Bogie Based on Multi-domain Fusion CNN. *J. Syst. Simul.* **2018**, *30*, 4492–4497.
55. Zhu, Z.; Wu, S.; Fu, K. Characteristic Analysis of High-Speed Train Vibration Based on Entropy Feature. *J. Vib. Meas. Diagn.* **2015**, *35*, 381–387.
56. Liu, H.; Meng, Q.; Zhao, Y.; Luo, J. Dual Tree Complex Wavelet Based Fault Characteristic Analysis for High-speed Trains. *Control Eng. China* **2018**, *25*, 1386–1392.
57. Zhao, Y.; Tan, X. Applications of Feature Matrix Construction Method in Fault Diagnoise of High-speed Train. *Comput. Eng.* **2017**, *43*, 21–25.
58. Ye, Y.; Zhang, Y.; Wang, Q.; Wang, Z.; Teng, Z.; Zhang, H. Fault diagnosis of high-speed train suspension systems using multiscale permutation entropy and linear local tangent space alignment. *Mech. Syst. Signal Proc.* **2020**, *138*, 106565. [CrossRef]
59. Wu, Y.; Jiang, B.; Lu, N.; Zhou, D. ToMFIR-based incipient fault detection and estimation for high-speed rail vehicle suspension system. *J. Frankl. Inst.* **2015**, *352*, 1672–1692. [CrossRef]
60. Kaiser, I.; Strano, S.; Terzo, M.; Tordela, C. Anti-yaw damping monitoring of railway secondary suspension through a nonlinear constrained approach integrated with a randomly variable wheel-rail interaction. *Mech. Syst. Signal Proc.* **2021**, *146*, 107040. [CrossRef]
61. Zoljic-Beglerovic, S.; Stettinger, G.; Luber, B.; Horn, M. Railway Suspension System Fault Diagnosis using Cubature Kalman Filter Techniques. *IFAC-PapersOnLine* **2018**, *51*, 1330–1335. [CrossRef]
62. Aravanis, T.C.I.; Sakellariou, J.S.; Fassois, S.D. A stochastic Functional Model based method for random vibration based robust fault detection under variable non–measurable operating conditions with application to railway vehicle suspensions. *J. Sound Vibr.* **2020**, *466*, 115006. [CrossRef]
63. Skarlatos, D.; Karakasis, K.; Trochidis, A. Railway wheel fault diagnosis using a fuzzy-logic method. *Appl. Acoust.* **2004**, *65*, 951–966. [CrossRef]
64. Wang, Y.W.; Ni, Y.Q.; Wang, X. Real-time defect detection of high-speed train wheels by using Bayesian forecasting and dynamic model. *Mech. Syst. Signal Proc.* **2020**, *139*, 106654. [CrossRef]

65. Chi, Z.; Lin, J.; Chen, R.; Huang, S. Data-driven approach to study the polygonization of high-speed railway train wheel-sets using field data of China's HSR train. *Measurement* **2020**, *149*, 107022. [CrossRef]
66. Li, Y.; Zuo, M.J.; Lin, J.; Liu, J. Fault detection method for railway wheel flat using an adaptive multiscale morphological filter. *Mech. Syst. Signal Proc.* **2017**, *84*, 642–658. [CrossRef]
67. Zeng, Y.; Song, D.; Zhang, W.; Zhou, B.; Xie, M.; Tang, X. A new physics-based data-driven guideline for wear modelling and prediction of train wheels. *Wear* **2020**, *456–457*, 203355. [CrossRef]
68. Chellaswamy, C.; Krishnasamy, M.; Balaji, L.; Dhanalakshmi, A.; Ramesh, R. Optimized railway track health monitoring system based on dynamic differential evolution algorithm. *Measurement* **2020**, *152*, 107332. [CrossRef]
69. Liang, B.; Iwnicki, S.; Ball, A.; Young, A.E. Adaptive noise cancelling and time–frequency techniques for rail surface defect detection. *Mech. Syst. Signal Proc.* **2015**, *54–55*, 41–51. [CrossRef]
70. Ward, C.P.; Goodall, R.M.; Dixon, R. Contact Force Estimation in the Railway Vehicle Wheel-Rail Interface. *IFAC Proc. Vol.* **2011**, *44*, 4398–4403. [CrossRef]
71. Lingamanaik, S.N.; Thompson, C.; Nadarajah, N.; Ravitharan, R.; Widyastuti, H.; Chiu, W.K. Using Instrumented Revenue Vehicles to Inspect Track Integrity and Rolling Stock Performance in a Passenger Network during Peak Times. *Procedia Eng.* **2017**, *188*, 424–431. [CrossRef]
72. Zhou, Y.; Lin, L.; Wang, D.; He, M.; He, D. A new method to classify railway vehicle axle fatigue crack AE signal. *Appl. Acoust.* **2018**, *131*, 174–185. [CrossRef]
73. Chen, H.; Jiang, B.; Lu, N.; Chen, W. Real-time incipient fault detection for electrical traction systems of CRH2. *Neurocomputing* **2018**, *306*, 119–129. [CrossRef]
74. Wang, X.; Jiang, B.; Lu, N.; Cocquempot, V. Accurate Prediction of RUL under Uncertainty Conditions: Application to the Traction System of a High-speed Train. *IFAC-PapersOnLine* **2018**, *51*, 401–406. [CrossRef]
75. Tian, Y.; Zhang, K.; Jiang, B.; Yan, X. Interval observer and unknown input observer-based sensor fault estimation for high-speed railway traction motor. *J. Frankl. Inst.* **2020**, *357*, 1137–1154. [CrossRef]
76. Wu, Y.; Jiang, B.; Lu, N.; Yang, H.; Zhou, Y. Multiple incipient sensor faults diagnosis with application to high-speed railway traction devices. *ISA Trans.* **2017**, *67*, 183–192. [CrossRef]
77. Fei, M.; Ning, L.; Huiyu, M.; Yi, P.; Haoyuan, S.; Jianyong, Z. On-line fault diagnosis model for locomotive traction inverter based on wavelet transform and support vector machine. *Microelectron. Reliab.* **2018**, *88–90*, 1274–1280. [CrossRef]
78. Yang, C.; Peng, T.; Yang, C.; Chen, Z.; Gui, W. Fault Testing and Validation Simulation Platform for Traction Drive System of High-speed Trains. *Acta Autom. Sin.* **2019**, *45*, 2218–2232.
79. Jiang, B.; Chen, H.; Yi, H.; Lu, N. Data-driven fault diagnosis for dynamic traction systems in high-speed trains. *Sci. China (Inf. Sci.)* **2020**, *50*, 496–510.
80. Zhang, K.; Jiang, B.; Chen, F.; An, C.; Ren, F. Time-varying model identified based coupled fault diagnosis for high speed trains. *Control Decis.* **2019**, *34*, 274–278.
81. Tao, T.; Xu, H. Immersion and invariance fault-tolerant control for a class high-speed trains. *J. Jilin Univ. (Eng. Technol. Ed.)* **2015**, *45*, 554–561.
82. Ji, H.; Zhou, D. Incipient fault detection of the high-speed train air brake system with a combined index. *Control Eng. Pract.* **2020**, *100*, 104425. [CrossRef]
83. Liu, J.; Li, Y.; Zio, E. A SVM framework for fault detection of the braking system in a high speed train. *Mech. Syst. Signal Proc.* **2017**, *87*, 401–409. [CrossRef]
84. Sang, J.; Zhang, J.; Guo, T.; Zhou, D.; Chen, M.; Tai, X. Detection of incipient faults in EMU braking system based on data domain description and variable control limit. *Neurocomputing* **2020**, *383*, 348–358. [CrossRef]
85. Guo, T.; Tai, X.; Chen, M.; Zhou, D. A Convex Hull Vertices-Based Fault Diagnosis Algorithm for EMU Braking System. *Control Eng. China* **2019**, *26*, 1011–1014.
86. Wu, Y.; Jiang, B.; Wang, Y. Incipient winding fault detection and diagnosis for squirrel-cage induction motors equipped on CRH trains. *ISA Trans.* **2020**, *99*, 488–495. [CrossRef]
87. Bai, W.; Dong, H.; Yao, X.; Ning, B. Robust fault detection for the dynamics of high-speed train with multi-source finite frequency interference. *ISA Trans.* **2018**, *75*, 76–87. [CrossRef]
88. Yang, C.; Liu, Z.; Liu, K.; Zhong, J.; Han, Z. A Loose Default Diagnosis Method for Oblique Bracing Wire in High-Speed Railway. *IFAC-PapersOnLine* **2019**, *52*, 18–23. [CrossRef]
89. Yao, Z.; He, D.; Chen, Y.; Liu, B.; Miao, J.; Deng, J.; Shan, S. Inspection of exterior substance on high-speed train bottom based on improved deep learning method. *Measurement* **2020**, *163*, 108013. [CrossRef]
90. Han, Y.; Liu, Z.; Lyu, Y.; Liu, K.; Li, C.; Zhang, W. Deep learning-based visual ensemble method for high-speed railway catenary clevis fracture detection. *Neurocomputing* **2020**, *396*, 556–568. [CrossRef]
91. Liu, C.; Yang, S.; Cui, Y.; Yang, Y. An improved risk assessment method based on a comprehensive weighting algorithm in railway signaling safety analysis. *Saf. Sci.* **2020**, *128*, 104768. [CrossRef]
92. Oukhellou, L.; Debiolles, A.; Denœux, T.; Aknin, P. Fault diagnosis in railway track circuits using Dempster–Shafer classifier fusion. *Eng. Appl. Artif. Intell.* **2010**, *23*, 117–128. [CrossRef]
93. Liu, J.; Li, Q.; Chen, W.; Cao, T. A discrete hidden Markov model fault diagnosis strategy based on K-means clustering dedicated to PEM fuel cell systems of tramways. *Int. J. Hydrogen Energy* **2018**, *43*, 12428–12441. [CrossRef]

94. Xu, Q.; Zhang, P.; Liu, W.; Liu, Q.; Liu, C.; Wang, L.; Toprac, A.; Joe Qin, S. A Platform for Fault Diagnosis of High-Speed Train based on Big Data. *IFAC-PapersOnLine* **2018**, *51*, 309–314. [CrossRef]
95. Niu, G.; Xiong, L.; Qin, X.; Pecht, M. Fault detection isolation and diagnosis of multi-axle speed sensors for high-speed trains. *Mech. Syst. Signal Proc.* **2019**, *131*, 183–198. [CrossRef]
96. Cai, G.; Zhao, J.; Song, Q.; Zhou, M. System architecture of a train sensor network for automatic train safety monitoring. *Comput. Ind. Eng.* **2019**, *127*, 1183–1192. [CrossRef]
97. Huang, P.; Wen, C.; Fu, L.; Peng, Q.; Li, Z. A hybrid model to improve the train running time prediction ability during high-speed railway disruptions. *Saf. Sci.* **2020**, *122*, 104510. [CrossRef]
98. Yin, J.; Zhao, W. Fault diagnosis network design for vehicle on-board equipments of high-speed railway: A deep learning approach. *Eng. Appl. Artif. Intell.* **2016**, *56*, 250–259. [CrossRef]
99. Liu, P.; Yang, L.; Gao, Z.; Li, S.; Gao, Y. Fault tree analysis combined with quantitative analysis for high-speed railway accidents. *Saf. Sci.* **2015**, *79*, 344–357. [CrossRef]
100. Huang, W.; Liu, Y.; Zhang, Y.; Zhang, R.; Xu, M.; De Dieu, G.J.; Antwi, E.; Shuai, B. Fault Tree and Fuzzy D-S Evidential Reasoning combined approach: An application in railway dangerous goods transportation system accident analysis. *Inf. Sci.* **2020**, *520*, 117–129. [CrossRef]
101. Zhou, Y.; Tao, X.; Yu, Z.; Fujita, H. Train-movement situation recognition for safety justification using moving-horizon TBM-based multisensor data fusion. *Knowl. Based Syst.* **2019**, *177*, 117–126. [CrossRef]

 *applied
sciences*

Article

Analysis of Lateral Forces for Assessment of Safety against Derailment of the Specialized Train Composition for the Transportation of Long Rails

Valeri Stoilov [1], Petko Sinapov [2], Svetoslav Slavchev [1], Vladislav Maznichki [1] and Sanel Purgic [1,*]

[1] Faculty of Transport, Department of Railway Engineering, Technical University Sofia, 1000 Sofia, Bulgaria; vms123@tu-sofia.bg (V.S.); slavchev_s_s@tu-sofia.bg (S.S.); v.maznichki@tu-sofia.bg (V.M.)
[2] Faculty of Transport, Department of Mechanics, Technical University Sofia, 1000 Sofia, Bulgaria; p_sinapov@tu-sofia.bg
* Correspondence: s_purgic@tu-sofia.bg; Tel.: +359-898-813-435

Abstract: This study proposes a theoretical method for evaluating the "safety against derailment" indicator of a specialized train composition for the transportation of very long rails. A composition of nine wagons, suitable for the transportation of rails with a length of 120 m in three layers, is considered. For the remaining recommended rail lengths, the number of wagons is reduced or increased, with the calculation model being modified depending on the required configuration. When the composition is in a curve with the minimum radius (R = 150 m), the rails bend, and some of them come into contact with the vertical stanchions of the wagon and cause additional lateral forces. These forces are then transferred through the wagon body, central pivot, bogie frame, and wheels and act on the wheel–rail contact points. They could potentially lead to derailment of the train composition. The goal of this study is to determine the additional lateral forces that arise because of the bent rails. For the purposes of this study, the finite element method was used. Based on the displacements of the support points of the rails (caused by the geometry of the curve), the bending line of the elastic load is determined and the forces in the supports are calculated. The resulting forces are considered when determining the derailment safety criterion. The analysis of the results shows that the wagon with fixing blocks is the most at risk of derailment. The front and intermediate wagons have criterion values very close to that of the empty wagon. This shows that the emerging horizontal elastic forces do not significantly influence the derailment process. The obtained results show that the transportation of long rails with specialized train composition can be realized on four layers. This will significantly increase the efficiency of delivering new long rails.

Keywords: railway; derailment; long rails transportation; FEM

Citation: Stoilov, V.; Sinapov, P.; Slavchev, S.; Maznichki, V.; Purgic, S. Analysis of Lateral Forces for Assessment of Safety against Derailment of the Specialized Train Composition for the Transportation of Long Rails. *Appl. Sci.* **2024**, *14*, 860. https://doi.org/10.3390/app14020860

Academic Editor: Suchao Xie

Received: 30 November 2023
Revised: 13 January 2024
Accepted: 17 January 2024
Published: 19 January 2024

1. Introduction

The European Commission's road map [1] (with a time horizon of 2050) identifies 10 goals in defining transport policy and offers a list of 40 concrete initiatives to achieve the goals. One of them concerns passenger and freight transport, with the following aims:

- Reducing the share of road transport, with 50% of all medium-distance transport being carried out through rail and water transport;
- By 2030, 30% of freight transport over 300 km should be carried out through rail or water transport, and by 2050, over 50%;
- By 2050, most of the passenger transport over more than 300 km should be carried out through rail transport.

The main reasons for the strategic plans formulated in this way are dictated by the huge advantages that rail and water transport have over road and air transport: developed infrastructure, high speed of movement, the possibility of transporting loads of large volume and mass, low cost per unit of work, minimal staff per unit of work, environmental friendliness, high level of safety, and others.

In terms of safety (as well as personnel and environmental indicators), rail transport is the undisputed leader among other modes of transport. This stems from the strict regulatory documents for the admission of new vehicles into operation and the regulation of a strict system for operation and maintenance.

A special place for guaranteeing accident-free transport is occupied by the wheel–rail interaction. This is the reason why these elements are the subject of a few current studies [2–5], in which analytical, numerical, and experimental methods related to wheel and rail wear and the derailment phenomenon have been proposed and analyzed. A significant concern in modern publications is the elucidation of the derailment mechanism [6–13], in which new factors related to the process are introduced. Their reasons are logical since the phenomenon of derailment is a not infrequent event in the modern reality of railway transport and is associated with serious material damage and casualties. Derailments can occur due to the infrastructure or to rolling stock failures as stated in [14]. Some of the publications [8,9,11,15–17] investigate specific elements (turnouts, guardrails, switches, track defects, lubricators, crossings) or operational indicators (speed, train length, location of wagons carrying dangerous loads) affecting the wheel–rail system and contributing to the derailment of rolling stock. To better investigate the complex phenomenon of derailment, some research activities were devoted to studying the problem from an experimental point of view [18]. Many authors [8,9,19] propose the use of numerical methods to theoretically determine the "safety against derailment" indicator. In [20], a new type of finite element for FEM analysis is proposed, which more accurately reflects the interaction between the wheel and the rail.

The analysis made above clearly shows that there are many theoretically unexplained problems for the occurrence of the derailment process. This is clearly stated in [21], where a critical reading of the currently effective regulatory documents for the admission of new railway vehicles into operation of the European Union, CIS, Great Britain, the USA, China, South Korea, and Japan was made. The authors provide specific explanations of the most common trade-offs, point out the advantages and disadvantages of the methods from the relevant papers, and provide useful recommendations on using simulation methods to assess safety against derailment.

The extensive analysis of available publications allows for the determination of the following conclusions:

- The derailment phenomenon has a distinctly stochastic character. It depends on many factors, which at a certain moment in a complex combination can lead to its occurrence. Current research addresses the issue of derailment from the point of view of theory enrichment by accounting for one or more additional factors that could influence the occurrence of the adverse event. This effort to clarify the mechanism of the process is undeniably positive for two main reasons: first, although the wagons commissioned into operation meet the requirements of the regulatory documents, in practice, railway accidents caused by derailments often occur; and second, it could lead to an adjustment of national or international standards, which would reduce the likelihood of derailment.

- Although the proposed criteria are generally valid for the cases of examination of railway vehicles in a loaded or empty state, the safety assessment against derailment is carried out on an unloaded wagon. The reasons for this are logical—with a full wagon, the vertical force Q increases significantly (especially for freight wagons up to 3–4 times), while the lateral force Y increases less or in the same order under the influence of centrifugal forces.

- Currently, there are regulatory documents that, regardless of their shortcomings, should be respected to reduce the likelihood of derailment.

- In most analyzed publications and in all normative documents related to the derailment process, the evaluation is reduced to the application of Nadal's criterion [21]. The limit value of the ratio (Y/Q) at a coefficient of friction equal to 0.36 was determined via Nadal's criterion using Equation (1):

$$\left(\frac{Y}{Q}\right)_{lim} = \frac{\tan\gamma - 0.36}{1 + 0.36\,\tan\gamma} \tag{1}$$

where γ is the angle of inclination of the wheel flange.

Based on the third conclusion, a theoretical method for evaluating the safety against derailment of a specialized rolling stock for the transport of long rails is proposed in the present study. In this regard, the second conclusion has relevance to the problem: when transporting long rails, the load is placed not on one wagon, but on a whole specialized train composition (Figure 1). When passing through a curved section of the track, the elastic rails are deformed and describe a complex curved line, the shape of which depends on the radius of the curve, the number of rails, the type of rails, the method of fixing the rails, the number of vertical columns limiting the lateral displacements inside the gauge of the wagon, etc.

(a)

(b)

Figure 1. Train composition for the transport of long rails in a curve: (**a**) middle view; (**b**) front view (source: personal archive).

As a result of the interaction of the elastic rails with the vertical columns of the wagon, enormous lateral forces arise between them, which are transmitted through the vehicle structure and balanced on the rail track. This leads to an increase in the lateral forces Y. The mechanism described above necessitates an assessment of safety against derailment not only in a single empty wagon (the case when a new wagon is commissioned into service), but also in the whole train composition, which is loaded to the permissible values. There is a very small number of studies dealing with this issue. One of them is [22], which uses a multibody dynamics simulation for the assessment of safety against derailment when transporting long rails. In this case, rails are "only" 50 m long, loaded in one layer, using four wagons. This is not a common case according to European standards, where longer rails (mostly 120 m long) in three layers are transported in one train composition, which dramatically increases the efficiency of transport work.

Based on the above considerations, the study of the safety criterion against derailment should be carried out through two distinct stages:

- Determination of the lateral forces acting on the central bearings of the bogies, obtained because of the elastic deformations of the long rails transported.

- Application of the method for theoretical assessment of safety against derailment, considering the permissible vertical load and the total values of lateral forces resulting from the movement of the train composition in a curved section of the track and from the elastic deformation of the transported rails.

The first stage takes place in the following sequence:

1. Development of a computational model for the research object;
2. Numerical solution for determining the interaction forces between the deformed rails and the vertical columns of the wagon;
3. Determination of the additional forces acting on the central bearings of the bogies.

2. Computational Model

The task is complex and complicated since the dimensions of the studied objects are large, and the deformations are significant. In this case, non-linear effects occur that do not allow us to obtain an accurate solution even when using specialized software SolidWorks 2014 SP5. In the literature, there are studies investigating beams where the deformations are large [23–27]. In them, the following task is solved: at a given load, the bending line of the beam is sought. In the present work, the reverse task should be solved: determining the forces that arise at the support points at a given deformation of the investigated beam. For this purpose, a linear analysis method was used, and the research methods from [23–27] can be used to assess the accuracy of the solution.

The transportation of long rails can be carried out with wagons that have different parameters (length, base, width, own weight, load capacity, etc.). After analyzing the mentioned parameters, the authors propose as the most efficient version of the specialized train composition, one that should consist of nine wagons with precisely defined parameters. The idea is to obtain maximum use of both the wagons' carrying capacity and their length, and to have the necessary safety factor against derailment. Therefore, the specific technical parameters of the wagon, which in our opinion are the most efficient, are described. When using another modification in the wagons, the mentioned efficiency cannot be guaranteed.

Basic data and prerequisites for solving the task at these conditions are:

- In accordance with [28], the train composition is in a curve with radius $R = 150$ m (Figure 2a). The locomotive is not shown in the figure.
- The composition consists of nine wagons; the rails are located on support frames (Figure 2b) and are grouped in separate sections of five rails each.
- The middle wagon is equipped with a system to guarantee the immobility of the rails (so called "fixing block") (Figure 2b), and the load/pressure forces are shown in Figure 2c.
- On the remaining wagons (not equipped with fixing blocks), three or two (for the end wagons) support frames [29] (Figure 2b) are installed, depending on the length of the transported rails and the number of wagons.
- In the support frames, the rails are placed freely, and unlimited longitudinal displacement is allowed as well as limited lateral displacement within Δh, (Figure 2d). Movements along the vertical direction are limited by the supporting beams of the frames and the clearances Δv (Figure 2d).
- It is assumed that the first four wagons have entered the curve.

The last premise allows for only one half of the train composition to be considered (Figure 3). Element A_0 defines the position of the fixing block in the middle wagon, and the remaining elements (A_1 to A_{12}) define the position of the support frames. The front wagon is equipped with two support frames (A_{11} and A_{12}) in accordance with the requirements of [29]. The locomotive is not illustrated in Figure 3.

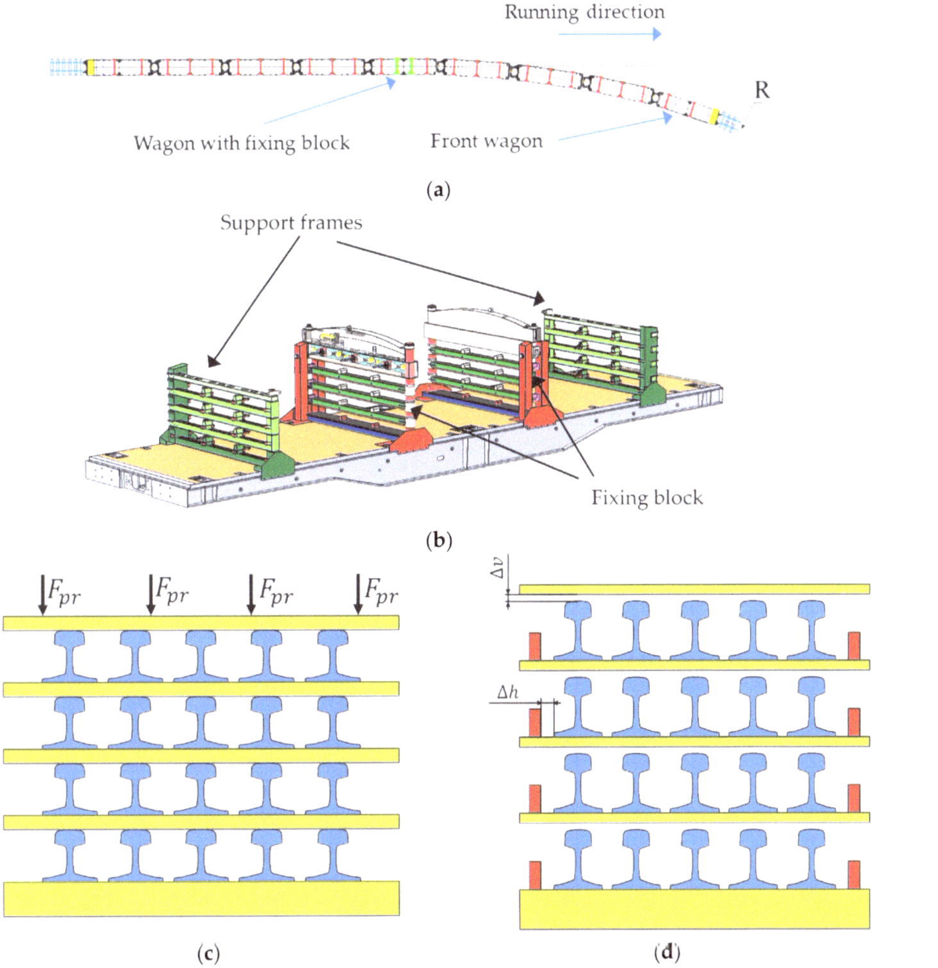

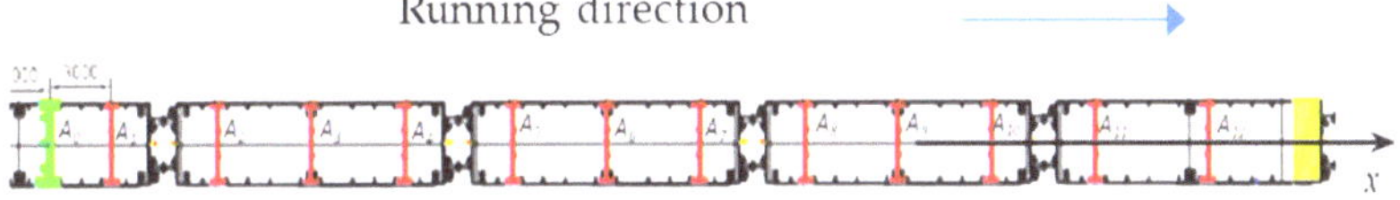

Figure 2. Train composition for transportation of long rails: (**a**) composition in the curve; (**b**) middle wagon with frames; (**c**) load/pressure forces in the fixing block; (**d**) movement limitations in support frames.

Figure 3. Half of the train composition with elements needed for calculation.

The following additional assumptions were made during the compilation of the computational model:

- The bending of a group of five rails (corresponding to one separate section) in the plane Oxy, referred to as "group of rails" or "beam" for brevity (Figures 4 and 5);
- Each rail group has a moment of inertia $I_{zi} = 0.00000486$ m^4 [30,31]. The total moment of inertia for five rails is $I_z = 5 . I_{zi}$;

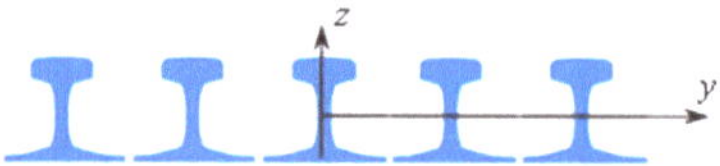

Figure 4. Cross section of one rail group.

- The friction between the rails is neglected due to the peculiarities of the investigated structure.
- In the wagon with the fixing blocks, it is assumed that the rails are immovably fixed, i.e., in the fixing block, it is assumed that we have a fixed beam (Figure 5a);
- At the locations of the supports A_i, it is assumed that unknown forces caused by the deformation of the beam act on the lateral direction (y axis) as shown in Figure 5b. The forces are directed perpendicular to the beam axis (x axis);
- Because of the large dimensions and deformations of the system and considering that the task is statically indeterminate, it is solved with a linear analysis method for the approximate determination of the forces in the supports. The obtained results contain some inaccuracies, which can be estimated by the studies presented in [25,27];
- When moving in a curve with radius $R = 150$ m, the speed of the train composition is small; therefore, the inertial forces are neglected.

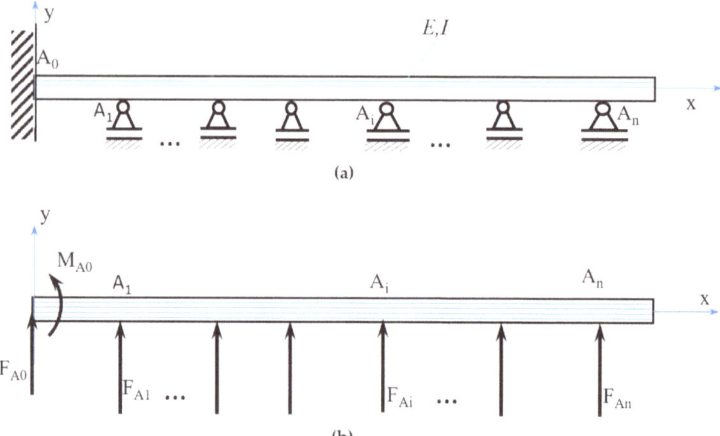

Figure 5. Calculation model of the fixed beam and lateral forces acting on it: (**a**) fixed beam in fixing block; (**b**) unknown forces acting on the lateral direction.

The classical form of the differential equation of the bending line in the given coordinate system and for each section of the beam is given in Equation (2) [32]:

$$\frac{d^2}{dx^2}\left(EI_z\frac{d^2u_y}{dx^2}\right) = q_y \tag{2}$$

For the analytical solution, it is necessary to integrate Equation (2), where a significant number of integration constants appear, depending on the number of sections, which must subsequently be determined by the boundary conditions. For this purpose, the finite element method was applied [32,33]. The large dimensions of the beams make it difficult to use precise models like those shown in [2,34,35]. The beam is divided into n-number of finite elements, and each of them is located between two adjacent supports A_i (Figure 5). The type of finite element is shown in Figure 6 [32].

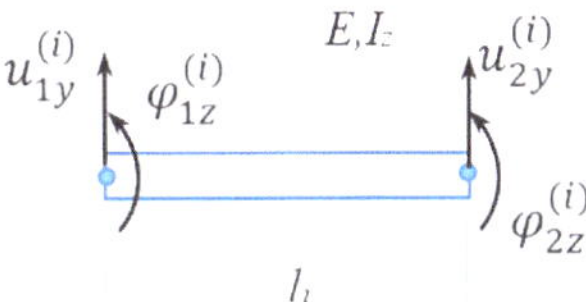

Figure 6. Finite beam element [28].

The position of each element is described by four local coordinates—two displacements and two rotations ($u_{1y}^{(i)}, \varphi_{1z}^{(i)}, u_{2y}^{(i)}, \varphi_{2z}^{(i)}$) of each of the nodes [32].

The stiffness matrix [32,33] of the element i is given in Equation (3):

$$\left[K^{(e)}\right] = \frac{2EI_z}{l_i^3} \begin{bmatrix} 6 & 3l_i & -6 & 3l_i \\ 3l_i & 2l_i^2 & -3l_i & l_i^2 \\ -6 & -3l_i & 6 & -3l_i \\ 3l_i & l_i^2 & -3l_i & 2l_i^2 \end{bmatrix} \tag{3}$$

where l_i is the length of the element and E is the modulus of elasticity.

The discretized model has n number of elements (n = 12 in Figure 7) with different lengths l_i and correspondingly different stiffness matrices. Its position is described by $2n + 2$ global coordinates (Figure 7a). The forces act on the nodes, and the moment acts in node A_0 (Figure 7b).

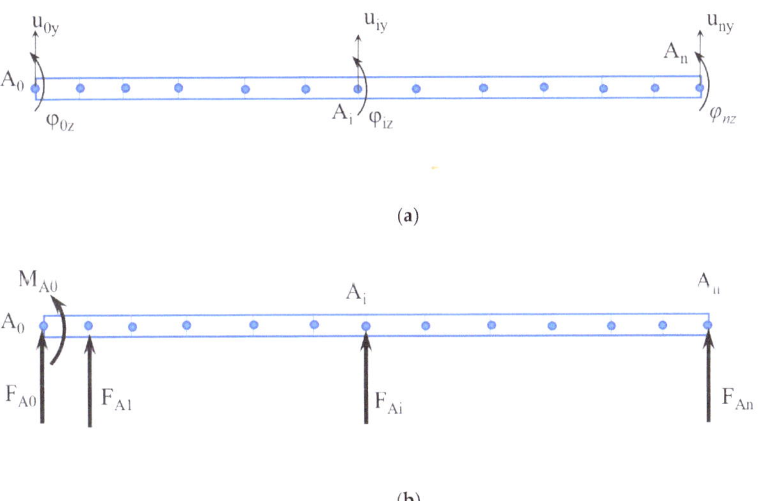

Figure 7. Discretized model of the beam: (**a**) with defined nodes coordinates; (**b**) with defined loads in nodes.

For the beam in Figure 7, Equation (4) is valid:

$$[K_0]\{U_0\} = \{F_0\} \tag{4}$$

where $\{U_0\} = \{u_{0y}, \varphi_{0z}, u_{1y}, \varphi_{1z}......, u_{iy}, \varphi_{iz}......u_{ny}, \varphi_{nz},\}^T$ is a vector of global coordinates with dimension $(2n + 2) \times 1$ and $\{F_0\} = \{F_{A0}, M_{A0}, F_{A1}, 0, F_{A2}, 0....F_{Ai}, 0..F_{An}, 0\}^T$ is a vector of node loads with dimension $(2n + 2) \times 1$. $\{F_0\}$ contains forces and moments applied at the nodes. Except for A0, there are no applied moments in all other nodes. $[K_0]$

is the global stiffness matrix, with dimension (2n + 2) $\times$ (2n + 2). Considering that $u_{0y} = 0$ and $\phi_{0z} = 0$, Equation (4) is reduced to Equation (5):

$$[K]\{U\} = \{F\} \tag{5}$$

where $[K] = [K_0\,(3{:}2n + 2,\,3{:}2n + 2)]$ is the stiffness matrix after considering the boundary conditions and has a dimension of $2n \times 2n$ (the first two rows and columns of the matrix $[K_0]$ are removed), $\{U\} = \{u_{1y}, \varphi_{1z}......, u_{iy}, \varphi_{iz}, ..., u_{ny}, \varphi_{nz,}\}^{T}$ is a vector of global coordinates after considering the boundary conditions and has a dimension of $2n \times 1$, and $\{F\} = \{F_{A1}, 0, F_{A2}, 0....F_{Ai}, 0..F_{An}, 0\}^{T}$ is a vector of node loads after considering the boundary conditions and has a dimension of $2n \times 1$. F_{A0} and M_{A0} are determined using Equation (6):

$$[K_1]\{U\} = \left\{ \begin{array}{c} F_{A0} \\ M_{A0} \end{array} \right\} \tag{6}$$

where $[K_1] = [K_0\,(1{:}2,\,3{:}2n + 2)]$ is a $2 \times 2n$ matrix formed by the first two rows and columns from 3 to $2n + 2$ of the matrix $[K]$. Classical FEM analysis was used up to this step. In Equation (5) (unlike the classical problem), the displacements u_{iy} are known, but the forces F_{Ai} and rotations φ_{iz} of the nodes are not known. To make the equations easier to solve (with MATLAB R2020b software), it is necessary to discretize this continuous mechanical system. To determine the unknowns, the system is transformed into the form given in Equation (7):

$$[X]\{p\} = \{Z\} \Rightarrow \{p\} = [X]^{-1}\{Z\} \tag{7}$$

In Equation (7), the term $\{p\}$ has a dimension of $2n \times 1$ and contains the unknowns F_{Ai} and φ_{iz}. It is determined using Equation (8):

$$\{p\} = \{F_{A1}, F_{A2}, ...F_{An}, \varphi_{1z}, \varphi_{2z}......\varphi_{nz}\}^{T} \tag{8}$$

The term $\{Z\}$ in Equation (7) has a dimension of $2n \times 1$ and contains the predetermined displacements u_{iy} of the nodes (coordinates of A_i along the y-axis direction) and elements of the matrix $[K]$. It is determined using Equation (9):

$$\{Z\} = \left\{ \begin{array}{l} K_{1,1}u_{1y} + K_{1,3}u_{2y}....... + K_{1,2n-1}u_{ny} \\ K_{3,1}u_{1y} + K_{3,3}u_{2y}....... + K_{3,2n-1}u_{ny} \\ ... \\ .... \\ K_{2n-1,1}u_{1y} + K_{2n-1,3}u_{2y}....... + K_{2n-1,2n-1}u_{ny} \\ K_{2,1}u_{1y} + K_{2,3}u_{2y}....... + K_{2,2n-1}u_{ny} \\ K_{4,1}u_{1y} + K_{4,3}u_{2y}....... + K_{4,2n-1}u_{ny} \\ ... \\ .... \\ K_{2n,1}u_{1y} + K_{2n,3}u_{2y}....... + K_{2n,2n-1}u_{ny} \end{array} \right\} \tag{9}$$

The matrix $[X]$ has a dimension of $2n \times 2n$, and contains the null matrices and the elements of matrix $[K]$. It is determined using Equation (10):

$$[X] = \begin{bmatrix} 1, O_{1,n-1}, -K_{1,2}, -K_{1,4}, -K_{1,6}\cdots\cdots, -K_{1,2n} \\ 0, 1, O_{1,n-2}, -K_{3,2}, -K_{3,4}, -K_{3,6}\cdots\cdots, -K_{3,2n} \\ \cdots \\ \cdots \\ O_{1,n-1}, 1, -K_{2n-1,2}, -K_{2n-1,4}, -K_{2n-1,6}\cdots\cdots, -K_{2n-1,2n} \\ O_{1,n}, -K_{2,2}, -K_{2,4}, -K_{2,6}\cdots\cdots, -K_{2,2n} \\ O_{1,n}, -K_{4,2}, -K_{4,4}, -K_{4,6}\cdots\cdots, -K_{4,2n} \\ \cdots \\ \cdots \\ O_{1,n}, -K_{2n,2}, -K_{2n,4}, -K_{2n,6}\cdots\cdots, -K_{2n,2n} \end{bmatrix} \tag{10}$$

In this way, the unknowns from Equation (8) can easily be solved with MATLAB software. After determining the unknowns $F_{A1}, F_{A2}, ...F_{An}, \phi_{1z}, \phi_{2z}......\phi_{nz}$ (contained in $\{p\}$), M_{A0} and F_{A0} can be determined from Equation (6) or from the two equilibrium conditions of the entire beam.

3. Results from the Numerical Calculations

As mentioned above, in numerical calculations, the first step is to determine the interaction forces between the deformed rails and the vertical columns of the wagon. The second step is the determination of the additional forces acting on the central bearings of the bogies from the forces calculated in the first step. These lateral forces are transferred via central bearings to the bogie frame and then on to the wheels, acting on wheel–rail contact and causing additional lateral force. This force is used for the assessment of safety against derailment according to Equation (1).

3.1. Results from Calculation of Lateral Forces between Rails and Vertical Columns of the Wagon

To determine the coordinates of the points A_i (and the displacements u_{iy}, respectively) along the y-axis, it is assumed that the composition is in a curve with radius R (Figure 2a). It is assumed that the points A_1, A_2, A_4, A_5, A_7, A_8, A_{10} and A_{11} lie on the theoretical arc of a circle because that is where the central bearings of the wagons are positioned. The remaining points A_3, A_6, A_9 and A_{12} are located on the chords between the above points (Figure 3). Their position is determined by the line connecting the central bearings of the wagon. Point A_1 is located at the beginning of the curve.

The theoretical curve of the bending line, described with the coordinates from Table 1, is shown in Figure 8.

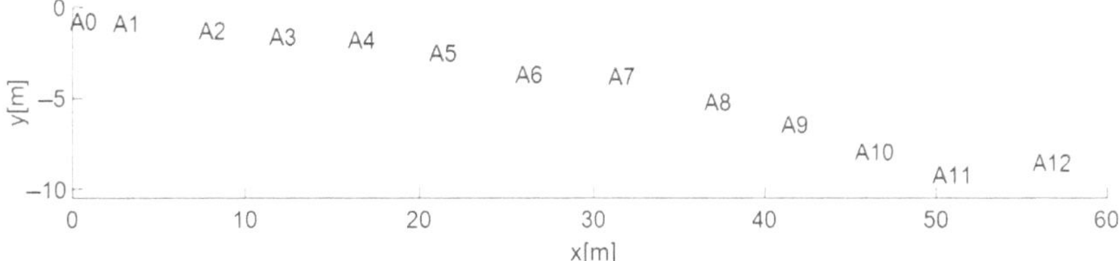

Figure 8. Bending line with coordinates of points A_i.

The coordinate system $A0xy$ is introduced so that the x-axis is in the direction of line A_0–A_1, and its direction relative to the curve is determined by the two central bearings of the wagon with the fixing blocks (Figure 2a).

The determination of the unknown forces F_{Ai} and the moment M_{A0} is carried out in the MATLAB program environment using the displacements from Table 1 (third column). Table 1 (column 5) also contains the obtained values for force reactions in points A_i. It should be noted that the values are obtained under the condition that there is no horizontal clearance between the rails and the support stanchions, i.e., $\Delta h = 0$ (Figure 2d).

Table 1. Coordinates of points A_i along the bending line and values of reaction forces (and moment M_{A0}) in points A_i without horizontal clearance between the rails and the support stanchions ($\Delta h = 0$).

Point	Coordinate x (m)	Coordinate $y = u_{iy}$ (m)	Reaction Identifier (Unit)	Value
A_0	0	0	M_{A0} (kN·m)	−20.9627
A_0	0	0	F_{A0} (kN)	−20.9627
A_1	3.000	0	F_{A1} (kN)	15.4067
A_2	8.190	−0.168	F_{A2} (kN)	28.7388
A_3	12.690	−0.525	F_{A3} (kN)	−45.0839
A_4	17.190	−0.883	F_{A4} (kN)	22.1165
A_5	22.380	−1.540	F_{A5} (kN)	21.0936
A_6	26.880	−2.320	F_{A6} (kN)	−42.6837
A_7	31.380	−3.100	F_{A7} (kN)	21.6934
A_8	36.570	−4.241	F_{A8} (kN)	20.7446
A_9	41.070	−5.436	F_{A9} (kN)	−43.0170
A_{10}	45.570	−6.631	F_{A10} (kN)	25.6134
A_{11}	50.760	−8.245	F_{A11} (kN)	4.9837
A_{12}	56.260	−10.200	F_{A12} (kN)	−8.6432

During the calculations, the presence of large forces was found (Table 1, last column), especially in the points that do not lie on the theoretical arc of a circle (A_3, A_6 and A_9). This is logical and is dictated by the fact that the clearance Δh is not provided in the model, i.e., $\Delta h = 0$ (Figure 2d). To increase the adequacy of the model, a limited horizontal displacement of the points was allowed due to the available constructively provided clearances between the rails and the support frames. With successive iterations based on a possible displacement of the bending line in the support points, new results for the y-coordinates of points A_i and values for the force reactions in points A_i were obtained and are presented in Table 2.

Some of them have been adjusted relative to their original position up to 90 mm (constructively provided clearance between the rails and the support frames) in the lateral direction. In this case, the group of rails does not necessarily rest against each of the side supports, but settles freely, hindered by the minimal frictional forces at the base of the rail. This fact was established during real measurements carried out on another specialized train composition for the transport of long rails. The clearance and possible settlement of the beam at a given point was obtained by iterations through a certain step with a numerical experiment until the actual measured values were obtained.

The iterations of horizontal clearance between the rails and limiting pins Δh (Figure 2d) show that the reaction forces decrease because the bending line changes.

With new coordinates from Table 2 (third column), new values of reaction forces are determined and are presented in the last column of Table 2.

An analysis of the data from Table 2 shows that the maximum lateral forces are at points A_0 and A_1 of the wagon with fixing block and at points A_{11} and A_{12} of the front (closest to the locomotive) wagon.

These lateral forces are transferred via central bearings to the bogie frame and then on the wheels, acting on wheel–rail contact and causing additional lateral force. The results from the calculation of these additional forces for different wagons of a specialized train composition are presented in the next subsection.

Table 2. Adjusted coordinates of points A_i along the bending line and values of reaction forces (and moment M_{A0}) in points A_i with horizontal clearance between the rails and the support stanchions ($\Delta h = 90$ mm).

Point	Coordinate x (m)	Coordinate $y = u_{iy}$ (m)	Reaction Identifier (Unit)	Value
A_0	0	0	M_{A0} (kN·m)	−21.8257
A_0	0	0	F_{A0} (kN)	−21.8257
A_1	3.000	0	F_{A1} (kN)	22.8997
A_2	8.190	−0.1436	F_{A2} (kN)	−1.0583
A_3	12.690	−0.4350	F_{A3} (kN)	1.3180
A_4	17.190	−0.8733	F_{A4} (kN)	−1.0934
A_5	22.380	−1.5400	F_{A5} (kN)	−1.1026
A_6	26.880	−2.2530	F_{A6} (kN)	1.1627
A_7	31.380	−3.1000	F_{A7} (kN)	−0.0952
A_8	36.570	−4.2410	F_{A8} (kN)	−0.2744
A_9	41.070	−5.3690	F_{A9} (kN)	−0.4341
A_{10}	45.570	−6.6274	F_{A10} (kN)	1.2717
A_{11}	50.760	−8.2450	F_{A11} (kN)	4.8339
A_{12}	56.260	−10.1100	F_{A12} (kN)	−5.6022

3.2. Results from Calculation of Additional Lateral Forces in Wheel–Rail Contact

The next step is to determine the additional forces acting on the central bearings of the bogies from forces calculated in the previous subsection. They are used to assess the safety against derailment of the entire specialized train composition. The calculation is carried out only for the wagon with the fixing blocks and the front wagon.

The forces at points A_0 and A_1 of the wagon with fixing block with their respective values from Table 2 are applied to the wagon, as shown in Figure 9.

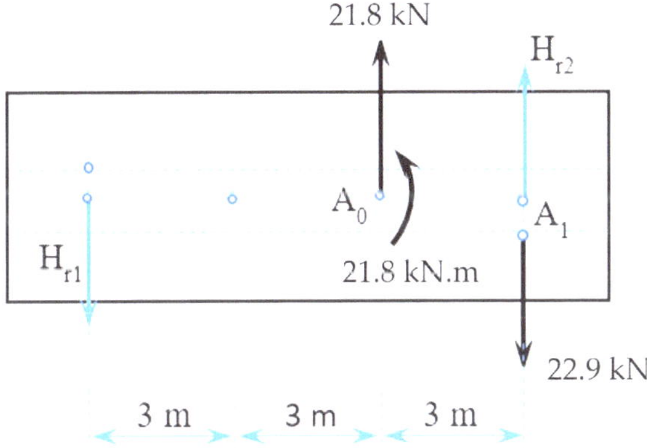

Figure 9. Forces and moment acting on a wagon with fixing block when entering the curve.

The reason for analyzing the front wagon is that it is at risk of derailment due to the incomplete vertical load from the rails' weight (leading to a reduction in the vertical force Q from Equation (1)) and due to the asymmetric distribution of the load in the structure resulting from the placement of the support frames on the wagon (Figure 3). The forces at points A_{11} and A_{12} of the front wagon with their respective values from Table 2 are applied to the wagon, as shown in Figure 10.

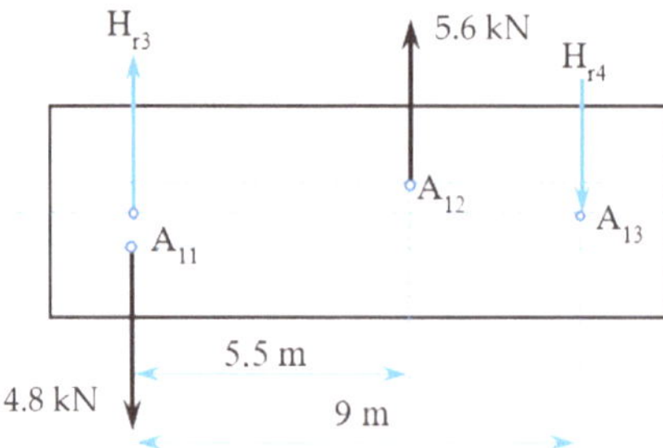

Figure 10. Forces acting on a front wagon.

The determination of the additional horizontal forces H_{ri} is trivial, and their values are given in Table 3.

Table 3. Values of additional lateral forces H_{ri}.

Wagon with Fixing Block	Front Wagon	Unit	Value
H_{r1}	-	(kN)	4.84
H_{r2}	-	(kN)	5.94
-	H_{r3}	(kN)	2.62
-	H_{r4}	(kN)	3.42

4. Assessment of Safety against Derailment

The assessment of safety against derailment of the specialized train composition for transportation of long rails was performed theoretically in full accordance with the method proposed in [36]. Briefly, this is an analytical method developed according to EN 14363:2019 [28], Section 6.1, Method 2 (paragraph 6.1.5.2). The method proposed in [36] has been verified by comparing the results of theoretical and experimental studies and shows a very good compliance.

For assessment purposes, the following corrections were made in the present method compared to that in [36]:

- The vertical load from the weight of 60 (4 rows × 15 rails per row) rails of type 60E1 with a length of 120 m (on the entire train composition) was applied;
- The horizontal forces from Table 3 are applied in the central bearings of the wagon.

It is assumed that the specialized train composition for transportation of long rails is composed of nine flat wagons of the Smmns series with the following parameters:

- The tare weight of each wagon is 21 tons;
- The distance between pivots is 9000 mm;
- The maximum load capacity is 69 tons;
- The maximum permissible load capacity when transporting rails is 0.8 × 69 tons = 49.6 tons according to [29];
- The wagon gauge is G1;
- The weight of two fixing blocks at the middle wagon is 2 × 2.5 tons = 5 tons;
- The weight of one support frame is 1 ton.

The calculations were performed for a horizontal track without overhang at a friction coefficient $\mu = 0.36$.

Three types of wagons from the loaded train composition were studied:

- The wagon equipped with fixing block to limit the longitudinal movements of the rails during transportation—the reason for this is that this is the wagon with the maximum tare weight (taking into account the weight of the installed frames) and in it, the lateral forces F_{A0} and F_{A1} (Table 2), respectively, and the lateral force H_{ri} have a maximum values;

- Front wagon with protective walls. It has a reduced weight of the payload, which is placed asymmetrically on the wagon;
- Intermediate wagon (heaviest loaded wagon of all the wagons discussed above).

Additionally, an assessment was also performed for the empty Smmns series wagon in accordance with the requirements of [28], to assess the influence of the load on the derailment process. Detailed data from the simulation are presented in Table 4 only for the first type of wagon—wagon with fixing block—and for the other types of wagons only the final assessments are given separately in Table 5.

Table 4. Results from calculation of parameters used for assessment of safety against derailment of wagon with fixing block.

Parameter Description	Symbol/Equation	Value
Total reaction force of the attacking wheel of wheelset [1]	Y_{1a} [1]	122.814 kN
Total reaction force of the non-attacking wheel of wheelset [1]	Y_{1i} [1]	−70.732 kN
Minimum deflection of the bogie frame to be reached during tests of the wagon [28]	$\Delta z_{jk}^{+} = g_{lim}^{+}.2a^{+}$	0.0076 m
Twist of the bogie frame during test ($2a^{+}$ is distance between wheelsets) [28]	$g_{lim}^{+} = 7 - 5/2a^{+}$	4.2222‰
Minimum deflection of the wagon frame to be reached during tests of the wagon [28]	$\Delta z* = g_{lim}^{*}.2a$	33.2 mm
Twist of the wagon frame during test ($2a$ is the pivot distance) [28]	$g_{lim}^{*} = \frac{15}{2a} + 2$	3.6483‰
Load force during torsion test	ΔF_{p}	10 kN
Deflection of the wagon frame under ΔF_{p}	Δz_{p}	88.22 mm
Force needed to achieve $\Delta z*$	$\Delta F_{z*} = \frac{\Delta F_{p}.\Delta z^{*}}{\Delta z_{p}}$	3.763 kN
Maximum force additionally loading bogie frame under ΔF_{z*}	$\Delta F'_{z*,max} = \frac{\Delta F_{z*}.(b_{1F}+b_{s})}{2b_{1F}}$	3.481 kN
Minimum force additionally loading bogie frame under ΔF_{z*}	$\Delta F'_{z*,min} = \Delta F_{z*} - \Delta F'_{z*,max}$	0.282 kN
Maximum force overloading first wheel	$\Delta F'_{1z*,max} = \frac{\Delta F'_{z*,max}}{2}$	1.7405 kN
Minimum force overloading first wheel	$\Delta F'_{1z*,min} = \frac{\Delta F'_{z*,min}}{2}$	0.141 kN
Additional maximum vertical reaction force in wheels (Equation (25) of [36])	$\Delta Q_{1,max}$	2.007 kN
Additional minimum vertical reaction force in wheels (Equation (25) of [36])	$\Delta Q_{1,min}$	−0.125 kN
Nominal vertical force loading the wheels	$Q_{nom} = Q/N$	93.401 kN
Minimum vertical wheel reaction force	$Q_{jk,min} = Q_{nom} + \Delta Q_{1,min}$	93.276 kN
Minimum vertical wheel reaction force	$Q_{jk,max} = Q_{nom} + \Delta Q_{1,max}$	95.408 kN

[1] Calculated according to methodology described in [36].

With the data from Table 5, the final assessment of safety against derailment can be conducted. This is performed using Equation (11) and the calculated value is equal to 1.0698. According to [28], when using the theoretical assessment methods, the limit value of 1.2 is reduced by 10%, which means that the limit value of Nadal's criterion should be set to 1.08 and compared with the calculated value, as shown in Equation (12).

$$\left(\frac{Y}{Q}\right)_{ja} = \frac{Y_{ja}}{Q_{jk,min}+\Delta Q_{j}}; \qquad \frac{Y_{ja}}{Q_{jk,min}+\Delta Q_{j}} \leq \left(\frac{Y}{Q}\right)_{lim} \tag{11}$$

$$\left(\frac{Y}{Q}\right)_{ja} \leq \left(\frac{Y}{Q}\right)_{lim}; \qquad 1.0698 \leq 1.08 \tag{12}$$

Data for the safety against derailment evaluation criterion for all wagon types considered are given in Table 5.

Table 5. Safety criterion for different wagon types.

Criterion	Wagon with Fixing Block	Front Wagon	Intermediate Wagon	Empty Wagon
$\left(\frac{Y}{Q}\right)_{ja} \leq 1.08$	1.0698	0.8024	0.8295	0.8268

The obtained values of the safety criterion for all wagon types considered are lower than the limit value of 1.08, which means that for specialized train composition for transportation of long rails, the requirement for safety against derailment is fulfilled.

The developed model allows us to evaluate the influence of individual factors like pivot distance, own weight, number of rails, movement speed, overhang of the outer rail of the track, friction coefficient, etc., on the criterion. It was found that the friction coefficient has the most serious influence. By increasing its value by only one hundredth from 0.36 to 0.37, the criterion increases from 1.0698 to 1.0874 and exceeds the permissible limits for theoretical evaluation.

The remaining parameters have relatively less influence on the change in intensity of the criterion. For example, when reducing the number of rails from 60 (four rows with 15 rails each) to 45 (three rows with 15 rails each) as recommended in [29] (the reduction by 25%), the criterion decreases from 1.0698 to 1.0450, i.e., by only 2.37%, which is within the computational error. This makes it inexpedient to use an approach from [29] with only three layers of long rails in the composition.

5. Conclusions

The present study proposes a theoretical method for assessment of safety against derailment of a specialized train composition for the transportation of long rails. For this purpose, a method was developed to determine the horizontal forces in the vertical stanchions of the wagon, occurring when they contact the transported rails in a curved section of the track. The forces are used in the computational model for the study of safety against derailment and an evaluation is made according to Nadal's criterion.

- Four different types of train composition wagons were considered: a wagon with fixing blocks, front wagon, intermediate wagon, and an empty wagon. The analysis of the results shows that the wagon with fixing blocks is most at risk of derailment. The obtained value (Table 4) fully corresponds to the requirements of the normative documents not only for real tests (<1.2), but also for theoretical studies (<1.08).
- The front and intermediate wagons have criterion values very close to that of the empty wagon. This shows that the emerging horizontal elastic forces do not significantly influence the derailment process.
- The developed model allows us to evaluate the influence of individual factors like pivot distance, own weight, number of rails, movement speed, overhang of the outer rail of the track, friction coefficient, etc., on the safety against derailment criterion.
- The transportation of long rails can be carried out with wagons that have different parameters (length, base, width, own weight, load capacity, etc.). After analyzing the mentioned parameters, the authors propose the most efficient version of the specialized train composition for 120 m long rails. When using another modification in the wagons or another rail length, the mentioned efficiency cannot be guaranteed.
- The obtained results show that the transportation of long rails with specialized train composition can be realized on four levels. This will significantly increase the efficiency of operators when delivering long new rails.

The transportation of rails on more than three levels is not new in global practice and is widely used in India (a record was set there for the transportation of rails with a length of 260 m, in five rows with 12 rails each), the USA, Russia, etc. (in four rows with 12–15 rails each).

For future research, it would be necessary to conduct a physical validation of the model. This will certainly be carried out if there is a proposal from companies carrying out the transportation of long rails.

Author Contributions: Conceptualization, V.S. and P.S.; methodology, V.S. and P.S.; software, S.S., P.S. and V.M.; validation, V.S. and S.P.; formal analysis, P.S., S.S. and V.M.; investigation, P.S. and S.S.; resources, S.S. and S.P.; data curation, S.S. and V.M.; writing—original draft preparation, V.S., P.S. and S.P.; writing—review and editing, V.S., P.S. and S.P.; visualization, S.S. and V.M.; supervision, V.S.; project administration, V.S.; funding acquisition, V.S. All authors have read and agreed to the published version of the manuscript.

Funding: This research was supported by the European Regional Development Fund within the Operational Programme "Science and Education for Smart Growth 2014–2020" under the Project CoE "National center of mechatronics and clean technologies" under Grant BG05M2OP001−1.001−0008-C01.

Institutional Review Board Statement: Not applicable.

Informed Consent Statement: Not applicable.

Data Availability Statement: Data are contained within this article.

Acknowledgments: The authors would like to thank the Research and Development Sector at the Technical University of Sofia for the financial and technical support.

Conflicts of Interest: The authors declare no conflicts of interest.

References

1. An Official Website of the European Union. Roadmap to a Single European Transport Area: Towards a Competitive and Resource-Efficient Transport System. Available online: https://eur-lex.europa.eu/EN/legal-content/summary/roadmap-to-a-single-european-transport-area-towards-a-competitive-and-resource-efficient-transport-system.html (accessed on 24 November 2023).
2. Jagadeep, B.; Kumar, P.-K.; Subbaiah, K.-V. Stress Analysis on Rail Wheel Contact. *Int. J. Res. Eng.* **2018**, *1*, 47–52.
3. Bosso, N.; Magelli, M.; Zampieri, N. Simulation of wheel and rail profile wear: A review of numerical models. *Rail. Eng. Sci.* **2022**, *30*, 403–436. [CrossRef]
4. Gao, Y.; Wang, P.; Wang, K.; Xu, J.; Dong, Z. Damage tolerance of fractured rails on continuous welded rail track for high-speed railways. *Rail. Eng. Sci.* **2021**, *29*, 59–73. [CrossRef]
5. Milković, D.; Simić, G.; Radulović, S.; Lučanin, V.; Kostić, A. Experimental approach to assessment of safety against derailment of freight wagons. In *Experimental Research and Numerical Simulation in Applied Sciences. CNNTech 2022*; Lecture Notes in Networks and Systems, 564; Mitrovic, N., Mladenovic, G., Mitrovic, A., Eds.; Springer: Cham, Switzerland, 2023.
6. Zeng, J.; Wu, P. Study on the wheel/rail interaction and derailment safety. *Wear* **2008**, *265*, 1452–1459. [CrossRef]
7. Matsumoto, A.; Michitsuji, Y.; Ichiyanagi, Y.; Sato, Y.; Ohno, H.; Tanimoto, M.; Iwamoto, A.; Nakai, T. Safety measures against flange-climb derailment in sharp curve-considering friction coefficient between wheel and rail. *Wear* **2019**, *432–433*, 202931. [CrossRef]
8. Lai, J.; Xu, J.; Liao, T.; Zheng, Z.; Chen, R.; Wang, P. Investigation on train dynamic derailment in railway turnouts caused by track failure. *Eng. Fail. Anal.* **2022**, *134*, 106050. [CrossRef]
9. Yang, Z.; Li, Z. Wheel-rail dynamic interaction. In *Woodhead Publishing Series in Civil and Structural Engineering—Rail Infrastructure Resilience*; Calçada, R., Kaewunruen, S., Eds.; Woodhead Publishing: Cambridge, UK, 2022; pp. 111–135, ISBN 9780128210420. [CrossRef]
10. Huang, J.-Y. Nadal's Limit (L/V) to Wheel Climb and Two Derailment Modes. *Eng. Phys.* **2021**, *5*, 8–14. [CrossRef]
11. Atmadzhova, D. Analysis of mathematical expressions for determine the criteria derailment of railway wheelset. *Mech. Transp. Comm.* **2011**, *3*, 42–49.
12. Song, I.-H.; Koo, J.-S.; Shim, J.-S.; Bae, H.-U.; Lim, N.-H. Theoretical Prediction of Impact Force Acting on Derailment Containment Provisions (DCPs). *Appl. Sci.* **2023**, *13*, 3899. [CrossRef]
13. Santamaria, J.; Vadillo, E.-G.; Gomez, J. Influence of creep forces on the risk of derailment of railway vehicles. *Veh. Syst. Dyn.* **2009**, *47*, 721–752. [CrossRef]
14. Liu, X.; Saat, R.; Barkan, C. Analysis of Causes of Major Train Derailment and Their Effect on Accident Rates. *Trans. Res. Rec. J. Trans. Res. Board.* **2012**, *2289*, 154–163. [CrossRef]
15. Ishida, M.; Ban, T.; Iida, K.; Ishida, H.; Aoki, F. Effect of moderating friction of wheel/rail interface on vehicle/track dynamic behaviour. *Wear* **2008**, *265*, 1497–1503. [CrossRef]
16. Sala, A.J.; Felez, J.; de Dios Sanz, J.; Gonzalez, J. New Anti-Derailment System in Railway Crossings. *Machines* **2022**, *10*, 1224. [CrossRef]

17. Anderson, R.-T.; Barkan, C.-P.-L. Derailment Probability Analyses and Modeling of Mainline Freight Trains. In Proceedings of the 8th International Heavy Haul Conference, Rio de Janeiro, Brazil, 14–16 June 2005; pp. 491–497.
18. Diana, G.; Sabbioni, E.; Somaschini, C.; Tarsitano, D.; Cavicchi, P.; Di Mario, M.; Labbadia, L. Full-scale derailment tests on freight wagons. *Veh. Syst. Dyn.* **2021**, *60*, 1849–1866. [CrossRef]
19. Kalivoda, J.; Neduzha, L. Simulation of Safety Against Derailment Tests of an Electric Locomotive. In Proceedings of 25th International Conference Engineering Mechanics, Svratka, Czech Republic, 13–16 May 2019; pp. 177–180.
20. Ju, S.-H.; Ro, T.-I. Vibration and Derailment Analyses of Trains Moving on Curved and Can't Rails. *Appl. Sci.* **2021**, *11*, 5106. [CrossRef]
21. Wilson, N.; Fries, R.; Wittea, M.; Haigermoser, A.; Wrang, M.; Evans, J.; Orlova, A. Assessment of safety against derailment using simulations and vehicle acceptance tests: A worldwide comparison of state-of-the-art assessment methods. *Veh. Syst. Dyn.* **2011**, *49*, 1113–1157. [CrossRef]
22. Wikaranadhi, P.; Handoko, Y.A. Curving Performance Analysis of a Freight Train Transporting 50-Meter-long Rail Using Multibody Dynamics Simulation. *J. Eng. Tech. Sci.* **2023**, *55*, 189–199. [CrossRef]
23. Shvartsman, B.-S. Direct method for analysis of flexible cantilever beam subjected to two follower forces. *Int. J. Non. Lin. Mech.* **2009**, *44*, 249–252. [CrossRef]
24. Ghuku, S.; Saha, K.-N. A theoretical and experimental study on geometric nonlinearity of initially curved cantilever beams. *Eng. Sci. Tech. Int. J.* **2016**, *19*, 135–146. [CrossRef]
25. Beléndez, T.; Neipp, C.; Beléndez, A. Large and small deflections of a cantilever beam. *Eur. J. Phys.* **2002**, *23*, 371–379. [CrossRef]
26. Kocatürk, T.; Akbaş, Ş.-D.; Şimşek, M. Large deflection static analysis of a cantilever beam subjected to a point load. *Int. J. Eng. Appl. Sci.* **2010**, *2*, 1–13.
27. Wang, C.-M.; Lam, K.-Y.; He, X.-Q.; Chucheepsakul, S. Large deflections of an end supported beam subjected to a point load. *Int. J. Non. Lin. Mech.* **1997**, *32*, 63–72. [CrossRef]
28. *EN 14363:2019*; Railway Applications—Testing and Simulation for the Acceptance of Running Characteristics of Railway Vehicles—Running Behaviour and Stationary Tests. European Committee for Standardization: Brussels, Belgium, 2019.
29. Loading Guidelines. *Code of Practice for the Loading and Securing of Goods on Railway Wagons, Goods*, 7th ed.; International Union of Railways (UIC): Paris, France, 2023; Volume 2.
30. *EN 13674-1:2011*; Railway Applications—Track—Rail—Part 1: Vignole Railway Rails 46 kg/m and Above. European Committee for Standardization: Brussels, Belgium, 2011.
31. Specifications Of UIC 60 Steel Rail. Available online: https://www.railroadpart.com/news/specifications-of-uic-60-steel-rail.html (accessed on 16 November 2023).
32. Merkel, M.; Öchsner, A. *Eindimensionale Finite Elemente*, 2nd ed.; Springer-Verlag: Berlin Heidelberg, Germany, 2015; pp. 79–83.
33. Rao, S.-S. *The Finite Element Method in Engineering*, 5th ed.; Butterworth-Heinemann: Oxford, UK, 2011; pp. 311–354.
34. Gardie, E.; Dubale, H.; Tefera, E.; Bezzie, Y.-M.; Amsalu, C. Numerical analysis of rail joint in a vertical applied load and determining the possible location of joints. *Forc. Mech.* **2022**, *6*, 100064. [CrossRef]
35. Kukulski, J.; Jacyna, M.; Gołębiowski, P. Finite Element Method in Assessing Strength Properties of a Railway Surface and Its Elements. *Symmetry* **2019**, *11*, 1014. [CrossRef]
36. Stoilov, V.; Slavchev, S.; Maznichki, V.; Purgic, S. Method for Theoretical Assessment of Safety against Derailment of New Freight Wagons. *Appl. Sci.* **2023**, *13*, 12698. [CrossRef]

 *applied sciences*

Article

Method for Theoretical Assessment of Safety against Derailment of New Freight Wagons

Valeri Stoilov, Svetoslav Slavchev, Vladislav Maznichki and Sanel Purgic *

Faculty of Transport, Department of Railway Engineering, Technical University Sofia, 1000 Sofia, Bulgaria; vms123@tu-sofia.bg (V.S.); slavchev_s_s@tu-sofia.bg (S.S.); v.maznichki@tu-sofia.bg (V.M.)
* Correspondence: s_purgic@tu-sofia.bg; Tel.: +359-898-813-435

Abstract: The assessment of safety against freight wagons' derailment is a mandatory element of the documents provided to the EU notifying authorities of the entry into service of new freight wagons. The assessment methodology is presented in EN 14363:2019. It is mainly aimed at the experimental measurement of certain parameters, and the data are used to calculate the safety criterion. The practical implementation of the tests is accompanied by many difficulties: finding a track with a proper radius, ensuring free access to the railway infrastructure for a long period of time, waiting for suitable metrological conditions, preparing the curve and the test wagon, etc. These difficulties are well known to the European legislators, and as a solution, they propose a large set of reference wagons that have undergone real tests. It is sufficient to demonstrate that the parameters of the new wagon relate to some of the reference wagon parameters to avoid such a requirement. Proving the "convergence" of the parameters of the new and the reference wagons is a lengthy, complex, and, in many cases, subjective process. To introduce an objective assessment, the authors set themselves the task of developing a theoretical method to assess safety against derailment.

Keywords: railway; freight wagon; derailment; safety; assessment

Citation: Stoilov, V.; Slavchev, S.; Maznichki, V.; Purgic, S. Method for Theoretical Assessment of Safety against Derailment of New Freight Wagons. *Appl. Sci.* **2023**, *13*, 12698. https://doi.org/10.3390/app132312698

Academic Editor: Suchao Xie

Received: 2 October 2023
Revised: 11 November 2023
Accepted: 13 November 2023
Published: 27 November 2023

1. Introduction

The methods for assessing safety against derailment for new vehicle acceptance are mostly performed using some combination of testing and theoretical mathematical calculations or simulations. Most methods are based on dynamic tests, which can be performed as controlled tests in a laboratory or as long runs under representative service conditions or under the described test conditions, using exactly defined track characteristics and statistical assessment methods. All three test types have their advantages and disadvantages, as described in [1].

There are many methods used for evaluating safety against derailments worldwide. A good overview of these methods is given in [1]. Theoretical mathematical calculations have a long tradition since the first attempts from Hertz [2], Klingel [3], and Nadal [4] to mathematically describe the processes occurring in the wheel–rail interface. In particular, the Nadal criterion [4] sets the basics for further developing theoretical methods to assess safety against derailment [5–8]. Today, it is applied in different international and European normative documents [9,10].

On the other hand, the experimental measurement of certain parameters relevant to derailment safety is still a standard procedure for approving new freight wagons for operational commissioning. The measured values of these parameters are then used to calculate the safety criterion; if the limit values are not exceeded, the new wagon is approved. The performance of these tests is strictly regulated by international and European authorities in standards and regulatory works [9,10]. Even if these test procedures and conditions are strictly described, some new approaches and methods for measuring forces and displacements are used nowadays [11,12]. However, the practical implementation of the tests is accompanied by many difficulties: finding a track with a proper radius, ensuring free access to the railway infrastructure for a long period of time, waiting for suitable metrological conditions, preparing the curve and the test wagon, etc. For the wagon manufacturers, these difficulties mean long-term approval procedures and higher expenses. This is why most wagon manufacturers attempt to avoid dynamic tests since this is possible according to EN 14363:2019 [9]. The required evidence and procedure for acceptance are well defined in this European standard, and, thus, it is possible to compare the resulting parameters from the theoretical calculations or simulations of safety against derailment with parameters of reference wagons that have undergone real tests. This approach is allowed according to clause 6.1.5.2.6 of EN 14363:2019 [9] when using Method 2 for the assessment. This method is described in detail in Section 2 of this paper.

When using Method 2 of EN 14363:2019 [9], it is necessary to determine the vertical and transversal forces acting in wheel–rail contact. In tests, these parameters are measured directly. It is not explicitly pointed out which theoretical method or type of simulation should be used. The only condition is that the method used for calculation or simulation delivers sufficiently reliable results that are close enough to the results from dynamic tests. Depending on the mechanism that causes derailment of wagons, many methods that determine forces and displacements exist. As mentioned above, a very good overview of these methods is given in [1].

Many methods are based on the flange climb in the wheel–rail contact [13–16], which is the most frequent reason for derailment. Many researchers in this field deal with geometrical properties of wheel–rail contact on straight tracks or in curves [13,15], including dynamic properties of suspension, while others propose methods to control these processes [14,16].

Some researchers deal with derailment caused by other mechanisms: passing through an S-shaped curve [17], passing through turnouts and switches [18], or due to track failure [19]. There is also research on the effects of wind loads on dynamic behavior and derailment of the train composition [20]. New methodologies are also presented with some non-standard approaches, like a response surface methodology [21].

Over the last 70 years, computer development has allowed their very intense use in numerical and multibody simulations, as well as in testing and analysis techniques. There is much research using multibody dynamic approaches to assess safety against derailment [22,23]. Also, other types of simulations are used for this purpose [24,25]. Many simulation methods have their advantages and problems [1,24–26]. The main issue for simulations and theoretical calculations in the last decades is that a vehicle model must be validated in tests; it is still unclear what can be determined as a validated model. In [10], computer simulations were introduced to supplement or replace testing and to be used as an alternative assessment method. The same thing happened in the last version of [9], but with some restrictions regarding the validation procedure: in Europe, for every new freight wagon that should enter the service across European railways, mandatory documents provided to the EU notifying authorities is an assessment of safety against derailment. According to Method 2 of EN 14363:2019 [9], it is sufficient to demonstrate that the parameters of the new wagon relate to some of the reference wagon parameters to avoid such a requirement. Proving the "convergence" of the parameters of the new and the reference wagons is a lengthy, complex, and, in many cases, subjective process.

The final decision is met by the notified bodies, independent of the method used for theoretical calculations or simulations. The notified body will accept a new freight wagon for commissioning if safety against derailment is proven and assessed with appropriate methods or simulation. Moreover, the multibody approaches and different simulations and theoretical methods can be very complex, as well as costly and time-consuming. From the perspective of the wagon manufacturers, the used method must be objective enough, fast in fulfillment, not expensive, and capable of delivering reliable results.

This work introduces an objective theoretical method to assess safety against derailment. Consequently, it is possible to shorten the time and the costs of the approval procedure. The results obtained from theoretical analyses, as described in Section 3, were compared and validated with results from experimental tests carried out on real objects to achieve a sufficient degree of validation.

2. Method for Assessment of Theoretical Safety against Freight Wagon Derailment

In accordance with EN 14363:2019 [9], Section 6.1, Method 2 (paragraph 6.1.5.2), when analyzing the safety criterion against derailment, Equation (1) must be considered:

$$\left(\frac{Y}{Q}\right)_{ja} = \frac{Y_{ja}}{Q_{jk,\min}+\Delta Q_{jH}}; \qquad \frac{Y_{ja}}{Q_{jk,\min}+\Delta Q_{jH}} \leq \left(\frac{Y}{Q}\right)_{\lim} \tag{1}$$

where Y_{ja} is the reaction of the rail at its contact with the attacking wheel, $Q_{jk,\min}$ is the lowest value of the vertical reaction of the wheel calculated when the frame of the wagon is twisted, and ΔQ_{jH} is a load on the wheel from the moment of the forces acting on the 2 wheels of the examined wheel axle (Figure 1). The value of $(Y/Q)_{\lim}$ is equal to 1.2, and the railway vehicle is safe against derailment if $(Y/Q)_{ja} \leq 1.2$.

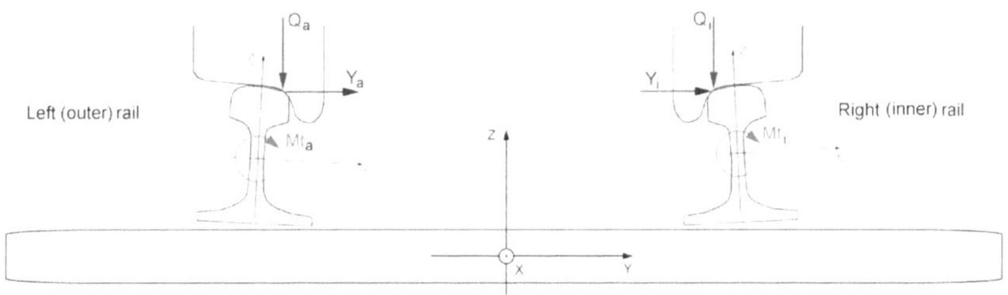

Figure 1. Forces acting in wheel–rail contact.

The component ΔQ_{jH} is determined using Equation (2):

$$\Delta Q_{jH} = (Y_{ja} + Y_{ji}) \cdot \frac{h}{2b_0} \tag{2}$$

where Y_{ja} is the total reaction of the rail at its contact with the attacking (outer) wheel; h is the effective height above the rail of the journal box suspension (for the most commonly used bogie, Y25, $h = 365$ mm is assumed); Y_{ji} is horizontal load force between the inner (non-attacking) wheel of the examined axle and the inner rail (Figure 1); $2b_0$ is the nominal transverse distance between the contact points of the wheels ($2b_0 = 1500$ mm is assumed); j is index (number) of the examined axle; a is index (number) of the outer wheel; and i is index of the inner wheel.

For the theoretical assessment of the criterion from Equation (1), it is necessary to apply appropriate methods for the theoretical determination of the following parameters:

- Y_{ja}—the total reaction of the rail in contact with the attacking (outer) wheel. The parameter is involved in Equations (1) and (2).
- Y_{ji}—the horizontal load force between the inner (non-attacking) wheel of the examined track axle and the inner rail. The parameter is involved in Equation (2).
- $Q_{jk,\min}$—the lowest value of the vertical reaction of the wheel calculated when the frame of the wagon is twisted. The parameter is involved in Equation (1).

The methods for determining these three important parameters are given in the next three subsections and represent the main goal of our paper.

2.1. Methodology for Theoretical Determination of Leading Forces, Y_a, on Axles of Railway Vehicles with Bogies

In [27], a theoretical method for determining the total reaction of the rail on the wheels of a bogie is proposed. Briefly, the method consists of the following: when moving on a curved section of the rail track, the wagon performs two movements—translational and rotational. The rotation occurs around the center of rotation, M (Figure 2), characterized by the pole distance, x, which can be determined using Equation (3):

$$x = l + \frac{R.\sigma_b}{2.l} \tag{3}$$

where $2.l$ is the distance between wheelsets of the bogie (for compliance with European standards, it should be noted that $2.l = 2a^+$), R is the radius of the calculation curve, and σ_b is the current coordinate.

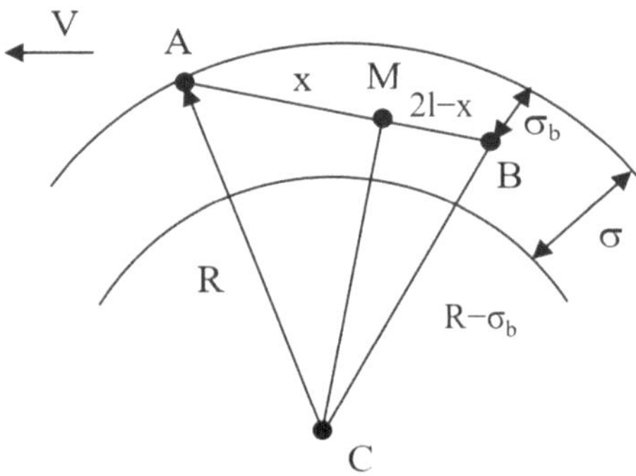

Figure 2. Movement of bogie in curved section of the track.

The current coordinate, σ_b, depends on the position of the bogie when passing a curved section of the track. In Figure 3, the bogie is represented by section AB (AB′ or AB″). For this purpose, the transverse dimensions of the track with gauge $2s$ and of the bogie are reduced by the constant amount, $2d$, defining the transverse distance between the bases of the wheel flanges of the same axle.

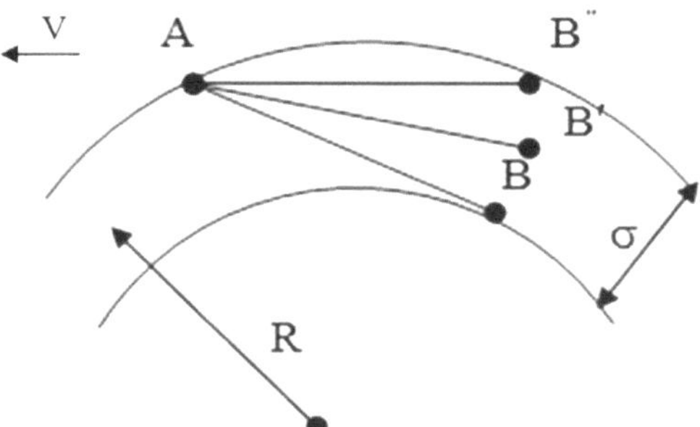

Figure 3. Representation of bogie moving in curved section of the track.

The bases of the flanges of the two wheels merge and are represented in Figure 3 as points A and B (B′ and B″). The same points depict the attacking (A) and the non-attacking (B) axle, respectively.

The reduced gauge is determined via Equation (4):

$$\sigma = 2s - 2d = \Delta + \delta \tag{4}$$

where Δ is the total clearance between the flange and the rail, equal to 0.01 m; and δ is an additional expansion of the rail track in a curved section depending on the radius of the calculation curve, as determined with data from Table 1 [27].

Table 1. Additional expansion of the rail track in a curved section depends on the radius of the calculation curve [13].

Radius R (m)	δ (mm)
125–150	20
150–180	15
180–250	10
250–300	5
Over 300	0

In this way, the calculated value represents the maximum total clearance between the rails and the axle in a curved section of the rail track. In Figure 3, the attacking axle (point A) always contacts the outer rail. Depending on the movement speed and the radius of the curve, the second axle (point B or B′ or B″) can take one of the following positions:

- AB—maximum crossing ($\sigma_b = \sigma = \Delta + \delta$);
- AB′—free settling ($0 \leq \sigma_b \leq \delta$);
- AB″—maximum displacement ($\sigma_b = 0$).

When moving in a curve, the following forces act on the bogie (Figure 4):

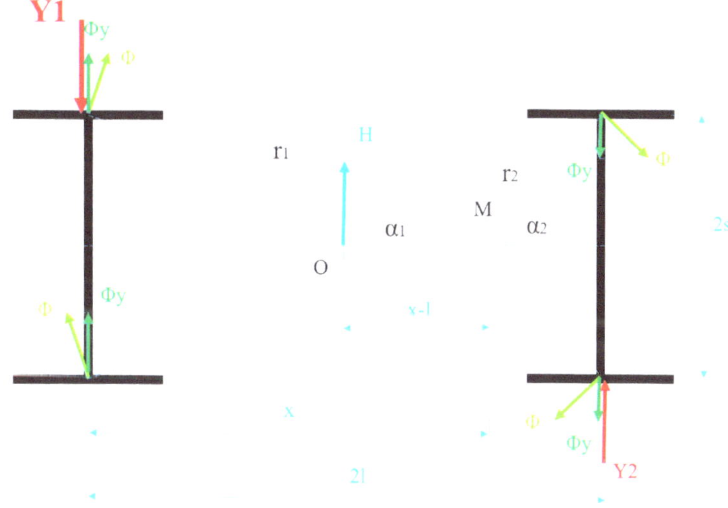

Figure 4. Forces acting on the bogie while moving in a curve.

1. The transverse force, H, is induced by the centrifugal (H_c) and wind (H_w) forces. It is applied at the mass center of the wagon and is determined via Equation (5):

$$H = H_c + H_w \tag{5}$$

2. The centrifugal force is defined using Equation (6):

$$H_c = P \cdot \left(\frac{v^2}{R \cdot g} - \frac{h}{2 \cdot s} \right) \tag{6}$$

where P (in [N]) is the weight of the wagon, v (in [m/s]) is the movement speed, g (in [m/s^2]) is the gravitational acceleration, R (in [m]) is the curve radius, $2 \cdot s$ (in [mm]) is the distance between wheels rolling circles (normal track width $2.s = 1500$ mm), and h (in [mm]) is the overhang of the outer rail determined from the table in Figure 5 [27].

R (m)	Speed v (m/s)													
	25	30	35	40	45	50	55	60	65	70	75	80	85	90
150	50	70	100	125	150	150	150							
160	50	70	90	120	150	150	150							
170	45	65	85	115	140	150	150	150						
180	40	60	80	105	135	150	150	150						
190	40	55	80	100	125	150	150	150						
200	40	55	75	95	120	150	150	150	150					
225	35	50	65	85	110	135	150	150	150					
250	30	45	60	75	100	120	145	150	150	150				
275	30	40	55	70	90	110	130	150	150	150	150			
300	25	35	50	65	80	100	120	140	150	150	150	150		
325	25	35	45	60	75	90	110	130	150	150	150	150		
350	25	30	45	55	70	85	105	125	145	150	150	150	150	
375	20	30	40	50	65	80	95	115	135	150	150	150	150	
400	20	30	40	50	60	75	90	110	125	145	150	150	150	150
450	20	25	35	45	55	65	80	95	110	130	150	150	150	150
500		25	30	40	50	60	75	85	100	115	135	150	150	150
550		20	30	35	45	55	65	80	90	105	120	140	150	150
600		20	25	35	40	50	60	70	85	100	110	125	145	150

Figure 5. Overhang of the outer rail, depending on radius of curve and movement speed [27].

3. The wind force is determined via Equation (7):

$$H_w = F.W \tag{7}$$

where F is the surface of the wagon on which the wind is acting (in [m^2]), and W is the wind pressure (in [N/m^2]).

4. The frictional forces Φ obtained because of the rotation around the pole M are determined using Equation (8):

$$\Phi = \mu.N_{st} \tag{8}$$

where μ is the coefficient of friction between the wheel and the rail, and N_{st} is the static vertical load on one wheel, as determined using Equation (9):

$$N_{st} = \frac{P}{N} \tag{9}$$

where N is the number of wheels. For compliance with European standards, it should be noted that $N_{st} = Q_{nom}$.

5. The total reaction Y_i from rails on the wheelset i are obtained from the equilibrium conditions $\Sigma Y = 0$ and $\Sigma M_M = 0$, according to Equation (10):

$$\left| \begin{array}{l} Y_1 - Y_2 - H - 2.\Phi.cos\alpha_1 + 2\Phi cos\alpha_2 = 0 \\ Y_1 x + Y_2(2l - x) - H(x - l) - 2\Phi.r_1 - 2\Phi.r_2 = 0 \end{array} \right. \tag{10}$$

where Φ_{yi} is the component of force Φ along the y-axis, and r_i is the distance from pole M to the corresponding contact point between the rail and wheel of the i-th wheel axle.

In Equation (10), there are four unknown terms implicitly set by the centrifugal forces: Y_1, Y_2, x, and speed v. Therefore, the total reactions, Y_1 and Y_2, are determined according to the following methodology, and the graphical representation is shown in Figure 6.

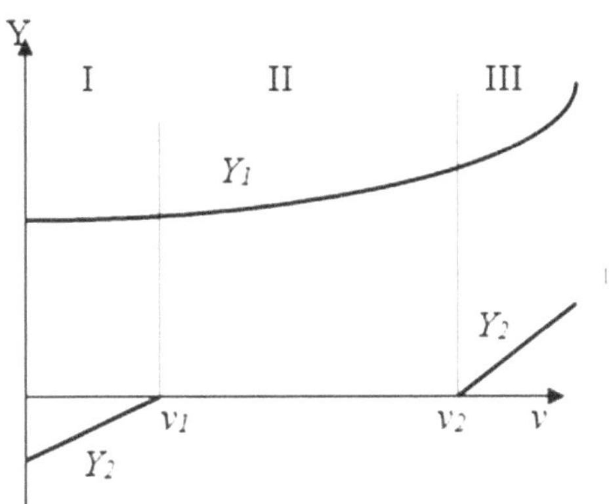

Figure 6. Graphical representation of dependencies $Y_1 = Y_1(v)$ and $Y_2 = Y_2(v)$ [13].

Step 1. It is assumed that the bogie is in the limit state between "maximum overshoot" and "free settling". It is possible at a precisely determined but unknown speed, v_1. From the condition for the considered boundary condition, it follows that the distance from pole M, according to Equation (11), is

$$x = x_{max} = l + \frac{R.\sigma_b}{2.l} \tag{11}$$

which is typical for the "maximum overshoot" position. It also follows that the total reaction of the second wheel axle is zero, e.g.,

$$Y_2 = 0 \tag{12}$$

which is typical for the "free settling" position.

This allows Equation (10) to be solved and to obtain specific values for Y_1 and v_1, which are typical for the limited state. When solving Equation (10), it is possible for Y_1 or for v_1 to obtain negative values. This indicates that the boundary condition is not valid for the specified track and bogie parameters. In this case, it is necessary to go to Step 3 of the current methodology.

Step 2. When the bogie is in the "maximum overshoot" state, it will move with a speed in the interval from 0 to v_1, and the pole distance will be $x = x_{max}$. Therefore, the system of Equation (10) can be solved concerning Y_1 and Y_2 by setting discrete movement speed values in the specified interval.

Step 3. It is assumed that the bogie is in a limited state between free settling and maximum displacement. From this, the next conditions are shown in the following Equation (13):

$$x = x_{min} = l \quad \text{and} \quad Y_2 = 0 \tag{13}$$

Movement speed v_2 and force Y_1, in this case, can be found by solving Equation (10) under the conditions of Equation (13).

Step 4. If the bogie is freely fixed ($Y_2 = 0$), then, in Equation (10), there are three unknowns: Y_1, v, and x. In this case, the condition in Equation (14) is relevant:

$$v_1 < v < v_2 \quad \text{and} \quad x_{min} < x < x_{max} \tag{14}$$

Therefore, it is possible to obtain the remaining two unknowns by setting discrete values of v or x in Equation (10). The calculation process is greatly simplified when setting values of the parameter x.

Step 5. If the design speed of the wagon, v_k, is higher than v_2, it is necessary to build the third zone of the horizontal dynamic calculations, i.e., the zone of maximum displacement. In this case, the condition in Equation (15) is valid:

$$x = x_{min} = l \tag{15}$$

Therefore, by setting movement speed values in the interval between v_2 and v_k, the full reaction forces, Y_1 and Y_2, can be determined.

The methodology proposed above allows us to determine the full reactions, Y_1 and Y_2, of the first- and second-wheel axles of each bogie at different speeds, curve radii, specific track parameters, different wheel loads, different bogie wheel axle distances, and other parameters.

2.2. Methodology for Theoretical Determination of the Horizontal Load Force between the Inner (Non-Attacking) Wheel Y_{ji} of the Investigated Wheel Axle and the Inner Rail

The inner wheel of the examined wheel axle does not contact its flange with the corresponding rail. Therefore, the horizontal force acting between them arises from the frictional forces, which are determined via Equation (16):

$$Y_{ji} = \mu.Q_{ji} = \mu.\left(2.Q_{nom} - Q_{ja,min}\right) \tag{16}$$

where μ is the coefficient of friction between the wheel and the rail, assumed to be equal to 0.4 for clean and dry rails, Q_{ji} is the vertical load force of the inner wheel (index i) on axle j, Q_{nom} is the nominal vertical load force of the wagon wheels, and $Q_{ja,min}$ is the minimum vertical force acting on the outer (attacking) wheel of axle j. It is determined in accordance with the methodology given in Section 2.3. of this paper. Q_{nom} is determined using the ratio of the force from the weight of the wagon Q and the number of wheels of the vehicle N, as given in Equation (17):

$$Q_{nom} = \frac{Q}{N} \tag{17}$$

2.3. Methodology for Theoretical Determination of the Smallest Value of the Vertical Reaction of the Wheel, $Q_{jk,,min}$, Calculated during Torsion of the Wagon Frame

The proposed methodology for the theoretical determination of the minimum value of the vertical reaction of the wheels, $Q_{jk,,min}$, allows us to obtain the corresponding maximum value of this parameter, $Q_{jk,,max}$. The calculations are carried out in the following sequence:

1. The frame of the wagon is loaded with an arbitrary force, ΔF_p (Figure 7), according to UIC Leaflet 432 [28], and the deflection of the frame, Δz_p, in the area around the lateral supports is determined (Figure 8).

2. In accordance with EN 14363 [9], the minimum deflection of the frame Δz^* is determined, which should be reached during the real (in situ) testing of the wagon. It is determined via Equation (18), subject to the requirement in Equation (19). In this case, $2a^*$ is valid for wagon frames with pivot distances between 4 and 30 m.

$$\Delta z^* = g^* .2a^* \tag{18}$$

$$g^* = \frac{15}{2a^*} + 2 \tag{19}$$

3. Recalculation of the force ΔF_p from step 1 for loading the wagon frame to achieve the minimum deflection Δz^* according to Equation (20):

$$\Delta F_{z*} = \frac{\Delta F_p . \Delta z^*}{\Delta z_p} \tag{20}$$

The result of Equation (20) gives the force that acts on one side of the bogie in the area around the lateral support in Figure 8. This means that force ΔF_{z*} significantly loads the two unilaterally located wheels and significantly less for the other two.

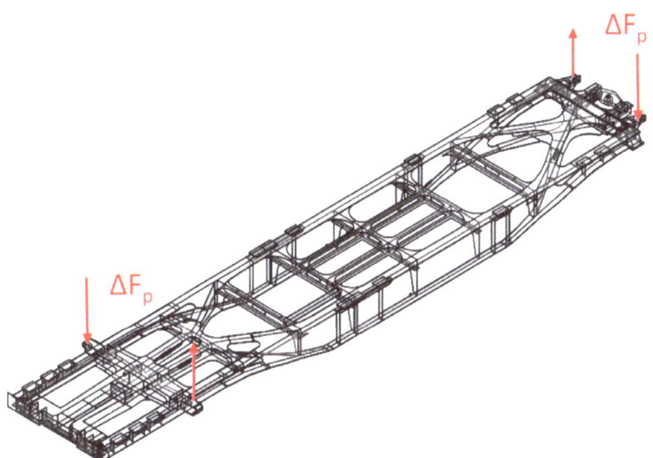

Figure 7. Forces acting on wagon frame during torsion tests.

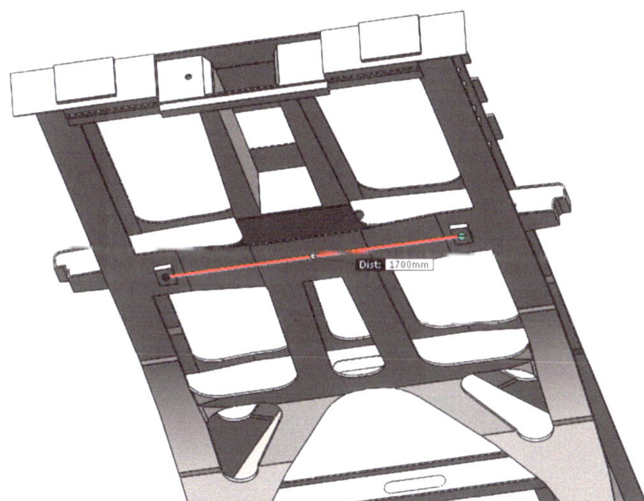

Figure 8. Area of lateral support in which the deflection of the frame is measured.

4. The force ΔF_{z*} is then transmitted from the lateral support to the side beams of the bogie with a value of $\Delta F'_{z*max}$ and $\Delta F'_{z*min}$ according to Equations (21) and (22). The corresponding distances, b_{1F} and b_s, are shown in Figure 9.

$$\Delta F'_{z*,max} = \frac{\Delta F_{z*} . (b_{1F} + b_s)}{2 b_{1F}} \tag{21}$$

$$\Delta F'_{z*,min} = \Delta F_{z*} - \Delta F'_{z*,max} \tag{22}$$

From the side beam, the forces $\Delta F'_{z*max}$ and $\Delta F'_{z*min}$ are distributed between the two axle journals of the overloaded and the two axle journals of the unloaded wheels, with the forces $\Delta F'_{z*max}$ and $\Delta F'_{z*min}$ acting on the first (attacking) wheel axle, defined using Equations (23) and (24):

$$\Delta F'_{1z*,max} = \frac{\Delta F'_{z*,max}}{2} \tag{23}$$

$$\Delta F'_{1z*,min} = \frac{\Delta F'_{z*,min}}{2} \tag{24}$$

From the corresponding axle journal, the forces from Equations (23) and (24) cause additional reactions in the two wheels, with values defined in Equations (25) and (26):

$$\Delta Q_{1,max} = \frac{\Delta F'_{1z*,max} \cdot (b_{1F} + b_0) - \Delta F'_{1z*,min} \cdot (b_{1F} - b_0)}{2b_0} \tag{25}$$

$$\Delta Q_{1,min} = \Delta F'_{1z*,max} + \Delta F'_{1z*,min} - \Delta Q_{1,max} \tag{26}$$

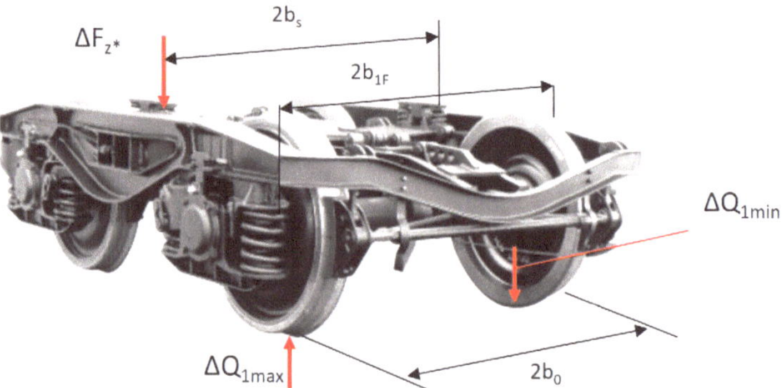

Figure 9. Forces acting on the standard Y25 bogie with distances used for calculations.

5. The minimum value of the wheel reaction, $Q_{jk,,min}$, is determined using Equation (27), and the maximum value is evaluated using Equation (28), respectively:

$$Q_{jk,min} = Q_{nom} + \Delta Q_{1,min} \tag{27}$$

$$Q_{jk,max} = Q_{nom} + \Delta Q_{1,max} \tag{28}$$

where the force Q_{nom} is determined via Equation (17).

3. Results from the Theoretical Derailment Safety Assessment

This study was conducted only for the first bogie axle of a Sggmrss series wagon (90 feet) in an unloaded condition equipped with a standard Y25 bogie. The reason for this is that theoretical analyses categorically state that the first wheel axle of an unloaded wagon is most at risk of derailment. This conclusion is also confirmed in the test results of all wagons from the reference list given in UIC Leaflet 530-2 [29]. This was also found during field tests of the same wagon [11,30]. The initial data used in the theoretical study are as follows:

- Tare weight of the wagon, 27.5 t;
- Curve radius, $R = 150$ m;
- Clearance between flanges and rail threads in a straight section of the track, equal to $\delta = 0.01$ m;

- Additional tracks widening in a curved section, $\delta = 0.002$ m (in accordance with the test data of the wagon [30]);
- Coefficient of friction between the rail and the wheel, $\mu = 0.4$;
- Wheel axle distance, $a^+ = 1.8$ m;
- Pivot distance (for one wagon section only), $a^* = 11.995$ m;
- Speed of passing through the curve, $v = 7$ km/h (in accordance with the test data of the wagon [30]);
- Wind pressure, $W = 0$ N/m^2 (in accordance with the test data of the wagon [30]);
- Distance between the rolling circles of the two wheels of the same axle, $2b_0 = 1.5$ m;
- Transverse distance between the axle journals, $2b_{jF} = 2.0$ m;
- Distance between the side supports on the bogie, $2b_s = 1.7$ m;
- Overhang of the outer rail, $h = 0.15$ m;
- Earth acceleration, $g = 9.81$ m/s^2.

For the theoretical determination of the safety criterion against derailment of a Sggmrss wagon (90 feet), the methods described in detail in Section 2 of this paper were applied. The results from the calculations conducted with the mentioned methodology are given in Table 2.

Table 2. Results from the calculation needed for determination of safety against derailment.

Parameter	Value	Remark
v_1	58.3 km/h	Methodology from Section 2.1.
$Y_1 = Y_{1a}$	24.718 kN	Methodology from Section 2.1.
Y_{1i}	-14.024 kN	Methodology from Section 2.2.
g^*	3.251‰	Equation (19)
Δz^*	39 mm	Equation (18)
ΔF_p	50 kN	The selected load value for the torsional stiffness test [30]
Δz_p	0.08265 mm	Deflection of the frame under the Load, ΔF_p, determined in [31]
ΔF_{z^*}	23.59 kN	Equation (20)
$\Delta F'_{z^*,max}$	21.82 kN	Equation (21)
$\Delta F'_{z^*,min}$	1.769 kN	Equation (22)
$\Delta F'_{1z^*,max}$	10.909 kN	Equation (23)
$\Delta F'_{1z^*,min}$	0.885 kN	Equation (24)
$\Delta Q_{1,max}$	12.58 kN	Equation (25)
$\Delta Q_{1,min}$	-0.7862 kN	Equation (26)
Q_{nom}	22.48 kN	Equation (17)
$Q_{jk,min}$	21.695 kN	Equation (27)
$Q_{jk,max}$	35.061 kN	Equation (28)

With the data from Table 2, the final assessment of safety against derailment can be conducted. This is performed using Equation (1), and the calculated value is equal to 1.017. According to [9], when using the theoretical assessment methods, the limit value of 1.2 is reduced by 10%, which means that the limit value of Nadal's criterion should be set to 1.08 and compared with the calculated value, as shown in Equation (29).

$$\left(\frac{Y}{Q} \right)_{ja} = 1.017; \qquad 1.017 \leq 1.08 \tag{29}$$

The obtained value of the safety criterion, 1.017, is lower than the limit value, 1.08, meaning that, for wagon series Sggmrss, the requirement for safety against derailment is fulfilled.

4. Discussion

The partial results from the experimental study are given in [11,30]. In [11], as well as in this paper, only an excerpt of the results is presented due to the sensitivity and confidentiality of the data contained in the report [30], as requested by the wagon manufacturer. For the final assessment in tests, not all parameters from Table 2 were determined, but only a few of them, which are necessary for the assessment and measured directly or indirectly. These parameters are Y_{1i}, Δz^*, Q_{nom}, and $Q_{jk,,min}$. A comparison of the results from tests and calculations is given in Table 3.

Table 3. Comparison of results from the calculation and test.

Parameter	Value from Calculation	Value from Test
Y_{1i}	-14.024 kN	-13.5 kN
Δz^*	39 mm	40.1 mm
Q_{nom}	22.48 kN	22.10 kN
$Q_{jk,min}$	21.695 kN	19.81 kN

The values for the lateral force, Y_{1i}, were measured in the test on the flat track with a radius of 150 m, while forces Q_{nom} and $Q_{jk,min}$ were measured on the twist test rig, as well as Δz^* [11,30]. It should be noted that all values of parameters measured in the tests are the average values from different numbers of tests (seven tests on twist rig and three on flat track) conducted. In [30], the measurement uncertainty was determined at 1.4% for vertical forces and 1.2% for displacements and twists. This, along with wagon imperfections caused by production, welding, and other influential factors, explains the deviation from the theoretical assumptions.

The results from the tests also confirm that the safety against derailment for this wagon fulfills the requirements. The value obtained in the tests is equal to 1.03 (Table 1 in [11]). It should be mentioned that the limit value in this case is set to 1.2, as stated in Equation (1) and Reference [9].

Table 3 shows that the values of measured and calculated parameters are close enough (in order of $\pm10\%$). This gives reason to claim that the proposed theoretical safety assessment method delivers very good results and can be used for the safety assessment of similar wagons.

The advantages of our method compared to similar methods are mainly the use of fewer input parameters, simplicity, and no need for complex simulations. On the other hand, the proposed method uses some initial parameters for which assumptions are made (the value of the coefficient of friction, μ) or for which their values are determined in tests (additional tracks widening in a curved section (δ) or wind pressure (W)). With other values for these parameters, safety against derailment would have values other than the calculated values, and the final fulfillment could be questionable. This is the reason why it would be necessary to conduct more assessments using the proposed method and on different wagon series for future research. This would help to additionally verify the results from calculations.

5. Conclusions

It is possible to save costs and time for the acceptance procedure by using the resulting parameters from the theoretical calculations of safety against derailment with the proposed method and by comparing them with the parameters of the reference wagons that have undergone real tests. This approach is allowed by clause 6.1.5.2.6 of EN 14363:2019 [9]. In this paper, we show that the proposed methodology for calculating safety against derailment gives good results and is verified in a test on a real object, the wagon Sggmrss series. For future research, it would be necessary to conduct more theoretical assessments using the proposed method on different wagon series, as doing so would help to verify the calculation results. The proposed theoretical safety assessment method can be used to study the safety regarding derailment for other new wagons and railway vehicles.

However, the final decision on whether the proposed method is appropriate for the assessment of safety in terms of preventing derailment lies with the notified body. If the authorities are satisfied with the results obtained in the theoretical calculation, this method could be allowed to be used for the assessment.

Author Contributions: Conceptualization, V.S. and S.P.; methodology, V.S. and S.P.; software, S.S. and V.M.; validation, V.S. and S.P.; formal analysis, S.S. and V.M.; investigation, S.S.; resources, S.S. and S.P.; data curation, S.S. and V.M.; writing—original draft preparation, V.S. and S.P.; writing—review and editing, V.S. and S.P.; visualization, V.M.; supervision, V.S.; project administration, V.S.; funding acquisition, V.S. All authors have read and agreed to the published version of the manuscript.

Funding: This research was supported by the European Regional Development Fund within the Operational Programme "Science and Education for Smart Growth 2014–2020", under the Project CoE "National center of mechatronics and clean technologies", under Grant BG05M2OP001-1.001-0008-C01.

Institutional Review Board Statement: Not applicable.

Informed Consent Statement: Not applicable.

Data Availability Statement: Data are contained within this article and References [11,30,31]. Reference [11] is openly available in [Springer Link] at [https://doi.org/10.1007/978-3-031-19499-3_13]. References [30,31] are not openly available because they contain sensitive and confidential data.

Acknowledgments: The authors would like to thank the Research and Development Sector at the Technical University of Sofia for the financial and technical support.

Conflicts of Interest: The authors declare no conflict of interest.

References

1. Wilson, N.; Fries, R.; Wittea, M.; Haigermoser, A.; Wrang, M.; Evans, J.; Orlova, A. Assessment of safety against derailment using simulations and vehicle acceptance tests: A worldwide comparison of state-of-the-art assessment methods. *Veh. Syst. Dyn.* **2011**, *49*, 1113–1157. [CrossRef]
2. Hertz, H. Über die Berührung fester elastischer Körper. *J. Reine Angew. Math.* **1881**, *92*, 156–171.
3. Klingel, W. Über den Lauf Eisenbahnwagen auf Gerarder Bahn. *Organ Fortscr. Eisenbahnwesens* **1883**, *20*, 113–123.
4. Nadal, M.J. *Locomotives a Vapeur*, 1st ed.; Collection Encyclopedie Scientifique: Paris, France, 1908.
5. Kardas-Cinal, E. Selected problems in railway vehicle dynamics related to running safety. *Arch. Trans.* **2014**, *31*, 37–45. [CrossRef]
6. Zeng, J.; Guan, Q. Study on flange climb derailment criteria of a railway wheelset. *Veh. Syst. Dyn.* **2008**, *46*, 239–251. [CrossRef]
7. Zeng, J.; Wu, P. Study on the wheel/rail interaction and derailment safety. *Wear* **2008**, *265*, 1452–1459. [CrossRef]
8. Konowrocki, R.; Chojnacki, A. Analysis of rail vehicles' operational reliability in the aspect of safety against derailment based on various methods of determining the assessment criterion. *Maint. Reliab.* **2020**, *22*, 73–85. [CrossRef]
9. EN 14363:2019; Railway Applications—Testing and Simulation for the Acceptance of Running Characteristics of Railway Vehicles—Running Behaviour and Stationary Tests. European Committee for Standardization: Brussels, Belgium, 2019.
10. UIC CODE 518; Testing and Approval of Railway Vehicles from the Point of View of Their Dynamic Behaviour—Safety—Track Fatigue—Running Behaviour, 5th ed. UIC—Railway Technical Publications: Paris, France, 2009.
11. Milković, D.; Simić, G.; Radulović, S.; Lučanin, V.; Kostić, A. Experimental approach to assessment of safety against derailment of freight wagons. In *Experimental Research and Numerical Simulation in Applied Sciences. CNNTech 2022*; Lecture Notes in Networks and Systems, 564; Mitrovic, N., Mladenovic, G., Mitrovic, A., Eds.; Springer: Cham, Switzerland, 2023.
12. Zeng, J.; Wie, L.; Wu, P. Safety evaluation for railway vehicles using an improved indirect measurement method of wheel–rail forces. *J. Mod. Transp.* **2016**, *24*, 114–123. [CrossRef]
13. Zeng, Y.; Wilson, N.; Lundberg, W.; Walker, R.; Shu, X.; Jones, M. Geometric Criterion for Flange Climb Derailment and IWS-Based Implementation. In *Advances in Dynamics of Vehicles on Roads and Tracks II. IAVSD 2021*; Lecture Notes in Mechanical Engineering; Orlova, A., Cole, D., Eds.; Springer: Cham, Switzerland, 2022.
14. Ishida, H.; Miyamoto, T.; Maebashi, E.; Doi, H.; Iida, K.; Furukawa, A. Safety Assessment for Flange Climb Derailment of Trains Running at Low Speeds on Sharp Curves. *Quart. Rep. RTRI* **2006**, *47*, 65–71. [CrossRef]
15. Molatefi, H.; Mazraeh, A. On the investigation of wheel flange climb derailment mechanism and methods to control it. *J. Theor. Appl. Mech.* **2016**, *54*, 541–550. [CrossRef]
16. Matsumoto, A.; Michitsuji, Y.; Ichiyanagi, Y.; Sato, Y.; Ohno, H.; Tanimoto, M.; Iwamoto, A.; Nakai, T. Safety measures against flange-climb derailment in sharp curve-considering friction coefficient between wheel and rail. *Wear* **2019**, *432–433*, 202931. [CrossRef]
17. Michalek, T.; Kohout, M. On the problems of lateral force effects of railway vehicles in S-curves. *Veh. Syst. Dyn.* **2022**, *60*, 2739–2757. [CrossRef]
18. Xu, J.; Zheng, Z.; Wang, P.; Wang, S. Influence of the motion conditions of wheelsets on dynamic derailment behavior of a bogie in railway turnouts. *Veh. Syst. Dyn.* **2022**, *60*, 3720–3742.

19. Lai, J.; Xu, J.; Liao, T.; Zheng, Z.; Chen, R.; Wang, P. Investigation on train dynamic derailment in railway turnouts caused by track failure. *Eng. Fail. Anal.* **2022**, *134*, 106050. [CrossRef]

20. Han, Y.; Zhang, X.; Wang, L.; Zhu, Z.; Cai, C.S.; He, X. Running Safety Assessment of a Train Traversing a Long-Span Bridge Under Sudden Changes in Wind Loads Owing to Damaged Wind Barriers. *Int. J. Struct. Stab. Dyn.* **2022**, *22*, 2241010. [CrossRef]

21. Pagaimo, J.; Magalhães, H.; Costa, J.N.; Ambrósio, J. Derailment study of railway cargo vehicles using a response surface methodology. *Veh. Syst. Dyn.* **2022**, *60*, 309–334. [CrossRef]

22. Eom, B.G.; Lee, H.S. Assessment of running safety of railway vehicles using multibody dynamics. *Int. J. Precis. Eng. Manuf.* **2010**, *11*, 315–320. [CrossRef]

23. Bruni, S.; Meijaard, J.P.; Rill, G.; Schwab, A.L. State-of-the-art and challenges of railway and road vehicle dynamics with multibody dynamics approaches. *Multibody Syst. Dyn.* **2020**, *49*, 1–32. [CrossRef]

24. Chudzikiewicz, A.; Opala, M. Application of Computer Simulation Methods for Running Safety Assessment of Railway Vehicles in Example of Freight Cars. *Appl. Mech. And Mat.* **2009**, *9*, 61–69.

25. Boronenko, Y.; Orlova, A.; Iofan, A.; Galperin, S. Effects that appear during the derailment of one wheelset in the freight wagon: Simulation and testing. *Veh. Syst. Dyn.* **2006**, *44*, 663–668. [CrossRef]

26. Evans, J.; Berg, M. Challenges in simulation of rail vehicle dynamics. *Int. J. Veh. Mech. Mob.* **2009**, *47*, 1023–1048. [CrossRef]

27. Stoliov, V.; Slavchev, S. *Wagons (in Bulgarian)*, 1st ed.; Technical University Sofia: Sofia, Bulgaria, 2014; pp. 150–156, ISBN 978-619-167-135-9.

28. *UIC CODE 432*; Wagons. Running Speeds. Technical Conditions to be Observed, 12th ed. UIC—Railway Technical Publications: Paris, France, 2008.

29. *UIC CODE 530-2*; Wagons—Running Safety, 7th ed. UIC—Railway Technical Publications: Paris, France, 2011.

30. *Test Report of Safety against Derailment of Sggmrss Wagon*; No: LSV 6/19 14.04.2019; Laboratory of Rail Vehicles, University of Belgrade: Belgrade, Serbia, 2019.

31. *Report "Strength Analysis of Wagon Series Sggmrss"*; Project-VS Ltd.: Sofia, Bulgaria, 2017.

Article

Influence of Variable Height of Piers on the Dynamic Characteristics of High-Speed Train–Track–Bridge Coupled Systems in Mountainous Areas

Yingying Zeng [1,2,3], Lizhong Jiang [1,2,3], Zhixiong Zhang [3], Han Zhao [3], Huifang Hu [3], Peng Zhang [3], Fang Tang [4] and Ping Xiang [1,3,*]

1. National Engineering Research Center of High-Speed Railway Construction Technology, Changsha 410075, China; lzhjiang@csu.edu.cn (L.J.)
2. China Railway Group Limited Corporation, Beijing 100039, China
3. School of Civil Engineering, Central South University, Changsha 410075, China
4. Nursing Department, Changsha Health Vocational College, Changsha 410075, China
* Correspondence: pxiang2-c@my.cityu.edu.hk

Abstract: With the increase in the occupancy ratio of bridges and the speed of trains, the probability of trains being located on bridges during earthquakes increases, and the risk of derailment increases. To investigate the influence of unequal-height piers on the dynamic response of high-speed railway train bridge systems, a seismic action model of high-speed train–track–bridge dynamic systems was established based on the in-house code using the finite element method and multi-body dynamics method. It is found that (1) compared to equal-height piers, the peak lateral dynamic response of unequal-height piers (with gradually increasing pier heights) decreases, while the peak vertical dynamic response increases; (2) the peak lateral dynamic response of unequal-height piers (with a steep increase in pier height) increases sharply, while the peak vertical dynamic response decreases; and (3) the safety indicators of equal-height piers are significantly superior to the two unequal-height pier operating conditions.

Keywords: high-speed railway; earthquake; bridge; coupling vibration; dynamic response; unequal-height pier

Citation: Zeng, Y.; Jiang, L.; Zhang, Z.; Zhao, H.; Hu, H.; Zhang, P.; Tang, F.; Xiang, P. Influence of Variable Height of Piers on the Dynamic Characteristics of High-Speed Train–Track–Bridge Coupled Systems in Mountainous Areas. *Appl. Sci.* **2023**, *13*, 10271. https://doi.org/10.3390/app131810271

Academic Editor: Jose Ramon Serrano

Received: 12 June 2023
Revised: 17 July 2023
Accepted: 27 July 2023
Published: 13 September 2023

1. Introduction

On 1 October 1964, with the opening of the Tokaido Shinkansen in Japan, the world's first high-speed railway was officially put into operation. There has subsequently been a wave of building high-speed railways around the world, and high-speed railways have developed rapidly [1–3]. Due to the limitations of computational methods, the initial research on vehicle bridge vibration problems mainly focused on independent analysis of vehicle and bridge models, and the models were too simplistic to best reflect the real vehicle bridge dynamic response [4]. The emergence of electronic computers and the development of finite element technology have further promoted the research on vehicle–bridge coupling vibration [5]. Matsuura Akio [6] used the energy method to derive the motion equation of the dynamic interaction between vehicles and bridges in high-speed railways. Chu K.H. et al. [7] first studied the vehicle–bridge coupling vibration system, considering that the vehicle body is a 3-degree of freedom rigid body, and established a vehicle–bridge coupling vibration calculation model. Dhar C.L. [8] established a vehicle model consisting of a vehicle body, a bogie, and a wheelset. The train–bridge coupled system was studied using a spring connection between the vehicle body and the bogie, and between the bogie and the wheelset, with the wheelset always in contact with the steel rail. Subsequently, train analysis models constructed with multiple rigid bodies were gradually recognized and adopted by scholars. This paper is also based on this to construct a 38-degree-of-freedom train model.

Since the 1960s, there have been multiple high-speed train derailments worldwide, which caused serious casualties and economic losses. One of the reasons for derailment is earthquakes [9–11]. Therefore, it is necessary to consider seismic effects in the study of the dynamic response of train–track–bridge coupled systems. Currently, many scholars have conducted extensive research on the dynamic response of vehicle–bridge coupling systems under seismic excitation. Lei et al. [12] used the relative motion method to solve seismic influences and analyzed the operational safety of high-speed trains passing over rigid-frame bridges with high piers under seismic influences. Xiang et al. [13,14] studied the dynamic response of trains and bridges under emergency braking during earthquakes, as well as the safety of train operation. Zeng [15], Han [16], Guo [17], Variyavwala J.P. [18] and others used cable-stayed bridges as examples to study the coupling dynamic interaction between trains and cable-stayed bridges under earthquake action, and obtained valuable conclusions.

In order to meet the requirements of smooth and stable operation of high-speed trains, land saving, and not affecting ground transportation, the proportion of high-speed railway bridges is constantly increasing. Many studies have shown that the characteristics of bridges significantly affect the dynamic response and safety of train–track–bridge coupled systems. It is necessary to study the effect of bridge characteristics on the train–track–bridge coupled system. Bridge characteristics include bridge structural stiffness, bridge deformation, abutment settlement, pier height, and so on. Zhai [19,20] analyzed the impact of bridge structural stiffness on the dynamic response of the coupling system. Their study shows that when the beam stiffness or lateral stiffness of bridge piers is insufficient, the main dynamic indicators of trains and bridges significantly increase with the decrease in stiffness. Fan et al. [21] found that the damping coefficient of bridges has great influence on the vertical acceleration and mid-span acceleration of vertical and horizontal beams. Chen et al. [22–24] theoretically derived an analytical expression for the mapping relationship between pier settlement and rail deformation in a dual block ballastless track–bridge system, and proposed a method for determining the safety threshold for high-speed railway pier settlement. Zhang et al. [25] studied the impact of differential settlement of high-speed railway bridge piers on various railway performance-related criteria. The wheel load reduction rate increases with differential settlement of bridge piers, and the vertical acceleration increases with differential settlement of bridge piers and train speed, respectively. Guo et al. [26] determined the mapping relationship between bridge deformation and train operation safety. This provides a convenient method for engineers to evaluate and maintain high-speed railway bridges. Feng et al. [27] studied the impact of uneven settlement of side piers on the stability and safety of train operation. Their study shows that under the condition of uneven settlement of side piers, the stability of train operation will be significantly affected.

At present, most bridge piers are of equal height, but in mountainous areas, due to geographic factors, an equal height of bridge piers cannot be guaranteed when constructing bridges, and there is an effect of pier height on the dynamic response of the system. Therefore, in this paper, three kinds of pier height conditions are set up, and the seismic effect is considered to investigate the influence law of unequal-height piers on the dynamic response of high-speed train–track–bridge coupled systems.

2. Train–Track–Bridge Coupled Vibration Model

2.1. Train Model

The single-section train model is shown in Figure 1. Each carriage consists of four wheelsets, a car body, and two bogies at the front and rear. It is assumed that the car body, wheel pairs and bogies are rigid bodies, and the car body and bogies, and bogies and wheelsets are simulated to be connected by linear spring dampers in longitudinal, lateral and vertical directions [26,28,29]. The car body contains six degrees of freedom, each wheel pair contains five degrees of freedom, and each bogie contains six degrees of freedom. That is, a single-section train model with 38 degrees of freedom is established, and the distribution of degrees of freedom is shown in Table 1.

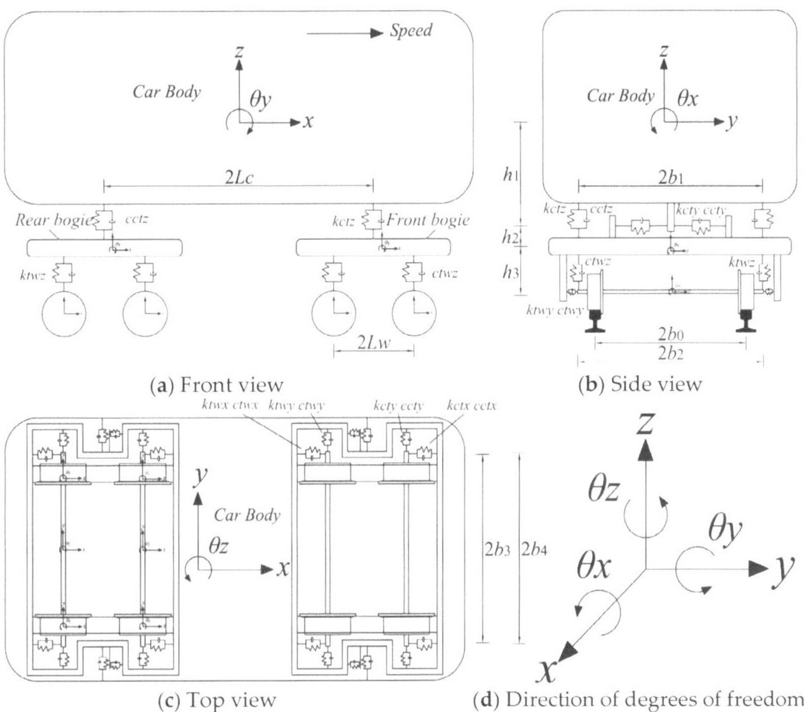

(**a**) Front view   (**b**) Side view

(**c**) Top view   (**d**) Direction of degrees of freedom

Figure 1. Single train model.

Table 1. Distribution of 38 degrees of freedom for a single train section.

	Vertical	Longitudinal	Lateral	Yaw	Roll	Pitch
Car body	z_c	x_c	y_c	θ_{zc}	θ_{xc}	θ_{yc}
Front bogie	z_{tf}	x_{tf}	y_{tf}	θ_{ztf}	θ_{xtf}	θ_{ytf}
Rear bogie	z_{tb}	x_{tb}	y_{tb}	θ_{ztb}	θ_{xtb}	θ_{ytb}
First wheel set	z_{w1}	x_{w1}	y_{w1}	θ_{zw1}	θ_{xw1}	-
Second wheel set	z_{w2}	x_{w2}	y_{w2}	θ_{zw2}	θ_{xw2}	-
Third wheel set	z_{w3}	x_{w3}	y_{w3}	θ_{zw3}	θ_{xw3}	-
Fourth wheel-set	z_{w4}	x_{w4}	y_{w4}	θ_{zw4}	θ_{xw4}	-

The subscript symbols c, wi, tf, and bf in Table 1 denote the train body, the i-th wheel set, and the front and rear bogies of the train, respectively. The mass $\mathbf{M}_{cc}$, stiffness $\mathbf{K}_{cc}$, and damping matrices $\mathbf{C}_{cc}$ of the train are obtained from the multi-rigid body dynamics.

2.2. Track and Bridge Model

This paper takes CRTS II slab ballastless track as an example to establish the track model. The track structure is mainly composed of a base, CA mortar layer, track board, elastic fasteners, rails and other components [30,31]. The steel rail is simulated as a beam element, and the track plate and base are both simulated as plate elements. The base is connected to the bridge, the track plate is connected to the base, and the steel rail is connected to the track plate as a linear spring damper, as shown in Figure 2.

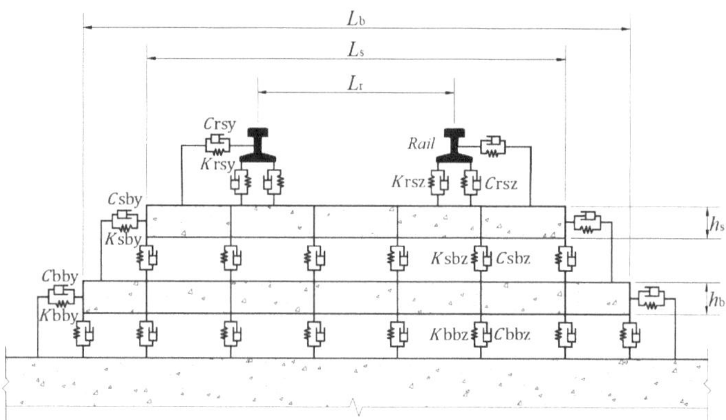

Figure 2. Cross section view of track model.

The bridge adopts a simply supported concrete box girder at both ends and a concrete pier, as shown in Figure 3. The connection between the steel beam and the steel rail is also considered as a linear spring damper connection. A bridge model is established based on the finite element method, and the pier and beam are simulated as Euler–Bernoulli beam elements. Each element has two nodes, and each node contains six degrees of freedom.

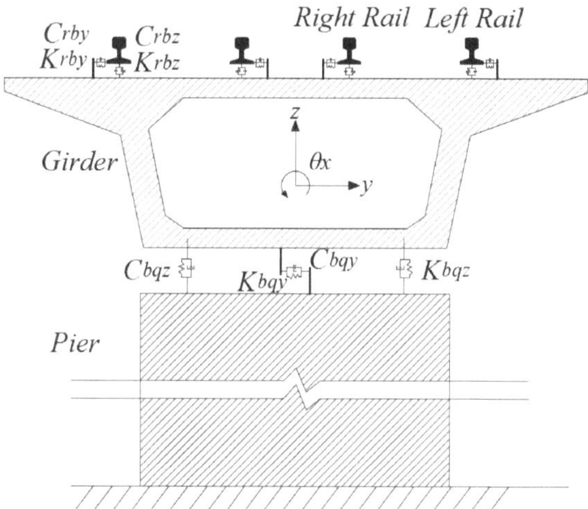

Figure 3. Cross section view of bridge model.

The mass matrix $\mathbf{M}_{rr}$, $\mathbf{M}_{bb}$, stiffness matrix $\mathbf{K}_{rr}$, $\mathbf{K}_{bb}$ and damping matrix $\mathbf{C}_{rr}$, $\mathbf{C}_{bb}$ of the bridge and track can be easily calculated via the finite element method [32,33]. The damping matrix $\mathbf{C}_{bb}$ of the bridge adopts Rayleigh damping, which can be composed of a mass matrix $\mathbf{M}_{bb}$ and stiffness matrix $\mathbf{K}_{bb}$ in a linear way.

$$\mathbf{C}_{bb} = \frac{2\omega_1\omega_2\zeta}{\omega_1 + \omega_2}\mathbf{M}_{bb} + \frac{2\zeta}{\omega_1 + \omega_2}\mathbf{K}_{bb} \tag{1}$$

where ω_1, ω_2 denote first-order and second-order natural frequencies (rad) of the bridge structure, respectively, and ζ denotes the bridge damping ratio. Here, it is assumed that the first-order and second-order corresponding bridge damping ratios are equal. Moreover, the stiffness and damping matrices ($\mathbf{K}_{rb}$, $\mathbf{K}_{br}$, $\mathbf{C}_{rb}$, $\mathbf{C}_{br}$) of the interaction between the track and the bridge are also easily obtained, with the following relationships:

$$\mathbf{K}_{br} = \mathbf{K}_{rb}^{\mathrm{T}}, \mathbf{C}_{br} = \mathbf{C}_{rb}^{\mathrm{T}} \tag{2}$$

2.3. Rail Irregularity

In reality, when trains pass over bridges, there are irregularities in the running surface of the steel rails, which can be divided into four types: high and low irregularities, track orientation irregularities, horizontal irregularities, and gauge irregularities [34]. In this paper, the low interference rail irregularity power spectral density (PSD) of Germany is used to describe the rail irregularity state of high-speed railways [29,35].

$$\begin{cases} S_v(\Omega) = \dfrac{A_v \cdot \Omega_c^2}{(\Omega^2 + \Omega_r^2)(\Omega^2 + \Omega_c^2)} \\[2mm] S_a(\Omega) = \dfrac{A_a \cdot \Omega_c^2}{(\Omega^2 + \Omega_r^2)(\Omega^2 + \Omega_c^2)} \\[2mm] S_c(\Omega) = \dfrac{A_v \cdot b_0^{-2} \cdot \Omega_c^2 \cdot \Omega^2}{(\Omega^2 + \Omega_r^2)(\Omega^2 + \Omega_c^2)(\Omega^2 + \Omega_s^2)} \\[2mm] S_s(\Omega) = \dfrac{A_g \cdot \Omega_c^2 \cdot \Omega^2}{(\Omega^2 + \Omega_r^2)(\Omega^2 + \Omega_c^2)(\Omega^2 + \Omega_s^2)} \end{cases} \tag{3}$$

where, $S_v(\Omega)$, $S_a(\Omega)$, $S_c(\Omega)$, $S_s(\Omega)$ denote the irregularity PSD of the vertical rail profile, rail cross-level, rail alignment and rail distance, respectively $(\mathrm{m}^2/(\mathrm{rad/m}))$, A_v, A_a, A_g denote the constants of roughness $(\mathrm{m}^2 \cdot \mathrm{rad/m})$, Ω_c, Ω_r, Ω_s denote the truncation frequency (rad/m), b_0 denotes the half distance between two rails (m), and the corresponding parameters are shown in Table 2.

Table 2. Parameters of PSD of rail irregularity.

Ω_c	Ω_r	Ω_s	A_g	A_v	A_a
0.8264	0.0206	0.438	5.32×10^{-8}	2.119×10^{-7}	4.032×10^{-7}

2.4. Train–Track–Bridge Coupled Vibration

According to the mass matrix, stiffness matrix and damping matrix obtained using the finite element method, multi-rigid body dynamics and other processing methods, based on the energy principle, the train–track–bridge coupled vibration equation can be derived, as shown below:

$$\begin{bmatrix} \mathbf{M}_{cc} & 0 & 0 \\ 0 & \mathbf{M}_{rr} & 0 \\ 0 & 0 & \mathbf{M}_{bb} \end{bmatrix} \begin{Bmatrix} \ddot{\mathbf{X}}_{cc} \\ \ddot{\mathbf{X}}_{rr} \\ \ddot{\mathbf{X}}_{bb} \end{Bmatrix} + \begin{bmatrix} \mathbf{C}_{cc} & \mathbf{C}_{cr} & 0 \\ \mathbf{C}_{rc} & \mathbf{C}_{rr} & \mathbf{C}_{rb} \\ 0 & \mathbf{C}_{br} & \mathbf{C}_{bb} \end{bmatrix} \begin{Bmatrix} \dot{\mathbf{X}}_{cc} \\ \dot{\mathbf{X}}_{rr} \\ \dot{\mathbf{X}}_{bb} \end{Bmatrix} + \begin{bmatrix} \mathbf{K}_{cc} & \mathbf{K}_{cr} & 0 \\ \mathbf{K}_{rc} & \mathbf{K}_{rr} & \mathbf{K}_{rb} \\ 0 & \mathbf{K}_{br} & \mathbf{K}_{bb} \end{bmatrix} \begin{Bmatrix} \mathbf{X}_{cc} \\ \mathbf{X}_{rr} \\ \mathbf{X}_{bb} \end{Bmatrix} = \begin{Bmatrix} \mathbf{Q}_{cc} \\ \mathbf{Q}_{rr} \\ 0 \end{Bmatrix} \tag{4}$$

where $\mathbf{X}_{cc}$, $\mathbf{X}_{rr}$ and $\mathbf{X}_{bb}$ denote the displacement vector of the train, rail and bridge, respectively, while $\mathbf{Q}_{cc}$, $\mathbf{Q}_{rr}$ denote the load train vector of the train and rail, respectively.

2.5. Equation Solving of System Coupled Vibration

Since the equations of motion are functional equations with respect to time, at present, numerical analysis methods are commonly used, such as the mean acceleration method, linear acceleration method, Wilson-θ method, Newmark's method, Runge–Kutta method, and Houbolt's method. In this paper, the MATLAB program solves the system dynamic response based on the Wilson-θ method.

The Wilson-θ method is an improvement on the linear acceleration method, which is one of the simplest and best methods to use, expanding the time step from Δt to $\theta \Delta t$, and unconditionally stabilizing at $\theta > 1.37$, and the MATLAB program in this paper takes $\theta = 1.4$. Since $\mathbf{M}$, $\mathbf{K}$, $\mathbf{C}$ vary with time, it is necessary to establish $\mathbf{M}_{t+\theta \Delta t}$, $\mathbf{K}_{t+\theta \Delta t}$, $\mathbf{C}_{t+\theta \Delta t}$ at each moment. At the moment t, the initial acceleration is calculated from the given initial velocity of the system and the initial displacement using the following equation:

$$\mathbf{M}_t \ddot{\mathbf{q}}_t + \mathbf{C}_t \dot{\mathbf{q}}_t + \mathbf{K}_t \mathbf{q}_t = \mathbf{Q}_t \tag{5}$$

Calculate $\mathbf{Q}_{t+\theta\Delta t}$ at each moment in time according to the linear law of change.

$$\mathbf{Q}_{t+\theta\Delta t} = \mathbf{Q}_t + \theta(\mathbf{Q}_{t+\Delta t} - \mathbf{Q}_t) \tag{6}$$

Calculate the equivalent stiffness matrix and the equivalent load matrix at time $t + \theta\Delta t$, respectively.

$$\begin{cases} \overline{\mathbf{K}}_{t+\theta\Delta t} = \mathbf{K}_{t+\theta\Delta t} + \dfrac{6}{(\theta\Delta t)^2}\mathbf{M}_{t+\theta\Delta t} + \dfrac{3}{\theta\Delta t}\mathbf{C}_{t+\theta\Delta t} \\[2ex] \overline{\mathbf{Q}}_{t+\theta\Delta t} = \mathbf{Q}_{t+\theta\Delta t} + \left(\dfrac{6\mathbf{q}_t}{(\theta\Delta t)^2} + \dfrac{6\dot{\mathbf{q}}_t}{\theta\Delta t} + 2\ddot{\mathbf{q}}_t \right)\mathbf{M}_{t+\theta\Delta t} + \left(\dfrac{3\mathbf{q}_t}{\theta\Delta t} + 2\dot{\mathbf{q}}_t + \dfrac{\theta\Delta t\ddot{\mathbf{q}}_t}{2} \right)\mathbf{C}_{t+\theta\Delta t} \end{cases} \tag{7}$$

Solve for $\mathbf{q}_{t+\theta\Delta t}$ at each moment using the following equation.

$$\overline{\mathbf{K}}_{t+\theta\Delta t} \cdot \mathbf{q}_{t+\theta\Delta t} = \overline{\mathbf{Q}}_{t+\theta\Delta t} \tag{8}$$

Finally, the acceleration, velocity, and displacement responses for each moment can be solved by the following equation:

$$\begin{cases} \ddot{\mathbf{q}}_{t+\Delta t} = \dfrac{6}{\theta(\theta\Delta t)^2}(\mathbf{q}_{t+\theta\Delta t} - \mathbf{q}_t) - \dfrac{6}{\theta^2\Delta t}\dot{\mathbf{q}}_t + \left(1 - \dfrac{3}{\theta}\right)\ddot{\mathbf{q}}_t \\[2ex] \dot{\mathbf{q}}_{t+\Delta t} = \dot{\mathbf{q}}_t + \dfrac{\Delta t}{2}\left(\ddot{\mathbf{q}}_{t+\Delta t} + \ddot{\mathbf{q}}_t\right) \\[2ex] \mathbf{q}_{t+\Delta t} = \mathbf{q}_t + \Delta t\dot{\mathbf{q}}_t + \dfrac{(\Delta t)^2}{6}\left(\ddot{\mathbf{q}}_{t+\Delta t} + 2\ddot{\mathbf{q}}_t\right) \end{cases} \tag{9}$$

3. Analysis and Calculation Parameters

3.1. Research Scenario

In order to analyze the influence of unequal-height piers on the dynamic characteristics of high-speed train–track–bridge coupling systems, this paper takes a three span simply supported box girder bridge as an example. As shown in Figure 4, the train speed is 350 km/h, the span length is 32 m, and the bridge consists of a round-ended solid pier with a height of 8 m and a cross-sectional area of 7.6 m², and a C35 cast-in-place concrete pier shaft. The structural damping ratio is 3%. The corresponding parameters of the bridge are shown in Table 3, and the track and wheel set parameters are detailed in the Ref. [36]. The train formation is: motor train + trailer × 6 + motor train. The corresponding parameters of the high-speed train and trailer are based on the German ICE3 train set parameters, as detailed in Ref. [37].

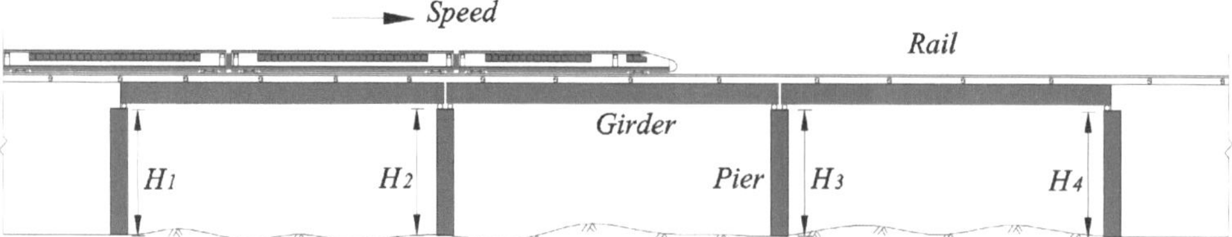

Figure 4. High-speed train–track–bridge coupled system.

Table 3. Some parameters of the bridge.

Parameters	Definitions	Values	Units
E_b	Elastic modulus	3.45×10^{10}	N/m^2
I	Mass moment of inertia of cross section	12.744	m^4
μ	Poisson's ratio	0.2	/
$\overline{m}_b$	Mass per unit length	2.972×10^4	kg/m
L_e	Length of element	3.2	m
ζ	Damping ratio	0.03	/

3.2. Pier Height Working Condition

In order to fully analyze the influence of unequal-height piers on the dynamic characteristics of a high-speed train–track–bridge coupling system, the equal-height pier condition is set, and two unequal-height pier conditions are considered, namely gradually increasing unequal-height piers and sharply increasing unequal-height piers. The pier height parameters are shown in Table 4 below:

Table 4. Pier height parameters under different working conditions.

Working Condition Type	H_1	H_2	H_3	H_4
Equal-height pier (m)	8	8	8	8
Gradually increasing (m)	8	10	12	14
Sharply increasing (m)	8	8	14	14

3.3. Earthquake Excitation Model

For high-speed railways, earthquakes not only directly affect the operation of high-speed trains and pose a threat of derailment to passengers, but also affect the construction and maintenance of high-speed railway lines. Once an earthquake occurs, it causes serious damage to infrastructures such as tracks and viaducts, greatly increasing maintenance costs and operational losses. Based on the above situation, it is of great significance to consider the seismic influence in vehicle–bridge vibration analysis.

Due to the limited number of existing seismic records, it is not enough to solely study the dynamic response of the system under earthquakes using existing records. Therefore, this article uses artificial seismic waves as seismic excitations. The mathematical model of seismic waves is simulated by using trigonometric series synthesis [35], and the artificial seismic wave is synthesized by using the simulated standard response spectrum. The acceleration equation of seismic ground motion is expressed in Equation (10):

$$\ddot{q}(t) = I(t) \cdot \sum_{k=1}^{n} A(\omega_k) \sin(\omega_k + \varphi_k) \tag{10}$$

$$I(t) = \begin{cases} \left(\frac{t}{t_1}\right)^2 & 0 \leq t \leq t_1 \\ 1 & t_1 \leq t \leq t_2 \\ e^{-c(t-t_2)} & t_2 \leq t \end{cases} \tag{11}$$

$$A(\omega_k) = \sqrt{4 S_g(\omega) \Delta\omega} \tag{12}$$

$$\omega_k = \omega_l + \left(k - \frac{1}{2}\right)\Delta\omega$$
$$\Delta\omega = \frac{\omega_u - \omega_l}{n}, k = 1, 2, \cdots n \tag{13}$$

where $I(t)$ denotes the intensity envelope function, which describes the process of earthquakes from appearance, enhancement, stationarity, and weakening, as shown in Figure 5. The values of each parameter of the intensity envelope function are referred to in Ref. [35]. c, $A(\omega_k)$, $S_g(\omega)$, ω_k, φ_k, ω_u, ω_l denote the attenuation coefficient of the weakening section, the amplitude of corresponding ω_k, the spectral density function of ground motion acceleration power, the k-th frequency, the random phase angle evenly distributed and mutually independent within $[0, 2\pi]$, and the upper and lower limits of frequency, respectively.

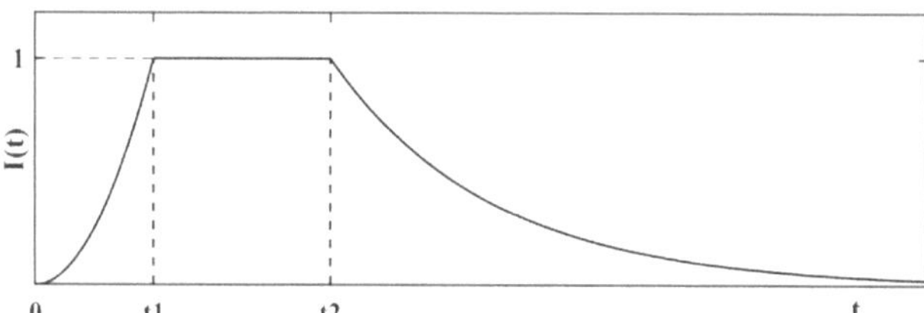

Figure 5. Strength envelope function diagram.

In this paper, the ground vibration power spectral density function model adopts the Clough–Penzien model that considers the low-frequency components.

$$S_g(\omega) = \left[\frac{\omega_g^4 + 4\zeta_g^4 \omega_g^2 \omega^2}{(\omega_g^2 - \omega^2)^2 + 4\zeta_g^2 \omega_g^2 \omega^2}\right] \times \left[\frac{\omega^4}{(\omega_f^2 - \omega^2)^2 + 4\zeta_f^2 \omega_f^2 \omega^2}\right] S_0 \tag{14}$$

where ω_g, ζ_g denote the site excellence frequency and site damping ratio, respectively, ω_f, ζ_f denote the ground filtering parameters for controlling low-frequency components of ground motion, and S_0 denotes the spectral intensity factor, as shown in Table 5.

Table 5. Seismic PSD parameter values.

Parameters	Site Classification			
	I	II	III	IV
ζ_g	0.728	0.775	0.822	0.868
ζ_f	0.411	0.557	1.140	2.208
$\omega_g(\mathrm{rad \cdot s^{-1}})$	24.763	18.656	13.491	9.848
$\omega_f(\mathrm{rad \cdot s^{-1}})$	0.453	0.355	0.154	0.082
$S_0(\mathrm{cm^2/s^2})$ (PGA = 0.1 g)	11.241	15.546	22.370	31.201
$S_0(\mathrm{cm^2/s^2})$ (PGA = 0.2 g)	43.028	59.512	85.598	119.304
$S_0(\mathrm{cm^2/s^2})$ (PGA = 0.4 g)	172.103	238.028	342.905	478.009

The site categories are divided into four classes, I, II, III and IV, according to the soil layer where the structure is located.

(1) Class I site soil: rocky, compact gravelly soil.
(2) Class II site soil: medium-dense, loose gravel soil, dense, medium-dense gravel, coarse and medium sand; clayey soil with foundation soil with a permissible bearing capacity $[\sigma_0] > 150$ kPa.
(3) Class III site soil: loose gravel, coarse and medium sand, dense and medium-dense fine and silty sand, clayey soil with a permissible bearing capacity $[\sigma 0] \leq 150$ kPa and fill soil with $[\sigma_0] \geq 130$ kPa.
(4) Class IV site soil: silty soil, loose fine and chalky sand, recently deposited clayey soil; fill with foundation soil with an allowable bearing capacity $[\sigma_0] < 130$ kPa.

Taking a Class II site with an eight-degree fortification and a 0.1 g seismic acceleration as an example, the design earthquake group is set as the second group. The seismic excitation adopts a combination of longitudinal, transverse, and vertical earthquakes $E_x + E_y + 0.65E_x$.

4. Dynamic Response of the System

This paper conducted a study on the influence of three pier height working conditions on the coupled vibration response and running safety of the vehicle bridge system at five speeds of 150, 200, 250, 300, and 350 km/h. The constant height pier working conditions were selected to explore the influence of vehicle speed and seismic action on the dynamic response of the high-speed train–track–bridge system.

4.1. Validation of the System

In order to verify the reliability and accuracy of the train–track–bridge coupled vibration system model established in this paper, an example of Ref. [38] is used for the reliability analysis, the parameters of the bridge and the train are set to be the same as those in Ref. [38], and the train is set to run at a speed of 240 km/h. Comparing the vertical displacement of the bridge spans, as can be seen in Figure 6, the vertical time-dependent response of the bridge spans obtained via the two models are very close and almost overlap, which means that the train–track–bridge coupled vibration system model is reliable.

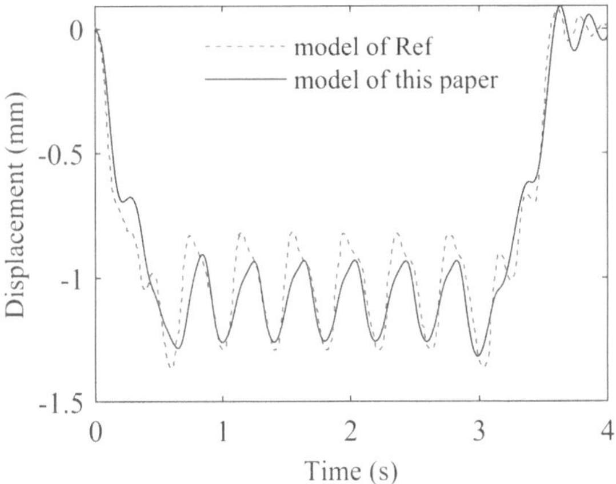

Figure 6. Model validation with the example in Ref. [38].

4.2. Train Running Safety Indicators

4.2.1. Derailment Coefficient

French scientist Nadal first began to study the wheel climbing phenomenon, and in 1896, according to the climbing wheel climbing tendency of static equilibrium conditions, deduced the minimum derailment coefficient Q/P [39]. The wheel–rail contact diagram as shown in Figure 7.

$$\frac{Q}{P} = \frac{\tan\alpha - \mu}{1 + \mu\tan\alpha} \tag{15}$$

where Q denotes the wheel rail lateral force, P denotes the wheel weight, α denotes the angle between tangent line and horizontal line at the contact point of wheel and rail, that is, rim contact angle; and μ denotes the coefficient of friction between wheel rim and rail side.

The derailment factor varies from country to country. According to Chinese standard TB10761-2013 [40], the derailment factor should meet the following requirements:

$$\begin{cases} Q/P \leq 0.8 \text{ (Allowable limit)} \\ Q/P \leq 1.0 \text{ (Danger limit)} \end{cases} \tag{16}$$

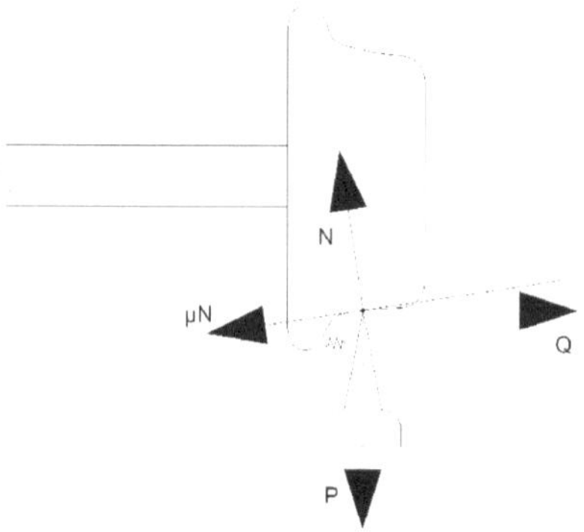

Figure 7. Wheel–rail contact diagram.

4.2.2. Rate of Wheel Load Reduction

In addition to the derailment coefficient, people also use the rate of wheel load reduction to determine the train running safety. When the wheels are substantially loaded down, that is, the vertical force on the wheels and rail is very small, the corresponding lateral force on the wheels and rail is often very small, and due to the impact of measurement error, it is difficult to calculate the derailment coefficient, especially when the wheel and rail are detached, which provides the derailment coefficient to assess the train running safety, addressing this serious problem.

Therefore, the increase in the rate of wheel load reduction can be used to comprehensively and effectively assess the derailment stability of the train's running. The rate of wheel load reduction is defined as the ratio of the wheel weight reduction ΔP on the reduced side to the average static wheel weight $\overline{P}$ of the wheelset. The rate of wheel load reduction standard stipulated by China is as follows [41]:

$$
\begin{cases}
\Delta P/\overline{P} \leq 0.65 \ (\text{First limit, eligibility criteria}) \\
\Delta P/\overline{P} \leq 0.6 \ (\text{Second limit, increased safety margin criteria})
\end{cases}
\tag{17}
$$

4.3. The Influence of Train Speed on System Dynamic Response

We compare the dynamic response of the system under no seismic excitation with that under seismic excitation, and analyze the influence of an earthquake on the dynamic response of the train–track–bridge system. The train passed over the bridge at speeds of 150, 200, 250, 300, and 350 km/h, respectively, and we recorded the peak dynamic response of the system, as shown in the following figures.

From Figures 8 and 9, it can be seen that seismic excitation has a significant impact on the dynamic response of bridges, with a particularly prominent impact on the acceleration response. The lateral dynamic response of the bridge increases more than the vertical dynamic response. This is because when there is no earthquake, the lateral dynamic response of the train mainly comes from the lateral track's irregularity, and the vertical dynamic response mainly comes from the train's gravity. The seismic excitation is input according to the vertical seismic wave strength, which is 0.65 times the horizontal seismic wave strength. Therefore, the seismic excitation has a more intense and sensitive impact on the lateral dynamic response of the bridge than the vertical dynamic response, and under earthquake excitation, some responses exceed the safety index limit, seriously reducing the smoothness and safety of the train. Overall, the dynamic response of the bridge increases with the increase in vehicle speed.

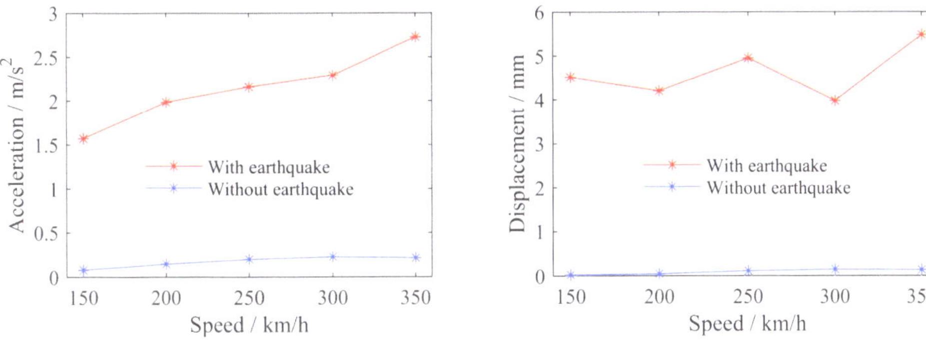

Figure 8. Peak lateral response of bridge midspan.

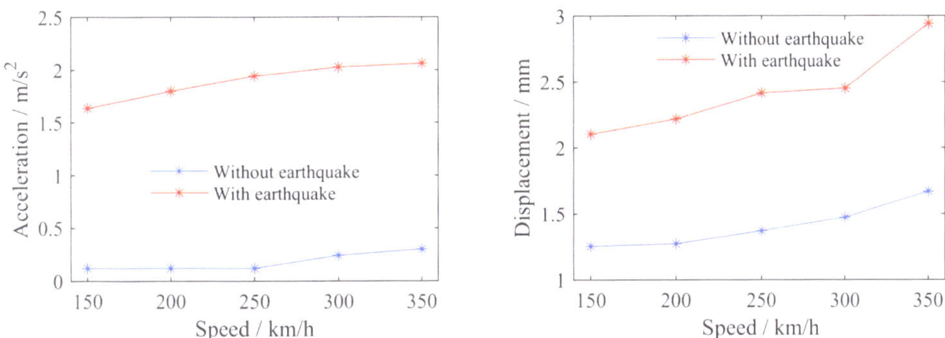

Figure 9. Peak vertical response of bridge midspan.

From Figure 10, it can be seen that the train response generally increases with the increase in operating speed, and the lateral response is more affected by seismic excitation than the vertical response, for the same reason as the above bridge response influence law. From Figure 11, it can be seen that under seismic excitation, the wheel rail load reduction rate and derailment coefficient of trains generally increase with the increase in vehicle speed. When the operating speed is 350 km/h, the two safety indicators of trains approach the limit values [42]. When the seismic acceleration is greater, the vehicle speed is faster, and the bridge structure is more flexible, the risk of train derailment will greatly increase, and the safety and comfort of trains will be greatly reduced.

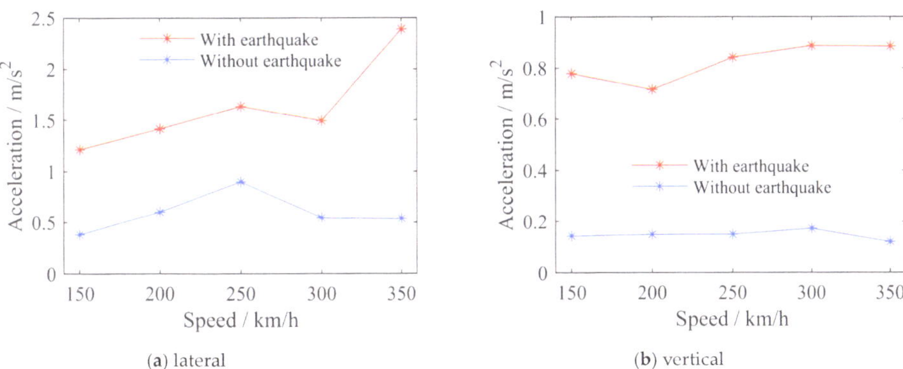

(a) lateral

(b) vertical

Figure 10. Peak train acceleration.

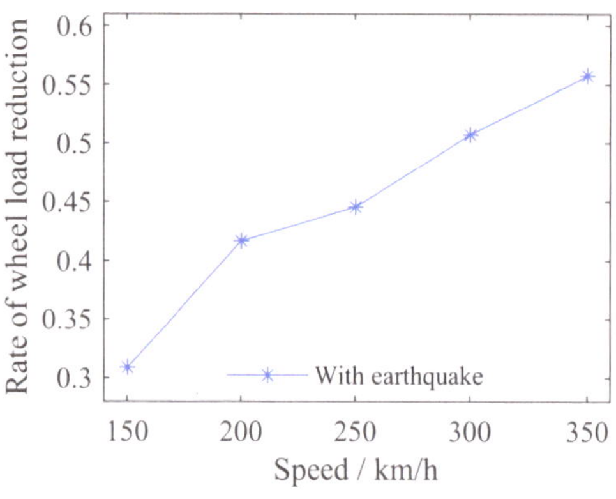
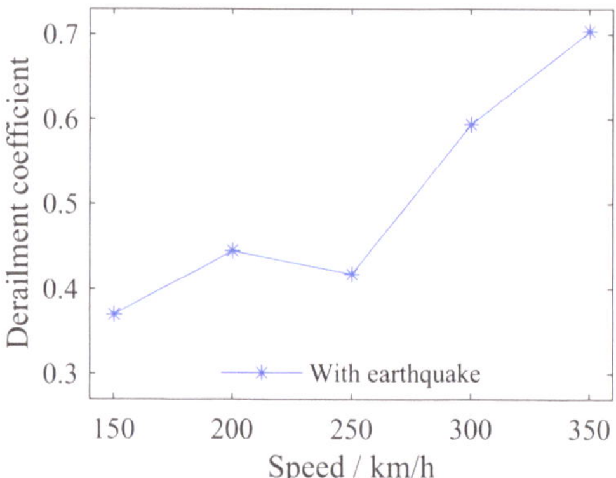

Figure 11. Train safety index.

4.4. *The Influence of Pier Height on System Dynamic Response*

This paper only analyzes the influence of unequal pier heights on the dynamic response and running safety of the high-speed train–track–bridge system under seismic excitation when the train passes over the bridge at a speed of 300 km/h. The first- to fourth-order natural frequencies of the bridge as shown in Table 6.

Table 6. First- to fourth-order natural frequencies of the bridge.

Working Condition Type	1st Modulus	2nd Modulus	3rd Modulus	4th Modulus
Equal-height pier	13.187	13.302	15.806	21.443
Gradually increasing	10.452	10.518	12.998	15.538
Sharply increasing	9.630	10.832	13.964	15.595

The working conditions of the bridge piers are shown in Table 4 in Section 3.2. From Figure 12, it can be seen that compared to equal-height piers, the peak lateral dynamic response of bridges with unequal-height piers (gradually increasing) decreases, while the peak vertical dynamic response of bridges increases. The peak lateral dynamic response of bridges with unequal-height piers (Sharply increasing) increases sharply, while the peak vertical dynamic response of bridges decreases. From Figure 13, it can be seen that compared to equal-height piers, the peak lateral dynamic response of trains with unequal-height piers (gradually increasing) decreases, which is beneficial for stabilizing the vehicle body. The peak lateral dynamic response of trains with unequal-height piers (sharply increasing) increases sharply, seriously reducing passenger comfort. Moreover, the vertical dynamic response of trains under both unequal-height pier conditions increases, and the safety indicators of equal-height piers are significantly better than those under the two unequal-height pier conditions. From the data in the above figures, it can be concluded that considering the comfort and safety of the train, the optimal choice is to design equal-height bridge piers, followed by gradually increasing pier heights, and avoiding steep increases in pier heights.

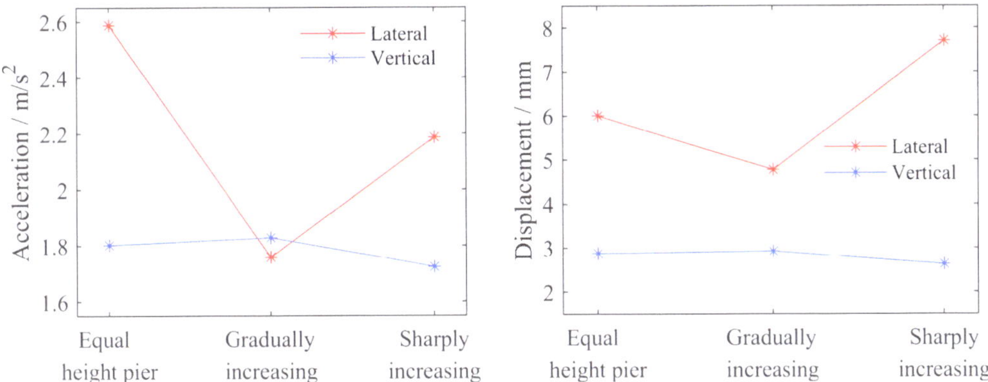

Figure 12. Peak response of bridge midspan.

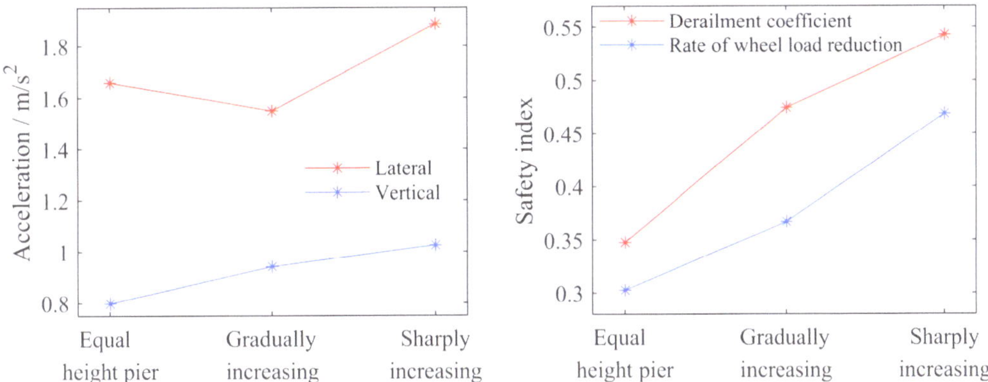

Figure 13. Train dynamic response.

5. Conclusions

In this study, based on the finite element method and multi-body dynamics method, a high-speed train–track–bridge coupling system model was established. Calculation and analysis were conducted via programming to explore the influence of unequal-height piers on the dynamic response of a high-speed railway vehicle bridge system. The following conclusion was obtained:

(1) The dynamic response of the system and the safety index of the train generally increase with the increase in the train running speed.

(2) Under seismic excitation, the dynamic response of the system is significantly increased, and the lateral dynamic response of the system is more affected by seismic excitation than the vertical response.

(3) Compared to equal-height piers, the peak lateral dynamic response of the system with unequal-height piers (gradually increasing) decreases, which is beneficial for stabilizing the vehicle body. The peak lateral dynamic response of the system with unequal-height piers (steep increase in pier height) increases sharply, seriously reducing passenger comfort.

(4) The vertical dynamic response of trains under two unequal-height pier conditions increases, and the safety indicators of equal-height piers are significantly better than the two unequal-height pier conditions. The recommended design is the optimal choice for equal-height bridge piers, followed by gradually increasing pier heights, and avoiding steep increases in pier heights.

Author Contributions: Data curation, Z.Z., H.Z., H.H., P.Z. and F.T.; Writing—original draft, Y.Z.; Writing—review & editing, L.J. and P.X. All authors have read and agreed to the published version of the manuscript.

Funding: This research work was jointly supported by the China Railway Corporation Limited Science and technology research and development program (2022-major-17), Science and Technology Research and Development Program Project of China railway group limited (2020-Key-02), the National Natural Science Foundation of China (Grant No. 11972379), the Hunan Natural Science Foundation Science and Education Joint Fund Project 2022JJ60109, Hunan Science Fund for Distinguished Young Scholars (2021JJ10061).

Data Availability Statement: Underlying research materials related to this paper can be accessed by requesting from the corresponding author.

Conflicts of Interest: The authors declare no conflict to interest.

References

1. Souza, E.F.; Bragança, C.; Meixedo, A.; Ribeiro, D.; Bittencourt, T.N.; Carvalho, H. Drive-by Methodologies Applied to Railway Infrastructure Subsystems: A Literature Review—Part I: Bridges and Viaducts. *Appl. Sci.* **2023**, *13*, 6940. [CrossRef]
2. Arvidsson, T.; Andersson, A.; Karoumi, R. International journal of rail transportation train running safety on non-ballasted bridges train running safety on non-ballasted bridges. *Int. J. Rail Transp.* **2019**, *7*, 1–22. [CrossRef]
3. Zhai, W.; Han, Z.; Chen, Z.; Ling, L.; Zhu, S. Train–track–bridge dynamic interaction: A state-of-the-art review. *Veh. Syst. Dyn.* **2019**, *57*, 984–1027. [CrossRef]
4. Biggs, J.M. *Introduction to Structural Dynamics*; McGraw-Hill Book Company: New York, NY, USA, 1964; pp. 23–32.
5. Liu, W.K.; Li, S.; Park, H.S. Eighty Years of the Finite Element Method: Birth, Evolution, and Future. *Arch. Comput. Methods Eng.* **2022**, *29*, 4431–4453. [CrossRef]
6. Matsuura, A. Dynamic behavior of bridge girder for high speed railway bridge. *Q. Rep. RTRI* **1979**, *20*, 70–76.
7. Chu, K.-H.; Dhar Chaman, L.; Garg Vijay, K. Railway-Bridge Impact: Simplified Train and Bridge Model. *J. Struct. Div.* **1979**, *105*, 1823–1844. [CrossRef]
8. Dhar, C.L. A method of computing railway bridge impact. *Diss. Abstr. Int.* **1978**, *39-08*, 3947.
9. Liu, H.; Wang, P.; Wei, X.; Xiao, J.; Chen, R. Longitudinal Seismic Response of Continuously Welded Track on Railway Arch Bridges. *Appl. Sci.* **2018**, *8*, 775. [CrossRef]
10. Zhao, H.; Wei, B.; Jiang, L.; Xiang, P.; Zhang, X.; Ma, H.; Xu, S.; Wang, L.; Wu, H.; Xie, X. A velocity-related running safety assessment index in seismic design for railway bridge. *Mech. Syst. Signal Process.* **2023**, *198*, 110305. [CrossRef]
11. Yang, Y.B.; Wu, Y.S. Dynamic stability of trains moving over bridges shaken by earthquakes. *J. Sound Vib.* **2002**, *258*, 65–94. [CrossRef]
12. Hujun, L.; Xiaozhen, L.; Yan, Z. Train running safety analysis of high-pier rigid frame bridge under earthquake action. In Proceedings of the 9th International Conference on Structural Dynamics, EURODYN 2014, Porto, Portugal, 30 June 2014; pp. 1267–1272.
13. Xiang, P.; Ma, H.; Zhao, H.; Jiang, L.; Xu, S.; Liu, X. Safety analysis of train-track-bridge coupled braking system under earthquake. *Structures* **2023**, *53*, 1519–1529. [CrossRef]
14. Xiang, P.; Huang, W.; Jiang, L.Z.; Lu, D.G.; Liu, X.; Zhang, Q. Investigations on the influence of prestressed concrete creep on train-track-bridge system. *Constr. Build. Mater.* **2021**, *293*, 123504. [CrossRef]
15. Zeng, Y.; Zheng, H.; Jiang, Y.; Ran, J.; He, X. Modal Analysis of a Steel Truss Girder Cable-Stayed Bridge with Single Tower and Single Cable Plane. *Appl. Sci.* **2022**, *12*, 7627. [CrossRef]
16. Han, Y.; Xia, H.; Guo, W.-W. Dynamic response of cable-stayed bridge to running trains and earthquakes. *Eng. Mech.* **2006**, *23*, 93.
17. Guo, J.; Zhong, J.; Dang, X.; Yuan, W. Seismic Responses of a Cable-Stayed Bridge with Consideration of Uniform Temperature Load. *Appl. Sci.* **2016**, *6*, 408. [CrossRef]
18. Variyavwala, J.P.; Desai, A.K. Evaluation of Train Running Safety on Railway Cable-Stayed Bridge During Seismic Excitation. *Int. J. Struct. Stab. Dyn.* **2023**, *23*, 2350066. [CrossRef]
19. Zhai, W.; Wang, S. Influence of bridge structure stiffness on the dynamic performance of high-speed train-track-bridge coupled system. *Zhongguo Tiedao Kexue/China Railw. Sci.* **2012**, *33*, 19–26. [CrossRef]
20. Chao, C.; Liang, L.; Zhai, W. Dynamic Response of High-Speed Train Subjected to Stiffness Degradation of Simply Supported Bridge. In Proceedings of the 2nd International Conference on Rail Transportation, ICRT 2021, Chengdu, China, 5–6 July 2021; pp. 134–145.
21. Fan, X.; Liu, L.; Wang, X.; Cao, J.; Cheng, W. Vibration Response Analysis of Overhead System Regarding Train-Track-Bridge Dynamic Interaction. *Appl. Sci.* **2022**, *12*, 9053. [CrossRef]
22. Chen, Z.; Zhai, W.; Cai, C.; Sun, Y. Safety threshold of high-speed railway pier settlement based on train-track-bridge dynamic interaction. *Sci. China Technol. Sci.* **2015**, *58*, 202–210. [CrossRef]

23. Chen, Z.; Zhai, W.; Tian, G. Study on the safe value of multi-pier settlement for simply supported girder bridges in high-speed railways. *Struct. Infrastruct. Eng.* **2018**, *14*, 400–410. [CrossRef]
24. Chen, Z.; Zhai, W.; Yin, Q. Analysis of structural stresses of tracks and vehicle dynamic responses in traintrackbridge system with pier settlement. *Proc. Inst. Mech. Eng. Part F J. Rail Rapid Transit* **2018**, *232*, 421–434. [CrossRef]
25. Zhang, X.; Shan, Y.; Yang, X. Effect of Bridge-Pier Differential Settlement on the Dynamic Response of a High-Speed Railway Train-Track-Bridge System. *Math. Probl. Eng.* **2017**, *2017*, 8960628. [CrossRef]
26. Gou, H.; Liu, C.; Hua, H.; Bao, Y.; Pu, Q. Mapping relationship between dynamic responses of high-speed trains and additional bridge deformations. *JVC/J. Vib. Control* **2021**, *27*, 1051–1062. [CrossRef]
27. Feng, Y.-L.; Jiang, L.-Z.; Chen, M.-C.; Zhou, W.-B.; Liu, X.; Zhang, Y.-T. Deformation compatibility relationship of track interlayer with uneven settlement of side pier of continuous girder bridge and its dynamic application. *Gongcheng Lixue/Eng. Mech.* **2021**, *38*, 179–190. [CrossRef]
28. Xu, L.; Yu, Z.; Shan, Z. Numerical simulation for traintrackbridge dynamic interaction considering damage constitutive relation of concrete tracks. *Arch. Civ. Mech. Eng.* **2021**, *21*, 116. [CrossRef]
29. Jiang, L.; Liu, X.; Xiang, P.; Zhou, W. Train-bridge system dynamics analysis with uncertain parameters based on new point estimate method. *Eng. Struct.* **2019**, *199*, 109454. [CrossRef]
30. Zhai, W. Development of vehicle-track coupled dynamics theory and engineering practice. *Kexue Tongbao/Chin. Sci. Bull.* **2022**, *67*, 3793–3807. [CrossRef]
31. Xu, L.; Lu, T. Influence of the finite element type of the sleeper on vehicle-track interaction: A numerical study. *Veh. Syst. Dyn.* **2021**, *59*, 1533–1556. [CrossRef]
32. Chen, Y.; Huang, X.-Q.; Ma, Y.-F. Coupling vibration of vehicle-bridge system. *Appl. Math. Mech.* **2004**, *25*, 390–395. [CrossRef]
33. Li, X.-Z.; Zhang, L.-M.; Zhang, J. State-of-the-art review and trend of studies on coupling vibration for vehicle and highway bridge system. *Gongcheng Lixue/Eng. Mech.* **2008**, *25*, 230–240.
34. Zhao, H.; Wei, B.; Guo, P.; Tan, J.; Xiang, P.; Jiang, L.; Fu, W.; Liu, X. Random analysis of train-bridge coupled system under non-uniform ground motion. *Adv. Struct. Eng.* **2023**. [CrossRef]
35. Gou, H.; Liu, C.; Xie, R.; Bao, Y.; Zhao, L.; Pu, Q. Running safety of high-speed train on deformed railway bridges with interlayer connection failure. *Steel Compos. Struct.* **2021**, *39*, 261–274. [CrossRef]
36. Yu, Z.-w.; Mao, J.-f. Probability analysis of train-track-bridge interactions using a random wheel/rail contact model. *Eng. Struct.* **2017**, *144*, 120–138. [CrossRef]
37. Zeng, Q.; Dimitrakopoulos, E.G. Seismic response analysis of an interacting curved bridge-train system under frequent earthquakes. *Earthq. Eng. Struct. Dyn.* **2016**, *45*, 1129–1148. [CrossRef]
38. Mao, J.; Yu, Z.; Xiao, Y.; Jin, C.; Bai, Y. Random dynamic analysis of a train-bridge coupled system involving random system parameters based on probability density evolution method. *Probabilistic Eng. Mech.* **2016**, *46*, 48–61. [CrossRef]
39. Zeng, Q.; Dimitrakopoulos, E.G. Derailment mechanism of trains running over bridges during strong earthquakes. *Procedia Eng.* **2017**, *199*, 2633–2638. [CrossRef]
40. China, S. *Technical Regulations for Dynamic Acceptance for High-Speed Railways Construction*; Standards Press of China: Beijing, China, 2013.
41. Zhao, H.; Wei, B.; Jiang, L.; Xiang, P. Seismic running safety assessment for stochastic vibration of train–bridge coupled system. *Arch. Civ. Mech. Eng.* **2022**, *22*, 180. [CrossRef]
42. Luo, X. Study on methodology for running safety assessment of trains in seismic design of railway structures. *Soil Dyn. Earthq. Eng.* **2005**, *25*, 79–91. [CrossRef]

applied sciences

Article

Flow-Induced Vibration Hybrid Modeling Method and Dynamic Characteristics of U-Section Rubber Outer Windshield System of High-Speed Trains

Yizheng Yu [1,2], Pengxiang Lv [1,3,4], Xiao Liu [1,3,4],* and Xiang Liu [1,3,4],*

[1] Key Laboratory of Traffic Safety on Track, School of Traffic & Transportation Engineering, Central South University, Ministry of Education, Changsha 410075, China; yuyizheng@cccar.com.cn (Y.Y.); lpx18613762267@163.com (P.L.)

[2] National Rail Vehicle Engineering R&D Center, CRRC Changchun Railway Vehicles Co., Ltd., Changchun 130062, China

[3] Joint International Research Laboratory of Key Technologies for Rail Traffic Safety, Changsha 410075, China

[4] National and Local Joint Engineering Research Center of Safety Technology for Rail Vehicle, Changsha 410075, China

* Correspondence: xiaoliu11@csu.edu.cn (X.L.); xiangliu@csu.edu.cn (X.L.); Tel.: +86-15-6751-04818 (X.L.); +86-13-8731-44366 (X.L.)

Abstract: The flow-induced vibration characteristic of the U-section rubber outer windshield structure of high-speed train is the key factor to limit its high-speed movement. Accurate and effective flow-induced vibration analysis of windshield structures is an important topic. In this paper, a hybrid modeling method for the analysis of flow-induced vibration of windshield structure is innovatively proposed for the U-section rubber windshield system of high-speed train. The method uses the external aerodynamic load obtained by aerodynamic simulation as the input condition of the flow-induced vibration model, and maps the aerodynamic load to the structural dynamics model characterized by the modal test data of the windshield structure. The flow-induced vibration model is established by means of modal superposition method and the time-domain response is effectively integrated by Runge Kutta method with variable step size. The results show that this method can effectively simulate the flow induced vibration of the wind baffle structure, and the real-time relationship between the aerodynamic load and the modal characteristics of the structure and the response of displacement and velocity can be obtained. On this basis, the comprehensive dynamic performance of the windshield system of high-speed trains at 400 km/h under external aerodynamic load is studied, that is, the force, displacement and velocity variation rules of the flexible structure are examined. It is determined that the displacement and velocity response curve of the measuring point near the lower side of the U-section rubber outer windshield is significantly higher than that of other parts. Moreover, the contribution of the first mode to the dynamic response of the structure is very obvious. This method provides an efficient calculation method for analyzing the flow-induced vibration characteristics of complex flexible structures.

Keywords: U-section rubber outer windshield structure of high-speed trains; aerodynamic simulation; modal test; modal superposition method; flow-induced vibration model

Citation: Yu, Y.; Lv, P.; Liu, X.; Liu, X. Flow-Induced Vibration Hybrid Modeling Method and Dynamic Characteristics of U-Section Rubber Outer Windshield System of High-Speed Trains. *Appl. Sci.* **2023**, *13*, 5813. https://doi.org/10.3390/app13095813

Academic Editor: Junhong Park

Received: 22 March 2023
Revised: 28 April 2023
Accepted: 1 May 2023
Published: 8 May 2023

1. Introduction

The windshield system is a flexible structure connecting the two car ends of the high-speed train, usually composed of outer windshield and inner windshield. The design of the windshield system reduces not only the aerodynamic resistance of train operation but also the aerodynamic noise of the connection between the two car ends [1–6]. According to the train operation and maintenance data, once the train operating speed increases, the aerodynamic effect surges, and the lightweight design requirements make the windshield system more sensitive to flow-induced vibration [7–13], which intensifies the flow-induced vibration phenomenon of the outer windshield structure, and the windshield is obviously turned out or even torn. Therefore, when designing a higher speed train, the flow-induced vibration characteristics of the windshield system are a key design consideration [3,14–20].

In the current study, the analysis methods related to flow-induced vibration are roughly divided into two categories [13,21–28]. (1) Means of monitoring analysis through experiments. Cai Jianming et al. [29] determined in the following test that the U-shaped rubber outer windshield of some high-speed trains had a gap between the windshield connection between the two cars due to the installation process problem. When the train runs at a high speed, high-frequency vibrations occur at the inner windshield at the end of the train where there is a gap in the outer windshield. Wang Haiyan et al. [30] conducted a modal test on the inner windshield of high-speed trains with a speed of 350 km/h, and obtained the vibration characteristics of the inner windshield. The modal test was combined with the real vehicle dynamic test to compare and analyze the dynamic influence of the outer windshield on the inner windshield under the condition of with or without gaps in the outer windshield. Li Suxuan et al. [31] studied the vibration and deformation causes of the outer windshield with a maximum speed of 350 km/h, and analyzed the abnormal vibration and high noise of the inner windshield in detail through experiments and data analysis. However, due to the relatively high cost of experimental testing, it is only suitable for testing of finished products, not for the design stage. (2) Modeling and simulation analysis with the help of commercial software. With the improvement of computer performance and the rapid development of computational mechanics, numerical calculation methods have gradually become a very important research method in train aerodynamics. The numerical calculation method can analyze the influence of a certain parameter separately, which is a very important research method in train aerodynamics. Numerical calculation methods are much more convenient, cheaper, and can consider a variety of situations separately, which is why they are the most widely used research methods. Many scholars have carried out extensive research on train aerodynamics based on numerical calculation methods. Liu Zhen et al. [32] conducted a fatigue life analysis under aerodynamic load on the CRH2-300 EMU rubber windshield, which was used to explore the problem of cracks in the rubber outer windshield. Wang et al. [33] used the FLDutil module of the fluid software SC/Tetra (version 11.0) and the input port of the analysis software Ansys (version 10.0) to analyze the load of the high-speed trains to obtain its response and analyze the vibration situation. Miao [34] used three-dimensional numerical methods to discuss the influence of four schemes: full enveloping windshield, top open windshield, bottom open windshield, and top end and low end simultaneous opening windshield on the aerodynamic performance of high-speed trains, and obtained the aerodynamic loads of the following trains under different schemes. Long [35] used the Naiver–Stokes equation and the k-ε turbulence model to numerically simulate and calculate the aerodynamic characteristics of high-speed trains on flat ground, bridges, cuttings and embankments, and obtained the aerodynamic load of high-speed trains in this driving environment. Wang [36] established a train aerodynamic model to study the aerodynamic characteristics of high-speed trains under the action of two different lateral winds: average wind and index wind. Ouyang [37] used hybrid LES/APE method to simulate the unsteady flow field and the near-field aerodynamic noise of a 1/25 scale eight-coach high-speed train in long tunnel to study the changes in sound pressure in different cases, including tunnel with fully reflective walls and tunnel with fully absorptive walls and open air. S. Maleki et al. [38] used various turbulence modeling approaches including ELES, SAS, URANS and RANS to predict the aerodynamic flow around a double-stacked freight wagon. A. Premoli [39] used computational fluid dynamics to investigate the effect of the relative motion between train and infrastructure scenario.

However, the windshield material, structure and connection are complex, which not only makes it difficult to model accurately, but also increases the number of modeling degrees of freedom. In addition, the calculation time and efficiency of this method are low, so it is difficult to apply it well to the dynamic response and fluid-structure interaction analysis of windshield system, and it is also difficult to apply it to the optimization design or selection stage. In summary, it is difficult to analyze the flow-induced vibration characteristics of the windshield system quickly and accurately according to the existing research methods.

The windshield structure in this paper is modeled by mode superposition method based on the results of modal test, which is different from the previous works based on finite element simulations. On the one hand, the current experiment-based model can represent the dynamic characteristics of the windshield structure accurately without the need for cumbersome and complicated structural modeling processes. On the other hand, by using the mode superposition method, the computational efficiency of flow-induced vibration simulation is greatly increased. At the same time, due to the application of the mode superposition method, the aerodynamic loads under different operating conditions can be mapped to the structural nodes easily. In general, this method provides an efficient and reliable tool for the simulation of flow-induced vibration of complex flexible structures.

In this paper, a method is proposed to establish a flow-induced vibration simulation model of windshield structure based on modal superposition method and then analyze the flow-induced vibration characteristics of windshield structure by using aerodynamic simulation data combined with structural modal test data as input, and the specific principle is shown in Figure 1.

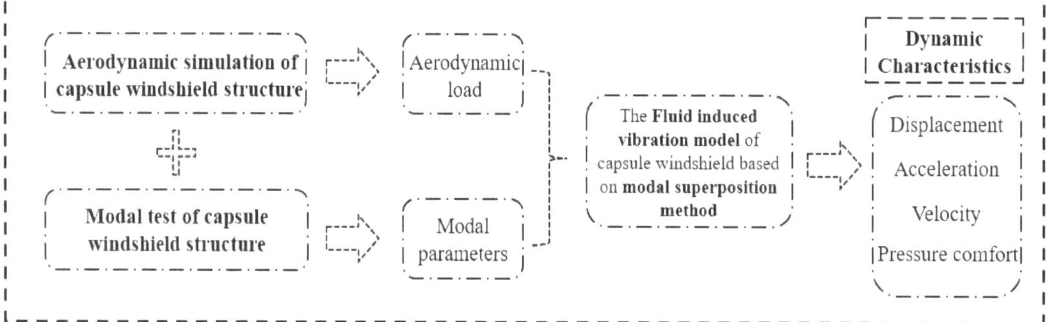

Figure 1. Schematic diagram of U-section rubber outer windshield structure flow-induced vibration analysis.

The rest of this paper is organized as follows: Section 2 introduces the aerodynamic simulation analysis and surface aerodynamic load extraction of U-section rubber windshield. In Section 3, the modal experiment of U-section rubber outer windshield structure and the analysis of the results are briefly summarized. In Section 4, the flow-induced vibration hybrid simulation method of U-section rubber outer windshield is introduced in detail. In Section 5, based on the simulation method proposed in this paper, the dynamic response of U-section rubber windshield was analyzed. Combined with the natural frequency of the outer windshield, the vibration mechanism and vibration characteristics of the outer windshield are summarized and discussed. Finally, Section 6 summarizes this research.

2. Extraction of Aerodynamic Load on U-Section Rubber Outer Windshield

In this section, an aerodynamic model of the U-section rubber outer windshield of high-speed train is developed, and the modelling process and the analysis of the aerodynamic load characteristics on the windshield surface are briefly described. In addition to this, an integration method for the aerodynamic loads is also presented.

2.1. Geometric Model and Computational Setup

To reduce the amount of calculation, the aerodynamic simulation model of the U-section rubber outer windshield adopts a three-car marshaling test train, including the head, middle and tail trains, and two groups of outer windshield structures. The middle section of the train remains unchanged, and the shortened model has relatively little effect on the flow field structure at the first windshield of the train [40]. During the numerical simulation, the modeling is carried out in strict accordance with the installation and size of the internal and outer windshields of the test train. In addition, the train's height H = 4.0 m is selected as the characteristic length, the width of the train is 0.8 H, and the length of the train is 20.8 H. The established model of the high-speed train is shown in Figure 2.

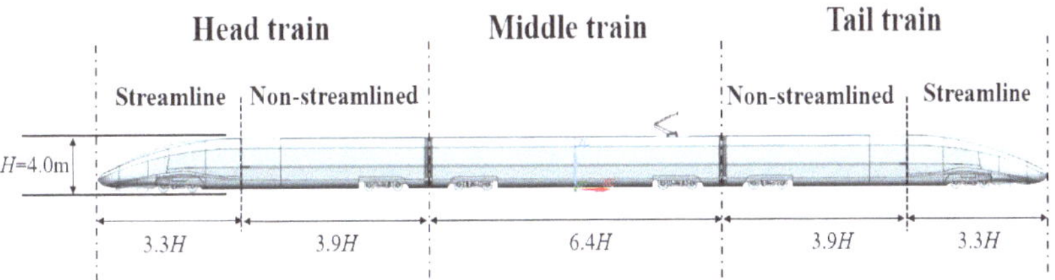

Figure 2. Geometric model of the train.

The high-speed train adopts a U-section rubber outer windshield with two windshield opening positions at the lower end and a total of four U-section rubber elements. The U-section rubber outer windshield structure is shown in Figure 3.

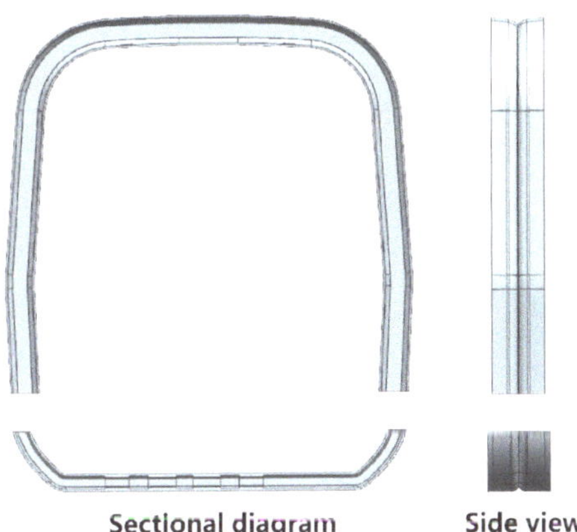

Figure 3. Geometric model of U-section rubber outer windshield.

The commercial software STRA-CCM+ (version 2020.3) is used to simulate the flow of the high-speed train in open line. The calculation area is shown in Figure 4. The calculation domain is 66.4 H long, 20 H wide and 15 H high. The nose tip of the head train is 18 H from the entrance of the calculation domain, and the nose tip of the tail train is 30 H from the exit of the calculation domain. The train is located in the middle of the calculation domain, and the lowest point of the wheelset is 0.05 H from the ground. The size ratio of the model to the actual high-speed train is 1:8.

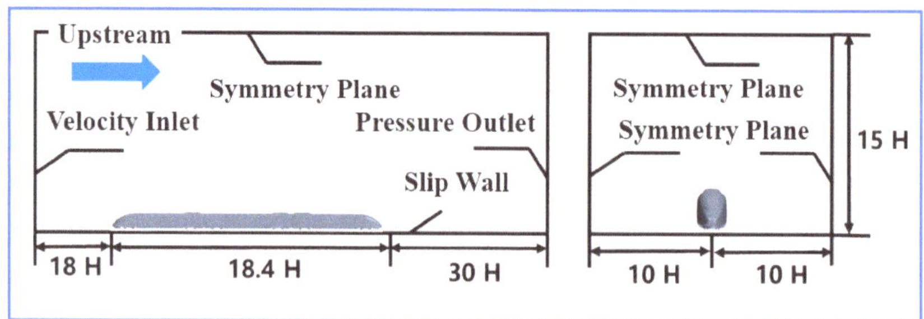

Figure 4. Computational domain.

The velocity inlet boundary condition is adopted at the inlet of the calculation domain, and the incoming flow velocity is set as 111 m/s. Pressure outlet boundary conditions are adopted at the outlet. The ground is set as a no-slip boundary condition, the velocity is consistent with the incoming flow velocity, and the rest are set to symmetry plane boundary conditions. The entire computing area grid is divided by tetrahedral mesh, and the total number of meshes is about 75 million; Figure 5 is a grid schematic. Three layers of nested encryption areas are set near the train's body, and the maximum grid scale is set to 200 mm, 400 mm, and 600 mm, respectively. Grid encryption is carried out on the train's body, bogie, windshield and other areas to ensure the accuracy of calculation. A 15-layer grid is set on the surface of the train body, windshield and pantograph. The first layer grid thickness is set to 0.02 mm, with a growth rate of 1.2, and the thickness of the first layer of the other surfaces is set to 0.05 mm, with a growth rate of 1.2. Overall, y+ is around 1, which meets the requirements of turbulence calculation [41–44].

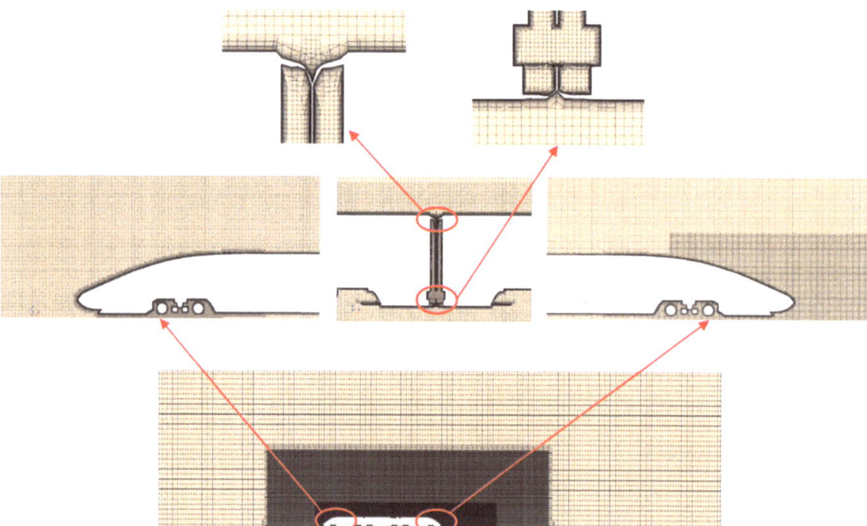

Figure 5. Computational grids.

For high-Reynolds-number flows, as the range for time and length scales that describe the flow depends on the Reynolds number, some level of turbulence modeling is required. The air flow around the train is highly turbulent, three-dimensional and time-dependent, and appropriate turbulence modeling is essential for accurate prediction. The model of train is 1:8 scaled, the implicit solution method pressure-based is selected for the constant flow field calculation, the $SSTk - \omega$ turbulence model is selected for numerical simulation, the SIMPLE algorithm is selected for pressure–velocity coupling, the pressure adopts

Standard discrete format, and the convection term and dissipation term are adopted in the second-order windward discrete format. The unsteady flow field calculation adopts the LES method, and the discrete momentum equation adopts the bounded center difference format, with a time step of 5×10^{-5} s, and each step is iterated 20 times, for a total of 10,000 time steps.

In order to obtain the time history data of pulsating wind pressure of the flow field in the windshield area, the measurement points are arranged on the outer windshield of the train, and the three-group train has a total of two windshields, and a total of 479 measurement points are arranged on the inside and outside of each windshield. The aerodynamic measurement point arrangement is shown in Figure 6.

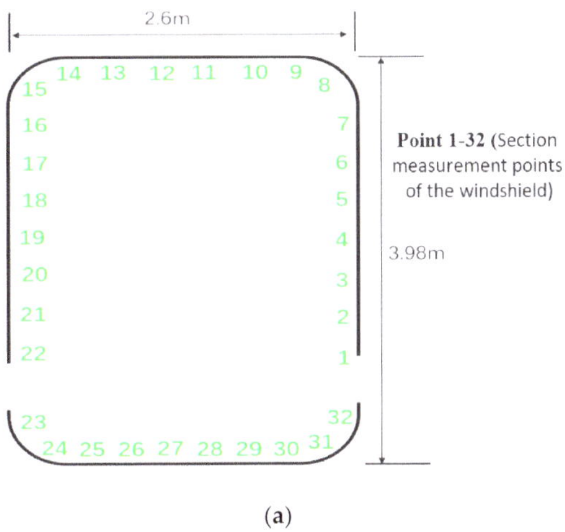

(a)

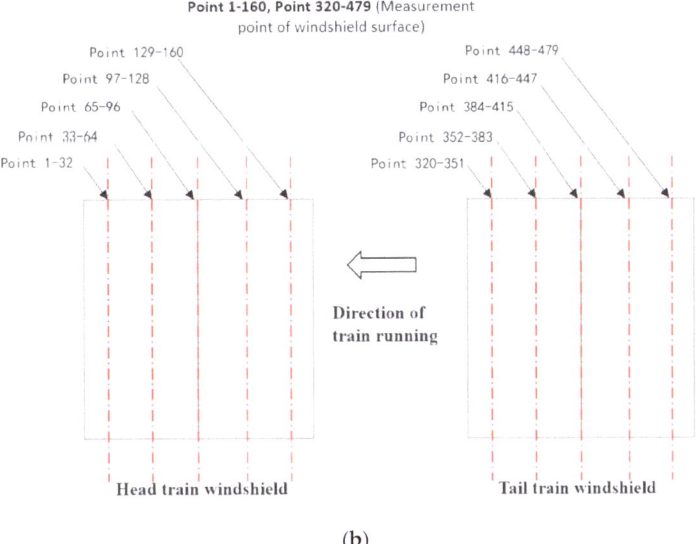

(b)

Figure 6. Schematic diagram of outer windshield measurement points. (a) Measurement points arranged circumferentially along the windshield section. (b) Lateral arrangement of measuring points on the outer surface of windshield.

2.2. Data Analysis

This section introduces the aerodynamic pressure distribution of the windshield surface of the train and the extraction of the aerodynamic load of the surface measurement point. Pressure cloud diagram of cross-section and surface of U-section rubber outer windshield when the velocity of the train is 400 km/h in open line is shown in Figures 7 and 8. Because the bottom of the windshield is close to the bogie area, and this area is similar to a groove, the speed of air decreases rapidly and the pressure value increases when the air flows through the windward side of bogie area. Through the arc connecting the wall and the train's body, the incoming flow separates, the speed begins to increase rapidly, causing its pressure to decrease to a negative value. The upper side of the tail train is also affected by the separation of incoming flow in the pantograph area, showing partial negative pressure in the area.

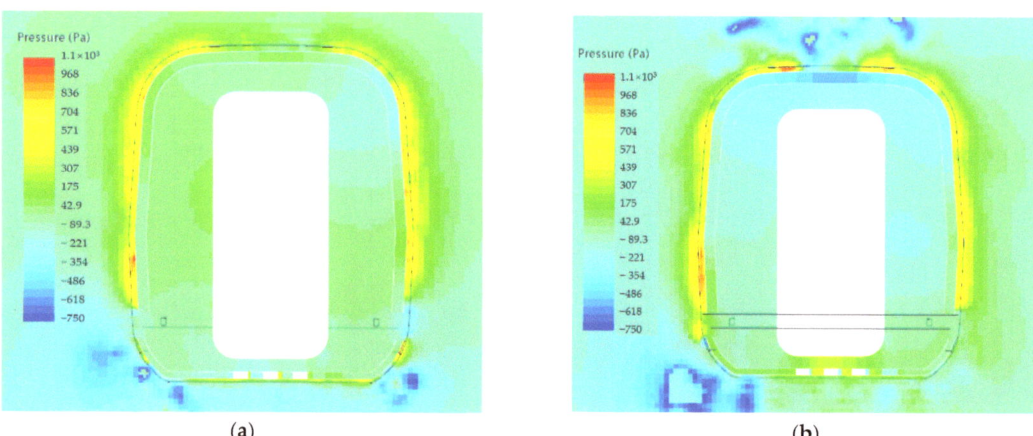

Figure 7. Pressure diagram of outer windshield section at 400 km/h. (**a**) Pressure diagram of head train windshield section. (**b**) Pressure diagram of tail train windshield section.

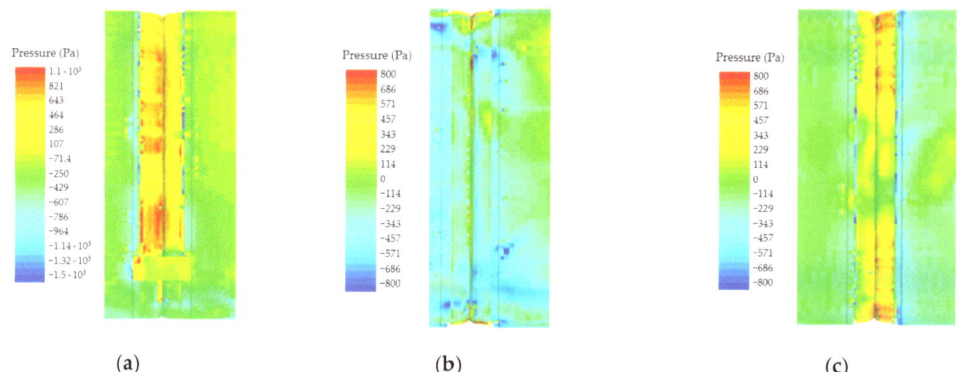

Figure 8. Pressure diagram of U-section rubber outer windshield surface at 400 km/h. (**a**) Side pressure cloud. (**b**) Side pressure cloud. (**c**) Top pressure cloud.

Through the flow field simulation calculation, the time history data of aerodynamic pressure of all the windshield surface measurement points shown in Figure 6 can be obtained, and the pressure change curve of the measurement points at different positions of the head train's windshield with time is shown in Figure 9, and it can be seen from the figure that the pressure amplitude of the bottom of the train is larger and the pressure amplitude of the top of the train is small. The pressure change curves of the measurement points at different positions of the windshield in the same horizontal plane are basically the same, and only the pressure amplitude at the peaks and troughs of the pressure wave is different.

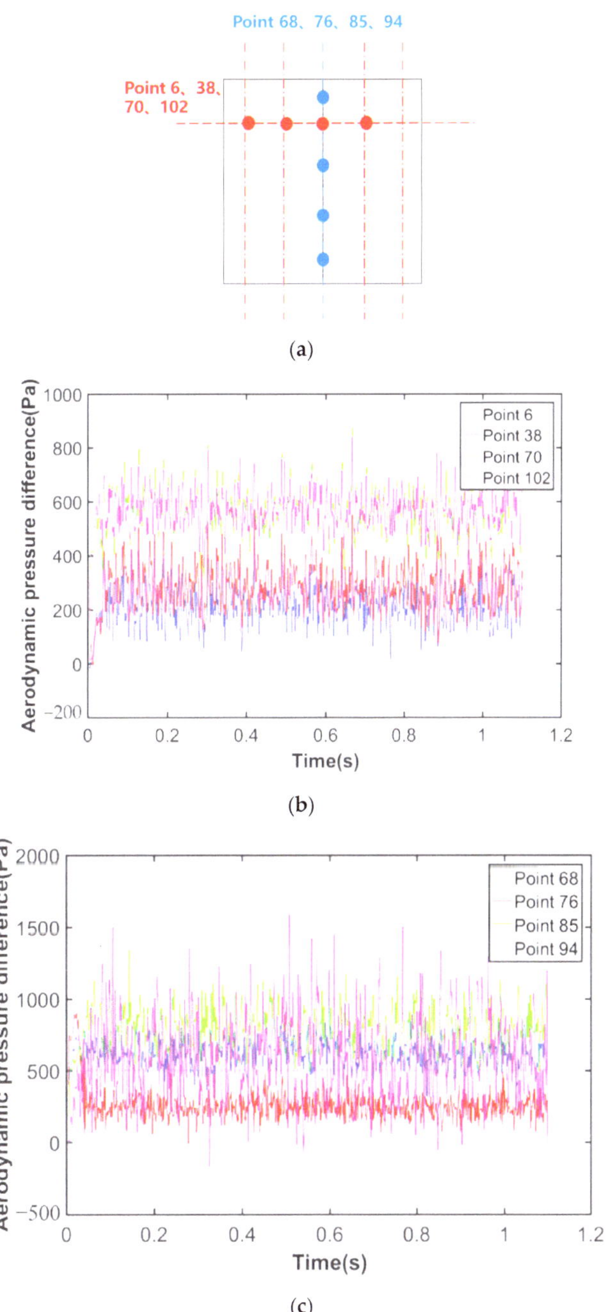

Figure 9. Aerodynamic pressure difference between inside and outside surface at different measuring points of head train windshield. (**a**) The layout diagram of horizontal and longitudinal measurement points. (**b**) Comparison of aerodynamic pressure difference at horizontal measurement points. (**c**) Comparison of aerodynamic pressure difference at longitudinal measurement points.

Next, all aerodynamic data of measurement points is extracted, and the aerodynamic pressure of the measurement points is mapped to the structural modal measurement points by the method of interpolation integration.

2.3. Aerodynamic Load Integration

Flow-induced vibration analysis needs to correspond the aerodynamic load of the measurement point in the flow field analysis to the aerodynamic load of the modal test. According to the measurement point arranged during the modal test to divide the windshield structure into multiple area units, in order to ensure a more accurate description of the aerodynamic load, it is necessary to use as many measurement points as possible to study the change in the aerodynamic load of the external flow field to the structure, and load the measurement point in the aerodynamic simulation to the measurement point of the modal test through the difference integration method. The mapping is shown in Figure 10. The pressure per area unit can be obtained from the aerodynamic load and the unit area. The detailed steps are described below.

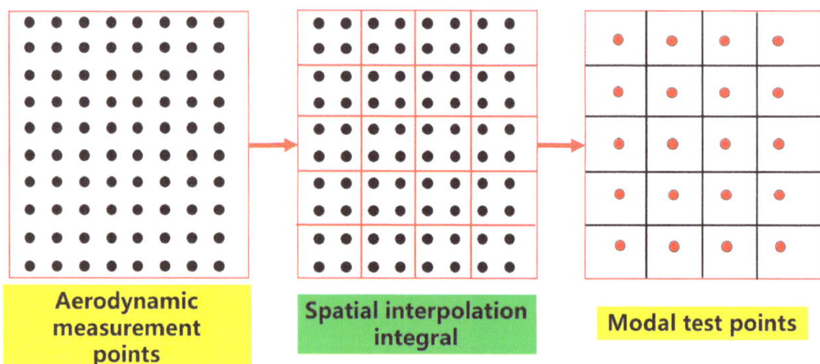

Figure 10. Mapping between aerodynamic measuring points and modal measuring points.

The inner and outer structure of the windshield can be divided into four parts: upper, lower, left and right. Each part of the structure establishes *x-y* local coordinates. The coordinate positions corresponding to the aerodynamic simulation measurement points and modal test measurement points of each part are ensured. Aerodynamic measurement points are used to interpolate each area unit at different positions. Assuming n aerodynamic pressure values are interpolated within an area unit, the average pressure for that area unit is

$$\overline{P} \times n = \sum_{i=1}^{n} p_i.\tag{1}$$

The pressure of each area unit can be expressed as

$$F = S \times \overline{P},\tag{2}$$

where S is the area of the area unit.

3. Modal Experiment and Analysis of U-Section Rubber Outer Windshield Structure

In this section, the modal test method for U-section rubber outer windshield structures is presented and, based on this method, the natural frequencies, damping ratios and modal shapes of the windshield structures are obtained.

3.1. Test Object, Measurement Point and Excitation Point Arrangement

Hammer method was used to test the structural modes, and signals were collected by force sensor (force hammer) and acceleration sensor (measuring point). The lateral part of the structure of the outer windshield was selected as the main test object in this test (see Figure 11a). The thickness of the root of the U-section rubber is 18 mm, the overall width of the U-section rubber is 240 mm, and there is a pretension between the two U-section rubbers. Twenty measuring points are arranged on the side of the windshield (see red numbers Figure 11d). Three sets of tests were carried out at three different excitation points, which were above point 7, in the middle of point 14 and 15, and on the upper side of point 10, as shown by the yellow five-pointed star in Figure 11d.

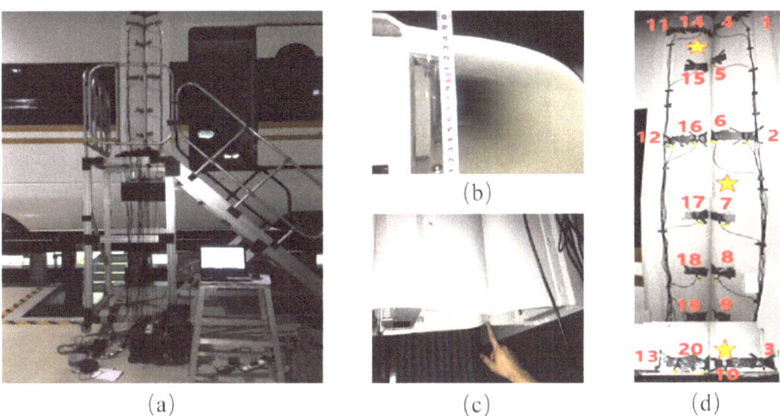

(a)   (b)   (c)   (d)

Figure 11. U-section rubber outer windshield structure and lateral part of the measurement points (red number) and excitation points (yellow five-pointed star) schematic. (**a**) The layout diagram of the modal experiment of the outer windshield. (**b**) The bottom section of the outer windshield. (**c**) The bottom of outer windshield. (**d**) The measurement points (red number) and excitation points (yellow five-pointed star).

3.2. Modal Analysis

After data of modal test collection, LMS PolyMAX module was used for mode extraction. The peaks with relatively concentrated S value (the mode with stable frequency, damping value and vector) were selected as the mode frequency, and the corresponding mode shapes of each order were extracted. In the modal test of the lateral part of the outer windshield structure, the results of the test under three different excitation points also have a good agreement. Based on the data measured at these excitation points, the first nine natural frequencies and damping ratios of the lateral part of the outer windshield can be obtained, as shown in Table 1, and the corresponding mode shapes of the first six modes are shown in Figure 12. The common point of the first four modes is large amplitude in the middle and small amplitude on both sides (connecting with the end wall of the train).

Table 1. Modal frequency and damping ratio of outer windshield.

Mode	Natural Frequency (Hz)	Damping Ratio (%)
1	12.02	6.25
2	14.25	6.92
3	18.31	6.37
4	23.66	8.63
5	26.72	6.21
6	28.41	5.36
7	31.78	5.72
8	33.46	5.70
9	39.57	4.10

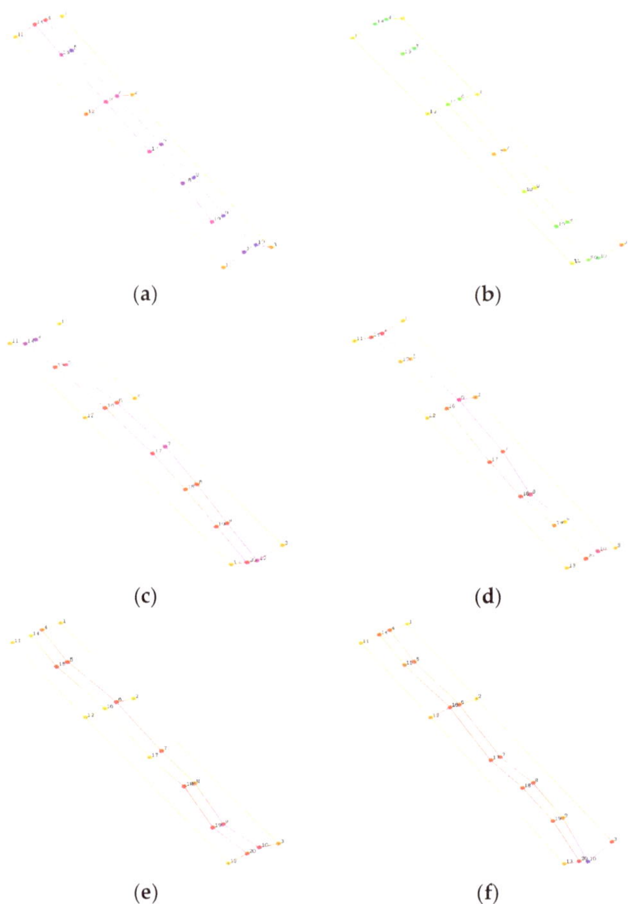

(a)

(b)

(c)

(d)

(e)

(f)

Figure 12. The first four mode shapes of the lateral part of the outer windshield structure. (**a**) ω_1 = 12.02 Hz. (**b**) ω_2 = 14.25 Hz. (**c**) ω_3 = 18.31 Hz. (**d**) ω_4 = 23.66 Hz. (**e**) ω_5 = 26.72 Hz. (**f**) ω_6 = 28.41 Hz. Nodes 1–20 correspond to the modal test points in Figure 11d and different colored lines represent the outer outline of the windshield structure.

4. Establishment of U-Section Rubber Outer Windshield Flow-Induced Vibration Response Analysis Method

In this section, based on the modal superposition method, the time course curves of the aerodynamic loads on the outer windshield surface in Section 2 and the modal parameters of the outer windshield structure in Section 3 are used as inputs to build a simulation model of the flow-induced vibration of the U-section rubber outer windshield. The dynamic response of the model is solved using the Runge–Kutta method. In addition to their mathematical formulation, the advantages of using each of these methods are also presented.

4.1. Flow-Induced Vibration Model Based on Mode Superposition Method of U-Section Rubber Outer Windshield Structure

The differential equation of motion for the outer windshield system can be expressed as

$$[M]\{\ddot{x}\} + [C]\{\dot{x}\} + [K]\{x\} = \{f\}, \tag{3}$$

where $[M]$ is the mass matrix, $[C]$ is the damping matrix, $[K]$ is the stiffness matrix, all of them are square matrices of order n, n is the degree of freedom of the system. $\{f\}$ is the generalized force vector, representing the force on each degree of freedom. $\{x\}$ is the generalized displacement matrix of the system, in which each parameter represents the generalized displacement on the corresponding degree of freedom of the system. $\{\dot{x}\}$ and $\{\ddot{x}\}$ represent the generalized velocity vector and the generalized acceleration vector, respectively.

Generally, the degrees of freedom of a multi-degree-of-freedom system are coupled to each other, its mass matrix, damping matrix and stiffness matrix are not diagonal matrices, and it is not possible to convert the set of differential equations into individual differential equations to solve them; they need to be converted into independent differential equations in modal coordinates to solve them through coordinate transformation. Coordinate transformation of $\{x\}$ is as follows:

$$\{x\} = [U] \times \{y\}, \tag{4}$$

where $[U]$ is the coordinate transformation matrix. Substituting Equation (4) into Equation (3), the differential equations of motion are converted into the form under modal coordinates:

$$[M][U]\{\ddot{y}\} + [C][U]\{\dot{y}\} + [K][U]\{y\} = \{f\}. \tag{5}$$

Both sides of Equation (5) are left multiplied by the modal transpose matrix $[U]^T$:

$$[U]^T[M][U]\{\ddot{y}\} + [U]^T[C][U]\{\dot{y}\} + [U]^T[K][U]\{y\} = [U]^T\{f\}. \tag{6}$$

The generalized mass matrix $[M_n]$, generalized damping matrix $[C_n]$, generalized stiffness matrix $[K_n]$ and generalized excitation $\{f_p\}$ are defined as

$$[M_n] = [U]^T[M][U], \tag{7}$$

$$[C_n] = [U]^T[C][U], \tag{8}$$

$$[K_n] = [U]^T[K][U], \tag{9}$$

$$\{f_p\} = [U]^T\{f\}. \tag{10}$$

Equation (6) can be expressed as

$$[M_n]\{\ddot{y}\} + [C_n]\{\dot{y}\} + [K_n]\{y\} = \{f_p\}. \tag{11}$$

To ensure that the transformed equations can be decoupled, $[M_n]$, $[C_n]$, $[K_n]$ are all diagonal matrices, and $[U]$ is the modal matrix. To obtain $[U]$, combining Equations (7) and (9) leads to

$$[K][U] = [M][U][M_n]^{-1}[K_n], \tag{12}$$

where both $[M_n]^{-1}$ and $[K]$ are diagonal matrices of order n. The matrixobtained by multiplying $[M_n]^{-1}$ and $[K]$ is also a diagonal matrix, and the ith element on the diagonal of this matrix is $\lambda_i = \frac{k_i}{m_i}$. Then, Equation (11) can be expressed as

$$[k_1 u_1, k_1 u_1, \ldots k_n u_n] = [\lambda_1 m_1 u_1, \lambda_2 m_2 u_2, \ldots, \lambda_n m_n u_n]. \tag{13}$$

That is, $\{u\}$ is the generalized eigenvector of the mass matrix $[M]$ and the stiffness matrix $[K]$, and the vector consisting of $[\lambda_1, \lambda_2, \ldots, \lambda_n]$ is its generalized eigenvalue. Equation (13) is the general eigenvalue problem if the mass matrix is the unit matrix, and therefore the modal vectors can be normalized. The generalized mass matrix corresponding to the ith order modal vector is

$$[M_i] = [u_i]^T [M][u_i]. \tag{14}$$

$[u_i]$ is ith order modal vector. Normalization of the modal vector results in

$$[\phi_i] = \frac{u_i}{\sqrt{M_i}}. \tag{15}$$

Equation (11) can be decomposed into n separate differential equations:

$$m_i \ddot{y}_i + c_i \dot{y}_i + k_i y_i = f_{pi}, (i = 1, 2, 3, \ldots, n). \tag{16}$$

Both sides of the equation are simultaneously divided by m_i and have

$$\frac{k_i}{m_i} = \omega_{ni}^2, \tag{17}$$

$$\frac{c_i}{m_i} = 2\zeta_i \omega_{ni}. \tag{18}$$

Thus, Equation (16) can be translated into

$$\ddot{y}_l + 2\zeta_i \omega_i \dot{y}_l + \omega_i^2 y_i = \frac{f_{pi}}{m_i}, (i = 1, 2, 3, \ldots, n). \tag{19}$$

Equation (18) is the differential equation of mode shape of the ith order. The modal superposition method based on the measured modal data of the structure can effectively restore the real mechanical properties of the windshield structure and avoid the problem of oversimplification of the modal simulation model of the structure.

4.2. Dynamic Response of U-Section Rubber Outer Windshield Structure

For the motion differential Equation (19) of a system with multiple degrees of freedom, it can be expressed as a matrix as follows:

$$\left\{ \begin{array}{c} \dot{Y} \\ \ddot{Y} \end{array} \right\} = \left[\begin{array}{cc} 0 & 1 \\ -\omega^2 & -2\delta\omega^2 \end{array} \right] \left\{ \begin{array}{c} Y \\ \dot{Y} \end{array} \right\} + \left\{ \begin{array}{c} 0 \\ F \end{array} \right\}. \tag{20}$$

Further, Equation (19) could write

$$\dot{X}(t) = AX(t) + U(t), \tag{21}$$

where

$$X(t) = \left\{ \begin{array}{c} Y \\ \dot{Y} \end{array} \right\}$$

$$A = \left[\begin{array}{cc} 0 & 1 \\ -\omega^2 & -2\delta\omega \end{array} \right], \tag{22}$$

$$U(t) = \left\{ \begin{array}{c} 0 \\ F \end{array} \right\}$$

where $X(t)$ is the state vector of the system, A is the system matrix, $U(t)$ is the external load vector. Equation (20) is the state equation of the system. The external load vector of each time step and the state vector of the system in the matrix form of the equation of motion can be obtained by the above derivation process.

The ode45 function in MATLAB uses the Runge–Kutta method to solve the differential equation of motion of the structure under aerodynamic load and calculate the displacement response. This method can solve nonlinear problems effectively and has the advantages of less computation and high precision. For the outer windshield structure, the displacement of each measuring point over time can be expressed as

$$W(x,y,t) = \sum_{i=1}^{N} \phi_i(x,y)q_i(t),\tag{23}$$

where $W(x,y,t)$ represents the displacement of a measuring point with coordinate (x,y) with time t, $\phi_i(x,y)$ represents the ith mode of a measuring point with coordinate (x,y), and $q_i(t)$ represents the displacement of a point with time t under generalized coordinates. N is the number of modes of the first N order. As the Runge–Kutta method is solved using variable step integration, the dynamic response curves obtained are not of the same time step in the time domain and the results can be post-processed using numerical interpolation methods.

5. Dynamic Response Analysis of U-Section Rubber Outer Windshield

Through the calculation and analysis of the U-section rubber outer windshield structure, it is determined that under the action of aerodynamic load, only the first three modes of the outer windshield have major contributions, and the contributions of the latter modes are negligible. In order to ensure accuracy, the first six modes are used for modal superposition in the flow-induced vibration analysis of windshield structure. Figure 13 is a schematic diagram of the selection of measuring points for the displacement response of the U-section rubber outer windshield. The left part corresponds to the left windshield in Figure 10d, and the right part corresponds to the right windshield in Figure 10d. A total of 12 measuring points are selected for analysis. On a common computer, Intel(R) Core(TM) i7-10750CPU is used. The calculation time of this example is 16.74 s, which has engineering availability.

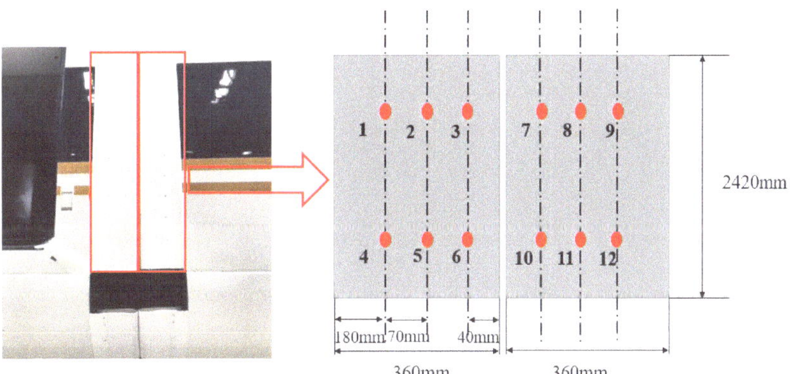

Figure 13. Schematic diagram of displacement response measurement point selection. The red circles are the displacement response measurement point 1–12.

Figures 14 and 15 show the displacement and velocity response curves of the measurement points on the left and right sides of the outer windshield within 6 s, respectively, positive in the direction toward the outside of the windshield. It can be seen from Figure 14 that the displacement and velocity amplitudes of measuring points 6 and 10 are the largest. This is due to the opening on the lower side of the outer windshield structure at points 6 and 10, which leads to lower stiffness compared with other measuring points and more obvious vibration there. On the other hand, measurement points 1, 4, 9 and 12 are closer to the connection between outer windshield and end wall. Due to the limitation of boundary conditions, their stiffness is larger than that of other measurement points, resulting in significantly smaller displacement and velocity response amplitudes than those of other measurement points.

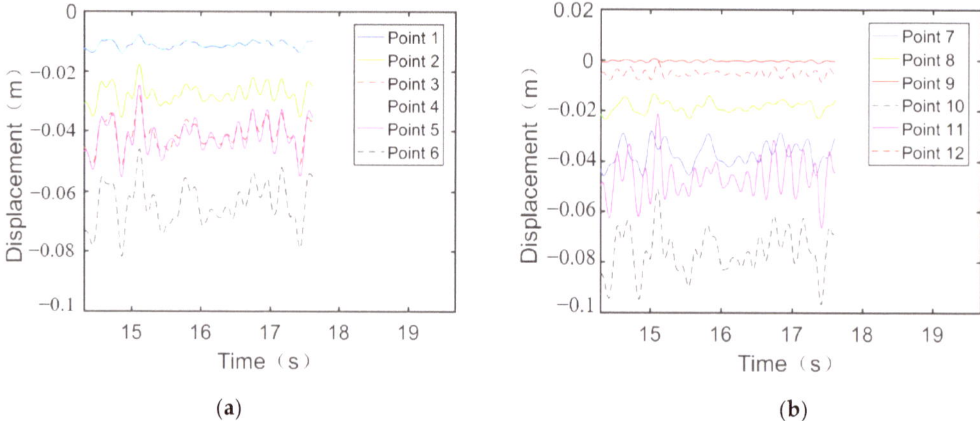

Figure 14. The displacement response of measurement points 1–6 (**a**) and 7–12 (**b**) on both sides of the U-section rubber outer windshield.

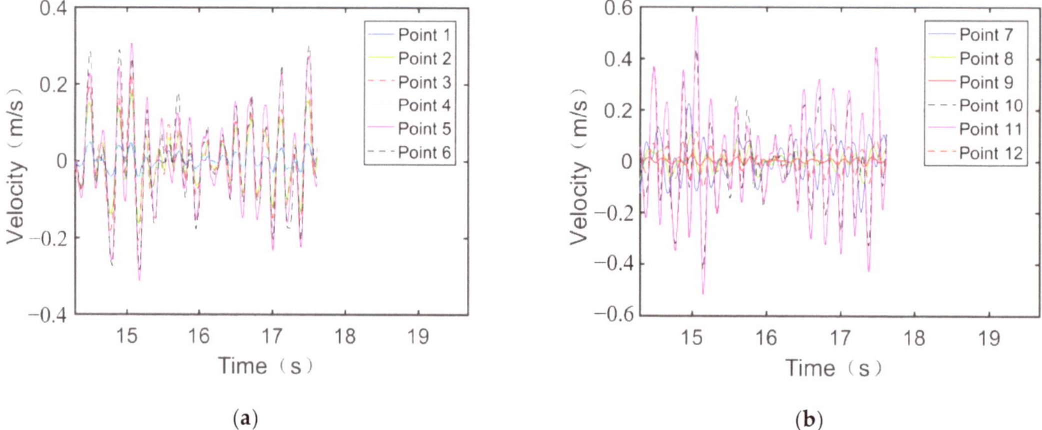

Figure 15. The velocity response of measurement points 1–6 (**a**) and 7–12 (**b**) on both sides of the U-section rubber outer windshield.

Figure 16 shows the modal contribution of displacement at measurement points 6 (left) and 10 (right). The black solid line is the actual displacement at measurement point of outer windshield, and the dashed lines in other colors represent the contribution of each mode to the displacement of outer windshield, respectively. It can be observed that for measurement points 6 and 10, the first-order mode has a strong influence on the displacement of the outer windshield, while the other modes contribute relatively little to the displacement response.

Next, in order to investigate the dynamic response of the measurement points for different positions of the windshield, the train is divided by direction of travel. Windshield 1 between the head car and the middle car and windshield 2 between the middle car and the tail car were selected for comparison in this paper, as shown in Figure 17. Each windshield was selected for comparison at the same measurement point locations as in Figure 13.

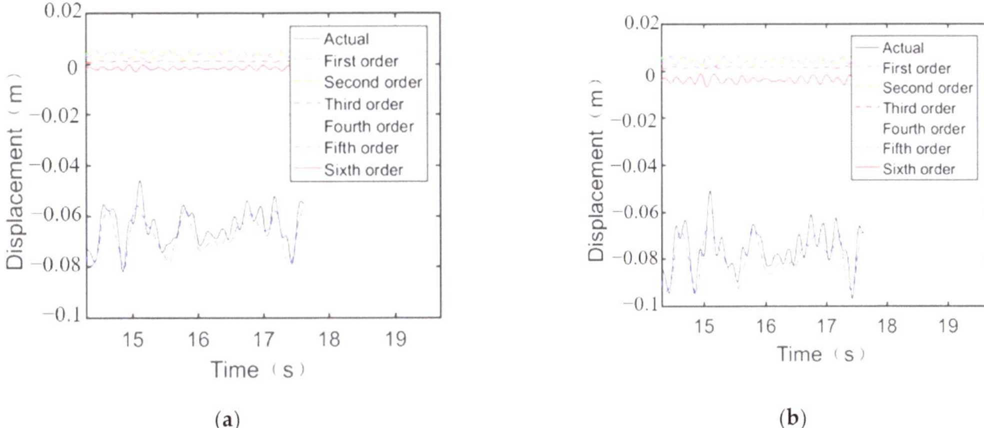

(a) (b)

Figure 16. The displacement modal contribution of U-section rubber outer windshield measuring point 6 (**a**) and measuring point 10 (**b**).

Figure 17. Diagram of the different positions of the windshield.

Figures 18 and 19 show a comparison of the maximum and mean values of the absolute values of each measurement point for different positions of the windshield at 400 km/h of the train. It can be seen that the amplitude of displacement vibrations at each measurement point on windshield 2 is higher than that of the measurement points on windshield 1, and the same is the case for the mean values. However, the difference is not very significant near the end of the car due to the constraints. It can be concluded that the outer windshield 2 vibrates more significantly under aerodynamic load excitation than windshield 1.

Furthermore, from Figure 13, it can be seen that the position of the measuring point on the lower side of the windshield is close to the windshield opening, and the constraint conditions at this place are weaker than the position of the measuring point on the upper side, and the stiffness is smaller. At the same time, for the measuring points on the lower side of the windshield, since both sides of the outer windshield are fixed at the end wall, points 4 and 12 near the end wall are more constrained than other measuring points at the same horizontal line (points 5, 6, 10, 11), while points 6 and 10 are more constrained than other measuring points at the same horizontal line. On the other hand, due to the opening on the lower side of the outer windshield, the airflow inside and outside the windshield is exchanged through the opening position, which increases the aerodynamic force on the surface at the lower side. Therefore, it can be seen from Figure 18 that the displacements of lower points are larger than those of upper points and displacements of points 6 and 10 are the largest.

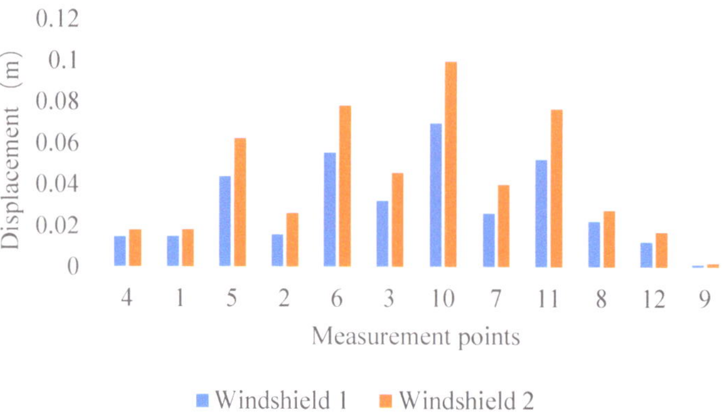

Figure 18. Comparison of the maximum values of the displacement of windshield 1 and windshield 2.

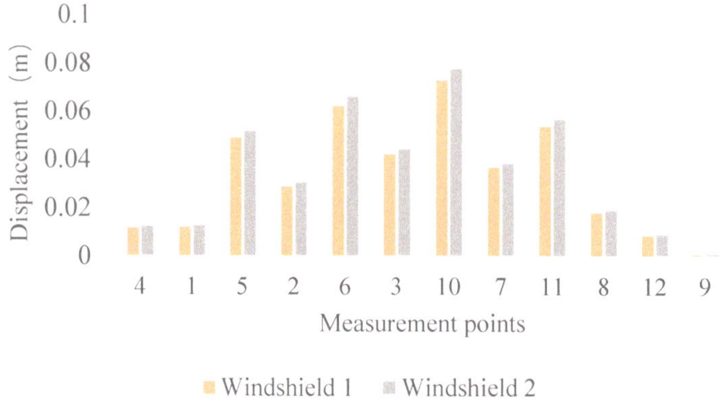

Figure 19. Comparison of the mean values of the displacement of windshield 1 and windshield 2.

6. Conclusions

In this paper, a new method is proposed to build a U-section rubber windshield structure flow-induced vibration model by using the external aerodynamic load and the measured modal data of the outer windshield structure as input, combined with the mode superposition method. This method can be used to analyze the dynamic response of U-section rubber outer windshield structure efficiently.

Since the aerodynamic load on the windshield surface as the input condition does not need to be iterated repeatedly, it is not limited by the computing resources required and can handle the dynamic response simulation calculation under various working conditions such as open line operation, open line intersection or tunnel operation.

The use of measured structural modal data can avoid the problem of inaccurate dynamic characteristics caused by oversimplification of U-section rubber outer windshield simulation modeling. This method is suitable for dynamic characterization of complex flexible structures.

Through the flow-induced vibration analysis of U-section rubber outer windshield structure of a 400 km/h train, it is determined that the displacement and velocity response curves of the measuring point near the windshield opening position are significantly higher than those of other parts due to the influence of the opening position. This area has a great potential for vibration reduction, and attention should be paid to the influence of the structure's first-order mode on its response.

Appl. Sci. **2023**, *13*, 5813

Author Contributions: Conceptualization, X.L. (Xiang Liu); methodology, Y.Y. and X.L. (Xiang Liu); software, X.L. (Xiao Liu); formal analysis, Y.Y., P.L. and X.L. (Xiao Liu); writing—original draft preparation, Y.Y., P.L. and X.L. (Xiao Liu); writing review and editing, Y.Y., P.L., X.L. (Xiao Liu) and X.L. (Xiang Liu); supervision, X.L. (Xiang Liu); funding acquisition, X.L. (Xiang Liu). All authors have read and agreed to the published version of the manuscript.

Funding: This research was funded by Science and Technology Research and Development Program of China National Railway Group Co., Ltd., grant number P2021J037. National Natural Science Foundation, grant number 11802345 and the Fundamental Research Funds for the Central Universities of Central South University, grant number 1053320216395.

Institutional Review Board Statement: Not applicable.

Informed Consent Statement: Not applicable.

Data Availability Statement: Not applicable.

Conflicts of Interest: The authors declare no conflict of interest.

References

1. Shen, Z. Dynamic Environment of High-speed Train and Its Distinguished Technology. *J. China Railw. Soc.* **2006**, *28*, 1–5.
2. Yin, T.; Cheng, H.; Shi, Z.; Dong, J. Analysis of the influence of the relative position of the end of the EMU on the internal windshield linkage. *Shandong Ind. Technol.* **2019**, 1.
3. Schetz, J.A. Aerodynamics of High-Speed Trains. *Annu. Rev. Fluid Mech.* **2001**, *33*, 371–414. [CrossRef]
4. Baker, C. The flow around high speed trains. *J. Wind Eng. Ind. Aerodyn.* **2012**, *98*, 277–298. [CrossRef]
5. Baker, C.J.; Dalley, S.J.; Johnson, T.; Quinn, A.; Wright, N.G. The slipstream and wake of a high-speed train. *Proc. Inst. Mech. Eng. Part F J. Rail Rapid Transit* **2001**, *215*, 83–99. [CrossRef]
6. Schulte-Werning. Research of European railway operators to reduce the environmental impact of high-speed trains. *Proc. Inst. Mech. Eng. Part F J. Rail Rapid Transit* **2003**, *217*, 249–257. [CrossRef]
7. Liu, H.; Wei, X.; Zeng, J.; Wang, Y. Effect of vestibule diaphragm device on train dynamic performance. *J. Traffic Transp. Eng.* **2003**, 3, 5.
8. Niu, J.; Liang, X.; Xiong, X.; Liu, F. Effect of outside vehicle windshield on aerodynamic performance of high-speed train under crosswind. *J. Shandong Univ. Eng. Sci. Ed.* **2016**, *8*, 108–115.
9. Tan, L.; Su, R.; Huang, L.; Cheng, Z.; Zhu, T.; Mu, H. Investigation of white EPDM based vestibule diaphragm for high-speed train. *Electr. Locomot. Urban Rail Veh.* **2015**, *38*, 4.
10. Tang, M.; Xiong, X.; Zhong, M.; Wang, Z. Influence of installation spacing of external vestibule diaphragm of high-speed train on aerodynamic characteristics of the vestibule diaphragm. *J. Railw. Sci. Eng.* **2019**, *16*, 10.
11. Yang, J.; Jiang, C.; Gao, Z.; Lv, Y.; Zhang, J.; Li, C. Influence of Inter-car Wind-shield Schemes on Aerodynamic Performance of High-speed Trains. *J. China Railw. Soc.* **2012**, *34*, 7.
12. Fujii, K.; Ogawa, T. Aerodynamics of high speed trains passing by each other. *Comput. Fluids* **1995**, *24*, 897–908. [CrossRef]
13. Niu, J.; Wang, Y.; Zhou, D. Effect of the outer windshield schemes on aerodynamic characteristics around the car-connecting parts and train aerodynamic performance. *Mech. Syst. Signal Process.* **2019**, *130*, 1–16. [CrossRef]
14. Kurita, T.; Mizushima, F. Environmental Measures along Shinkansen Lines with FASTECH360 High-Speed Test Trains. *JR East Tech. Rev.* **2010**, *16*, 47–55.
15. Li, X.B.; Chen, G.; Wang, Z.; Xiong, X.H.; Yin, J. Dynamic analysis of the flow fields around single- and double-unit trains. *J. Wind Eng. Ind. Aerodyn.* **2019**, *188*, 136–150. [CrossRef]
16. Peters, J.L. Optimising aerodynamics to raise IC performance. *Railw. Gaz. Int.* **1982**, *138*, 817–819.
17. Bell, J.R.; Burton, D.; Thompson, M.C.; Herbst, A.H.; Sheridan, J. The effect of tail geometry on the slipstream and unsteady wake structure of high-speed trains. *Exp. Therm. Fluid Sci. Int. J. Exp. Heat Transf. Thermodyn. Fluid Mech.* **2017**, *83*, 215–230. [CrossRef]
18. Shiraishi, H. Improvement of Smooth Covers between Vehicles for Shinkansen High Speed Test Trains. *Foreign Roll. Stock* 2010.
19. Tian, H.Q. Review of research on high-speed railway aerodynamics in China. *Transp. Saf. Environ.* **2019**, *1*, 1–21. [CrossRef]
20. Xia, Y.; Liu, T.; Gu, H.; Guo, Z.; Li, L. Aerodynamic effects of the gap spacing between adjacent vehicles on wind tunnel train models. *Eng. Appl. Comput. Fluid Mech.* **2020**, *14*, 835–852. [CrossRef]
21. Bathe, K.-J.; Hou, Z.; Ji, S. Finite element analysis of fluid flows fully coupled with structural interactions. *Comput. Struct.* **1999**, *72*, 1–16. [CrossRef]
22. Benney, M.R.J.; Stein, M.K.R.; Kalro, V.; Tezduyar, T.E.; Leonard, J.W.; Accorsi, M.L. Parachute performance simulations: A 3D fluid-structure interaction model. *Comput. Methods Appl. Mech. Eng.* **1992**, *94*, 353–371.
23. Horiuchi, M. Outline of Shinkansen high-speed test train type E955 (FASTECH360Z). *JR East Tech. Rev. Ser.* **2006**, *8*, 6–10.
24. Loon, R.V.; Anderson, P.D.; Vosse, F.; Sherwin, S.J. Comparison of various fluid-structure interaction methods for deformable bodies. *Comput. Struct.* **2007**, *85*, 833–843. [CrossRef]

25. Luo, L.L. Aerodynamic noise radiating from the inter-coach windshield region of a high-speed train. *J. Low Freq. Noise Vib. Act. Control.* **2018**, *37*, 590–610.

26. Mohmmed, A.O.; Al-Kayiem, H.H.; Sabir, O.; Ahmed, O. One-way coupled fluid-structure interaction of gas-liquid slug flow in a horizontal pipe: Experiments and simulations. *J. Fluids Struct.* **2020**, *97*, 103083. [CrossRef]

27. *EN 14067-4:2013*; Railway Applications—Aerodynamics—Part 4: Requirements and Test Procedures for Aerodynamics on Open Track. European Standards: Plzen, Czech Republic, 2013.

28. Wijesooriya, K.; Mohotti, D.; Amin, A.; Chauhan, K. Comparison between an uncoupled one-way and two-way fluid structure interaction simulation on a super-tall slender structure. *Eng. Struct.* **2021**, *229*, 111636. [CrossRef]

29. Cai, J.; Zhang, S.; Wang, H.; Xiong, X.; Tang, M. Modal finite element calculation and test analysis on U-shaped rubber outer windshield structure of high speed train. *J. Railw. Sci. Eng.* **2020**, *17*, 10.

30. Wang, H.; Sun, G.; She, P.; Tang, M. Experimental Study on Dynamic Characteristics of the Vehicle-end Inner Windshield in High-speed EMUs. *Electr. Drive Locomot.* **2020**, *4*, 6.

31. Li, S.; Wang, J. Research on Vibration of High speed Train Windshield. *Sci. Technol. Innov. Rev.* **2020**, *17*, 2.

32. Liu, Z. Research on Windshield Aerodynamic Characteristics and Pneumatic Fatigue Strength of High Speed Train. Master's Thesis, Central South University, Changsha, China, 2014.

33. Wang, Y.; Cai, X.; Zhao, W. FSI Technology and Its Application in High-Speed EMUs Structure Design. *J. Dalian Jiaotong Univ.* **2012**, *33*, 5.

34. Miao, X.; Gao, G.; He, K. Wish a hundred years; KONG Fabing. Aerodynamic shape optimization of windshields on freight high-speed trains with crosswind. *J. Cent. S. Univ. Nat. Sci. Ed.* **2021**, *52*, 9.

35. Long, S. Aerodynamic Characteristics Simulation of High-speed Train Running on Different Infrastructures under Cross Winds. *Mechatronics* **2017**, *34*, 3–8.

36. Wang, Z.; Li, T.; Zhang, J. Research on Aerodynamic Performance of High-speed Train Subjected to Different Types of Crosswind Enhanced Publishing. *J. Mech. Eng.* **2018**, *9*. [CrossRef]

37. Ouyang, M.; Chen, S.; Li, Q. Analysis of unsteady flow field and near-field aerodynamic noise of scale high-speed trains in long tunnel. *Appl. Acoust.* **2023**, *205*, 109261. [CrossRef]

38. Maleki, S.; Burton, D.; Thompson, M.C. Assessment of various turbulence models (ELES, SAS, URANS and RANS) for predicting the aerodynamics of freight train container wagons. *J. Wind Eng. Ind. Aerodyn.* **2017**, *170*, 68–80. [CrossRef]

39. Premoli, A.; Rocchi, D.; Schito, P.; Tomasini, G. Comparison between steady and moving railway vehicles subjected to crosswind by CFD analysis. *J. Wind Eng. Ind. Aerodyn.* **2016**, *156*, 29–40. [CrossRef]

40. Sun, Z.; Wang, T.; Wu, F. Numerical investigation of influence of pantograph parameters and train length on aerodynamic drag of high-speed train. *J. Cent. South Univ.* **2020**, *27*, 1334–1350. [CrossRef]

41. Dong, T.; Liang, X.; Krajnovi, S.; Xiong, X.; Zhou, W. Effects of simplifying train bogies on surrounding flow and aerodynamic forces. *J. Wind Eng. Ind. Aerodyn.* **2019**, *191*, 170–182. [CrossRef]

42. Wang, S.; Bell, J.R.; Burton, D.; Herbst, A.H.; Sheridan, J.; Thompson, M.C. The performance of different turbulence models (URANS, SAS and DES) for predicting high-speed train slipstream. *J. Wind Eng. Ind. Aerodyn.* **2017**, *165*, 46–57. [CrossRef]

43. Zhang, J.; Adamu, A.; Su, X.; Guo, Z.; Gao, G. Effect of simplifying bogie regions on aerodynamic performance of high-speed train. *J. Cent. South Univ.* **2022**, *29*, 1717–1734. [CrossRef]

44. Zhang, J.; Guo, Z.; Han, S.; Krajnović, S.; Sheridan, J.; Gao, G. An IDDES study of the near-wake flow topology of a simplified heavy vehicle. *Transp. Saf. Environ.* **2022**, *4*, tdac015. [CrossRef]

*applied
sciences*

Article

Influence of Wheel-Rail Contact Algorithms on Running Safety Assessment of Trains under Earthquakes

Guanmian Cai [1], Zhihui Zhu [1,2,3,*], Wei Gong [1], Gaoyang Zhou [1], Lizhong Jiang [1,2,3] and Bailong Ye [1,2,3]

[1] School of Civil Engineering, Central South University, Changsha 410075, China; 18793161283@163.com (G.C.); 13135317967@163.com (W.G.); 214807004@csu.edu.cn (G.Z.); lzhjiang@csu.edu.cn (L.J.); 13607481792@163.com (B.Y.)

[2] National Engineering Research Center of High-Speed Railway Construction Technology, Central South University, Changsha 410075, China

[3] Hunan Provincial Key Laboratory for Disaster Prevention and Mitigation of Rail Transit Engineering Structures, Central South University, Changsha 410075, China

*   Correspondence: zzhh0703@163.com

Abstract: Accurate running safety assessment of trains under earthquakes is crucial to ensuring the safety of line operation. Extreme contact behaviors such as wheel flange contact and wheel jump during earthquakes will directly affect the running safety of trains. To accurately simulate a wheel-rail extreme contact state, the calculation of the normal compression amount, the normal contact stiffness, and a number of contact points are crucial in wheel-rail space contact modeling. Hence, in order to clarify the applicable algorithms during earthquakes, this paper first introduces different algorithms in three aspects mentioned above. Taking a single CRH2 motor vehicle passing through a ballastless track structure under EI-Centro wave excitation as an example, a comparative analysis of wheel-rail contact dynamics and running safety was conducted. The results showed that adopting the normal compression algorithm based on vertical penetration and the consideration of only single-point contact will result in the maximum calculation error of wheel-rail contact force to reach 339.50% and 35.00%, respectively. This significantly affects the accuracy of train safety assessment, while using the empirical formula for wheel-rail normal contact stiffness has relatively less impact. To ensure the accuracy of running safety assessment of trains during an earthquake, it is recommended to adopt the normal compression algorithm based on normal penetration and consider the multi-point contact in wheel-rail contact modelling.

Keywords: running safety assessment; earthquake; wheel-rail contact; contact point; normal compression amount; normal contact stiffness

Citation: Cai, G.; Zhu, Z.; Gong, W.; Zhou, G.; Jiang, L.; Ye, B. Influence of Wheel-Rail Contact Algorithms on Running Safety Assessment of Trains under Earthquakes. *Appl. Sci.* **2023**, *13*, 5230. https://doi.org/10.3390/app13095230

Academic Editor: Diogo Ribeiro

Received: 15 March 2023
Revised: 20 April 2023
Accepted: 20 April 2023
Published: 22 April 2023

1. Introduction

In China, high-speed railways are increasingly crossing high-intensity earthquake-prone areas given their wide distribution, thus influencing the running safety of trains [1]. The running safety problems of trains under earthquake excitation has become an ongoing concern. Under earthquake excitation, extreme contact behaviors such as wheel flange contact and wheel jumping occur, directly impacting the running safety of trains [2]. Therefore, it is crucial to accurately simulate the wheel-rail extreme contact state.

Wheel-rail contact modeling under earthquake excitation needs to consider wheel-rail spatial rolling contact characteristics, and the typical wheel-rail contact model is associated with three problems: (i) wheel-rail contact geometry calculation, (ii) normal contact force calculation, and (iii) tangential contact force calculation [3]. In previous studies, the contact trace method [4], nonlinear Hertz theory, and Shen-Hedrick-Elkins theory [5] have been used to solve the above problems and establish a corresponding model [6–8]. Since this model can balance the requirements of wheel-rail contact modeling and computational efficiency [9,10], it is widely applied in vehicle dynamic simulations and running safety analyses under earthquakes [11–13].

However, in order to ensure the accuracy of the wheel-rail extreme contact state simulation, the algorithm used in wheel-rail contact modelling under earthquake excitation in the following aspects needs to be further clarified. First, the contact points are generally detected by the maximum wheel-rail vertical penetration in most cases; thus, only a single contact point can be determined [14]. Seismic excitation causes a large relative lateral displacement between the wheel and rail [15], leading to simultaneous elastic penetrations in different regions [16,17]. Hence, the consideration of the number of wheel-rail contact points under earthquake excitation needs to be further clarified.

The wheel-rail contact geometry relationship, as the basis for contact force calculation, mainly includes determining the contact point and the normal compression amount, and the wheel-rail searching method plays a decisive role. The vertical searching method [9], in which the contact point and the normal compression amount are determined according to the maximum wheel-rail vertical penetration, is widely adopted for the convenience of calculation. As the normal compression amount is a key parameter in calculating the contact normal force [17], it is essential to ensuring its calculation accuracy. The approximate calculation method based on vertical penetration is applicable to ordinary conditions [18], whereas it is not convincing under extreme operation conditions such as during earthquakes. The normal searching method [19–21], in which the contact point position and the normal compression amount are determined according to the maximum wheel-rail normal penetration, can ensure the reliability of the solution to the normal compression amount under all conditions. Hence, it is crucial to clarify the necessity of introducing the normal searching method for seismic conditions by comparing the two methods.

The normal contact stiffness, which is another key parameter in the calculation of the wheel-rail normal contact force, is calculated on the basis of the geometric parameters at wheel-rail contact points when using the Hertzian contact theory [22]. However, an empirical formula based on wheel-rail constants [23] is widely used in the wheel-rail contact force calculation, which is only applicable in cases where the contact points are located in a narrow region of the wheel tread and rail head [24]. Hence, the normal contact stiffness algorithm also needs to be further clarified for seismic conditions.

Therefore, in order to evaluate the algorithm of normal compression, the normal contact stiffness and the considerations of single-point or multi-point contact under earthquake excitations, various algorithms in the above three aspects were introduced in wheel-rail contact modelling. Taking a single CRH2 motor vehicle passing through a ballastless track structure under El-Centro wave excitation as an example, the wheel-rail contact dynamics and the running safety evaluation results calculated based on different algorithms were compared and analyzed.

2. Wheel-Rail Contact Model

2.1. Wheel-Rail Contact Geometry Calculation

To solve the wheel-rail contact geometric relationship, two types of methods have been developed currently: the spatial direct search method and the projection method. The spatial direct search method requires searching for wheel rail contact points in the spatial dimension and involves a considerable number of iterative calculations [25–27]. As a representative of the projection method, the contact trace method [28] uses the idea of constructing spatial traces to transform the three-dimensional search problem into a planar search problem [9], greatly improving computational efficiency and making it widely used in the calculation of wheel rail spatial geometric relationships. Thus, the contact trace method was employed in this study, as illustrated in Figure 1. Two coordinate systems are defined in the system, namely, the absolute coordinate system (O-XYZ) and the wheelset coordinate system (O_{w}-$X_{\mathrm{w}}Y_{\mathrm{w}}Z_{\mathrm{w}}$). In the absolute coordinate system, O is positioned at the center of the track, with the OX axis pointing in the direction of rolling, the OZ axis pointing vertically, and the OY axis aligning with the track transverse direction as determined by the right-hand rule. In the wheelset coordinate system, O_w is situated at the center of mass of the wheelset, and the $O_w Y_w$-axis aligns with the wheelset's axis of revolution. The basic idea of this method is as follows. First, the wheel profile is discretized along the transverse direction to obtain several rolling circles. A coordinate expression of the potential contact points on the wheel in the absolute coordinate system $O\text{-}XYZ$ is obtained:

$$\begin{cases} x = d_w l_x - l_x R_w \tan \delta_w \\ y = d_w l_y + Y_w + \frac{R_w}{1-l_x^2}\left(l_x^2 l_y l_z \tan \delta_w - l_z \sqrt{1 - l_x^2\left(1 + (\tan \delta_w)^2\right)} \right) \\ z = d_w l_z + Z_w + \frac{R_w}{1-l_x^2}\left(l_x^2 l_z \tan \delta_w + l_y \sqrt{1 - l_x^2\left(1 + (\tan \delta_w)^2\right)} \right) \end{cases} \tag{1}$$

where l_w is the lateral distance of the rolling circle from the wheelset center of mass; R_w is the rolling circle radius; δ_w is the contact angle at the wheel-rail contact point; Y_w and Z_w are the lateral and vertical displacements of the wheelset, respectively. l_x, l_y, and l_z are the elements of direction cosine matrix between the absolute coordinate system and the wheelset coordinate system, which can be expressed as:

$$\begin{cases} l_x = -\cos \phi_w \sin \psi_w \\ l_y = \cos \phi_w \cos \psi_w \\ l_z = \sin \phi_w \end{cases} \tag{2}$$

where ϕ_w and ψ_w represents the rolling angle and the yaw angle of the wheelset, respectively.

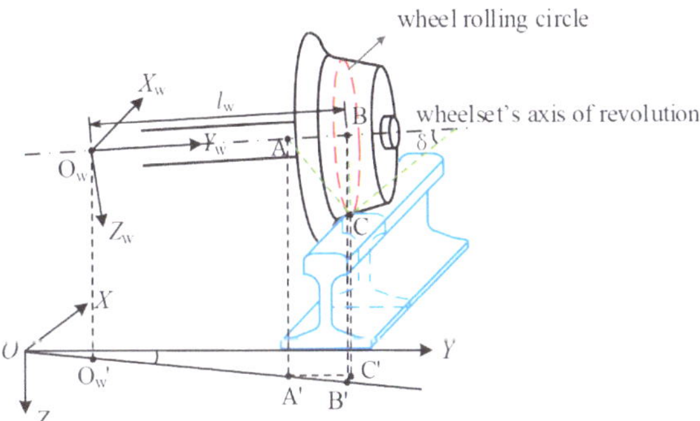

Figure 1. Calculation of wheel-rail contact geometric relationship.

The potential contact points of each rolling circle form a space curve, namely, the contact trace, and it is then projected onto the rail cross-section. To determine the contact point position and the normal compression amount in the projection plane, the calculation methods based on the vertical searching method or normal searching method are adopted, as shown in Figure 2. In the normal searching method, for each discrete point on the wheel and rail profiles, the distance from the normal to another profile curve is defined as the normal penetration amount. The maximum normal penetration is adopted for the judgement of the contact point position and the determination of the normal compression amount, which is characterized by the length of the red line segment shown in Figure 2. The mathematical expression can be written as:

$$\delta_{cn} = \max(d_n) \tag{3}$$

where d_n represents the normal penetration amount.

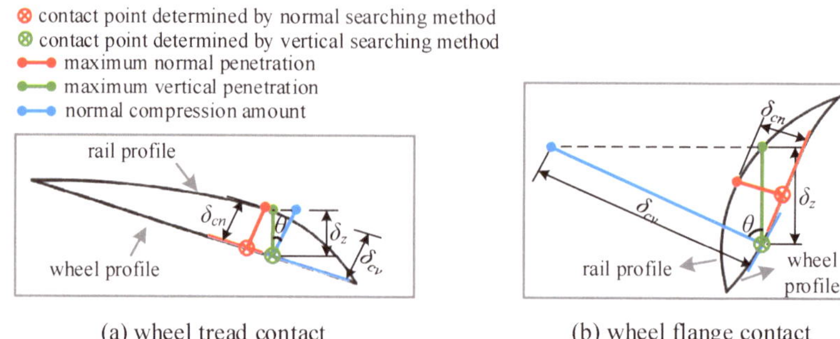

(a) wheel tread contact   (b) wheel flange contact

Figure 2. Determination of wheel-rail contact points and normal compression amounts based on different searching methods: (**a**) wheel-tread contact; (**b**) wheel-flange contact.

In the vertical searching method, the maximum wheel-rail vertical penetration is introduced as the judgement condition when determining the contact point, which is characterized by the length of the green line segment shown in Figure 2. The normal compression amount can be determined by constructing a right triangle, as characterized by the length of the blue line segment shown in Figure 2. The mathematical expression is given as:

$$\delta_{cv} = \delta_z / \cos\theta \tag{4}$$

where δ_z is the maximum vertical penetration amount, and θ is equal to $\delta_w - \phi_w$.

In the wheel tread penetration zone, as presented in Figure 2a, since θ is small, the difference between δ_{cn} and δ_{cv} is not evident, indicating that the algorithm based on the vertical penetration is applicable in this case. In the wheel flange penetration zone, as shown in Figure 2b, δ_{cv} is significantly different from δ_{cn} in this case due to the significant impact of θ. Hence, using the algorithm based on the vertical penetration tends to cause large calculation errors of the normal compression amount in this case.

In addition, for the consideration of the number of wheel-rail contact points, if only single-point contact is considered, the position corresponding to the maximum penetration in the whole range of wheel tread is generally regarded as the wheel-rail contact point. In contrast, to consider multipoint contact, the segment searching method can be introduced to determine each wheel-rail penetration zone, as illustrated in Figure 3. For each discrete point on the wheel profile, its vertical coordinate is denoted by Z_w, and the rail vertical coordinate corresponding to the vertical direction is denoted by Z_r. The difference between the two vertical coordinates is denoted by ΔZ and can be expressed as:

$$\Delta Z = Z_w - Z_r \tag{5}$$

Thus, each wheel-rail penetration zone can be determined when $\Delta Z < 0$. The contact point position and the normal compression amount can be determined in each penetration zone.

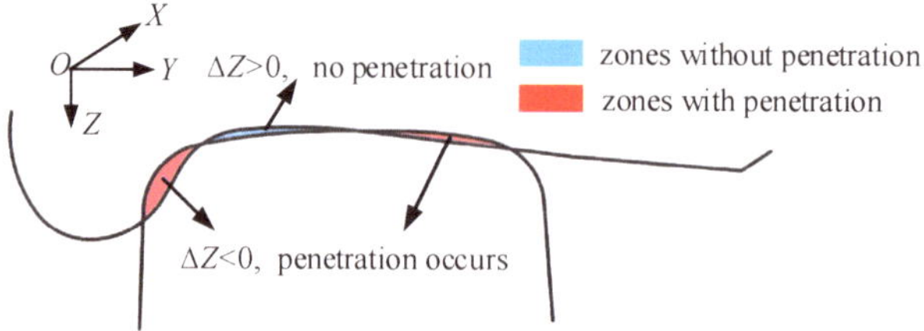

Figure 3. Wheel-rail segment searching method.

2.2. Wheel-Rail Contact Force Calculation

The wheel-rail normal contact force based on the Hertz elasticity theory is given as:

$$F_N = K_{nr}\delta_c^{3/2} \tag{6}$$

where K_{nr} is the wheel-rail normal contact stiffness, and δ_c is the wheel-rail normal compression amount.

According to an empirical formula, K_{nr} is expressed as:

$$K_{nr} = (1/G)^{3/2} \tag{7}$$

where G is the wheel-rail contact constant $\left(\mathrm{m/N}^{2/3}\right)$.

In contrast, the general expression for K_{nr} derived from the Hertz theory is given as [24]:

$$K_{nr} = \frac{2\sqrt{2}}{3} \frac{q_k}{(\delta_1 + \delta_2)} \left(\frac{\rho}{R_{wx}}\right)^{1/2} \sqrt{R_{wx}} \tag{8}$$

where δ_1 and δ_2 are the calculated parameters related to the Poisson's ratio and elastic modulus of the wheel and rail; ρ is the parameter related to R_{wy} and R_{ry} at the contact points; R_{wy} and R_{ry} are the curvature radii of the wheel and rail profiles; and R_{wx} is the rolling circle radius of the wheel. ρ/R_{wx} is obtained from the corresponding calculation table. The intermediate calculation parameter q_k is expressed as:

$$q_k = [m/2K(e)]^{3/2} \tag{9}$$

where m and $K(e)$ are the intermediate parameters in the solution obtained using the Hertz theory.

To clarify the difference between the empirical formula and the theoretical formula, the cases of wheel-tread contact and wheel-flange contact were analyzed. The different colored line segments on the wheel and rail profiles represent areas with different curvature radii in Figure 4. For wheel-tread contact, the contact points on the wheel and rail are located on lines AB and CD, respectively, as shown in Figure 4. In this case, the applicability of the empirical formula has been verified [24]. In contrast, for wheel-flange contact, the contact points are located on the lines EF and GH, respectively, as shown in Figure 4. In this case, R_{wy} and R_{ry} at the contact points are significantly different from that in wheel-tread contact. When using the theoretical formula, the calculation result of K_{nr} must be significantly different from that in wheel-tread contact according to Equation (8). However, K_{nr} will not change compared to that in wheel-tread contact when using the empirical formula according to Equation (7). Hence, using the empirical formula tends to cause large calculation errors of normal contact stiffness in this case.

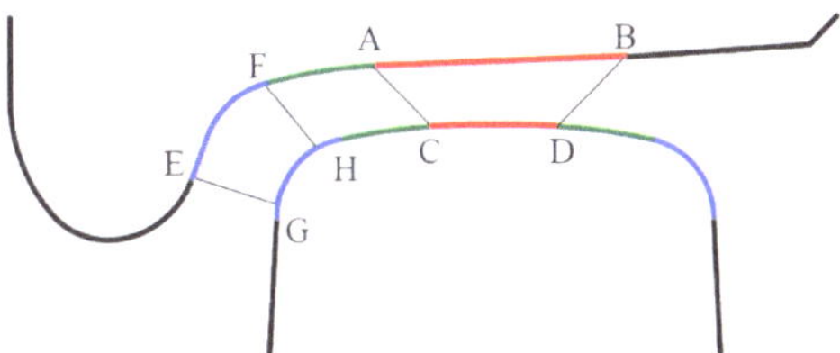

Figure 4. Distribution range of wheel-rail contact point positions in wheel-tread contact and flange contact.

After obtaining the normal contact force within each contact patch, the wheel-rail tangential force is first calculated using Kalker linear theory [29]. Considering the non-linear relationship between wheel-rail creepages and creep forces for large creepages, the

Shen-Hedrick-Elkins theory [5] is applied for the nonlinear correction of the tangential contact force.

Thus, the wheel-rail contact forces in each contact patch can be calculated, represented by blue line segments with arrows in Figure 5. Notably, the contact forces should be converted to an absolute coordinate system in dynamic analysis, which is given as:

$$F_Y = \sum F_{Yi}, \; F_Z = \sum F_{Zi}, \; M_X = \sum M_{Xi}, \; M_Z = \sum M_{Zi} (i = 1, 2, \ldots n) \tag{10}$$

where F_Y, F_Z, M_X and M_Z are composed by the wheel rail contact force and spin moment; i is the indicator of each wheel-rail contact point, and n is the number of contact points; and F_{Yi}, F_{Zi}, M_{Xi} and M_{Zi} can be expressed as:

$$\begin{aligned}
F_{Xi} &= F_{Ni} \cdot i + F_{xi} \cdot i + F_{yi} \cdot i, \; F_{Yi} = F_{Ni} \cdot j + F_{xi} \cdot j + F_{yi} \cdot j \\
F_{Zi} &= N_i \cdot k + F_{xi} \cdot k + F_{yi} \cdot k, \; M_{Xi} = M_i \cdot i, \; M_{Zi} = M_i \cdot k
\end{aligned} \tag{11}$$

where (i, j, k) is the unit vector in the absolute coordinate system; F_{Ni}, F_{xi}, F_{yi}, and M_i are the contact force and spin moment vectors at the ith wheel-rail contact point.

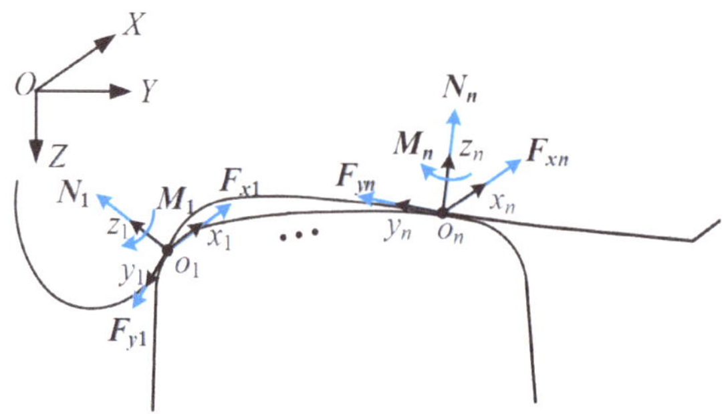

Figure 5. Wheel-rail contact force in the contact patch.

3. Dynamic Analysis Model of Train-Track Coupling System under Earthquakes

The dynamic analysis model of the train-track coupling system under the effect of earthquakes is used for analysis of the influence of different algorithms on wheel-rail contact dynamics and running safety, as presented in Figure 6. The coupling system is divided into two subsystems: a train subsystem and a track subsystem. The subsystems are coupled by wheel-rail interaction. Earthquake excitations are applied to the bottom of the track. The vibration equation of the coupling system can be written as:

$$\begin{bmatrix} M_v & 0 \\ 0 & M_r \end{bmatrix} \begin{Bmatrix} \ddot{X}_v \\ \ddot{X}_r \end{Bmatrix} + \begin{bmatrix} C_v & 0 \\ 0 & C_r \end{bmatrix} \begin{Bmatrix} \dot{X}_v \\ \dot{X}_r \end{Bmatrix} + \begin{bmatrix} K_v & 0 \\ 0 & K_r \end{bmatrix} \begin{Bmatrix} X_v \\ X_r \end{Bmatrix} = \begin{Bmatrix} F_{vg} + F_w \\ F_r + F_{EA} \end{Bmatrix} \tag{12}$$

where M_v, C_v, and K_v represent the mass, damping, and stiffness matrices of the train subsystem, respectively; X_v represents the displacement vector of the train subsystem; M_r, C_r, and K_r represent the mass, damping, and stiffness matrices of the track subsystem, respectively; X_r represents the displacement vector of the track subsystem; F_{vg} represents the self-weight load vector of the train system; F_w and F_r represent the load vectors of the wheel-rail force applied to the train subsystem and track subsystem, respectively; and F_{EA} is the earthquake load obtained using the large mass method [30].

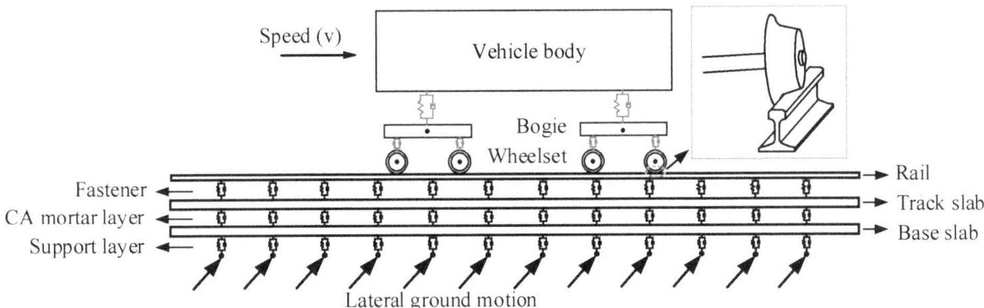

Figure 6. Train-track coupling system under earthquakes.

3.1. Train Subsystem

Each vehicle of a train can be simplified as a vibration system comprising vehicle body, bogies, wheelsets, and double-suspension systems, as shown in Figure 7. Four degrees of freedom, namely, vertical movement, lateral movement, rolling, and yawing, are considered for each wheelset, and for the vehicle body and each bogie, the degree of freedom of pitching is considered in addition to the former four degrees of freedom. To accurately simulate the motion behavior of vehicles under extreme conditions, such as during an earthquake, the nonlinearity of the secondary lateral stop, spring stiffness, and damping of the suspension systems [31–33], is considered in this paper. Thus, the vibration equation of the train subsystem can be given as:

$$M_v \ddot{X}_v + C_v \dot{X}_v + K_v X_v = F_{vg} + F_w \tag{13}$$

where the definitions of each variable are the same as that given for Equation (12).

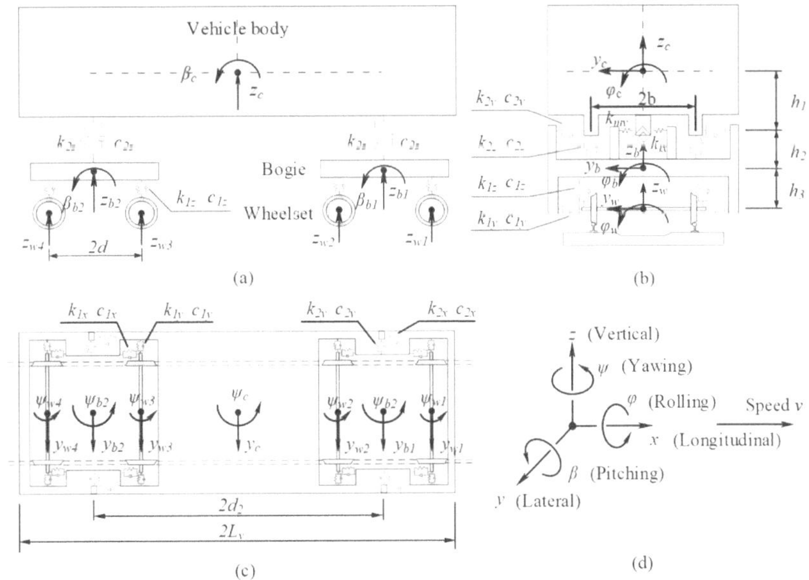

Figure 7. Vehicle model with corresponding DOFs: (**a**) Side view; (**b**) Front view; (**c**) Top view; (**d**) Sign convention of vehicle.

3.2. Track Subsystem

The China Railway Track System II (CRTS II) slab ballastless track structure was adopted in this study, the main construction and dimensions of which are shown in Figure 8. This type of ballastless track was derived from German Borg slab ballastless track structure and developed in China, which can provide a more stable, durable, and lower maintenance alternative to traditional ballasted track systems [34]. It has been widely used in China's high-speed rail line, including the Beijing-Tianjin Intercity Railway and the Beijing-Shanghai High-Speed Railway [35]. The track structure was established using the finite element (FE) method. For the modeling, the rail was built using beam element, the track plate and base plate were built using shell element, and the fastener was simulated by a spring damping element. The CA mortar layer and support layer were simulated by uniformly distributed spring damping element. The vibration equation of the track subsystem under earthquakes is given as:

$$M_r \ddot{X}_r + C_r \dot{X}_r + K_r X_r = F_r + F_{EA} \tag{14}$$

where the definitions of each variable are the same as that given for Equation (12).

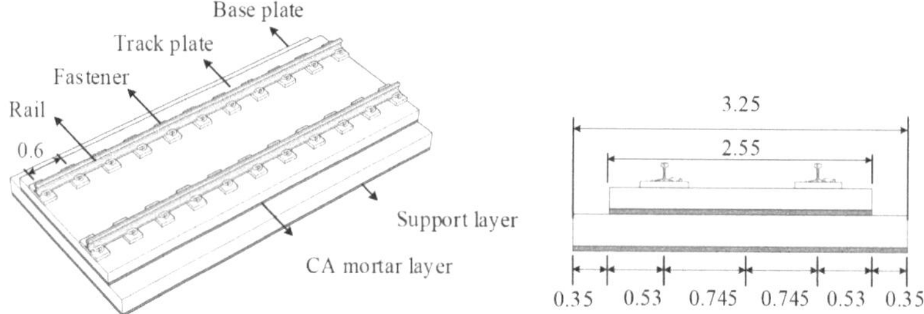

Figure 8. Diagram of CRTS II slab ballastless track (unit: m).

3.3. Train-Track Coupling

Figure 9 shows a scheme of the force state of each wheelset in the train subsystem. The loads at the wheelset centroid can be expressed as:

$$F_{Yw} = \sum F_{YLi} + \sum F_{YRi}, \quad F_{Zw} = \sum F_{ZLi} + \sum F_{ZRi},$$
$$M_{Xw} = -\sum F_{ZLi} \cdot l_{li} - \sum F_{YLi} \cdot r_{li} + \sum F_{ZRi} \cdot l_{ri} - \sum F_{YRi} \cdot r_{ri} + \sum M_{XLi} + \sum M_{XRi}$$
$$M_{Zw} = \sum F_{XLi} \cdot l_{li} + \sum F_{YLi} \cdot l_{li} \cdot \psi_w + \sum F_{XRi} \cdot l_{ri} + \sum F_{YLi} \cdot l_{li} \cdot \psi_w + \sum M_{ZLi} + \sum M_{ZRi}$$
$$(i = 1, 2, \ldots n) \tag{15}$$

where F_{Yw}, F_{Zw}, M_{Xw}, and M_{Zw} are the lateral force, vertical force, rolling moment, and yawing moment at the wheelset centroid, respectively; $F_{X(L,R)i}$, $F_{Y(L,R)i}$, $F_{Z(L,R)i}$, $M_{X(L,R)i}$, and $M_{Z(L,R)i}$ can be obtained by Equation (11).

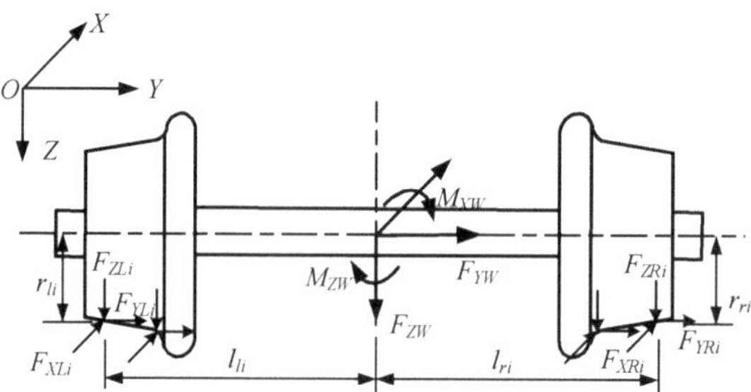

Figure 9. Wheelset force state.

F_w in Equation (13) can be written as:

$$F_w = \left[F_{w1}, F_{w2}, \cdots, F_{wj}\right]^T (j = 1, 2, \cdots, m) \tag{16}$$

where m is the number of vehicles, and F_{wj} is expressed as:

$$F_{wj(1\times31)} = \left[F_{cj}, F_{bj1}, F_{bj2}, F_{wj1}, \cdots, F_{wjk}\right]^T (k = 1, 2, 3, 4) \tag{17}$$

where F_{cj}, $F_{bj(1,2)}$, F_{wjk} are the load vectors corresponding to the vehicle body, bogies, and the kth wheelset of the jth vehicle. F_{wjk} can be written as:

$$F_{wjk(1\times4)} = \left[F_{Yw}, F_{Zw}, M_{Xw}, M_{Zw}\right]^T \tag{18}$$

In the track subsystem, taking the left rail as an example, the concentrated wheel-rail load applied to the rail can be expressed as:

$$F_{Yr} = -\sum F_{YLi}, \quad F_{Zr} = -\sum F_{ZLi} \tag{19}$$

Here, F_{Yr} and F_{Zr} represent the lateral and vertical forces, respectively; F_{YLi} and F_{ZLi} are the same as that in Equation (15).

The node load vector of the rail element is expressed as:

$$F_{re(1\times N_r)} = \left[F_{Yr} \; F_{Zr}\right]^T \cdot N(x) \tag{20}$$

where $N(x)$ represents the vector of the cubic Hermit form functions corresponding to the cell nodes, and N_r is the number of degrees of freedom of each node.

$F_{re(1\times N_r)}$ is applied to the corresponding degrees of freedom; thus, F_r in Equation (14) can be established.

4. Comparison of Different Wheel-Rail Contact Models

To analyze the impact of different algorithms on wheel-rail contact dynamics, different wheel-rail contact models were established based on the corresponding algorithms. By considering single-point contact, calculating normal compression amount based on vertical penetration and adopting the empirical formula for normal contact stiffness calculation, the corresponding wheel-rail contact model was established. This model was designated as Model 1. Similarly, different wheel-rail contact models were established, as listed in Table 1. The effect of considering multipoint contact versus single-point contact can be clarified by comparing Models 1 and 2. The effect of different algorithms for wheel-rail normal compression amount can be clarified by comparing Models 2 and 3, while the effect of different calculation formulae for the normal contact stiffness can be clarified by comparing Models 3 and 4.

Table 1. Setting of different wheel-rail contact models.

Wheel-Rail Contact Model	Consideration of Contact Point	Calculation Basis for Normal Compression Amount	Algorithm for Normal Contact Stiffness
Model 1	Single-point contact	Vertical penetration	Empirical formula
Model 2	Multipoint contact	Vertical penetration	Empirical formula
Model 3	Multipoint contact	Normal penetration	Empirical formula
Model 4	Multipoint contact	Normal penetration	Theoretical formula

A single CRH2 motor vehicle passing through the CRTS II slab ballastless track was taken as the example. The train running speed was set to 300 km/h. LM$_A$ and CHN60 were adopted as the wheel and rail profiles, respectively, as presented in Figure 10.

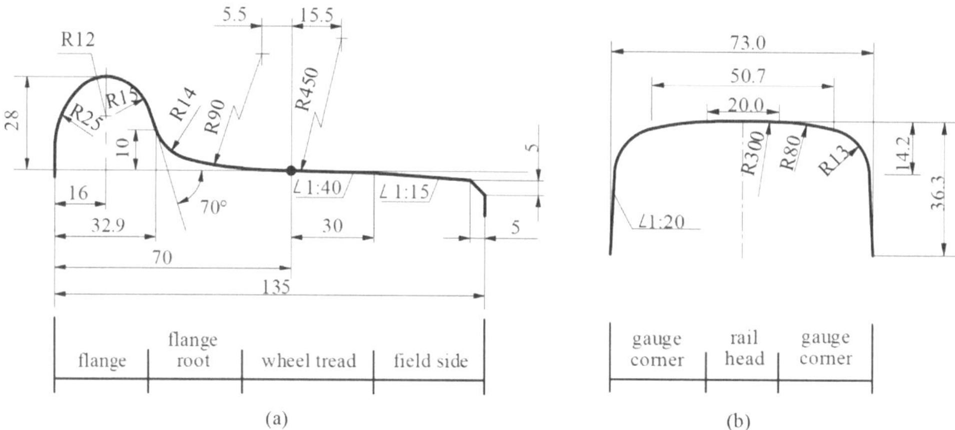

Figure 10. Profiles of the wheel and rail (units: mm): (**a**) wheel profile; (**b**) rail profile.

A comparative analysis under ordinary and seismic conditions was conducted. Under the seismic condition, the track irregularity excitation was considered; Figure 11 shows the irregularity samples. Since the wheel-rail dynamic response is mainly influenced by transverse excitations of earthquakes [36], only the transverse excitations were considered in this study. The El-Centro wave, generated by the 6.5 magnitude earthquake that took place in California, United States on 18 May 1957, has distinctive waveform characteristics that make it one of the frequently utilized seismic waves in the field of train-running safety under earthquakes [37,38]. The acceleration time history of this earthquake wave was obtained from the Pacific Earthquake Engineering Research (PEER) Ground Motion Database, and the peak acceleration was adjusted to 0.3 g as the ground motion input, as shown in Figure 12. For ordinary conditions, only track irregularity excitation was considered.

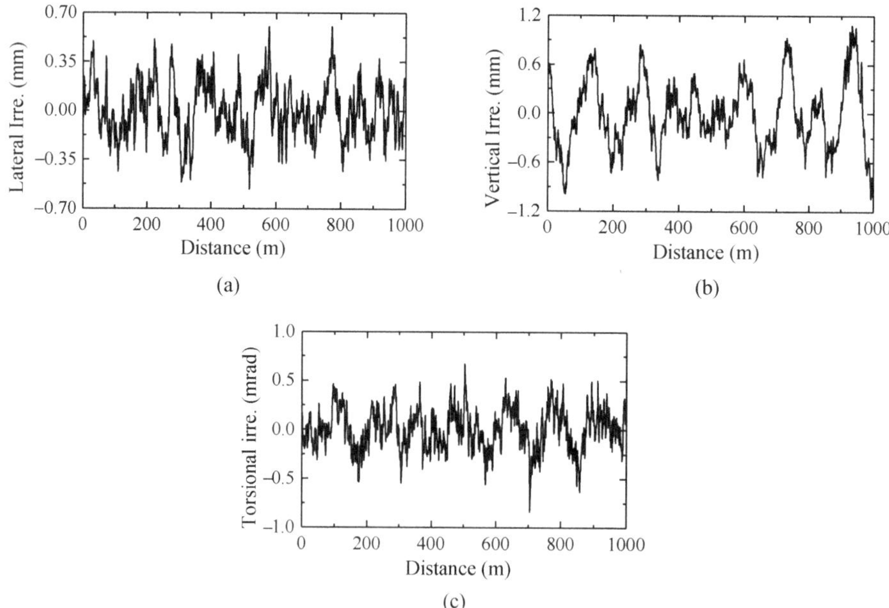

Figure 11. Samples of rail irregularities: (**a**) Lateral irregularities; (**b**) Vertical irregularities; (**c**) Torsional irregularities.

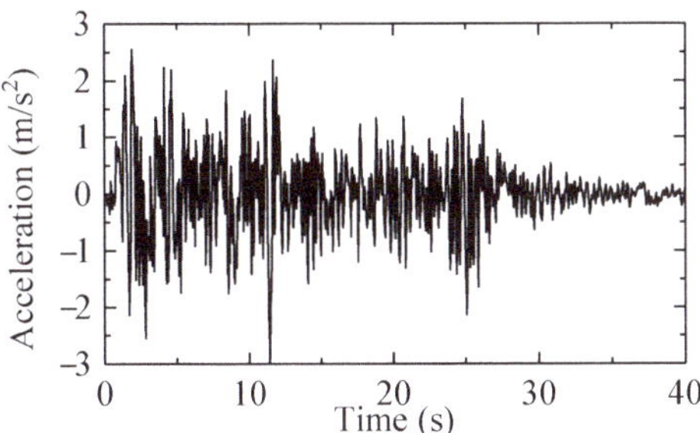

Figure 12. Acceleration-time history of the ground motion.

4.1. Comparison under Ordinary Conditions

To verify the correctness of the four models and to analyze the applicability of the wheel-rail contact algorithms under ordinary conditions, the wheel-rail contact dynamics on the left wheel of the first wheelset calculated based on the different models were compared, as shown in Figure 13.

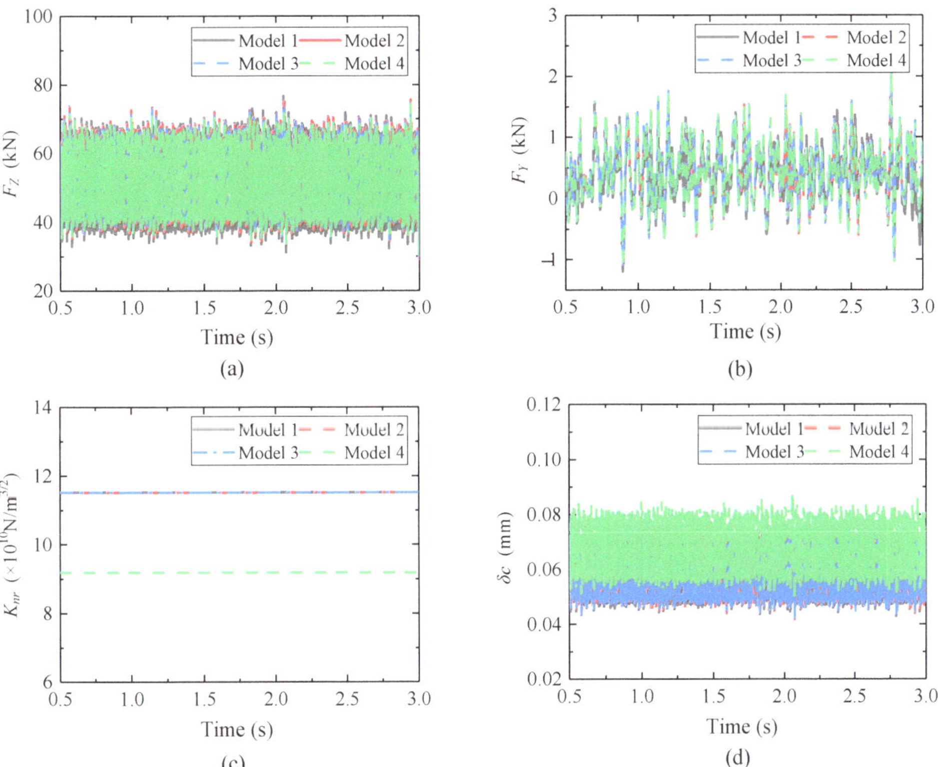

Figure 13. Time histories of wheel-rail dynamic responses based on different models: (**a**) Vertical wheel-rail force; (**b**) Lateral wheel-rail force; (**c**) Normal contact stiffness; (**d**) Normal compression amount.

Figure 13a,b show that the time histories of F_Y and F_Z based on the different models are in good agreement, thus verifying the correctness of the established models. As presented in Figure 13c, K_{nr} calculated based on Model 1 is maintained at 11.52×10^{10} N/m$^{3/2}$. In contrast, when calculating based on Model 4, this value is maintained at 9.20×10^{10} N/m$^{3/2}$

due to the straight segment of the wheel tread keeping in contact with the R300 arc segment of the rail head. Figure 13d shows a slight difference in δ_c calculated based on Models 1 and 4. Moreover, although the empirical formula causes a calculation error for K_{nr} and δ_c, it has largely no impact on the contact force. Thus, it can be concluded that under this condition, it is applicable to adopt the normal compression algorithm based on vertical penetration, the empirical formula for calculating the normal contact stiffness, and only consider single-point contact.

4.2. Comparison under Seismic Conditions

4.2.1. Wheel-Rail Contact Dynamics

The wheel-rail dynamic responses on the left side of the first wheelset calculated based on the different models were compared under seismic conditions firstly. The calculation results based on Models 1–4 are denoted by R_1, R_2, R_3, and R_4, respectively.

To clarify the effect of considering multipoint contact versus single-point contact, the wheel-rail dynamic responses calculated on the basis of Models 1 and 2 are compared, as shown in Figures 14 and 15. The relative calculation error between the two models can be expressed as $(R_1 - R_2)/R_2 \times 100\%$. Figure 14 shows that the differences in F_Y and F_Z that were calculated based on Models 1 and 2 mainly occur during the period when the contact force surges, i.e., when the wheel-flange contact occurs.

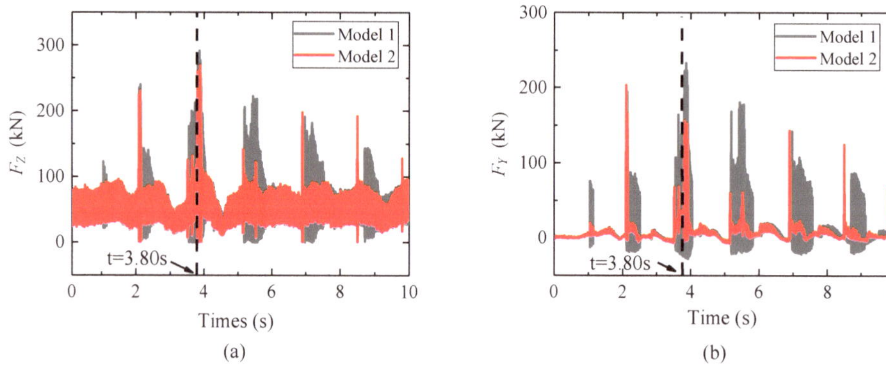

Figure 14. Time histories of wheel-rail contact forces based on Models 1 and 2: (**a**) Vertical wheel-rail force; (**b**) Lateral wheel-rail force.

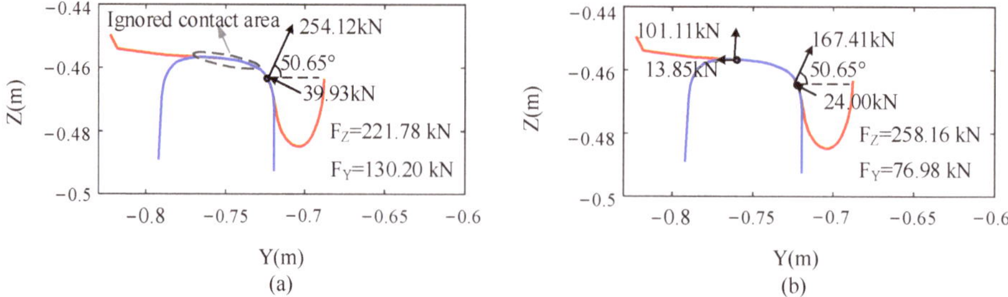

Figure 15. Wheel-rail contact state at time *t* = 3.80 s simulated based on Models 1 and 2: (**a**) Simulation result based on Model 1; (**b**) Simulation result based on Model 2.

Figure 15 shows the wheel-rail contact states obtained at a time of 3.80 s based on the two models. The wheel profile and rail profile are represented by red and blue curves respectively in Figure 15. In this case, the two-point contact state can be simulated based on Model 2 while the contact area existing in the wheel tread region is ignored when Model 1 is used. The solutions to F_Z based on Models 1 and 2 are 221.78 kN and 258.16 kN, respectively, with a calculated relative error of -14.09%, and the solutions to F_Y based on Models 1 and 2 are 130.20 kN and 76.98 kN, respectively, with a calculated relative error of 69.13%. Considering single point contact only will adversely affect the simulation of the wheel-rail contact state. Hence, multipoint contact should be considered under seismic conditions.

To clarify the effect of different algorithms for the wheel-rail normal compression amount, the wheel-rail dynamic responses calculated based on Models 2 and 3 are compared, as shown in Figure 16. Both wheel-tread contact region and flange-root contact region are considered as wheel-rail contact potential areas, which are recorded as region 1 and region 2, respectively. From Figure 16a,b, it can be found that the difference in δ_c based on the two models is mainly reflected in region 2, where the maximum values of δ_c obtained based on Models 2 and 3 are 0.186 mm and 0.083 mm, respectively, and the former is 2.24 times that of the latter. From Figure 16c,d, it is observed that the trend in F_N is consistent with δ_c since the same calculation formula for K_{nr} is adopted. The maximum values of F_N calculated based on Models 2 and 3, are 297.72 kN and 87.02 kN, respectively, in region 2, and the former is 3.42 times that of the latter. In addition, the wheel jumping amount, defined as the maximum vertical distance between the wheel and rail without wheel-rail contact occurring, was also introduced as an indicator of wheel rail dynamic response here. The responses of the wheel jump amount are given, as depicted in Figure 16e. Wheel jump is consistent with the occurrence of the contact in region 2 based on Model 2, which means that an unreasonable surge in the wheel-rail contact force will lead to an unreasonable wheel jump. Hence, the algorithm for the wheel-rail normal compression amount based on normal penetration should be introduced to avoid any unreasonable surge in the wheel-rail normal contact force.

To clarify the influence of different calculation formulae for the normal contact stiffness, the wheel-rail dynamic responses calculated based on Models 3 and 4 are compared, as plotted in Figure 17. For the contact in region 1, Figure 17a shows that based on Model 4, when contact occurs between the straight-line segment of the wheel tread and the R300 arc segment of the rail head, K_{nr} is maintained at 9.20×10^{10} N/m$^{3/2}$. When contact occurs between the R450 arc segment of the wheel tread and the R300 arc segment of the rail head, K_{nr} decreases to 8.21×10^{10} N/m$^{3/2}$. Figure 17c,e show that the theoretical formula also causes a difference in δ_c, but has no significant effect on F_N, which is consistent with the phenomenon observed under ordinary conditions. For the contact in region 2, the solutions to K_{nr} based on Models 3 and 4 are significantly different, as presented in Figure 17b. In this case, the contact mainly occurs between the R14 arc segment of the wheel flange root and the R13 arc segment of the rail gauge corner. The calculation result of K_{nr} based on the theoretical formula is 5.13×10^{10} N/m$^{3/2}$, which is approximately half of that based on the empirical formula. A different calculation formulae for K_{nr} will cause significant differences in δ_c, as shown in Figure 17d. Consequently, there is a noticeable difference after 8 s, as shown in Figure 17f. Therefore, to ensure the reliability of the normal contact force calculation, the theoretical calculation formula for the normal contact stiffness should be introduced.

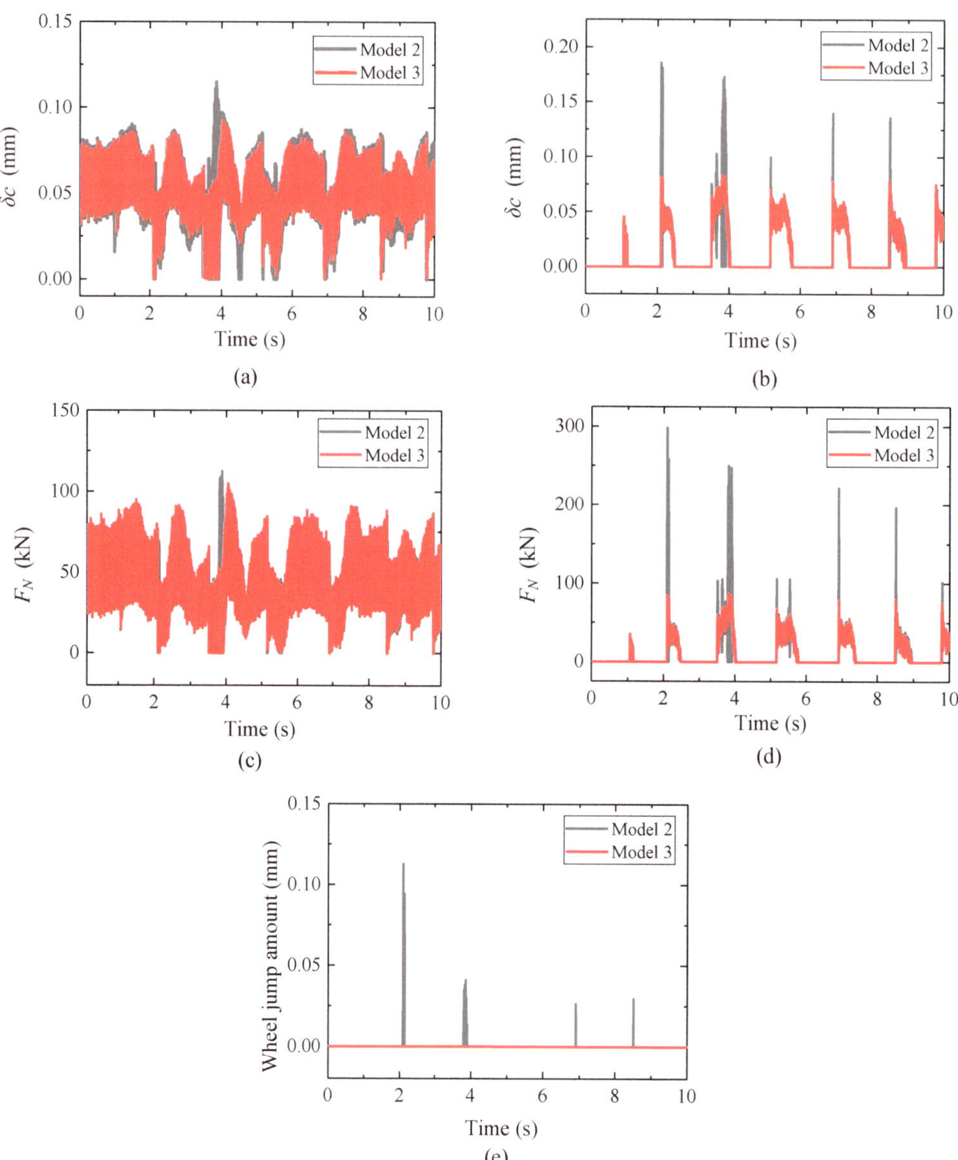

Figure 16. Time histories of wheel-rail dynamic responses based on Models 2 and 3: (**a**) Normal compression amount in region 1; (**b**) Normal compression amount in region 2; (**c**) Normal contact force in region 1; (**d**) Normal contact force in region 2; (**e**) Wheel jump amount.

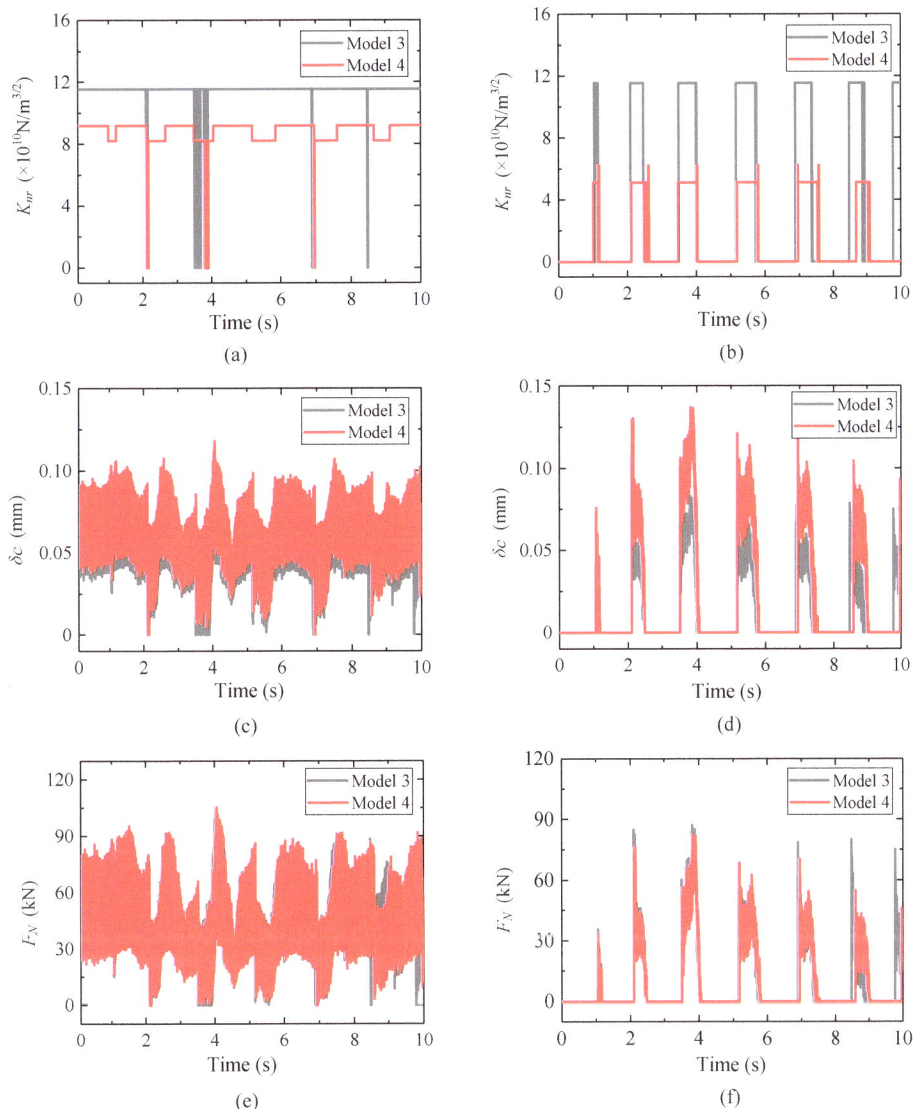

Figure 17. Time histories of wheel-rail dynamic responses based on Models 3 and 4: (**a**) Normal contact stiffness in region 1; (**b**) Normal contact stiffness in region 2; (**c**) Normal compression amount in region 1; (**d**) Normal compression amount in region 2; (**e**) Normal contact force in region 1; (**f**) Normal contact force in region 2.

To clarify the significance of the effect of the normal compression algorithms, the normal contact stiffness algorithms, and the considerations of the number of the contact point, the maximum values of wheel-rail forces on both sides of the four wheelsets calculated on the basis of the different models are presented in Figure 18. The relative calculation errors can be expressed as: $\Delta_{12} = (R_1 - R_2)/R_2 \times 100\%$, $\Delta_{23} = (R_2 - R_3)/R_3 \times 100\%$ and $\Delta_{34} = (R_3 - R_4)/R_4 \times 100\%$. As shown in Figure 18, the algorithm for the normal compression amount has the most significant impact on the F_Z and F_Y solutions, with Δ_{23} ranging from 48.97% to 339.50%. Regarding the impact of considering the wheel-rail contact point, considering only single-point contact may cause F_Z and F_Y to increase or decrease, with Δ_{12} ranging from -15.26% to 35.00%. The impact of the normal contact stiffness algorithm is minimal, with Δ_{34} ranging from 6.04% to 23.55%.

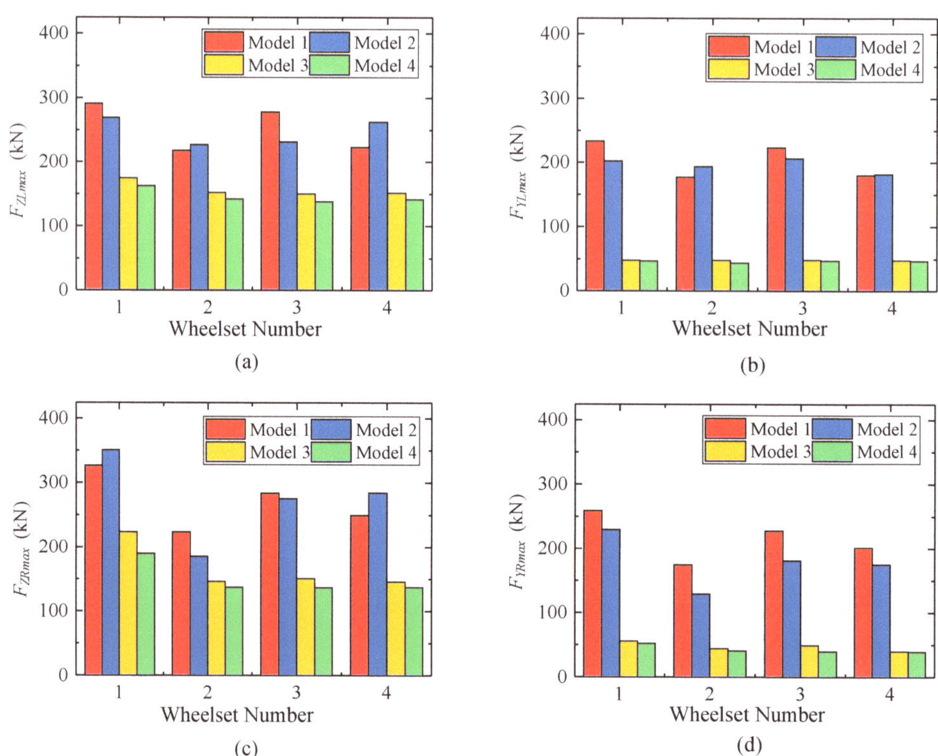

Figure 18. Maximum wheel-rail forces calculated based on different models: (**a**) Left vertical wheel-rail force; (**b**) Left lateral wheel-rail force; (**c**) Right vertical wheel-rail force; (**d**) Right lateral wheel-rail force.

4.2.2. Running Safety Assessment

In order to clarify the impact of different models on the running safety of trains during earthquakes, the calculation results of the running safety indices and running safety assessment results based on different models were compared. Three indices were adopted in this study: (i) derailment coefficient (Q/P), (ii) wheel unloading rate (ΔP/P), and (iii) wheelset lateral force ($\sum$Q). According to the Technical Code for Dynamic Acceptance of High-Speed Railway Engineering (TB10761-2013), the limit values of each index are listed in Table 2.

Table 2. Limit values of train running safety indices.

Running Safety Indices	Q/P	ΔP/P	$\sum$Q (kN)
Limit	0.8	0.8	55

Table 3 shows the maximum values of the three running safety indices calculated based on different models. The results demonstrate that the selection of algorithms in wheel-rail contact modelling has a significant effect on the safety assessment of trains during earthquakes. All the indices calculated based on Models 1 and 2 exceed their limits, indicating that the train has entered a dangerous state of operation. In comparison, all the indices calculated based on Models 3 and 4 are within the safe range, indicating that the train is in a safe state of operation. Comparing the results of Models 1 and 2, the relative calculation errors of Q/P, ΔP/P, and $\sum$Q reach to -4.60%, -5.0%, and -8.84%, respectively. This finding indicates that the running safety of trains during earthquakes is underestimated when considering only the single-point wheel-rail contact. Comparing the results of Models 2 and 3, the relative calculation error of Q/P, ΔP/P, and $\sum$Q reach to 11.54%, 20.99%, and 432.09%, respectively. This finding indicates that calculating the wheel-rail normal compression based on vertical penetration will significantly overestimate the running safety indices under earthquakes, leading to misjudgment of the running safety of trains. Comparing the results of Models 3 and 4, there is no significant difference in the indices, indicating that although using an empirical formula for wheel-rail normal contact stiffness will have a certain impact on the calculation accuracy of wheel-rail contact force, the influence on the calculation results of the running safety assessment of trains under earthquakes is small.

Table 3. Maximum values of the running safety indices based on different models.

Running Safety Indices	Model 1	Model 2	Model 3	Model 4
Q/P	0.83	0.87	0.78	0.78
ΔP/P	0.95	1.0	0.80	0.79
$\sum$Q (kN)	228.79	250.99	49.33	46.65

5. Model Validation

Through the analysis in Section 4, Model 4 should be the most reasonable choice as the wheel rail contact model under earthquake excitations. In order to verify the correctness of the model, the calculation results of wheel-rail contact force based on the model established in this paper were compared with those given in the study of Nishimura et al. [33]. Nishimura et al. simplified the vehicle as a system consisting of four rigid bodies, including a body, a frame, and two wheelsets. They considered the lateral, sinking, and rolling movements of each rigid body and assumed that the two wheelsets had the same motion state, establishing a 9-degree of freedom vehicle model. Furthermore, they also considered the lateral and vertical degrees of freedom of the rail, resulting in a comprehensive vehicle dynamic simulation model with a total of 13 degrees of freedom (as shown in Figure 19). The correctness of this model was verified through shaking table tests [39].

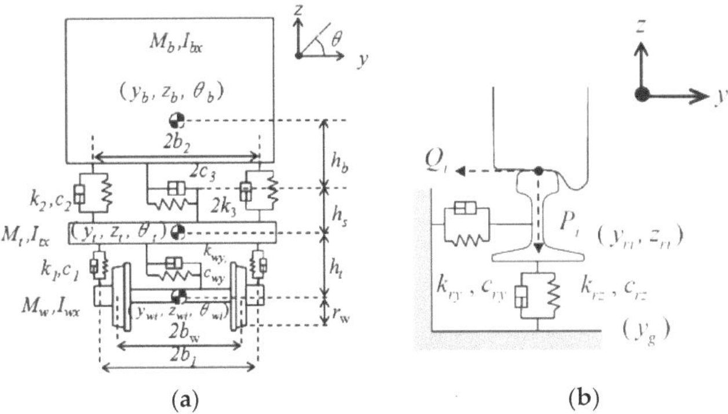

(a)   (b)

Figure 19. The vehicle dynamic analysis model established by Nishimura et al. [33]: (**a**) vehicle model; (**b**) track model.

Taking five-circle sine waves with frequencies of 0.5 Hz, 0.8 Hz, and 1.5 Hz as excitation, the wheel rail force calculation results are shown in Figures 20–22. It can be seen that the time histories of the vertical and lateral wheel-rail forces calculated based on this article are in good agreement with those presented in the literature, thus verifying the correctness and effectiveness of the model established in this article.

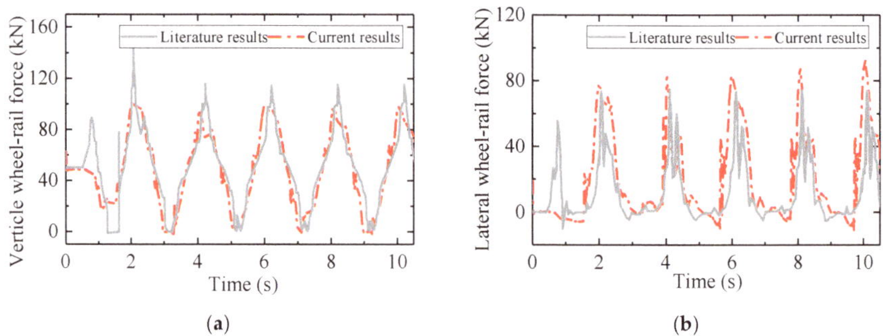

Figure 20. Time histories of wheel-rail contact force under excitation with frequency of 0.5 Hz and amplitude of 320 mm: (**a**) Vertical wheel-rail force; (**b**) Lateral wheel-rail force.

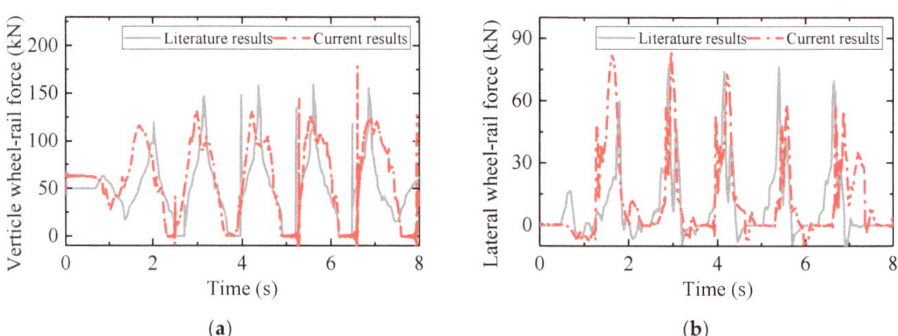

Figure 21. Time histories of wheel-rail contact force under excitation with frequency of 0.8 Hz and amplitude of 105 mm: (**a**) Vertical wheel-rail force; (**b**) Lateral wheel-rail force.

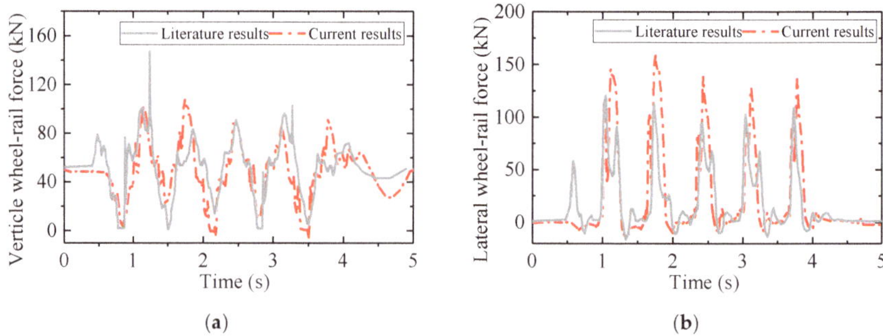

Figure 22. Time histories of wheel-rail contact force under excitation with frequency of 1.5 Hz and amplitude of 100 mm: (**a**) Vertical wheel-rail force; (**b**) Lateral wheel-rail force.

6. Conclusions

This paper investigated the effects of different normal compression algorithms, normal contact stiffness algorithms, and considerations of single-point/multi-point contact on the running safety assessment of trains under earthquakes. Through a case study of a single CRH2 train passing over a CRTS II type slab track structure under El-Centro wave excitation, a comparative analysis of the effect of different algorithms on the wheel-rail contact dynamics and running safety was conducted. The findings provided recommendations for the selection of wheel-rail contact algorithms under earthquakes. The main conclusions are as follows:

1. Using different wheel-rail contact algorithms will significantly affect the calculation accuracy of wheel-rail force in the case of flange-root contact under earthquakes. The most significant influence is due to the normal compression algorithm. Using an algorithm based on vertical penetration can lead to a maximum relative error of 339.50% for the case considered in this study. The consideration of the number of wheel-rail contact points also has a notable impact, with a maximum relative error of 35.00% caused by only considering single point contact. The influence of the normal contact stiffness algorithm is the least significant, with a maximum relative calculation error of 23.55% caused by using the empirical formula.

2. Using different wheel-rail contact algorithms will have a significant impact on the indices of running safety assessment under earthquakes. Using wheel-rail normal compression algorithm based on vertical penetration will significantly underestimating the train running safety margin, while only considering the wheel-rail single point contact will overestimate the train running safety margin, and using the wheel-rail normal contact empirical formula has little impact.

3. To ensure the accuracy of running safety assessment of trains under earthquakes, it is recommended to use the normal compression algorithm based on normal penetration and consider multi-point contact in wheel-rail contact modelling.

Author Contributions: Conceptualization, Z.Z.; methodology, G.C. and Z.Z.; formal analysis, G.C.; writing-original draft preparation, G.C.; writing-review and editing, W.G. and G.Z.; supervision, L.J. and B.Y.; funding acquisition, Z.Z. All authors have read and agreed to the published version of the manuscript.

Funding: This study was funded by the National Natural Science Foundation of China (Grant No. 52078498), Natural Science Foundation of Hunan Province of China (Grant number: 2022JJ30745), Frontier cross research project of Central South University (Grant number: 2023QYJC006), Science and Technology Research and Development Program Project of China railway group limited (Major Special Project, No.: 2021-Special-04-2) and Hunan Provincial Science and Technology Promotion Talent Project (Grant number: 2020TJ-Q19).

Institutional Review Board Statement: Not applicable.

Informed Consent Statement: Not applicable.

Data Availability Statement: No new data were created in this study.

References

1. Cao, L.C.; Yang, C.W.; Zhang, J.J. Derailment Behaviors of the Train-Ballasted Track-Subgrade System Subjected to Earthquake Using Shaking Table. *KSCE J. Civ. Eng.* **2020**, *24*, 2949–2960. [CrossRef]

2. Zeng, Q.; Dimitrakopoulos, E.G. Vehicle-bridge interaction analysis modeling derailment during earthquakes. *Nonlinear Dyn.* **2018**, *93*, 2315–2337. [CrossRef]

3. Meymand, S.Z.; Keylin, A.; Ahmadian, M. A survey of wheel-rail contact models for rail vehicles. *Veh. Syst. Dyn.* **2016**, *54*, 386–428. [CrossRef]

4. Chen, H.Y.; Jiang, L.Z.; Li, C.Q.; Li, J.; Shao, P.; He, W.K.; Liu, L.L. A semi-online spatial wheel-rail contact detection method. *Int. J. Rail Transp.* **2021**, *10*, 730–748. [CrossRef]

5. Shen, Z.Y.; Hedrick, J.K.; Elkins, J.A. A comparison of alternative creep force models for rail vehicle dynamic analysis. *Veh. Syst. Dyn.* **1983**, *12*, 79–83. [CrossRef]

6. Luo, J.; Zhu, S.Y.; Zhai, W.M. An advanced train-slab track spatially coupled dynamics model: Theoretical methodologies and numerical applications. *J. Sound Vib.* **2021**, *501*, 116059, 1–30. [CrossRef]

7. Zhai, W.M.; Han, Z.Y.; Chen, Z.W.; Ling, L.; Zhu, S.Y. Train-track-bridge dynamic interaction: A state-of-the-art review. *Veh. Syst. Dyn.* **2019**, *57*, 1–44. [CrossRef]

8. Xu, L.; Zhai, W.M. Train-track coupled dynamics analysis: System spatial variation on geometry, physics and mechanics. *Railw. Eng. Sci.* **2020**, *28*, 36–53. [CrossRef]

9. Liu, H.Y.; Yu, Z.W.; Guo, W.; Jiang, L.Z.; Kang, C.J. A novel method to search for the wheel-rail contact point. *Int. J. Struct. Stab. Dyn.* **2019**, *19*, 1950142, 1–21. [CrossRef]

10. Xu, J.M.; Wang, P.; Ma, X.C.; Xiao, J.L.; Chen, R. Comparison of calculation methods for wheel-switch rail normal and tangential contact. *Proc. Inst. Mech. Eng. Part F J. Rail Rapid Transit* **2016**, *231*, 148–161. [CrossRef]

11. Gong, W.; Zhu, Z.H.; Liu, Y.; Liu, R.T.; Tang, Y.J.; Jiang, L.Z. Running safety assessment of a train traversing a three-tower cable-stayed bridge under spatially varying ground motion. *Railw. Eng. Sci.* **2020**, *28*, 184–198. [CrossRef]

12. Chen, L.K.; Kurtulus, A.; Dong, Y.F.; Taciroglu, E.; Jiang, L.Z. Velocity pulse effects of near-fault earthquakes on a high-speed railway vehicle-ballastless track-benchmark bridge system. *Veh. Syst. Dyn.* **2021**, *60*, 2963–2987. [CrossRef]

13. Jin, Z.B.; Liu, W.Z. Fragility analysis for vehicle derailment on railway bridges under earthquakes. *Railw. Eng. Sci.* **2022**, *30*, 494–511. [CrossRef]

14. Ling, L.; Xiao, X.B.; Wu, L.; Jin, X.S. Study on dynamic responses and running safety boundary of high-speed train under seismic motions. *J. China Railw. Soc.* **2012**, *34*, 16–22. (In Chinese)

15. Zhao, H.; Wei, B.; Jiang, L.Z.; Xiang, P. Seismic running safety assessment for stochastic vibration of train-bridge coupled system. *Arch. Civ. Mech. Eng.* **2022**, *22*, 180, 1–24. [CrossRef]

16. Yamashita, S.; Sugiyama, H. Numerical procedure for dynamic simulation of two-point wheel/rail contact and flange climb derailment of railroad vehicles. *J. Comput. Nonlinear Dyn.* **2012**, *7*, 041012, 1–7. [CrossRef]

17. Ren, Z.S.; Iwnicki, S.D.; Xie, G. A new method for determining wheel-rail multi-point contact. *Veh. Syst. Dyn.* **2011**, *49*, 1533–1551. [CrossRef]

18. Xiao, X.B.; Jin, X.S.; Wen, Z.F.; Zhu, M.H.; Zhang, W.H. Effect of tangent track buckle on vehicle derailment. *Multibody Syst. Dyn.* **2010**, *25*, 1–41. [CrossRef]

19. Meli, E.; Ridolfi, A. An innovative wheel-rail contact model for railway vehicles under degraded adhesion conditions. *Multibody Syst. Dyn.* **2013**, *33*, 285–313. [CrossRef]

20. Montenegro, P.A.; Neves, S.G.M.; Calçada, R.; Tanabe, M.; Sogabe, M. Wheel-rail contact formulation for analyzing the lateral train-structure dynamic interaction. *Comput. Struct.* **2015**, *152*, 200–214. [CrossRef]

21. Magalhaes, H.; Marques, F.; Antunes, P.; Flores, P.; Pombo, J. Wheel-rail contact models in the presence of switches and crossings. *Veh. Syst. Dyn.* **2022**, *61*, 838–870. [CrossRef]

22. Shabana, A.A.; Zaazaa, K.E.; Escalona, J.L.; Sany, J.R. Development of elastic force model for wheel/rail contact problems. *J. Sound Vib.* **2004**, *269*, 295–325. [CrossRef]

23. Ling, L.; Jin, X.S. A 3D model for coupling dynamics analysis of high−speed train/track system. *J. Zhejiang Univ. Sci.-A* **2014**, *15*, 964–983. [CrossRef]

24. Guan, Q.H.; Zhao, X.; Wen, Z.F.; Jin, X.S. Calculation Method of Hertzian Normal Contact Stiffness. *J. Southwest Jiaotong Univ.* **2021**, *56*, 883–890. (In Chinese)

25. Shabana, A.A.; Berzeri, M.; Sany, J.R. Numerical Procedure for the Simulation of Wheel/Rail Contact Dynamics. *J. Dyn. Syst. Meas. Control.* **2001**, *123*, 168–178. [CrossRef]

26. Pombo, J.; Ambrósio, J.; Silva, M. A new wheel–rail contact model for railway dynamics. *Veh. Syst. Dyn.* **2007**, *45*, 165–189. [CrossRef]

27. Marques, F.; Magalhães, H.; Pombo, J.; Ambrósio, J.; Flores, P. A three-dimensional approach for contact detection between realistic wheel and rail surfaces for improved railway dynamic analysis. *Mech. Mach. Theory* **2020**, *149*, 1–28. [CrossRef]

28. Zhai, W.M. *Vehicle-Track Coupled Dynamics Theory and Applications*; Springer: Singapore, 2019; pp. 56–122.

29. Kalker, J.J. On the Rolling Contact of Two Elastic Bodies in the Presence of Dry Friction. Ph.D. Dissertation, Delft University of Technology, Delft, The Netherlands, 1967.

30. Zhang, N.; Xia, H.; De Roeck, G. Dynamic analysis of a train-bridge system under multi-support seismic excitations. *J. Mech. Sci. Technol.* **2010**, *24*, 2181–2188. [CrossRef]

31. Wu, X.W.; Chi, M.R.; Gao, H.; Ke, X.M.; Zeng, J.; Wu, P.B.; Zhu, M.H. Post-derailment dynamic behavior of railway vehicles travelling on a railway bridge during an earthquake. *Proc. Inst. Mech. Eng. Part F J. Rail Rapid Transit* **2014**, *230*, 418–439. [CrossRef]

32. Wang, W.; Li, G.X. Development of high-speed railway vehicle derailment simulation-Part I: A new wheel/rail contact method using the vehicle/rail coupled model. *Eng. Fail. Anal.* **2012**, *24*, 77–92. [CrossRef]

33. Nishimura, K.; Terumichi, Y.; Morimura, T.; Sogabe, K. Development of Vehicle Dynamics Simulation for Safety Analyses of Rail Vehicles on Excited Tracks. *J. Comput. Nonlinear Dyn.* **2009**, *4*, 011001, 1–9. [CrossRef]

34. Sun, L.; Chen, L.L.; Zelelew, H.H. Stress and Deflection Parametric Study of High-Speed Railway CRTS-II Ballastless Track Slab on Elevated Bridge Foundations. *J. Transp. Eng.* **2013**, *139*, 1224–1234. [CrossRef]

35. Xu, Y.D.; Yan, D.B.; Zhu, W.J.; Zhou, Y. Study on the mechanical performance and interface damage of CRTS II slab track with debonding repairment. *Constr. Build. Mater.* **2020**, *257*, 119600. [CrossRef]

36. Jin, Z.B.; Pei, S.L.; Li, X.Z.; Liu, H.Y.; Qiang, S.Z. Effect of vertical ground motion on earthquake-induced derailment of railway vehicles over simply-supported bridges. *J. Sound Vib.* **2016**, *383*, 277–294. [CrossRef]

37. Jiang, L.Z.; Zhou, T.; Liu, X.; Xiang, P.; Zhang, Y.T. An Efficient Model for Train-Track-Bridge-Coupled System under Seismic Excitation. *Shock. Vib.* **2021**, *2021*, 1–14. [CrossRef]

38. Li, M.; Liu, J.W.; Zhang, G.C. Safety Analysis of the Running Train under Earthquake Dynamic Disturbance. *Shock. Vib.* **2021**, *2021*, 1–21. [CrossRef]
39. Nishimura, K.; Terumichi, Y.; Morimura, T.; Adachi, M.; Morishita, Y.; Miwa, M. Using Full Scale Experiments to Verify a Simulation Used to Analyze the Safety of Rail Vehicles During Large Earthquakes. *J. Comput. Nonlinear Dyn.* **2015**, *10*, 031013, 1–9. [CrossRef]

Article

A New Car-Body Structure Design for High-Speed EMUs Based on the Topology Optimization Method

Chunyan Liu [1,2], Kai Ma [2,3], Tao Zhu [1,*], Haoxu Ding [1], Mou Sun [1] and Pingbo Wu [1]

[1]   State Key Laboratory of Rail Transit Vehicle System, Southwest Jiaotong University, Chengdu 610031, China; lcy0528@126.com (C.L.); dinghaoxu@my.swjtu.edu.cn (H.D.); d1023480113@163.com (M.S.); wupingbo@163.com (P.W.)

[2]   CRRC Changchun Railway Vehicles Co., Ltd., Changchun 130062, China; 013200020512@crrcgc.cc

[3]   School of Mechanical Engineering, Southwest Jiaotong University, Chengdu 610031, China

*   Correspondence: zhutao034@swjtu.edu.cn

Abstract: In recent years, the research and development of high-speed trains has advanced rapidly. The main development trends of high-speed trains are higher speeds, lower energy consumption, higher safety, and better environmental protection. The realization of a lightweight high-speed car body is one of the key features in the development trend of high-speed trains. Firstly, the basic dimensions of the car body's geometric model are determined according to the external dimensions of the body of a CRH EMU, and the specific topology optimization design domain is selected to establish the finite element analysis model; secondly, the strength and modal analyses of the topology optimization design domain are carried out to check the accuracy of the design domain and provide a comparative analysis for subsequent design. Then, the variables, constraints, and objective functions of the topology optimization design are determined to establish the mathematical model of topology optimization, and the design domain is calculated for topology optimization under single and multiple conditions, respectively. Finally, based on the topology optimization calculation results, truss-type reconstruction modeling is carried out for the car body's side walls, roof, underframe, end walls, and other parts. Compared with the conventional EMU body structure, the weight of the reconstructed body structure is reduced by about 18%. The results of the finite element analysis of the reconstructed car-body structure prove the reliability and safety of the structure, indicating that the reconstructed car-body scheme meets the corresponding performance indicators.

Keywords: high-speed car body; lightweight; truss body; topology optimization; structural reconstruction

Citation: Liu, C.; Ma, K.; Zhu, T.; Ding, H.; Sun, M.; Wu, P. A New Car-Body Structure Design for High-Speed EMUs Based on the Topology Optimization Method. *Appl. Sci.* **2024**, *14*, 1074. https://doi.org/10.3390/app14031074

Academic Editor: Suchao Xie

Received: 11 December 2023
Revised: 15 January 2024
Accepted: 17 January 2024
Published: 26 January 2024

1. Introduction

In recent years, the research and development of high-speed trains has advanced rapidly. The main development trends of high-speed trains are higher speeds, lower energy consumption, higher safety, and better environmental protection. The realization of a lightweight high-speed car body is one of the key features in the development trend of high-speed trains. When a train has higher-speed operation, the car-body structure needs to bear many complex combined load conditions. The bearing structure of the body of electric multiple units (EMUs) is usually welded in a cylinder shape. In order to further reduce the air resistance, the contour of the head and the external frame of the train are streamlined [1].

In the design of high-speed trains' car-body structures, various factors should be integrated and coordinated to improve the performance of the body. As a large vehicle, the safety of the body structure of a high-speed train EMU has always been an important subject in the design of high-speed trains' car bodies [2–4]. As a complex mechanical structure, the EMU's body structure should be considered in the design of strength, mode, and other performance indicators because of its various structural forms and changeable load conditions. For the next generation of high-speed EMUs' body structure design, the traditional CAE/CAD design and analysis process has some shortcomings, such as high R&D costs, long time cycles, and insufficient optimization analysis. With the continuous development of optimization technology, structural optimization design gradually tends to combine multiple disciplines and objectives, which can significantly shorten the optimization design cycle, improve the reliability of the vehicle structure and the degree of optimization analysis, and make full use of materials.

In order to obtain a car-body structure that can meet many design requirements at the same time, extensive research has been carried out in the field of structure optimization. Harte et al. [5] divided the light rail body structure into different subregions based on the calculation and analysis results and then optimized the calculation and analysis of each subregion through size optimization. Chiandussi et al. [6] took the automobile chassis as the research object and realized the light weight of the automobile chassis through topology optimization design, and the dynamic performance of the chassis was also significantly improved. Chen et al. [7] took the train underframe as the main research object, analyzed the results of a material analysis for topological optimization under different loads, determined the optimal distribution position of the inner ribs of the underframe, and obtained the optimal shape of the underframe section. Zhang et al. [8] analyzed the design scheme of the whole vehicle structure of a tracked vehicle, determined the location of the maximum stress and strain point through static analysis results, used Optistruct to optimize its topology, and analyzed the stiffness and strength of the optimization results. Based on a certain type of China Railway High-Speed (CRH) EMU, Ji [9] performed a sectional analysis and comparison of the existing car-body sections; optimized the car-body model in the transverse, longitudinal, and transverse directions with Optistruct; reconstructed the car-body model according to the optimization results; and compared and analyzed the static changes in the car body before and after optimization. Zhang et al. [10,11] established a parameter model for high-speed trains and compared the aerodynamic performance of the front-end model before and after optimization with a crosswind, proving that optimization analysis can effectively improve the anti-crosswind performance of the front end. Many scholars have combined multidisciplinary optimization techniques with vehicle structure optimization to achieve lightweight structures and further improve the performance in terms of vehicle strength [12–14], vibration [15–17], collision [18–20], and other aspects.

To summarize, the existing studies mainly focus on the optimization of the section and local structure of the car body, while research on the topology optimization of the whole vehicle is scarce, which limits the design of the car body. At present, there is an urgent need to carry out the research and development of the next-generation high-speed EMU body. As the main research and design method, structural optimization technology has been highly valued by researchers at home and abroad. Due to its significant challenges, topology optimization technology has been a major focus of research. The purpose and objectives of this study include an investigation of the shortcomings of existing research, aiming to optimize the topology of the whole structure of the car body. Through a simulation analysis platform, the topology optimization design of the car-body structure design domain based on the CRH profile data is carried out. With the aim of developing the structural form of the next-generation high-speed EMU body, a comprehensive and detailed optimization design study is carried out, and the material distribution results of each part of the body are obtained. Based on these results, the main bearing truss structure of each part is established, which provides a certain reference for the research and development of the body structure of the next generation of high-speed EMUs.

2. Topology Optimization Method of Car-Body Structure

2.1. Homogenization Method

In 1978, Benssousan [21] put forward the theoretical basis of the homogenization method to study the relationship between the macroscopic characteristics and microstructures of composite materials. This method can correlate variables of different scales and thus transform macroscopic problems into microscopic ones, such as replacing periodic microscopic structures with single cells, which has been widely used in engineering practice for decades [22]. Guedes and other scholars combined the homogenization method and the finite element method to establish a finite element equation based on the progressive homogenization method, serving to simplify the solution process and expand the solvable range and the complexity of the solution [23]. At the same time, as a calculation method for periodic composite materials, the homogenization method can improve the analysis efficiency, reduce the workload, and significantly shorten the calculation time under the premise of known material properties.

2.2. Variable-Density Method

The variable-density method was developed from the homogenization method, which deals with the intermediate density. It is one of the mainstream topology optimization methods based on finite elements. The optimization criteria adopted by the variable density method have the characteristics of a fast convergence speed, few iterations, and small computation, which is the focus of current structural topology optimization methods. Unlike the homogenization method, the variable-density method mainly uses material description for topology optimization. Upon introducing a reverie material, the material density is between 0 and 1, where 0 represents the hollowed-out state, 1 represents the solid state, and a density between 0 and 1 represents the point between the hollowed-out and solid states. The material's physical characteristics and the element's relative density depend on the interpolation function. Considering the relative density of each unit as a design variable, the number of design variables can be significantly reduced, and the computation is also reduced accordingly.

The design domain of the variable-density method is discretized into a finite element set defined by the element set $N_X = \{1, 2, \ldots, |N_X|\}$ in the x-direction and the element set $N_Y = \{1, 2, \ldots, |N_Y|\}$ in the y-direction. The element density is taken as the design variable:

$$(0 \leq \rho_{(x,y)} \leq 1, (x, y) \in N_X \times N_Y) \tag{1}$$

According to the optimality criterion, the element stiffness is effectively controlled. Then, the overall stiffness of the structure can be reasonably regulated, and the materials can be redistributed within the design domain. Thus, the topology optimization structure with the optimum structural stiffness and material distribution can be obtained [24].

$$E = \rho_e E_0 \tag{2}$$

$$E = \frac{\rho_e}{1 + q(1 - \rho_e)} E_0 \tag{3}$$

Solid isotropic material with penalization (SIMP) is a commonly used density–stiffness interpolation model, a common technique in topology optimization problems. The model assumes that the material density is constant within the cell and takes it as a design variable. In order to simplify the calculation and improve the efficiency, the material properties are simulated by the exponential function of the cell density. The SIMP method introduces relative density ρ_e and penalty factor P. When $0 \leq \rho_e \leq 1$, the element density is limited by penalty factor P, so the structural elements' density is as close as possible to 0 or 1. If the element density is 0, the material can be deleted; if the element density is 1, the material should be filled. For the SIMP interpolation model, the larger the penalty factor is, the better it is. When the penalty factor takes different values, the penalty effect is also different. The element density can be expressed as follows.

$$\varphi(x_i) = x_i^P, x_i \in [x_{\min}, 1], i = 1, 2, 3 \ldots, x_n \tag{4}$$

The relation between element density and elastic modulus is

$$E(x_i) = E_{min} + \varphi(x_i)(E - E_{min}), i = 1, 2, 3 \ldots, n \qquad (5)$$

where $E(x_i)$ is the elastic modulus of the element, E_{min} is the elastic modulus of the low-strength material element, and x_i represents the relative density of each element. To ensure the stability of numerical calculation, usually, $E_{min} = E/1000$ and $0 < E_{min} \leq E(x_i) \leq E$. The general optimization mathematical model of structural topology optimization can be formulated as follows.

$$\begin{aligned} \text{Minimize}: \quad & f(X) = f(x_1, x_2, \ldots, x_n) \\ \text{Subject to}: \quad & g_j(X) \leq 0 \\ & j = 1, 2, 3, \ldots, m \end{aligned} \qquad (6)$$

where n represents the number of design variables, X is the optimization design variable, $f(X)$ is the optimization objective function, $g(X)$ represents the design response requiring constraints, and j is the number of constraint equations.

2.3. Progressive Structural Optimization Method

The progressive structural optimization method was first put forward by Xie and Steven [25]. This method involves gradually removing inefficient or ineffective materials from the initial design space so that the final topology optimization result achieved is optimal. In other words, as the iteration progresses, some of the design variables change from 1 to 0.

2.4. Topology Optimization Process of Car-Body Structure

Based on the above optimization methods, the topology optimization process of the new-type car-body structure for EMUs is shown in Figure 1.

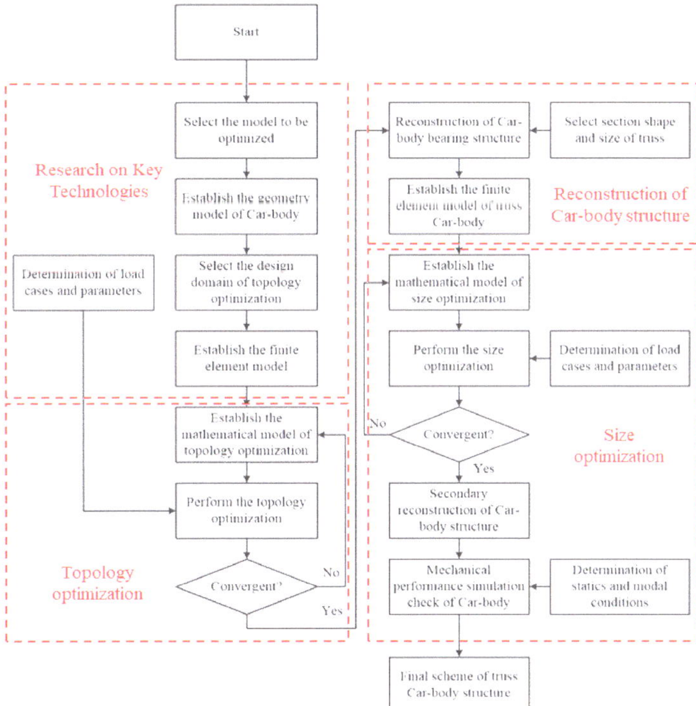

Figure 1. Vehicle body structure topology optimization process.

3. Selection of Topology Optimization Design Domain and Model Establishment

3.1. Selection of Topology Optimization Design Domain

The topology optimization design domain refers to the design space where the car-body structure can be optimized. A reasonable design domain can ensure the rationality and adaptivity of the topology optimization results, which is a crucial prerequisite for topology optimization calculation. In this paper, the topology optimization design domain is determined according to the outer contour size of the CRH vehicle car-body model, and the structure model is established. The car-body structure dimensions of a CRH vehicle are given in Table 1, including the body length, fixed distance, width, height, and other basic contour dimensions of the car-body structure.

Table 1. CRH car body's basic dimensions.

Structure Size	L/mm
Car-body length	25,000
Fixed distance	17,800
Car-body width	3360
Car-body height	4050
Height from coupler centerline to rail surface	950

Firstly, the initial topology optimization model is analyzed based on the existing car-body structure, and some structures suitable for topology optimization analysis (mainly considering the conventional extruded profile structure) are selected. Secondly, the selected parts of the car-body structure are classified, dividing it into the end wall, end, bottom frame, side wall, upper beam, and roof structure. Finally, it is necessary to facilitate the welding and manufacturing of the car-body structure and quickly match it with the vehicles already operating. Therefore, some of the necessary structural positions should be set aside, such as the coupler seat, car window, car door, etc. Based on the above analysis, the car-body structure model required for the initial topology optimization can be established. The geometric model of the car-body structure is shown in Figure 2.

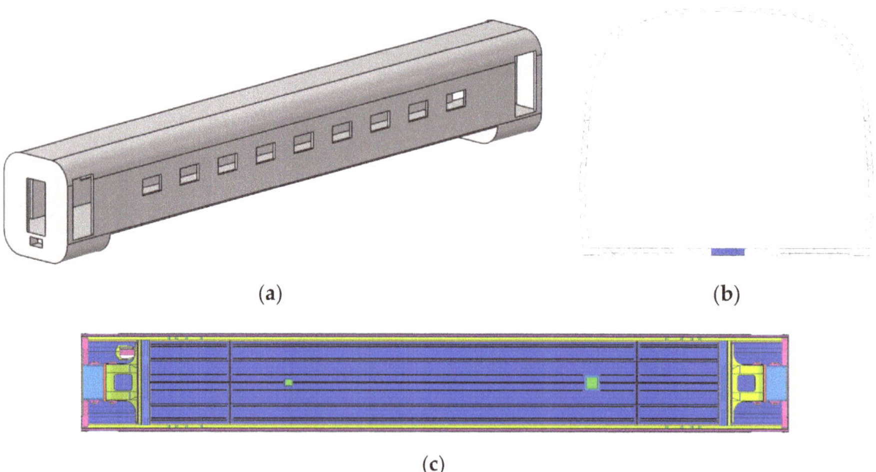

Figure 2. Geometric model of car-body structure: (**a**) axonometric drawing of car body; (**b**) cross-section of the car body; (**c**) the view of the car-body frame.

3.2. Establishment of Topology Optimization Design Domain Model

Since the topology optimization model is relatively simple and has few sharp corners or chamfering, the element size of the topology optimization design domain of the vehicle body is 50 mm with an eight-node hexahedral mesh [26]. The number of grid elements is determined by both the car-body structure parameters, as shown in Table 1, and the recommended mesh sizes in the topology optimization model, and the model has 390,620 meshes and 475,326 nodes after mesh division. Figure 3 shows the finite element model of the car-body structure. The selected material is aluminum 6005-T6, commonly used in the production of car-body structures. Its mechanical properties are as follows: the elastic modulus is 6.9×10^4 MPa, the density is 2.7×10^3 kg/m^3, and Poisson's ratio is 0.33. The working conditions and boundary conditions are shown in Table 2.

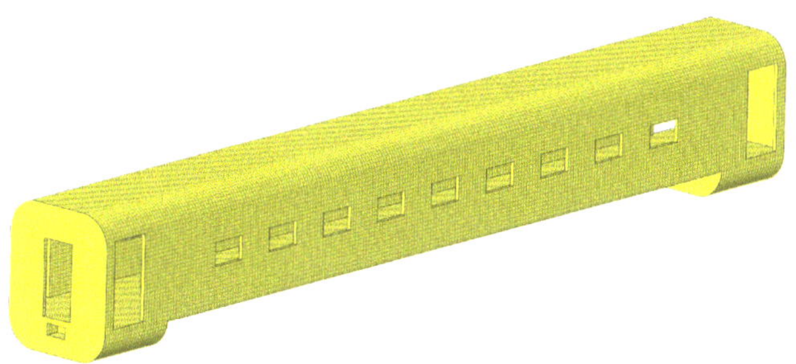

Figure 3. Finite element model of car-body structure.

Table 2. Working conditions and boundary conditions.

Working Condition	Load	Restraint	Condition Design Diagram
Longitudinal load	Compression force of 1500 kN for front-end coupler seat Compression forces of 300, 300, and 400 kN to the front-end wall near the roof, side wall, and chassis, respectively	Longitudinal constraint at the rear end wall	
Vertical load	1.3 times the weight of the car body	Vertical restraint at the secondary suspension	
Torsional load	Unit torsional load 1 kN·m	Full restraint at the secondary suspension	
Crosswind load	Unit wind pressure 450 Pa	Full restraint at the secondary suspension	

Table 2. *Cont.*

Working Condition	Load	Restraint	Condition Design Diagram
Three-point support load	-	Apply vertical displacement to constrained support points	

3.3. Results of Static Analysis

The design domain's static strength and natural vibration mode are preliminarily analyzed. This can better control the variable parameters of optimization, reduce the scope of constraints, and improve the spatial scope of the design domain. Theoretically, this can also make the structure optimization process more detailed, with better optimization results; secondly, it can also reduce the number of iterations and shorten the computation time. The results are shown in Figure 4 and Table 3.

Table 3. Results of finite element analysis in design domain.

Working Condition	Maximum Von Mises Stress/MPa
Longitudinal load	86.8
Vertical load	63.3
Torsional load	67.9
Crosswind load	45.0
Three-point support load	36.0

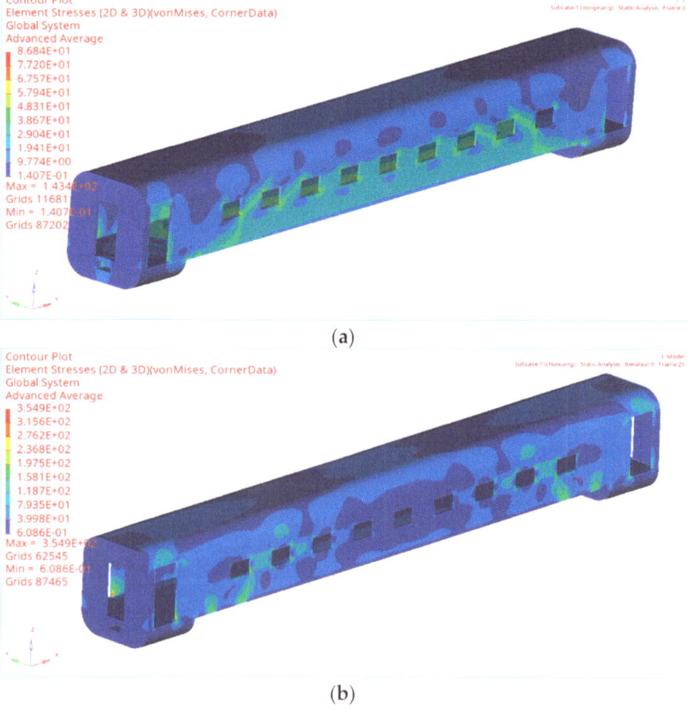

(a)

(b)

Figure 4. *Cont.*

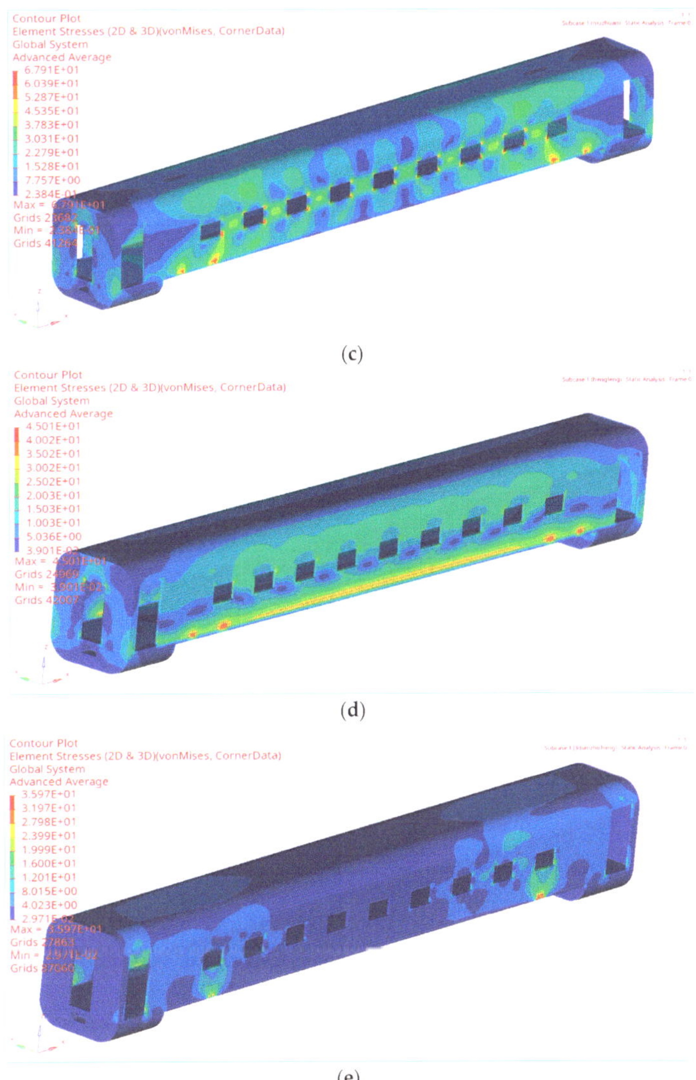

Figure 4. Finite element analysis of stress nephogram in design domain: (**a**) longitudinal load; (**b**) vertical load; (**c**) torsional load; (**d**) crosswind load; (**e**) three-point support load.

It can be seen from the calculation results that the maximum Von Mises stress in the design domain is 86.8 Mpa, which is far less than the allowable stress of the material and has ample space for optimization, and the design domain space is reasonable.

3.4. Results of Modal Analysis

In the general simulation analysis, the mass equipment is applied to the center of gravity by concentrating the mass points, and the rest of the mass of the equipment is loaded by uniformly distributing the mass points. Both structural and reconditioning modes are free vibration modes without any constraint. In the conceptual stage of car-body design, the influence of the equipment quality on the car-body mode should be considered in the topology optimization analysis and calculation. Table 4 and Figure 5 show the magnitude of the vibration frequency and corresponding mode shapes.

Table 4. Description of the natural frequency and mode shape of the car-body mode.

Modal Order Number	Mode Shape of the Car-Body Mode	
	Frequency/Hz	Vibration Mode
1	21.25	First vertical bending deformation
2	22.69	First diamond deformation
3	23.51	First transverse bending deformation
4	30.64	Breathing deformation

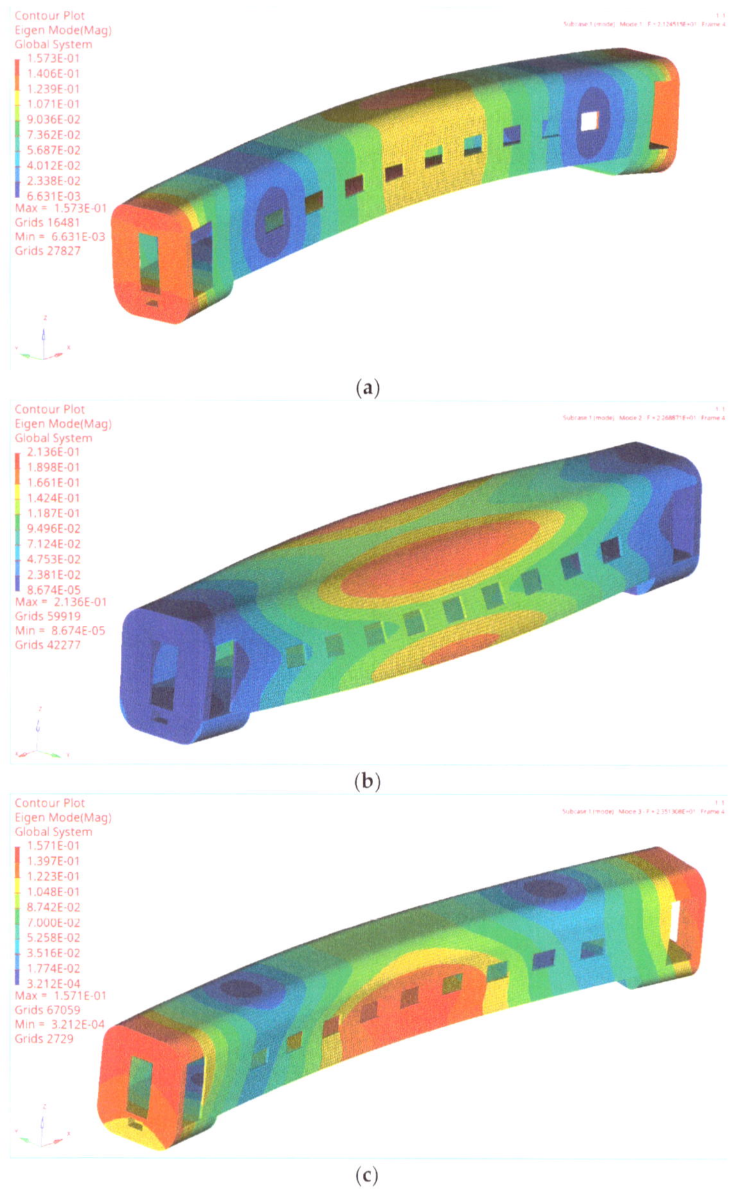

(a)

(b)

(c)

Figure 5. *Cont.*

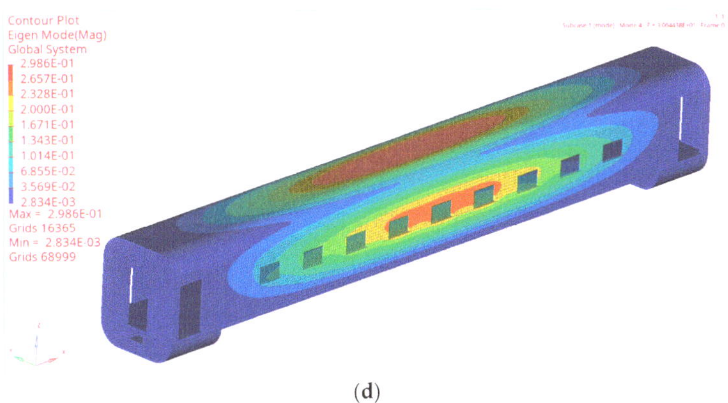

(d)

Figure 5. Mode shapes of each order in the state of the steel structure of the car body: (**a**) first vertical bending deformation; (**b**) first diamond deformation; (**c**) first transverse bending deformation; (**d**) breathing deformation.

4. Topology Optimization of Car-Body Structure

4.1. Topology Optimization Design

Before performing topology optimization, the mathematical model of topology optimization should be established. The design variables, constraints, and objective functions of topology optimization should be determined.

(1) Design variables

In order to ensure that the computer can complete the optimization calculation task and obtain the ideal optimization result, reasonable and appropriate design variables are indispensable. The more design variables, the more detailed the optimization results and the more in-depth the optimization degree, but too many design variables will also cause an increase in computing time. Therefore, when selecting design variables, the design variables should be reduced as much as possible to ensure the complete representation of the requirements. Based on the above principles, this paper takes the whole vehicle as a design variable with a minimum member size of 150 mm and a maximum member size of 400 mm.

(2) Constraints

Constraints are necessary to control the direction of the result generation in the optimization calculation. The constraint conditions can be divided into two types: size constraints and behavior constraints. Size constraints are mainly geometric restrictions on the design variables, while behavior constraints are used to characterize the state of the reaction structure, such as the frequency and intensity.

(1) Yield constraints

The yield strength is used as the constraint condition. The allowable yield utilization factor λ_{perm} is defined, and the finite elements meet the yield criteria as follows.

$$\lambda_y \leq \lambda_{\mathrm{perm}} \tag{7}$$

where λ_y is the yield utilization factor, $\lambda_y = 0.78\sigma_{\mathrm{vm}}/235$, and σ_{vm} is the Von Mises stress.

(2) Volume fraction constraints

Different volume fractions are used as response constraints to explore the influence of different volume fraction constraints on the optimization results and determine the value range of volume fractions.

This paper discusses only the constraints on the yield strength and first-order deformation frequency. The yield strength constraint limit is 215 MPa, the first-order deformation frequency constraint limit is 10 Hz, and the second-order deformation frequency constraint limit is 12 Hz.

(3) Objective function

This paper is a topological optimization solution analysis of the car-body structure, aiming to obtain the minimum material surplus in the design space to meet the design requirements. The optimization design should take the minimum volume of the residual material as the objective function. The formula for the volume as the objective function is as follows.

$$V = \sum_{i=1}^{n} \rho_{ij} V_i^0 \tag{8}$$

where ρ_{ij} represents the unit material density of the micro-element, and V_i^0 is the initial volume of the ith element.

For the car-body model in the initial state, the density of all unit materials is 1. After iterative calculation, if the unit material density is still 1, the unit material is more important for the car-body structure, and the unit is preserved. Meanwhile, when the unit material density is 0 after the iterative calculation is completed, the unit material is not important to the car-body structure, and the unit is deleted. Table 5 shows the specific topology optimization parameter settings.

Table 5. Topology optimization parameter settings.

Parameter	Description	Value
MINDIM	Minimum member size	150
MAXDIM	Maximum member size	400
OBJTOL	Tolerance of target function	0.005
CHECKER	Checkerboard parameter	1
DISCRETE	Discrete parameter	1

4.2. Topology Optimization Results

The force flow transfer path of the vehicle body is different under different loads, which leads to different topology optimization results under different loads. Based on different load conditions, this section describes the topology optimization analysis of single and multiple load conditions, respectively, to obtain the new main bearing structure of the vehicle body, satisfying multiple load conditions simultaneously.

(1) Topology optimization results of single working condition

After 75 iterations of analysis and calculation, the topology optimization calculation for the longitudinal load condition is terminated. Figure 6 lists the topological density cloud images with different thresholds.

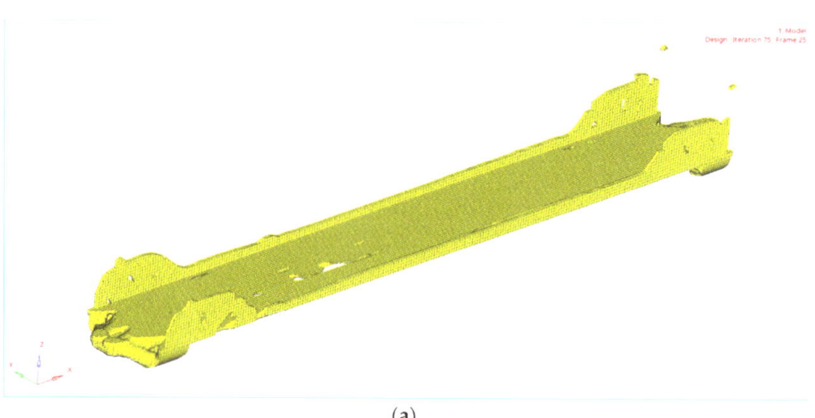

(a)

Figure 6. *Cont.*

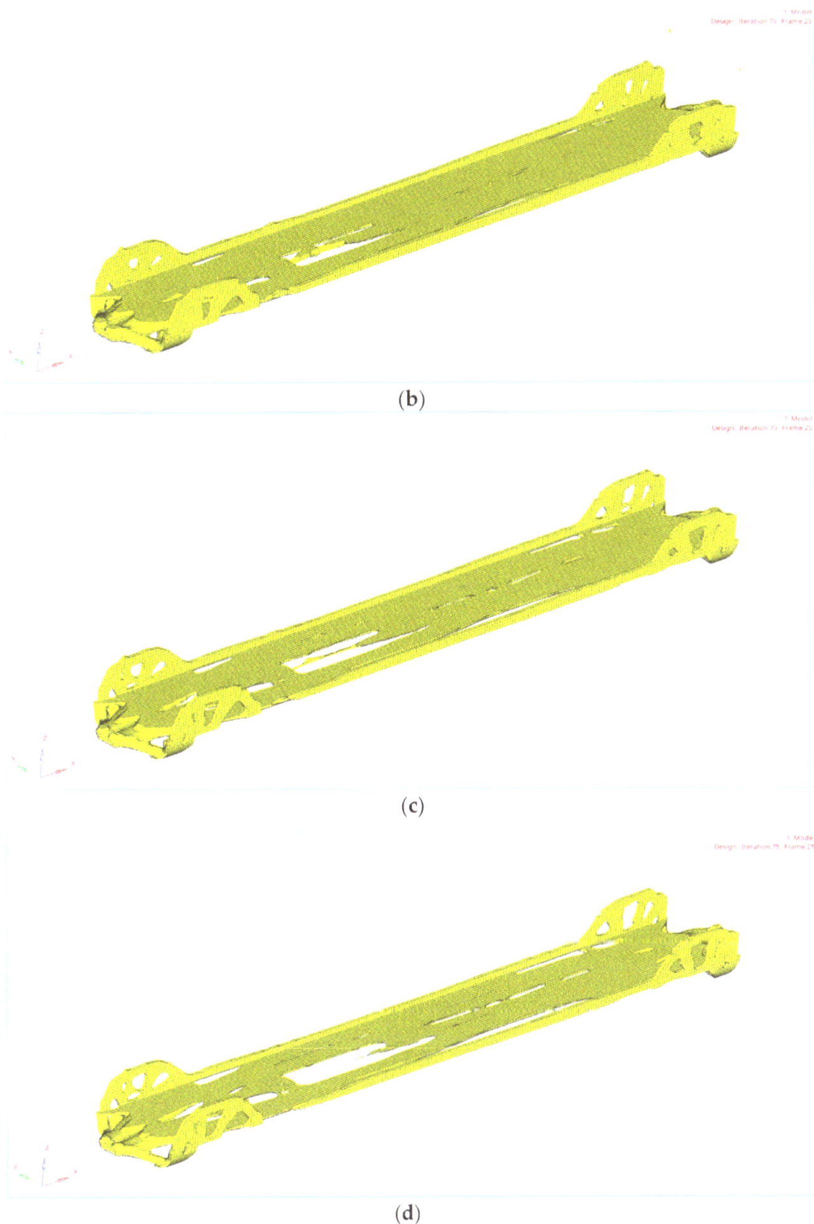

(**b**)

(**c**)

(**d**)

Figure 6. *Cont.*

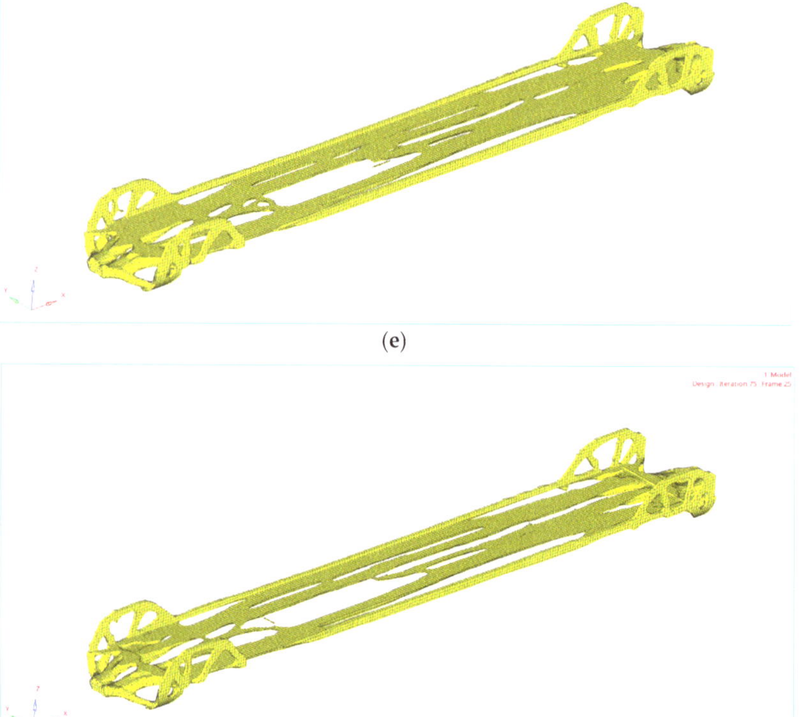

(e)

(f)

Figure 6. The optimization results of 75 iterations with different thresholds were obtained: (**a**) threshold 0.05; (**b**) threshold 0.1; (**c**) threshold 0.2; (**d**) threshold 0.3; (**e**) threshold 0.4; (**f**) threshold 0.5.

After 106 iterations of analysis and calculation, topology optimization calculation for vertical load conditions is terminated. Figure 7 lists the topological density cloud images with different thresholds.

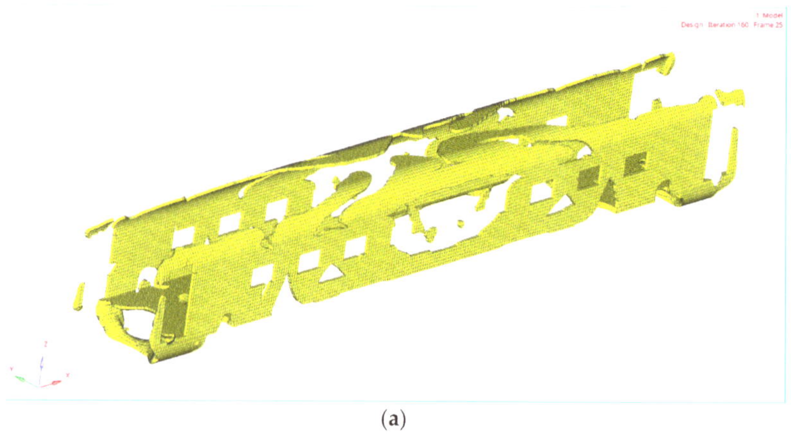

(a)

Figure 7. *Cont.*

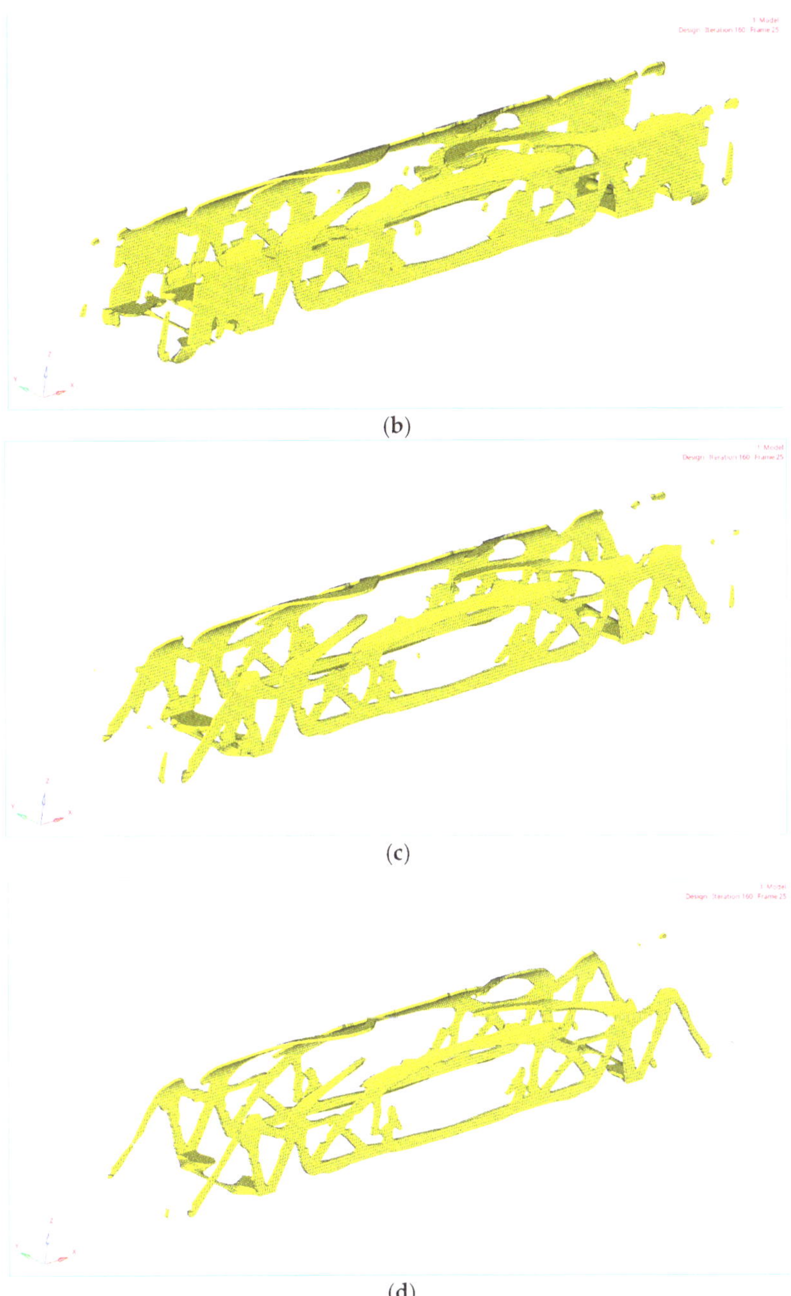

(b)

(c)

(d)

Figure 7. *Cont.*

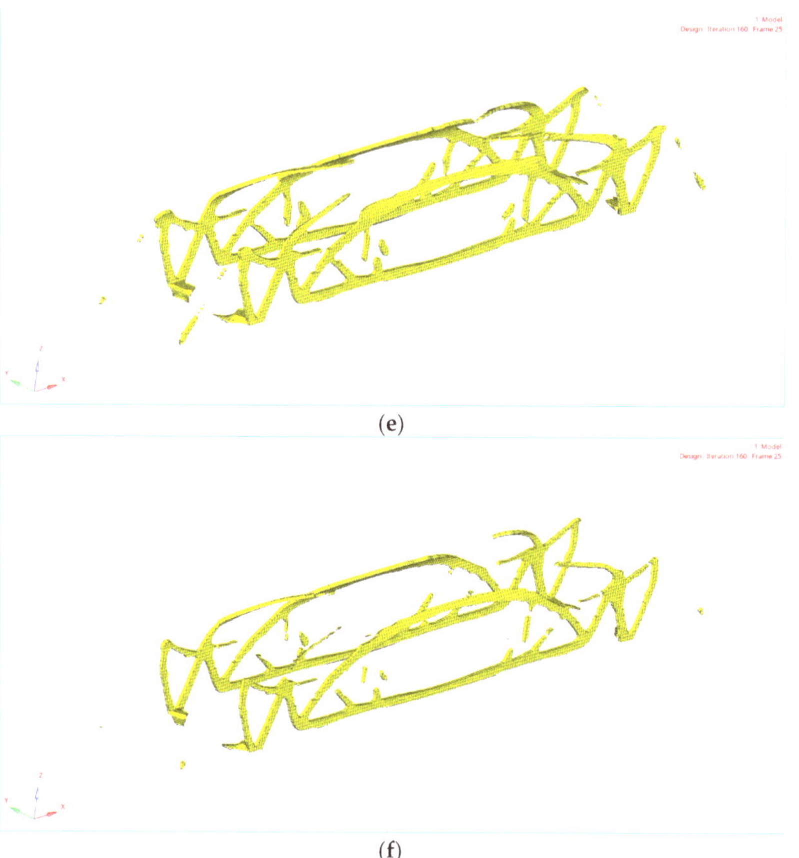

(e)

(f)

Figure 7. The optimization results of 106 iterations with different thresholds were obtained: (**a**) threshold 0.05; (**b**) threshold 0.1; (**c**) threshold 0.2; (**d**) threshold 0.3; (**e**) threshold 0.4; (**f**) threshold 0.5.

After 139 iterations of analysis and calculation, topology optimization calculation for torsional load conditions is terminated. Figure 8 lists the topological density cloud images with different thresholds.

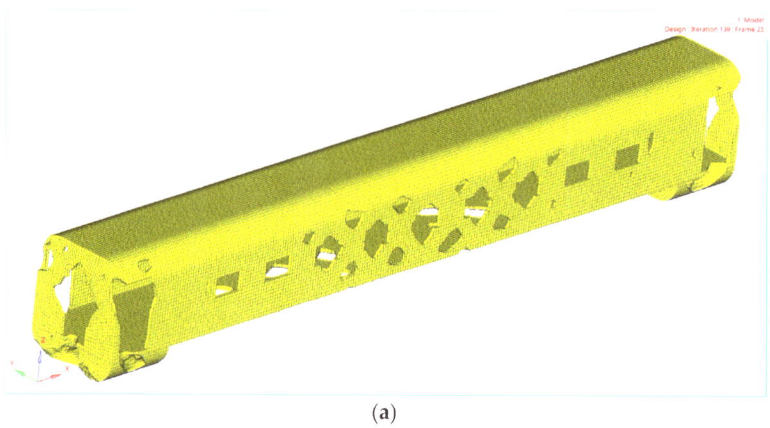

(a)

Figure 8. *Cont.*

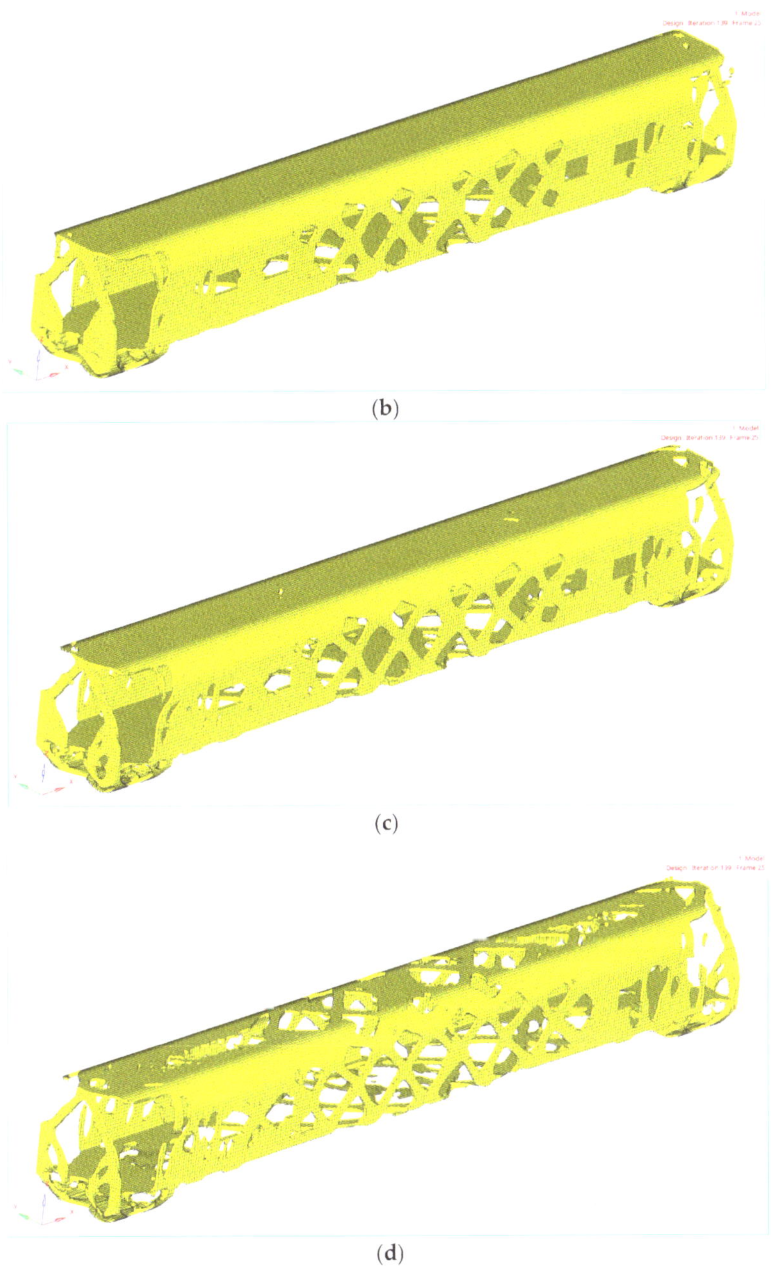

Figure 8. *Cont.*

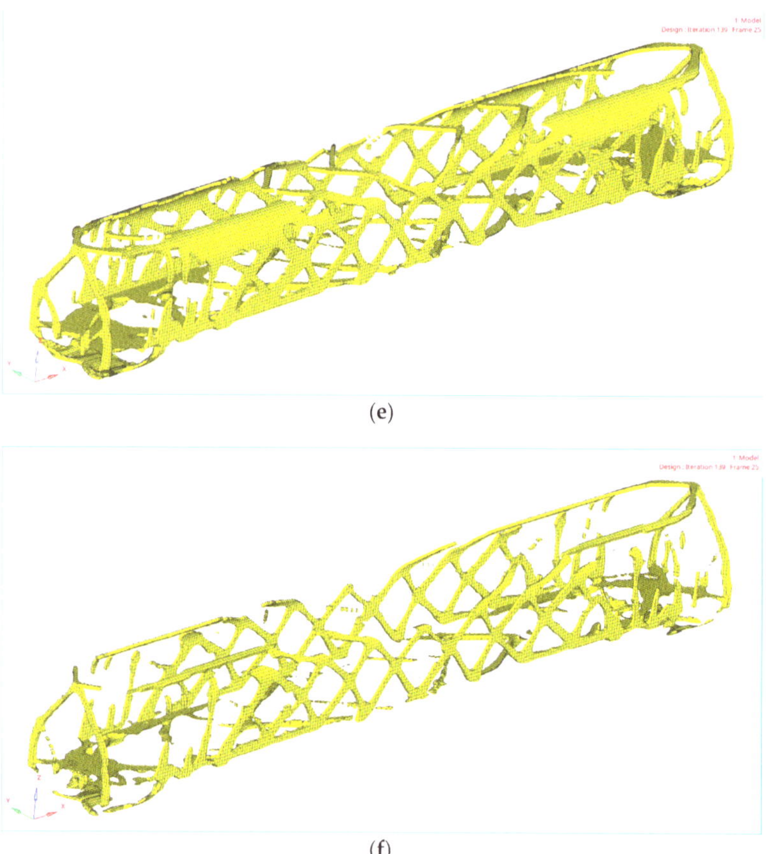

(e)

(f)

Figure 8. The optimization results of 139 iterations with different thresholds were obtained: (**a**) threshold 0.05; (**b**) threshold 0.1; (**c**) threshold 0.2; (**d**) threshold 0.3; (**e**) threshold 0.4; (**f**) threshold 0.5.

After 155 iterations of analysis and calculation, the topology optimization calculation of crosswind load conditions is terminated. Figure 9 lists the topological density cloud maps with different thresholds.

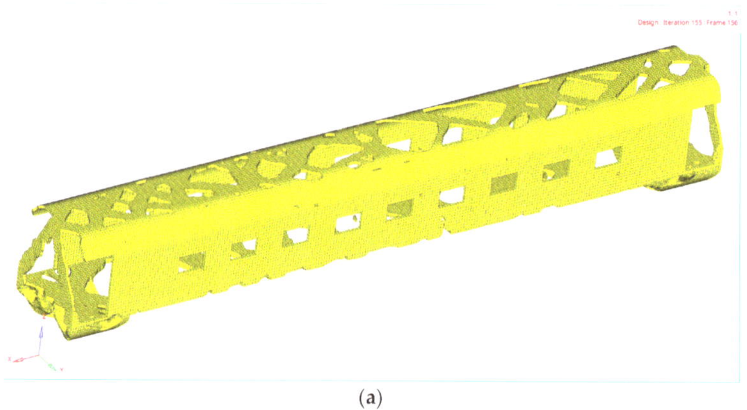

(a)

Figure 9. *Cont.*

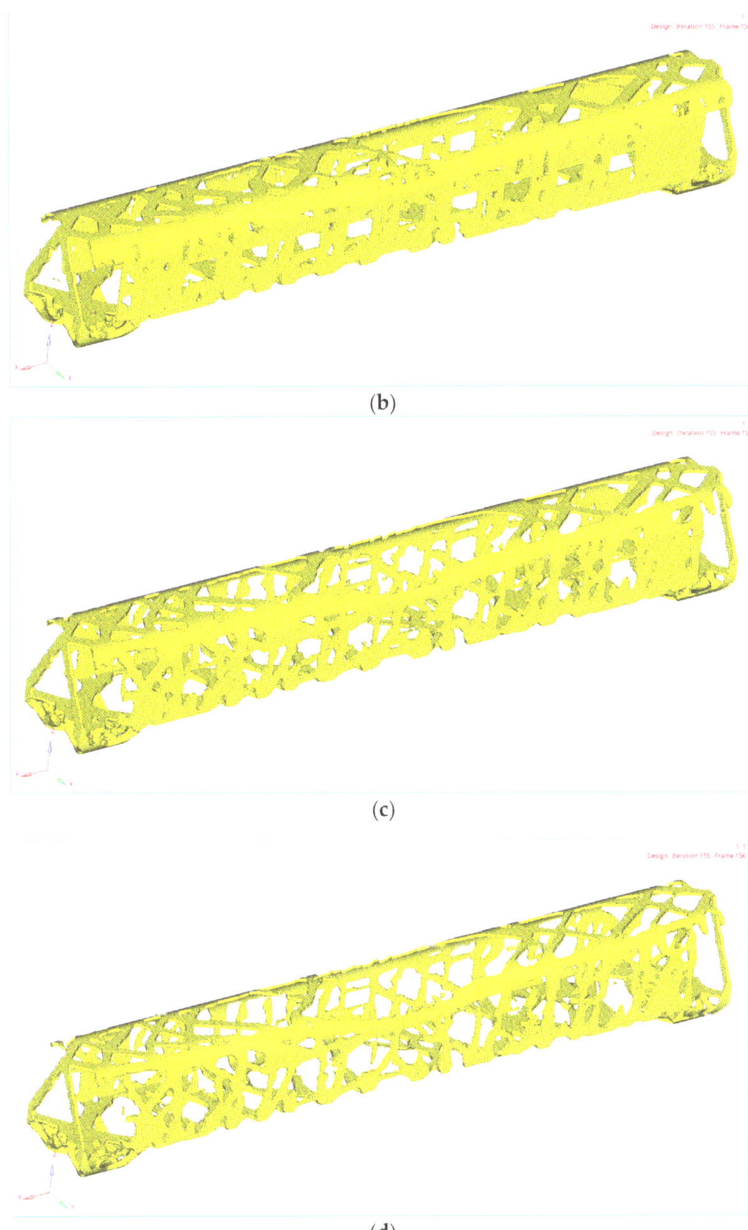

(**b**)

(**c**)

(**d**)

Figure 9. *Cont.*

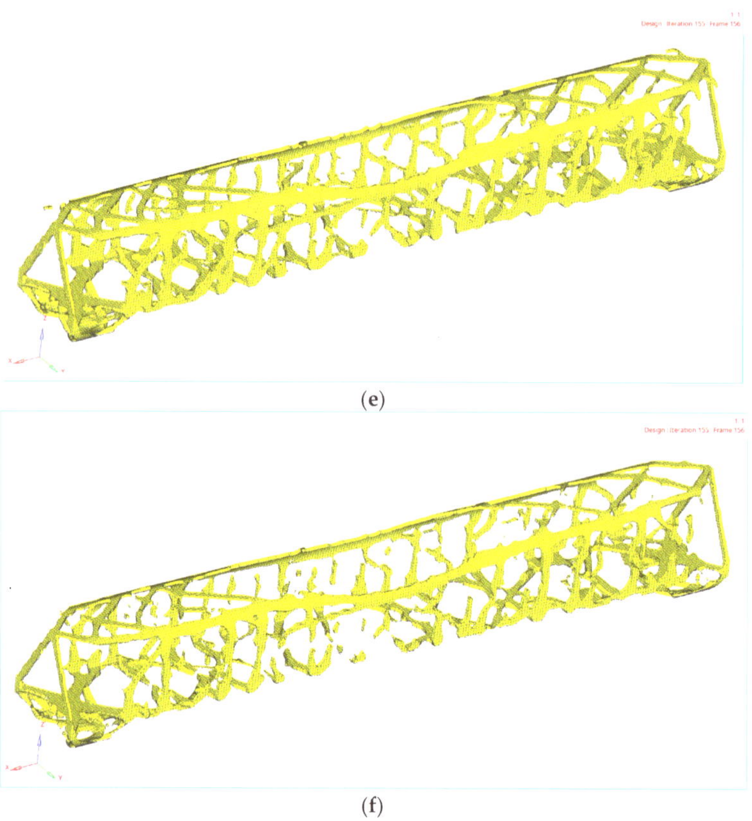

(e)

(f)

Figure 9. The optimization results of 155 iterations with different thresholds were obtained: (**a**) threshold 0.05; (**b**) threshold 0.1; (**c**) threshold 0.2; (**d**) threshold 0.3; (**e**) threshold 0.4; (**f**) threshold 0.5.

After 200 iterations of analysis and calculation, topology optimization calculation under three-point support load conditions is terminated. Figure 10 lists the topological density cloud diagrams with different thresholds.

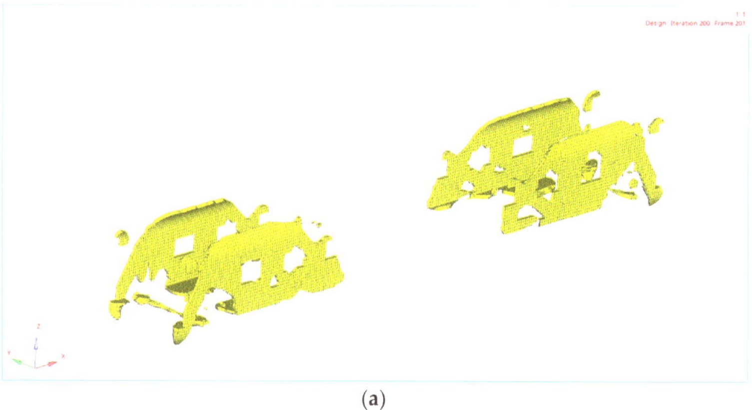

(a)

Figure 10. *Cont.*

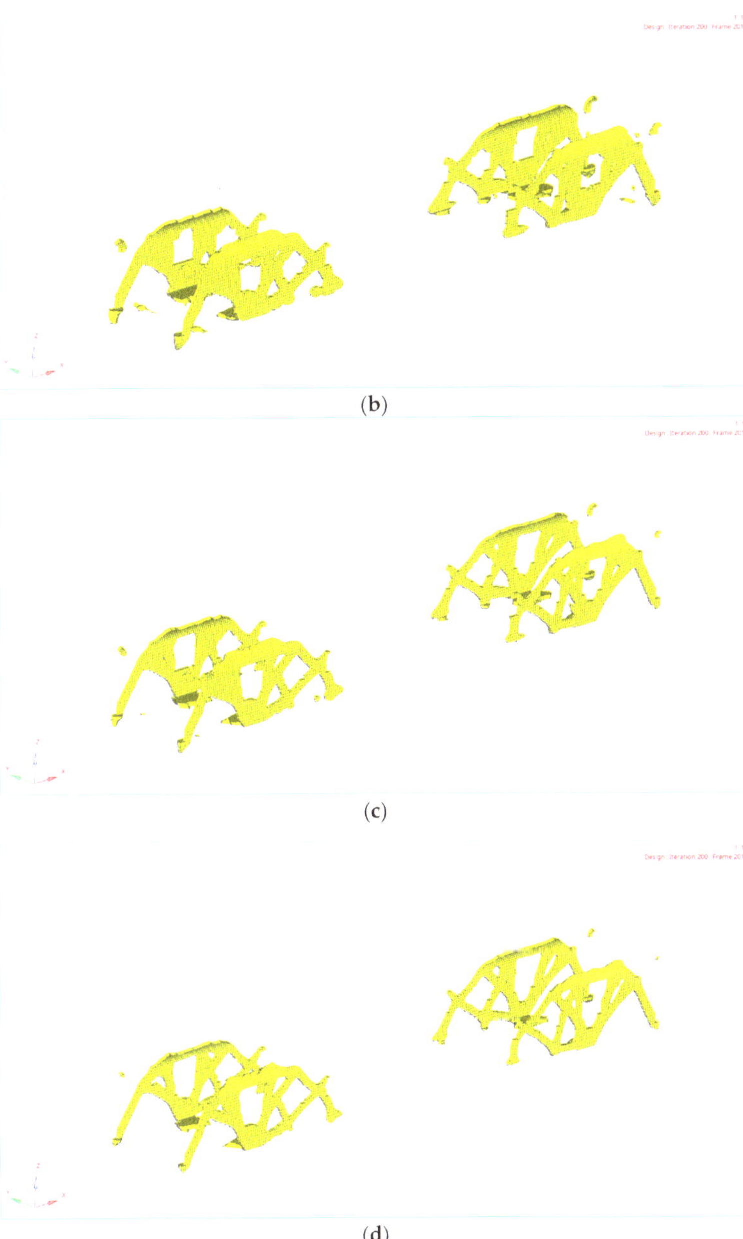

(**b**)

(**c**)

(**d**)

Figure 10. *Cont.*

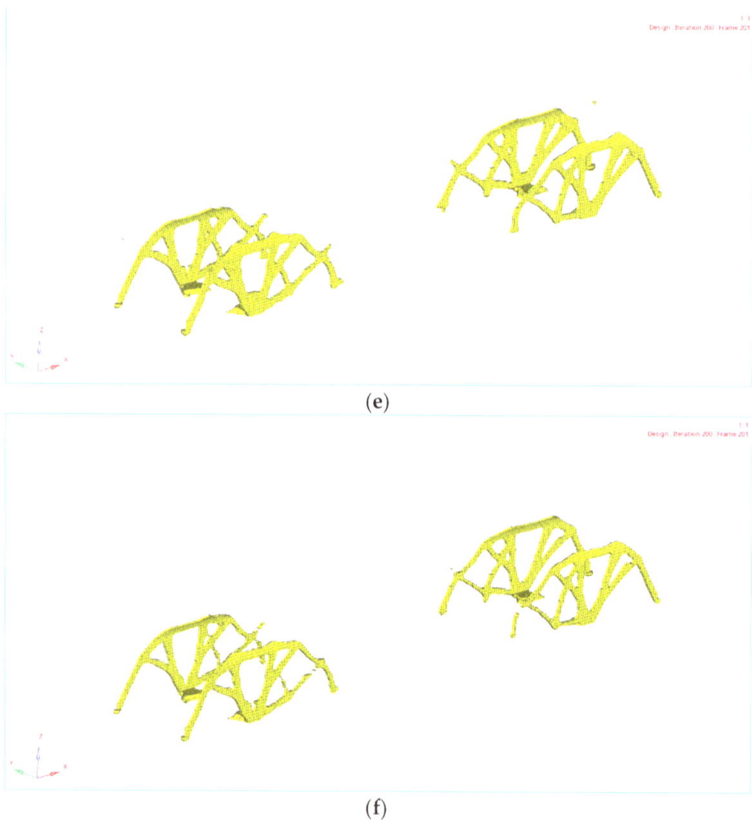

Figure 10. The optimization results of 200 iterations with different thresholds were obtained: (**a**) threshold 0.05; (**b**) threshold 0.1; (**c**) threshold 0.2; (**d**) threshold 0.3; (**e**) threshold 0.4; (**f**) threshold 0.5.

(2) Topology optimization results of multiple working conditions

In the multiple working condition topology optimization of the structure, the optimal topology optimization structure corresponding to different conditions may also differ. A material element deleted in one working condition may be retained in another. In other words, there may be conflicts between the deletion and retaining of material elements under different working conditions. To obtain the comprehensive optimal solution under various working conditions, the linear weighted relationship between the topology optimization results of each single load working condition is considered. By assigning different weight coefficients to each working condition, the complex multi-working condition optimization problem can be simplified to a single working condition optimization problem. The mathematical model is as follows.

$$\begin{cases} F(X) = \sum_{j=1}^{m} \omega_j f_j(X) = \omega_1 f_1(X) + \omega_2 f_2(X) + \ldots + \omega_m f_m(X) \\ \sum_{i=1}^{P} \omega_i = 1 \ (i = 1, 2, \ldots, \mathrm{p}) \end{cases} \tag{9}$$

where $F(X)$ represents the equivalent objective function under multiple working conditions, m represents the number of optimized working conditions, ω corresponds to the weight coefficient of each working condition, and $f(X)$ represents the input load under the single load condition.

When creating the response in the simulation software, it is necessary to set the corresponding proportion of each working condition. However, due to the different frequencies of each working condition during the operation of the vehicle body, the proportion of each working condition in the topology optimization is different, so two groups of weight coefficients are selected for comparative analysis.

(1) Topology optimization scheme with the same weight coefficients

In this group, the specific gravity of the five working conditions is set as 0.2 for topology optimization and submitted for calculation. Figures 11 and 12 show the changing trend of the objective function under topology optimization with the same weight coefficient after 139 iterations.

The optimization results of multiple working conditions with the same weight coefficient show that the residual material distribution of the car-body structure after the topology optimization of multiple working conditions is relatively straightforward. With the increase in the number of iterations, the truss structure in the middle of the car body is apparent. However, there are still more materials on both sides of the car body, and no apparent truss structure is generated.

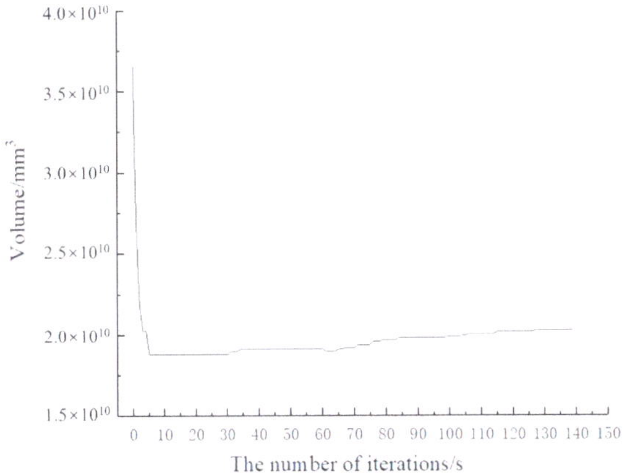

Figure 11. Volume change trend under topology optimization of the same weight coefficient.

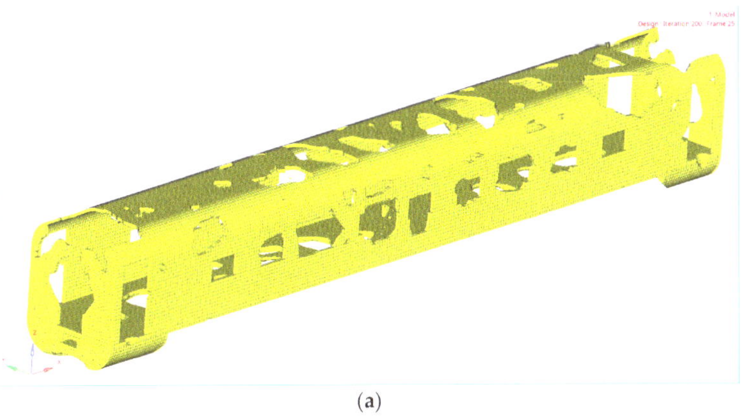

(**a**)

Figure 12. *Cont.*

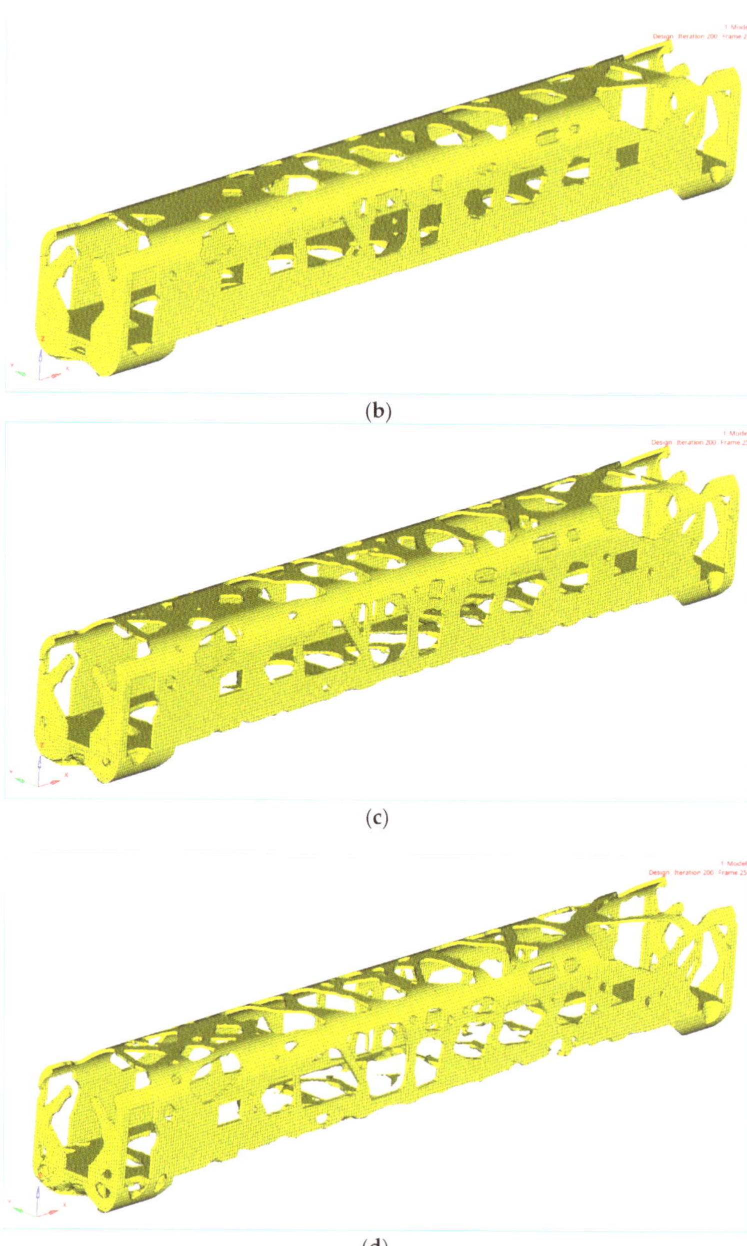

(b)

(c)

(d)

Figure 12. *Cont.*

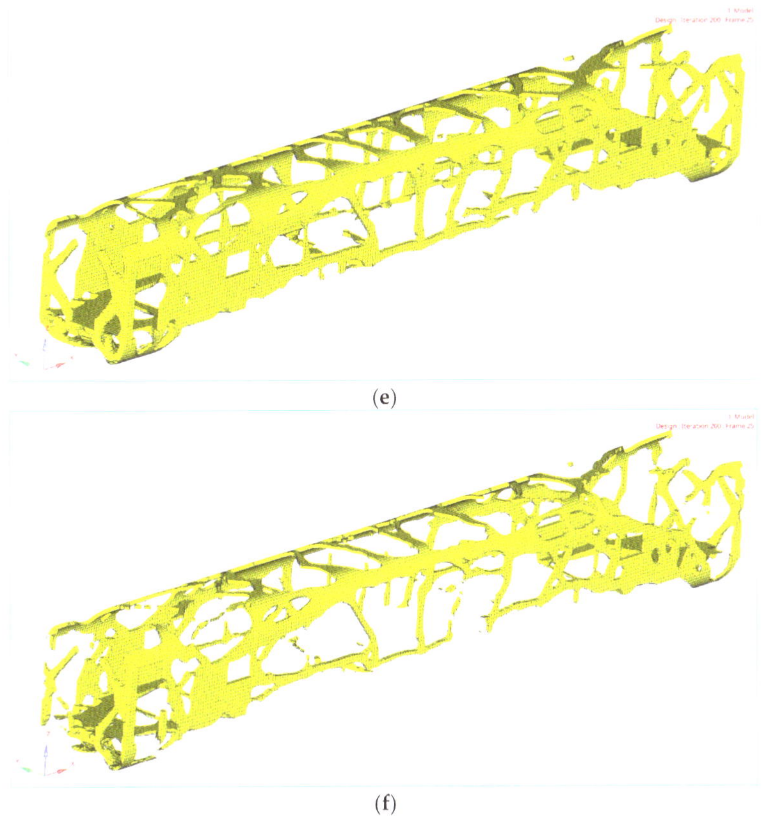

(e)

(f)

Figure 12. Optimization results with different thresholds in 139 iterations: (**a**) threshold 0.05; (**b**) threshold 0.1; (**c**) threshold 0.2; (**d**) threshold 0.3; (**e**) threshold 0.4; (**f**) threshold 0.5.

(2) Topology optimization scheme with different weight coefficients

A vertical load is always present in the daily operation of high-speed EMUs. The torsional load and transverse wind load are the most frequent loads of trains entering and exiting curves, intersections, and tunnels, while the longitudinal load and three-point support load are less frequent. According to the frequency of the five working conditions, the specific gravity of the five topology optimizations in this group is set as 0.4, 0.2, 0.2, 0.1, 0.1, including the vertical load of 0.4, the torsional load and the transverse wind load of 0.2, and the longitudinal load and the three-point support load of 0.1, and the calculation is performed based on the above optimization settings. Figures 13 and 14 show the changing trend of the objective function under topology optimization with different weight coefficients after 75 iterations.

The multi-condition optimization results with different weight coefficients show that the distribution of residual structural materials of the car body is clear after the multi-condition topology optimization. With the increase in the number of iterations, the material of the end walls on both sides is removed from the early stage of calculation, the structure of the bottom beam of the frame gradually becomes clear, and the cross-type grid structure of the side wall and the roof also gradually becomes prominent.

The multiple condition optimization with different weight coefficients is adopted by comprehensively comparing the above two schemes. This scheme can better integrate the characteristics of the topology optimization results of each single condition, and the optimization results are reasonable. The truss structure is evident, which can provide critical guiding suggestions for the establishment of the truss vehicle body structure in the later stage.

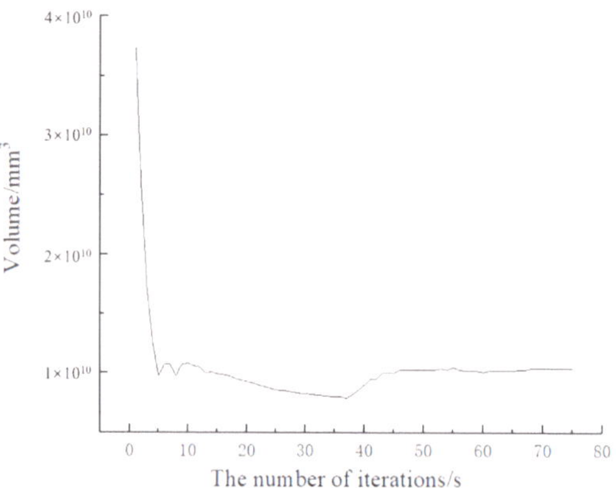

Figure 13. Volume change trend under topology optimization with different weight coefficients.

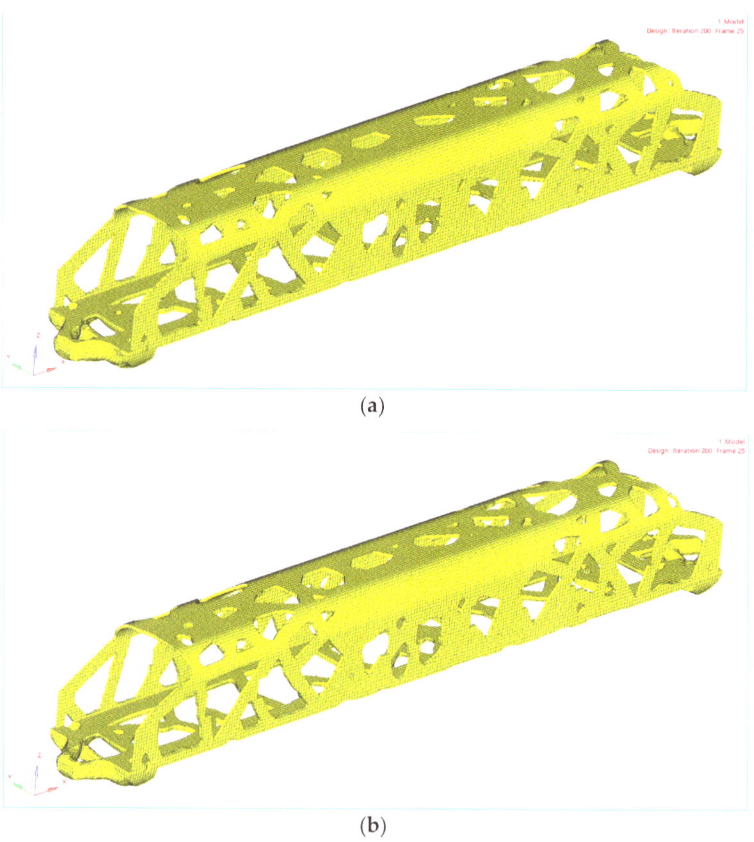

(a)

(b)

Figure 14. *Cont.*

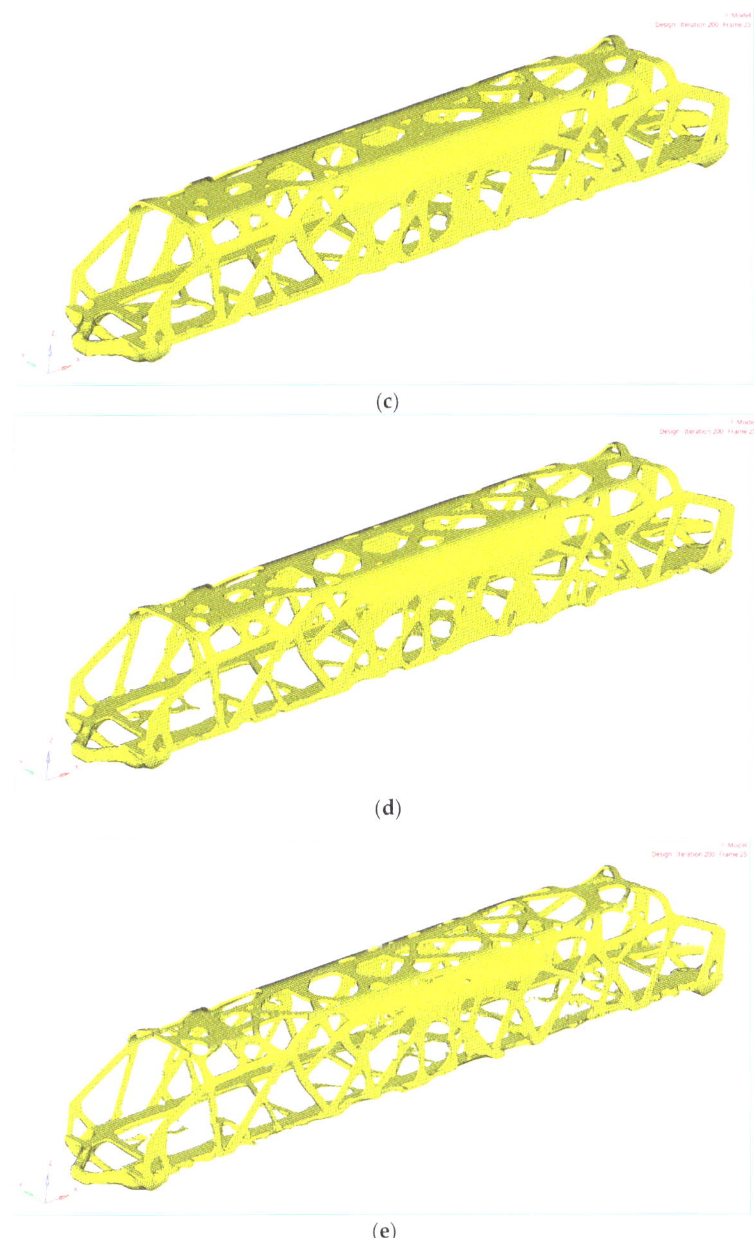

(**c**)

(**d**)

(**e**)

Figure 14. *Cont.*

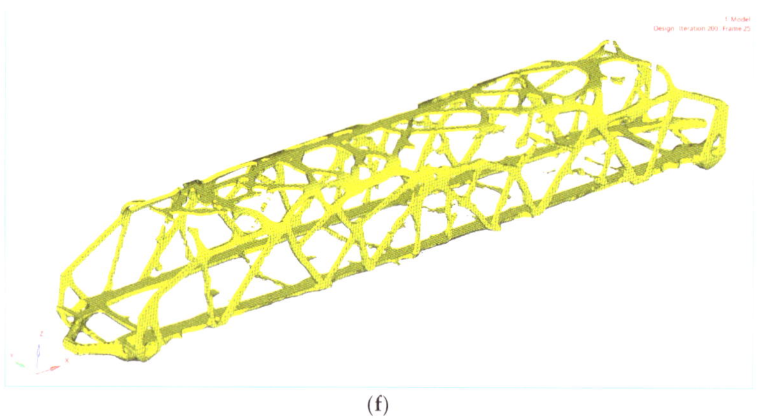

(**f**)

Figure 14. The optimization results of 75 iterations with different thresholds were obtained: (**a**) threshold 0.05; (**b**) threshold 0.1; (**c**) threshold 0.2; (**d**) threshold 0.3; (**e**) threshold 0.4; (**f**) threshold 0.5.

5. Reconstruction of Car-Body Bearing Structure

Although topology optimization can reflect the load transfer path by optimizing the resulting material distribution, this result cannot be directly applied to machining and manufacturing, and it can only provide ideas for subsequent design. In this section, the reconstruction scheme of the car-body bearing structure is determined, and the dynamic performance is checked.

5.1. Design Method of Truss Car-Body Structure Reconstruction

The bearing structure of high-speed EMUs mainly comprises the roof, side wall, bottom frame, end wall, end part, etc. By extracting the topology optimization results of each part of the car body, the basic shape of the new car-body structure of high-speed EMUs is obtained. The basic design principles of each car-body part are as follows. The end wall mainly adopts the triangular bearing structure; the end part is the coupler–seat rear inclined beam structure; the bottom frame mainly considers the longitudinal beam structure in the middle and both sides of the bottom frame; the side wall and the roof adopt the cross-beam structure; and the structure near the supporting point of the car body needs to be strengthened locally. In order to ensure that the vibration mode frequency meets the design requirements, the side wall and upper beam structure are strengthened.

The topology optimization results contain the residual distribution range of materials under different working conditions. However, this result is limited by the conceptual design stage of topology optimization, which does not consider various complex situations in vehicle body operation, such as the counterweight of cables, ventilation ducts, toilets, etc. Therefore, the topology optimization results are more suitable to guide the subsequent detailed design stage.

This process includes measuring various working conditions in the finite element software, examining the surplus material in the topology optimization results, determining each part's hole position and its sizes and conditions, evaluating multiple conditions considering the multiple and single topological optimization results, and incorporating the welding manufacture process, including the simplification of irregular holes and the identification of each part of the body after obtaining the neat hole position and its size.

5.2. Establishment of Car-Body Geometry Model and Finite Element Model

Many factors should be considered in the reconstruction of the truss car-body model. This paper establishes the main bearing structure model of the truss car body based on the basic data of the CRH vehicle contour and topology optimization results. In order to facilitate the subsequent simulation calculation, only a quarter of the car-body model is established, and the comparative modeling of each part of the car body is shown in Figures 15–21; different colors are used to distinguish the components.

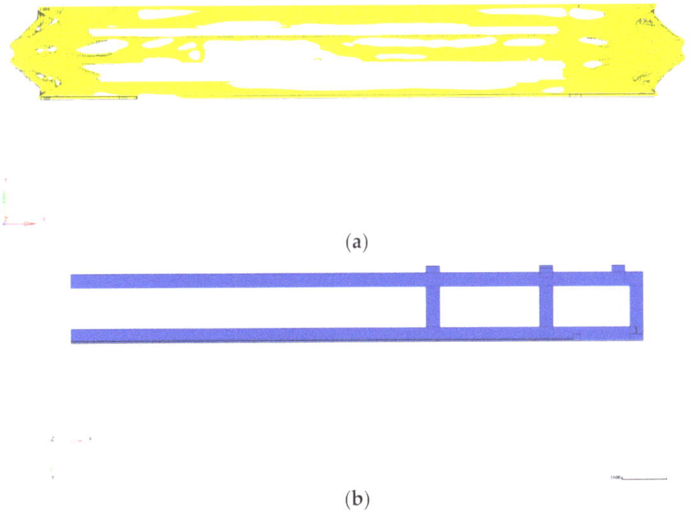

(a)

(b)

Figure 15. Contrast modeling of underframe structure: (**a**) results of topology optimization of underframe structure; (**b**) geometric modeling of underframe structure.

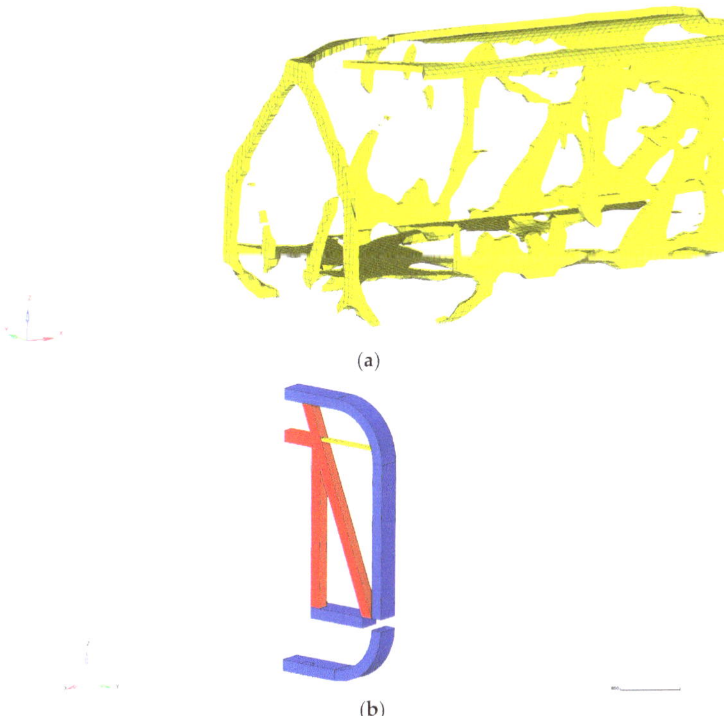

(a)

(b)

Figure 16. Contrast modeling of end-wall structure: (**a**) results of topology optimization of end-wall structure; (**b**) geometric modeling of end-wall structure.

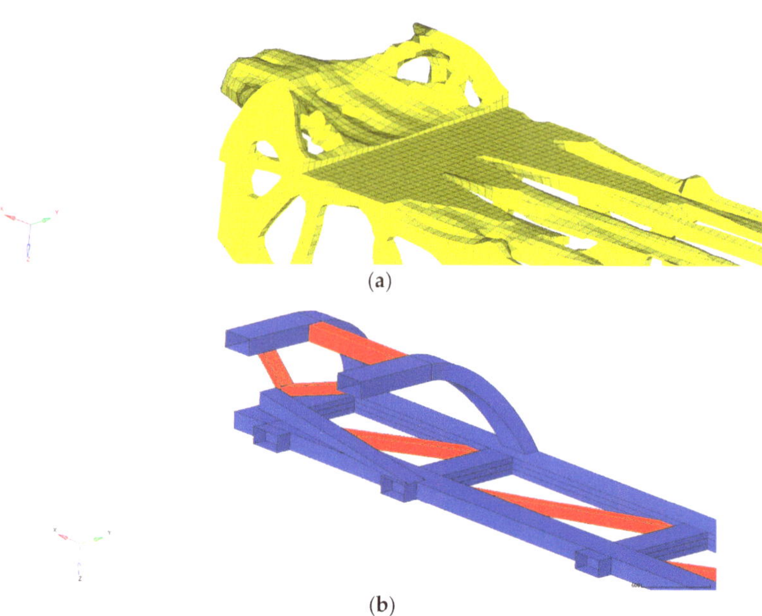

Figure 17. Contrast modeling of results of end structure: (**a**) results of topology optimization of results of end structure; (**b**) geometric modeling of results of end structure.

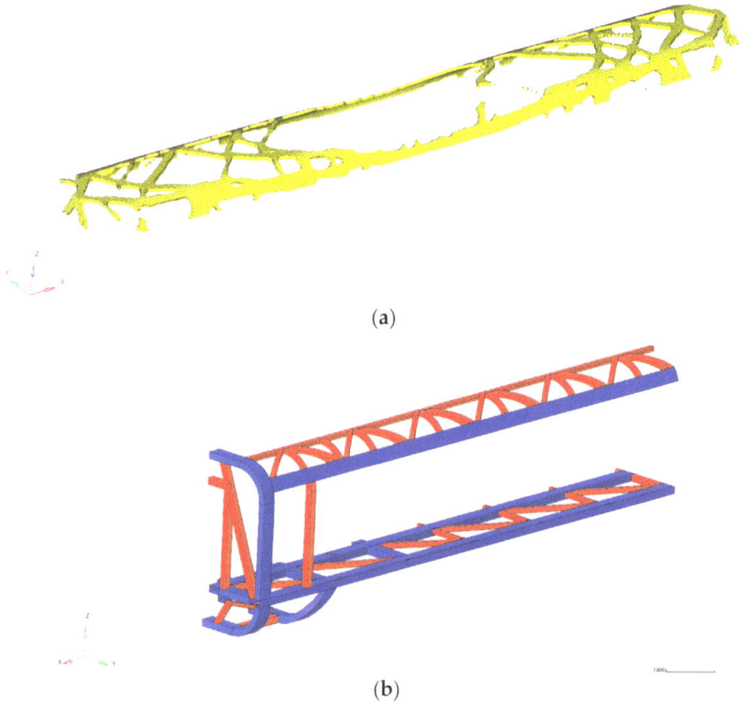

Figure 18. Contrast modeling of side beam structure (**a**) results of topology optimization of side beam structure; (**b**) geometric modeling of side beam structure.

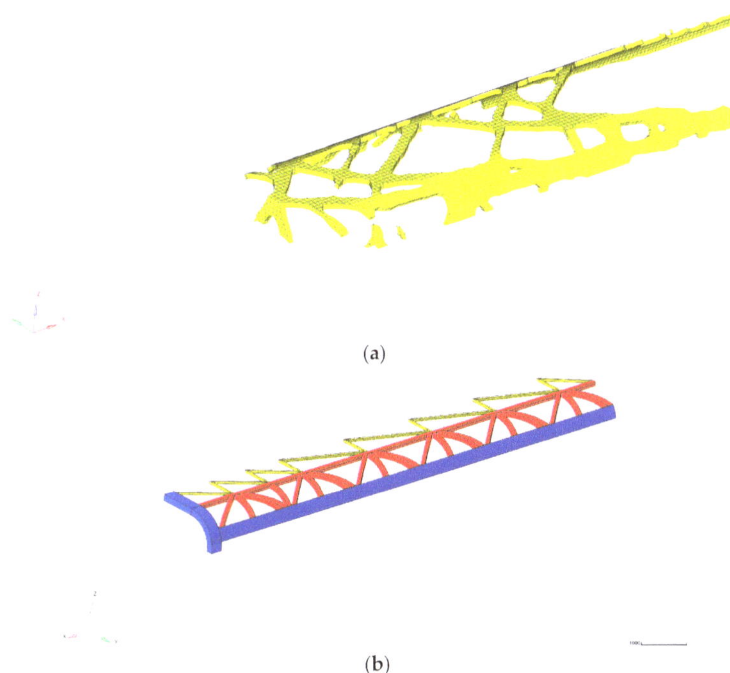

Figure 19. Contrast modeling of roof structure: (**a**) results of topology optimization of side wall structure; (**b**) geometric modeling of roof structure.

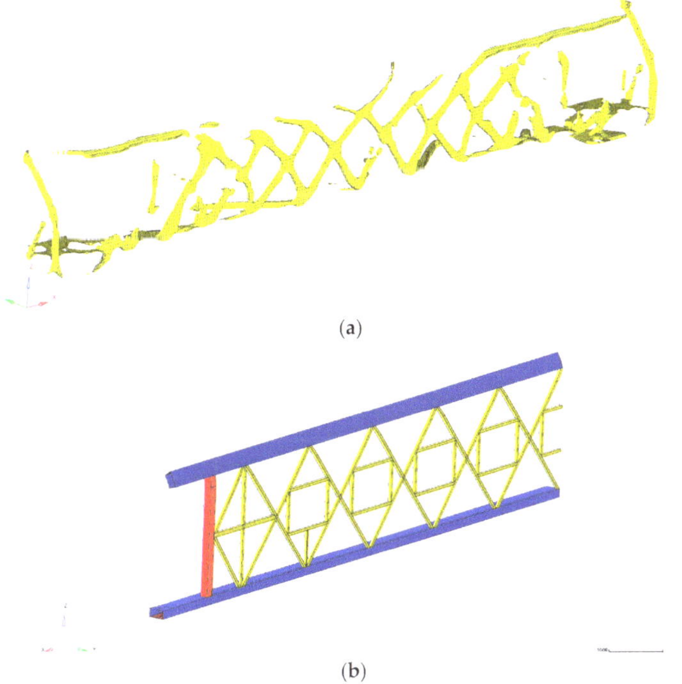

Figure 20. Contrast modeling of side wall structure: (**a**) results of topology optimization of side wall structure; (**b**) geometric modeling of side wall structure.

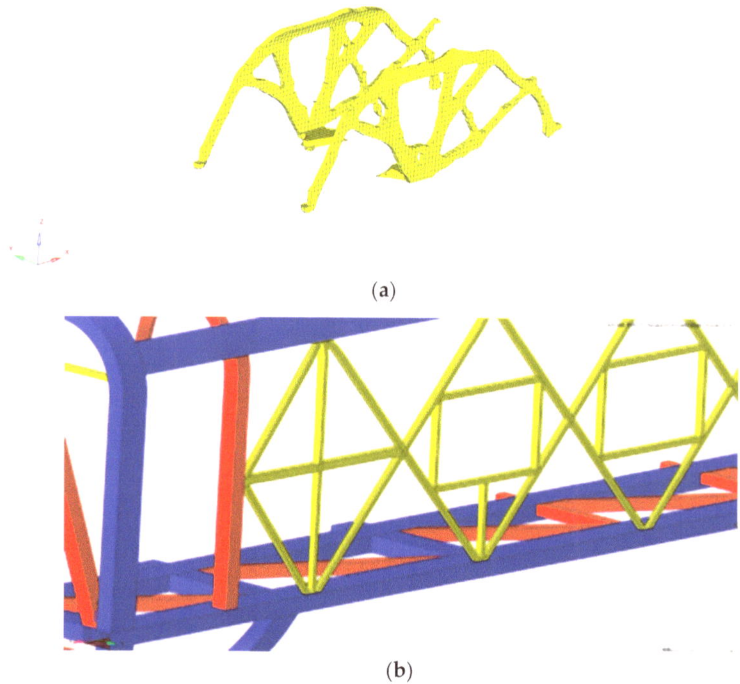

(**a**)

(**b**)

Figure 21. Contrast modeling of supporting structure: (**a**) results of topology optimization of supporting structure; (**b**) geometric modeling of supporting structure.

Each part's topology optimization modeling results are combined to obtain the truss body structure of high-speed EMUs after topology optimization. Figure 22 shows the model diagram of the quarter-truss vehicle. At the same time, to further improve the strength and stiffness of the car body and consider the tightness of the car-body structure, a layer of aluminum alloy skin with a thickness of 2 mm is added to the inner and outer surfaces of the car body. The quarter-body model with the skin added is shown in Figure 23. Different colors are used to distinguish the components.

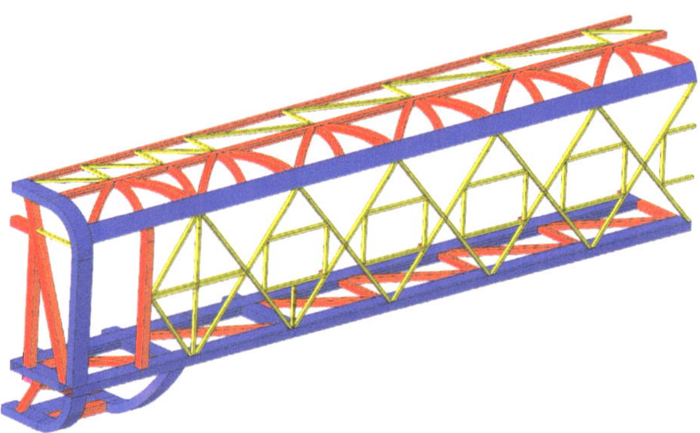

Figure 22. A quarter-truss model of the vehicle.

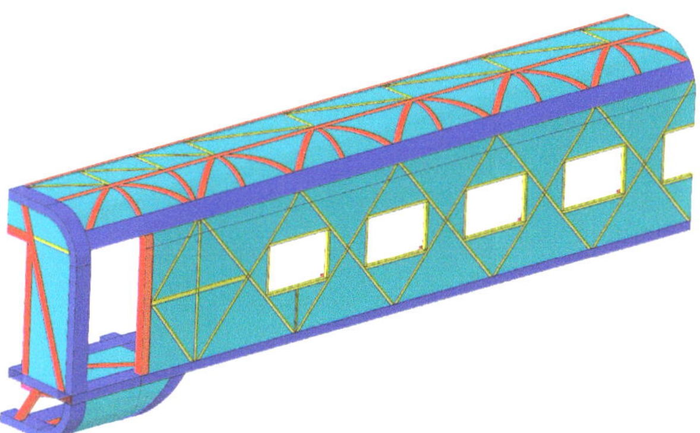

Figure 23. A quarter-truss model of the vehicle (skin).

The 3D model of the vehicle is imported into the finite element modeling software, and the car-body structure model is meshed. The mesh size is 20 mm × 20 mm, and the mesh shape is dominated by four-node thin-shell elements, totaling 1,865,412 elements and 1,772,788 nodes. The vehicle body model after finite element dispersion is shown in Figure 24. According to the measurement, the weight of the car body after topology optimization is 8.62 t. According to the relevant information, the weight of a CRH EMU's body is about 10.5 t. Compared with the CRH EMU's body, the weight of the reconstructed car body is reduced by 1.88 t from the original 10.5 t, which is about 18%.

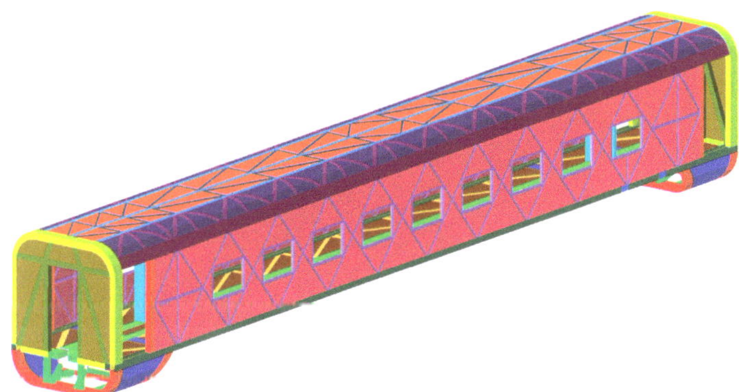

Figure 24. Finite element model of vehicle body after topology optimization.

5.3. Finite Element Analysis of the Reconstructed Model

In the above simulation calculation, the weight lifted by the off-board equipment and the weight of the passengers are not considered, and the traction transformer, auxiliary converter, sewage box, compressor, exhaust air cylinder, brake module, and other suspension equipment are not set separately. In this simulation analysis, it is necessary to consider the counterweights of cables, lighting equipment, ventilation ducts, toilets, inner end walls, tea rooms, seats, and other car components. However, the car-body structure design is only in the conceptual design stage, and the suspension position of the mass equipment has yet to be determined, so the mass equipment and the cable and other equipment are set by uniformly distributing the mass points. The reference design quality of the car-body structure is shown in Table 6.

A load and constraint are applied to the reconstructed finite element model of the car-body structure, and the simulation solution is carried out. The stress cloud diagram of the truss vehicle structure under different working conditions is obtained. The calculation results are shown in Figure 25.

Table 6. Body reference design quality.

Sequence	Description		Number/t
1	Weight of vehicle maintenance equipment (excluding bogie and car-body structure)		25.46
2	Weight of bogie		8.0
3	Passengers (80 kg/per person)	Capacity: 85	6.8
		Overcrowding: 120	9.6
4	Servicing equipment		0.4
5	Weight of vehicle with capacity passengers (excluding bogie and car-body structure)		32.26
6	Weight of vehicle with overcrowding passengers (excluding bogie and car-body structure)		35.06

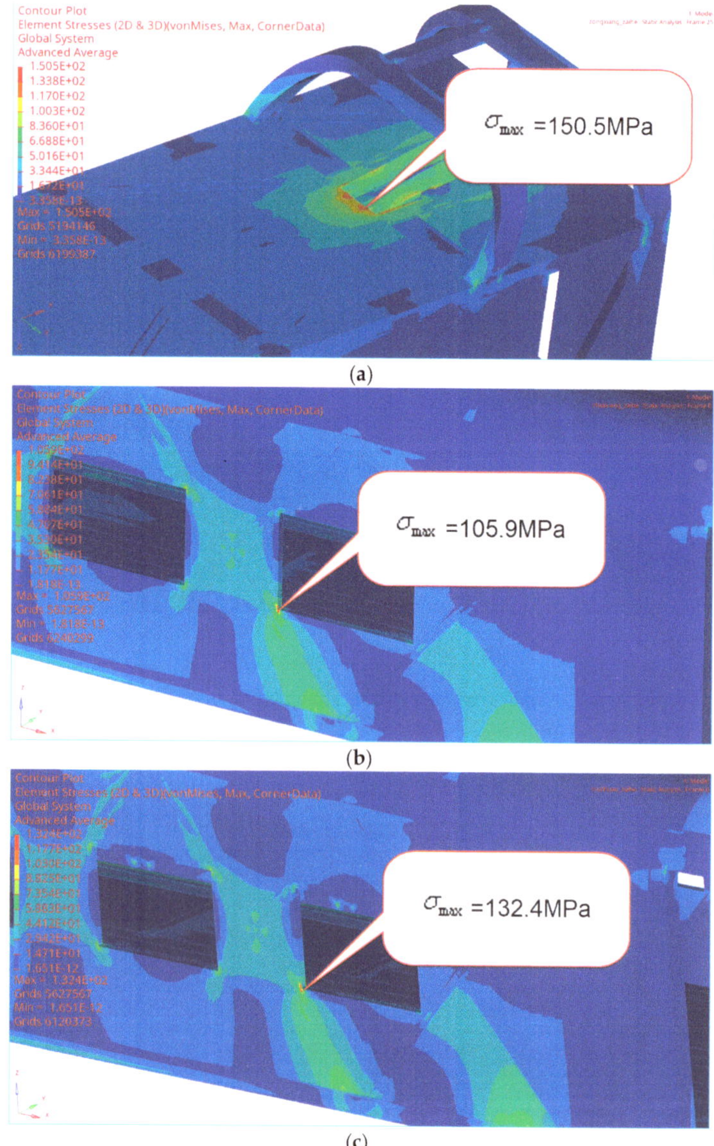

(a)

(b)

(c)

Figure 25. *Cont.*

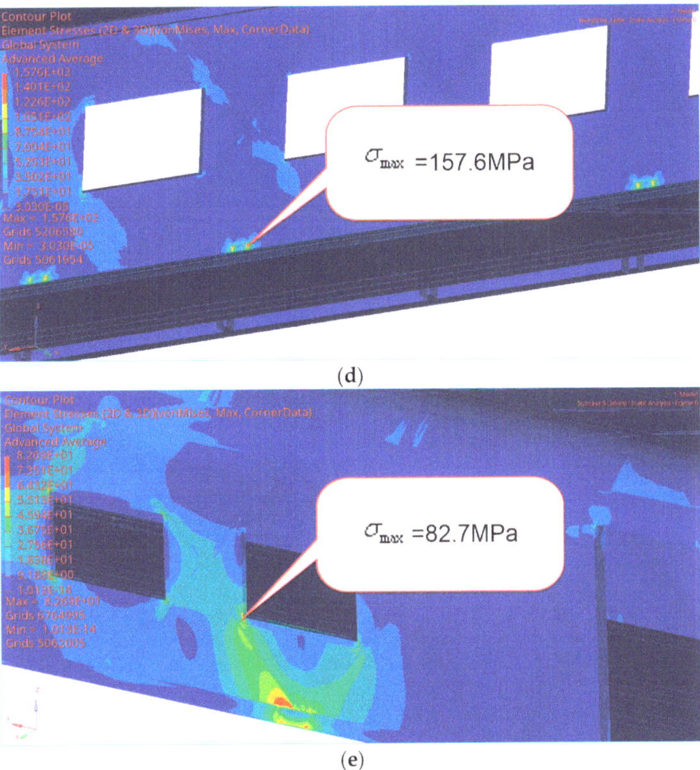

Figure 25. The stress nephogram of car-body model is reconstructed by finite element method: (**a**) stress cloud diagram (longitudinal load); (**b**) stress cloud diagram (vertical load); (**c**) stress cloud diagram (torsional load); (**d**) stress cloud diagram (crosswind load); (**e**) stress cloud diagram (three-point support load).

According to the calculation results, the maximum Von Mises stress of the car-body structure under these five working conditions is 157.6 MPa, less than the allowable stress of the material of 215 MPa. The maximum stress points are mainly concentrated at the window corner, the joint between the rear seat of the couplers and the bottom frame, and the corner of the side wall.

6. Conclusions

This paper uses topology optimization design for the car-body structure design domain based on CRH profile data. The following conclusions are drawn from the research analysis.

(1) The geometry model is established based on the contour size of the car-body structure. The design space of the topology optimization design domain is determined, and the finite element model of the vehicle body is established. The finite element analysis of the initial design domain is completed, and the correctness of the design space is checked, which provides a comparative reference for the subsequent topology optimization calculation results.

(2) The input parameter values of the design variables, constraints, and objective functions of topology optimization are determined, and the mathematical model of topology optimization is established. In the topology optimization design domain of the car-body structure, the topology configuration of the car-body structure is obtained by performing the topology optimization of the single and multiple operating conditions, respectively.

(3) The truss-body bearing structure is reconstructed based on the topology optimization results. The weight of the reconstructed structure is 8.6 t, which is about 18% lower than that of the current EMU structure in operation. The finite element analysis of the reconstructed truss car-body structure shows that the strength of the structure meets the requirements of the corresponding standards.

Author Contributions: C.L. and T.Z. were responsible for the whole experiment; K.M. wrote the manuscript; H.D., M.S. and P.W. carried out the topology optimization method and the numerical simulations. All authors have read and agreed to the published version of the manuscript.

Funding: This study was supported by the Science and Technology Research and Development Plan Project of China National Railway Group Co., Ltd. (P2020J024), the Sichuan Outstanding Youth Fund (2022JDJQ0025), and the Science and Technology Research and Development Plan Project of China National Railway Group Co., Ltd. (2022CDA001).

Institutional Review Board Statement: Not applicable.

Informed Consent Statement: Not applicable.

Data Availability Statement: Data available on request due to restrictions.

Conflicts of Interest: Authors C.L. and K.M. are employed by the company CRRC Changchun Railway Vehicles Co., Ltd. The remaining authors declare that the research was conducted in the absence of any commercial or financial relationships that could be construed as a potential conflict of interest.

References

1. Ding, S.S. Research on Key Technology of High-Speed Train Body Design. Ph.D. Thesis, Beijing Jiaotong University, Beijing, China, 2016.
2. Zhang, Q.G.; Qiao, K.; Cao, Y. The development of foreign high-speed double decker EMU and its inspiration to China. *China Railw.* **2018**, *7*, 103–107. [CrossRef]
3. Liu, C.Q.; Wang, L. 400 km/h cross-country interconnection high-speed EMUs. *Electr. Drive Locomot.* **2020**, *2*, 1–6. [CrossRef]
4. Deng, H.; Zhang, G.Q.; Zhang, Y.; Wang, S.B.; Wang, C. Development Concept of Next Generation High-speed Intelligent EMU. *Urban Mass Transit* **2022**, *25*, 11–15.
5. Harte, A.M.; McNamara, J.F.; Roddy, I.D. A multilevel approach to the optimisation of a composite light rail vehicle bodyshell. *Compos. Struct.* **2003**, *63*, 447–453. [CrossRef]
6. Chiandussi, G.; Gaviglio, I.; Ibba, A. Topology optimisation of an automotive component without final volume constraint specification. *Adv. Eng. Softw.* **2014**, *35*, 609–617. [CrossRef]
7. Chen, B.Z.; Zhang, X.Q.; Qiu, G.Y. Topology Optimization of Underframe for 400km/h High-Speed Train. *Mach. Des. Manuf.* **2021**, *7*, 272–275. [CrossRef]
8. Zhang, Q.; Sun, Q.Z.; Liu, G.F. Finite Element Analysis and Topological Optimization of a Tracked Vehicle Body Based on Hypermesh. *J. Jiamusi Univ. (Nat. Sci. Ed.)* **2020**, *38*, 123–126.
9. Ji, B.K. Research on Topology Optimization Design of High-Speed Train Body Structure Based on CRH EMU. Master's Thesis, Sichuan University of Science & Engneering, Zigong, China, 2019.
10. Zhang, L.; Zhang, J.Y.; Li, T.; Zhang, Y.D. Multi-objective aerodynamic optimization design of high-speed train head shape. *J. Zhejiang Univ.-Sci. A* **2017**, *18*, 841–854. [CrossRef]
11. Zhang, L.; Dai, Z.Y.; Li, T.; Zhang, J.Y. Multi-objective aerodynamic shape optimization of a streamlined high-speed train using Kriging model. *J. Zhejiang Univ.-Sci. A* **2022**, *23*, 225–245. [CrossRef]
12. Deb, A.; Chou, C.; Dutta, U.; Gunti, S. Practical Versus RSM-Based MDO in Vehicle Body Design. *SAE Int. J. Passeng. Cars-Mech. Syst.* **2012**, *5*, 110–119. [CrossRef]
13. Chen, X.; Wang, Y.Q.; Sun, K.; Feng, Z.Y.; Zhao, W.H.; Lu, B.H. Multi-Objective Optimization for Section of Sandwich Plate Applied to High-Speed Train Compartments. *J. Xi'an Jiaotong Univ. Nat. Sci. Ed.* **2013**, *47*, 62–67.
14. Sahib, M.M.; Kovács, G. Elaboration of a Multi-Objective Optimization Method for High-Speed Train Floors Using Composite Sandwich Structures. *Appl. Sci.* **2023**, *13*, 3876. [CrossRef]
15. Miao, B.R.; Zhang, L.M.; Zhang, W.H.; Yin, H.T.; Jin, D.C. High-speed Train Carbody Structure Fatigue Simulation Based on Dynamic Characteristics of the Overall Vehilce. *J. China Railw. Soc.* **2010**, *32*, 101–108. [CrossRef]
16. Tang, Q.C.; Ma, L.; Zhao, D.; Lei, J.Y.; Wang, Y.H. A multi-objective cross-entropy optimization algorithm and its application in high-speed train lateral control. *Appl. Soft Comput.* **2022**, *115*, 108151. [CrossRef]
17. Wang, Q.S.; Zeng, J.; Shi, H.L.; Jiang, X.S. Parameter optimization of multi-suspended equipment to suppress carbody vibration of high-speed railway vehicles: A comparative study. *Int. J. Rail Transp.* **2023**, *2023*, 2291202. [CrossRef]

18. Wang, S.M.; Peng, Y.; Wang, T.T.; Che, Q.W.; Xu, P. Collision performance and multi-objective robust optimization of a combined multi-cell thin-walled structure for high speed train. *Thin-Walled Struct.* **2019**, *135*, 341–355. [CrossRef]
19. Peng, Y.; Hou, L.; Che, Q.W.; Xu, P.; Li, F. Multi-objective robust optimization design of a front-end underframe structure for a high-speed train. *Eng. Optim.* **2019**, *51*, 753–774. [CrossRef]
20. Lu, S.S.; Xu, P.; Yan, K.P.; Yao, S.G.; Li, B.H. A force/stiffness equivalence method for the scaled modelling of a highspeed train head car. *Thin-Walled Struct.* **2019**, *137*, 129–142. [CrossRef]
21. Wang, R.; Gao, S.M.; Wu, H.Y. Progress in Hexahedral Mesh Generation and Optimization. *J. Comput.-Aided Des. Comput. Graph.* **2020**, *32*, 693–708.
22. Papanicolau, G.; Bensoussan, A.; Lions, J.L. *Asymptotic Analysis for Periodic Structures*; Elsevier: New York, NY, USA, 1978; ISBN 978-0-444-85172-7.
23. Wang, J. Application and Analysis of Asymptotic Homogenization Method in Masonry. Master's Thesis, Xiangtan University, Xiangtan, China, 2013.
24. Guedes, J.; Kikuchi, N. Preprocessing and postprocessing for materials based on the homogenization method with adaptive finite element methods. *Comput. Method. Appl. M* **1990**, *83*, 143–198. [CrossRef]
25. Zhang, G.F.; Xu, L.; Li, D.S.; Yu, F.C. Research on Sensitivity Filtering of Continuum Topology Optimization. *Modul. Mach. Tool Autom. Manuf. Tech.* **2021**, *06*, 29–32. [CrossRef]
26. Xie, Y.M.; Steven, G.P. *Evolutionary Structural Optimization*; Berlin Springer-Verlag: London, UK, 1997; ISBN 978-3-540-76153-2.

 *applied sciences*

Article

Dynamic Characteristics of Urban Rail Train in Multivehicle Marshaling under Traction Conditions

Yichao Zhang, Jianwei Yang *, Jinhai Wang and Yue Zhao

Beijing Key Laboratory of Performance Guarantee on Urban Rail Transit Vehicles, Beijing University of Civil Engineering and Architecture, Beijing 100044, China
* Correspondence: yangjianwei@bucea.edu.cn

Abstract: In recent years, urban rail transportation has rapidly developed in China and become one of the most important modes of travel. Most existing studies on the dynamic characteristics of urban rail trains have been based on single-section trains, and there have been fewer studies on marshaling urban rail trains that incorporate traction transmission systems. The dynamic performance of each carriage directly affects the operational reliability and even the running safety of urban rail trains. For this reason, in this paper, a marshaling urban rail train model with a traction transmission system was established and its accuracy was validated by field tests. This dynamics model enables the consideration of the coupling interactions between the gear transmission motion, the vertical, the lateral and the longitudinal motions of the vehicle. First, the model accuracy was validated by field tests. Then, the relationship between the motor torque and the running time of the urban rail train under traction conditions was calculated. Finally, the dynamic performance of each car of the marshaling train was studied. The research results show that there is a clear difference between the dynamics of the motor car and the trailer, and that the motor car is significantly inferior to the trailer. Among the four motor cars, the dynamic performances of the first and last moving cars were worse than those of the other motor cars. Among the two trailers, the trailer at the back was worse than the trailer at the front. The traction transmission system has a greater impact on the vertical and lateral vibration of the train bogie frame and wheelset, but the impact on the vibration of the car body is negligible. This paper provides theoretical support for the research one train dynamic performance optimization and operation safety.

Keywords: urban rail train; multivehicle marshaling; traction transmission system; traction condition; dynamic characteristics

Citation: Zhang, Y.; Yang, J.; Wang, J.; Zhao, Y. Dynamic Characteristics of Urban Rail Train in Multivehicle Marshaling under Traction Conditions. *Appl. Sci.* **2023**, *13*, 3022. https://doi.org/10.3390/app13053022

Academic Editor: Suchao Xie

Received: 30 January 2023
Revised: 16 February 2023
Accepted: 21 February 2023
Published: 26 February 2023

1. Introduction

In China, with the rapid development of cities, urban rail transit has increasingly become the first choice for people to travel. Compared with other modes of transportation, urban rail transit has the advantages of being economic, safe, convenient, environmentally friendly, highly efficient and able to fundamentally improve the development of public transportation in the process of urban development. In the actual operation process of urban rail trains, they will experience rapid and frequent traction and braking, and each motor car and trailer will be subject to different degrees of shock and vibration. The above phenomenon occurs between different cars will directly affect the operational safety of urban rail trains, as well as their stability and stability. Therefore, it is particularly important that a multicar marshaling model of urban rail trains is established, considering the interaction between the carriages and the traction transmission system, and to analyze its dynamic characteristics.

For a long time, the operational safety problems caused by the dynamic characteristics of urban rail vehicles have attracted extensive attention from scholars. Alexander et al. [1] established a fine finite element model of the car body and the bogie system and studied and analyzed the vibration transmission path of the bogie structure caused by the excitation of the wheel and rail. Simson et al. [2] studied the three-axle bogie locomotive, and the bending performance of the bogie under different track excitations is compared and analyzed. Ribeiro et al. [3] conducted modal test analysis on the closing part of the high-speed train bogie by the modal test method and corrected the finite element model according to the analysis results. Wang et al. [4] studied the time domain response reconstruction method based on using the wavelet transform to obtain the responses of multiple positions, and the vibration transmission paths between the train bogie and the car body are sorted. Ling et al. [5] studied the influence of rail corrugation on the vibration of subway bogies and proposed measures to reduce the vibration of subway bogies and reduce the failure rate of primary suspension springs, which provided a basis for the development of new subway bogies and track maintenance. The traction transmission system plays the role of transmitting force and motion in urban rail trains, and its dynamic performance has a significant impact on the stability and safety of train operations. Traction transmission failures can even lead to catastrophic consequences such as derailment or train collisions. Yang et al. [6,7] researched the time-domain and time–frequency domain responses of the gear system of a vehicle with the wheel flat under variable speed conditions, and the influence of wear parameters on the nonlinear dynamics of gear systems. Wang et al. [8] presented a novel analytical model of TVMS for profile-shifted spur gears, and carried out numerical simulations to analyze the individual and compound shifts. Wang et al. [9] established a gear system model of a railway vehicle, which they proposed in consideration of its time-varying mesh stiffness, nonlinear backlash, transmission error, time-varying external excitation and rail irregularity to investigate the nonlinear dynamics of a time-varying gear system. Wang et al. [10] studied the inner/outer races failure mechanism of axle box bearing. Zhang et al. [11] studied the dynamic responses of the electromechanical coupling model under variable conditions, and the dynamic characteristics of the traction motor were revealed by numerical analysis. Li et al. [12] studied the contribution of each vibration path of the gearbox, determined that the right bearing of the gearbox was the main transmission path of vibration through vibration tests and obtained the transmission law of the vibration transmission path of the gearbox. Zhang et al. [13] established a dynamic model considering the traction motor and gearbox. By comparing the simulation results with the experimental results in the time domain and time–frequency domain and analyzing the gear meshing force and wheel–rail force, it was found that compared with the wheelset, the gear meshing force has a greater impact on the dynamics of the traction motor and gearbox. Wang et al. [14] established the vehicle dynamics model that integrated the flexible gearbox housing, time-varying mesh stiffness and nonlinear gear backlash. The effect of wheel out-of-roundness on the dynamic response of the traction transmission system was studied, and the results show that wheel wear can cause the severe and complex torsional vibration of the gear drive system. Chen et al. [15–17] embedded the gear transmission system into the vehicle–track coupling model and considered complex excitations such as gear time-varying mesh stiffness, nonlinear backlash, nonlinear wheel–rail normal contact force and creep force. The vibration responses of the train and gear transmission system under braking conditions were in good agreement with the actual line test results. Zhang et al. [18] established a vehicle-track coupling model and compared and analyzed the models considering and not considering the gear transmission system. The gear system adopts spur gears and verifies the vertical and longitudinal vibration of the gear transmission system to the wheelset and traction motor. There is an obvious effect, but the effect on the lateral vibration of the bogie frame is not obvious. Wang et al. [19] conducted a non-smooth dynamic analysis of the wheelset system. The relationship between the time–frequency dynamic characteristics, slip velocity and nonlinear interaction force between the wheel and rail was studied. Zhou et al. [20,21] established an electromechanical coupling model and verified that the electromechanical coupling effect can be applied to monitor the faults of rotating components in the transmission path of the train transmission system. In addition to heavy-duty locomotives, Wang et al. [22] conducted similar research in the field of high-speed EMUs. Zhang et al. [23] established fully translational and fully rotational vehicle–track coupling models, studied the gear vibration mechanism through the time–frequency analysis method and analyzed the composition of the frequency band components considering gear excitation. Wang et al. [24] established a plane vibration model of the train gear system considering the effect of adhesion and studied the bifurcation characteristics of the gear system at different speeds. Huang et al. [25] established the torsional vibration model of a high-speed train gear system and studied the superharmonic resonance problem and the Hopf bifurcation problem of the system by a semi-analytical method. Kia et al. [26] conducted experiments on a train traction system and found that the vibration characteristics of the traction motor, bearing and gearbox have a significant impact on the motor current. Huang et al. [27] established a high-speed train dynamics model based on the SIMPACK software. The research results show that gear meshing excitation has a significant impact on motor vibration. The running stability and safety of trains are also important research directions. Based on the flexible vehicle–track coupled dynamic model, Chen et al. [28] studied the ride comfort on the straight lines and curves under different excitations, as well as the vibration of the car body at different positions found that the riding comfort is very sensitive to longwave excitation but not to shortwave excitation, and it is not sufficiently accurate to evaluate the ride comfort of subway vehicles only based on the vibration of the center of the floor. Pradhan et al. [29] established a train dynamics model based on the ADAMS VI-Rail software, which included the first five modes of the car body, and studied the comfort of the train.

In addition, the single vehicle model cannot accurately reflect the dynamic characteristics of the train system, and therefore, Belforte et al. [30] established a numerical calculation method for studying the dynamics of heavy freight trains, and the influence of train composition on the operational safety was demonstrated through numerical analysis. Tao et al. [31] established a vehicle–track rigid-flexible coupling dynamic model considering two carriages, studied the influence of wheel–rail interaction caused by the out-of-roundness of the wheel of one wheelset on the wheel–rail interaction of the other seven wheelsets, and found that the wheel–rail interaction dynamic interactions can be transferred from an out-of-round wheelset to another wheelset of the same bogie via the rails. However, more research has focused on the coupler. Wu et al. [32] established a train model consisting of two eight-axle trains and one simplified carriage, and analyzed the working principles and differences between the two coupler systems. Ding et al. [33] established a nonlinear finite element model of the collision between the train and the detailed coupler and studied the collision characteristics of the coupler under different working conditions. Yadav et al. [34] established a simplified and degree-of-freedom model of the coupler to study the effect of the coupler clearance on the longitudinal dynamics of the train. Zou et al. [35] established different marshaling train models, compared the longitudinal dynamic characteristics of the train under different rescue conditions, and analyzed the effect of the longitudinal impulse on the coupler and vehicle dynamic performance. However, the influence of the traction transmission system on the dynamic characteristics of marshaling trains is still unclear at present, so it is difficult to obtain a more accurate dynamic response of marshaling trains under variable speed conditions, and it is difficult to provide an effective reference for traffic safety, optimal design and fault diagnosis.

As a result, in this paper, on the basis of considering the traction transmission, the dynamic model of the marshaling urban rail train is established, and the dynamic characteristics of each carriage under the traction conditions of the train are studied. The specific research contents are as follows: Section 2 introduces the dynamic model of the marshaling urban rail train. Section 3 validates the model. Section 4 analyzes and discusses the dynamic characteristics. Section 5 summarizes the full text.

2. Dynamic Modeling

One effective means of solving the dynamics problem of a complex mechanical system is to establish its physical model and analyze it. In this paper, the dynamic model of a domestic urban rail train is established by the SIMPACK software. The train adopts the form of six carriages, and the traction transmission system is considered. Figure 1a shows the connection mode and structural characteristics of the various components of the train. When building this model, all parts were treated as rigid bodies. Figure 1b shows the modeling process.

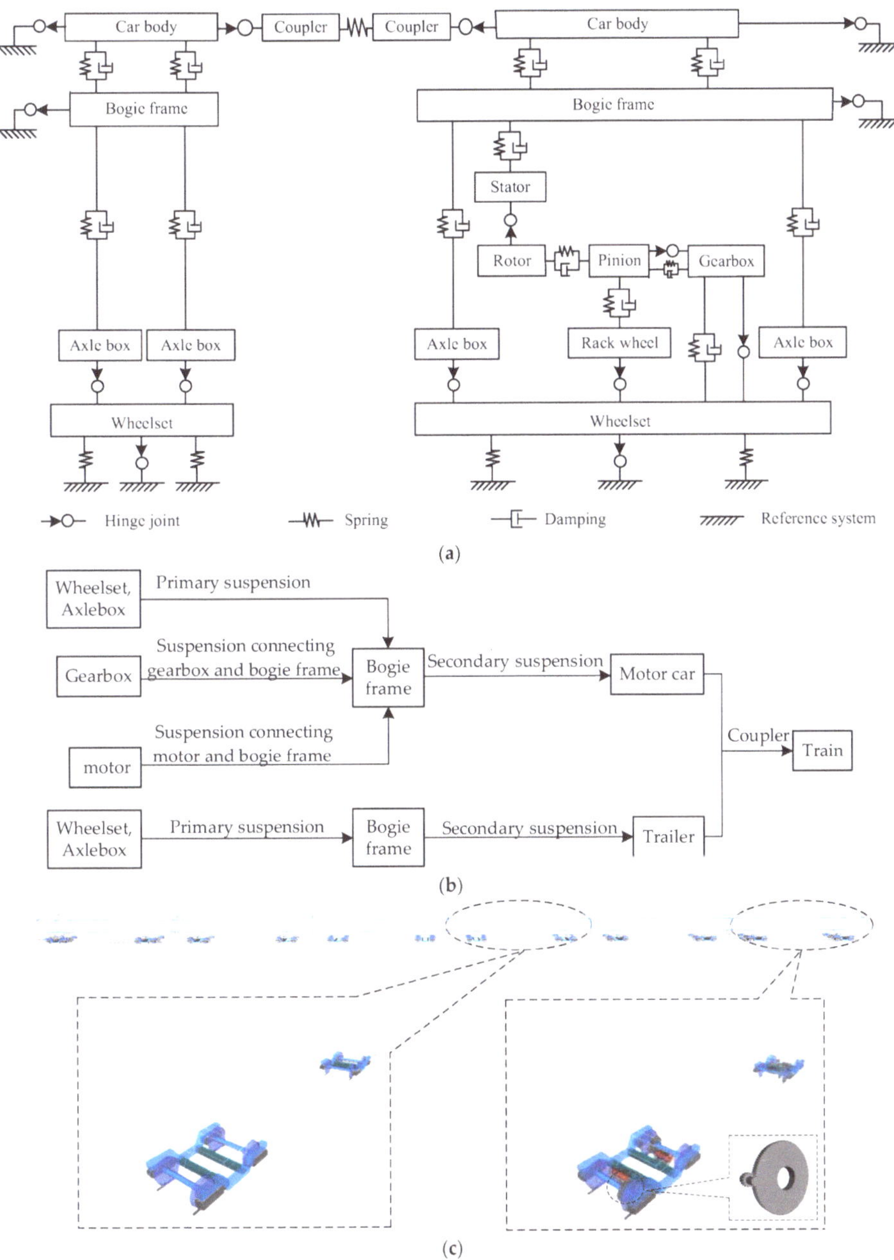

Figure 1. Marshaling urban rail train dynamics model with the traction transmission system: (**a**) Topology model of the urban rail train; (**b**) The modeling flow chart of the marshaling urban rail train; (**c**) The marshaling urban rail train dynamics model.

The trailer model consists of four wheelsets, eight axle boxes, two bogie frames and a car body. The wheelset is connected to the bogie frame through the primary suspension and then connected to the car body through the secondary suspension. The primary suspension and the secondary suspension were represented by springs and dampers in the model. Some parameters of the trailer are shown in Table 1.

Table 1. Main parameters of the urban rail train dynamics model.

Specification	Value
Car body mass (kg)	4.08×10^4
Bogie frame mass (kg)	3188
Wheelset mass (kg)	1640
Axle box mass (kg)	85.367
Gearwheel mass (kg)	53.15
Pinion mass (kg)	5.15
Gearbox mass (kg)	149.75
Rotor mass (kg)	178..36
Motor mass (kg)	422.82
Rotational inertia of car body $x/y/z$ (t·m^2)	75.06/2277.4/2277.4
Rotational inertia of bogie frame $x/y/z$ (kg·m^2)	2040/2710/3460
Rotational inertia of wheelset $x/y/z$ (kg·m^2)	725/100/725
Rotational inertia of axle box $x/y/z$ (kg·m^2)	1.455/2.448/2.011
Rotational inertia of gearwheel mass $x/y/z$ (kg·m^2)	4.55/4.85/4.555
Rotational inertia of pinion $x/y/z$ (kg·m^2)	0.006/0.007/0.006
Rotational inertia of gearbox $x/y/z$ (kg·m^2)	4.22/10.45/8.56
Rotational inertia of rotor $x/y/z$ (kg·m^2)	24.5/1.9/24.5
Rotational inertia of motor $x/y/z$ (kg·m^2)	77.5/24.7/75.2
Stiffness of primary suspension $x/y/z$ (N/m)	$9.2 \times 10^6/8 \times 10^6/1.5 \times 10^6$
Stiffness of secondary suspension $x/y/z$ (N/m)	$2.06 \times 10^5/2.06 \times 10^5/4.41 \times 10^5$
Stiffness between motor and bogie frame $x/y/z$ (N/m)	$3 \times 10^7/1 \times 10^7/3 \times 10^7$
Stiffness between gearbox and bogie frame $x/y/z$ (N/m)	$3 \times 10^6/3 \times 10^6/3 \times 10^6$
Damping coefficient of primary suspension $x/y/z$ (N·s/m)	5560/5560/1800
Vertical damping coefficient of secondary suspension (N·s/m)	6×10^4
Damping coefficient between motor and bogie frame $x/y/z$ (N·s/m)	$1 \times 10^3/1 \times 10^3/1 \times 10^3$
Damping coefficient between gearbox and bogie frame $x/y/z$ (N·s/m)	$3 \times 10^5/2 \times 10^5/3 \times 10^5$

The motor car model including four wheelsets, eight axle boxes, four gearboxes, four traction motors, two bogie frames and a car body. Compared with the trailer, the motor car has added a traction transmission system. The traction transmission system is composed of a traction motor, gearwheel, pinion and gearbox casing. The traction torque is first transmitted to the pinion through coupling and then transmitted to the gear through the meshing of the gearwheel. When establishing the gear system, the meshing of the gearwheel and the pinion is defined by force element No. 225, which can accurately describe the Young's modulus, Poisson's ratio and damping coefficient of the gear pair and establish the profile parameters. The geometric characteristic parameters are shown in Table 2. The traction motor consists of a stator and a rotor. The stator model is first established and connected to the bogie frame through the motor hanger. The rotor is then modeled and hinged to the stator. After that, the rotor is connected with the pinion through the coupling. The coupling is defined by force element No. 43, and reasonable stiffness and damping are set to ensure the torque transmission function of the coupling. The construction of the other parts of the motor car is the same as that of the trailer, and some parameters of the motor car are shown in Table 1.

Table 2. Main parameters of gear model.

Specification	Value
Tooth number of pinion/gear	16/107
Modification coefficient of pinion/gear (mm)	0.31449/−0.07614
Face width of pinion/gear (mm)	70/70
Module (mm)	85.367
Pressure angle (°)	20
Helix angle (°)	17
Poisson ratio	0.3
Young modulus (GN/m)	206
Damping coefficient (kN·s/m)	5

The marshaling train model is shown in Figure 1c, which adopts the marshaling form of "four motor cars + two trailers": M1–M2–T1–T2–M3–M4 (where M1, M2, M3 and M4 are motor cars and T1 and T2 are trailers). Here, the x axis is longitudinal, the y axis is lateral and the z axis is vertical. Each car is connected by a coupler, and the coupler is defined by force element No. 4.

3. Model Validation

To verify the accuracy and reliability of the urban rail train dynamic model, the measured tracking data and simulation data of a subway line in Beijing were used for comparative analysis, mainly to compare the vertical acceleration of the axle box and the gearbox box. In the actual test train, the fourth vehicle was selected as the test vehicle, the acceleration sensor was arranged on the axle box and the gearbox box for the tracking test, and the acceleration data of the corresponding components were obtained. The sensor uses a three-axis vibration acceleration sensor and the principle is piezoelectric. The gear box uses a vibration acceleration sensor with a range of 500 g, and the axle box uses a vibration acceleration sensor with a range of 50 g. The location of the sensor is above the axle box and to the left of the gear box case. The direction of the sensor: x axis is longitudinal, y axis is transverse and z axis is vertical. The sampling frequency is 10 K Hz. The scene photo is shown in Figure 2.

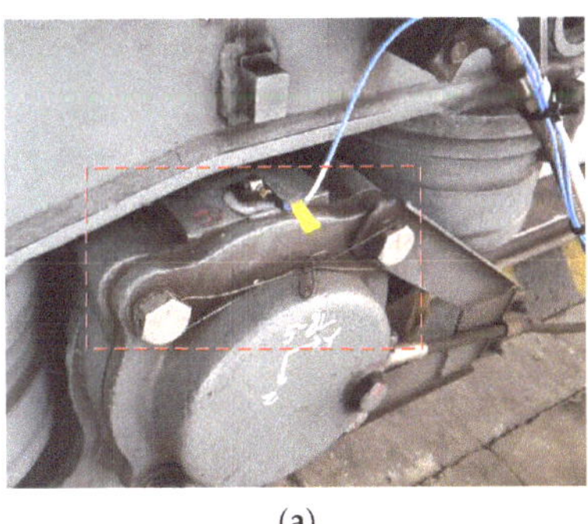
(a)

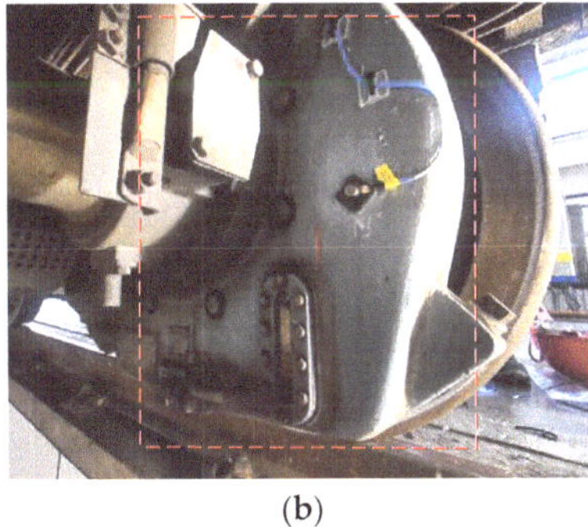
(b)

Figure 2. Sensor layout in the field test. (**a**) Layout of sensors on axle box, and the red circle indicates that the axle box; (**b**) Layout of sensors on gearbox, and the red circle indicates that the gearbox.

Compared with the actual operating conditions of urban rail trains, the simulated train parameters have certain differences which have a certain impact on the simulation results. The simulation model uses the measured track spectrum of a line in Beijing as the track excitation. The train simulation data and measured data are shown in Figure 3. Because the working conditions of urban rail trains in actual operation are very complex, they will experience continuous acceleration, coasting and braking, and will continue to pass through curves and turnouts, and the measured map will have a larger rail impact.

Therefore, the axle box, as the component with the closest contact distance with the wheel and rail, has the most obvious impact on the rail in the measured data. Since this situation is not considered in the simulation model, the overall vibration acceleration of the simulation model is smaller than the measured data. In this paper, the actual acquired vibration signals are low-pass filtered using a digital filter to filter out the high-frequency noise signals in the acquisition process. The comparison between the vertical dynamics simulation data of the urban rail train and the measured data is shown in the figure. Although there is a certain difference between the simulation data and the test data, the overall vibration characteristics are in good agreement with the measured data, so the established simulation model is more reasonable.

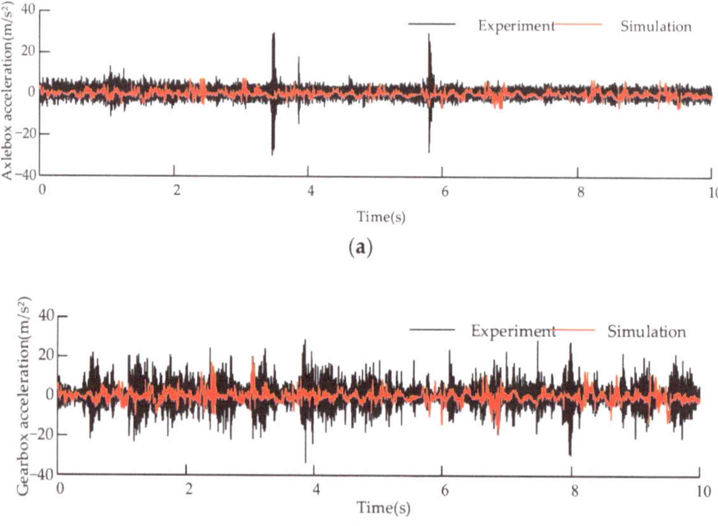

Figure 3. Comparison of the experimental data and simulation data in the vertical direction: the time series (**a**) on the axle box and (**b**) gearbox.

4. Analysis and Discussion

The working conditions studied in this paper adopt the linear traction working conditions, and the track spectrum is based on the measured data of a line in Beijing, the time is 40 s and the sampling frequency is 5000 Hz.

The traction motor of an urban rail train adopts an AC asynchronous motor, and the working speed range of the motor is divided into two parts: a constant torque area and a constant power area. When the traction motor works in the constant torque range, the output torque remains unchanged, and the working range is between the starting speed and the synchronous speed; when the traction motor works in the constant power range, the output power remains unchanged, and the output torque and speed have an inversely proportional relationship. According to the data provided by the traction motor, the motor output torque is obtained as a piecewise function:

$$T_s = \left\{ \begin{array}{ll} T_e & n \leq n_s \\ 9550\frac{P_e}{n} & n > n_s \end{array} \right. \tag{1}$$

where T_s is the driving torque; T_e is the rated torque; P_e is the rated power; n is the traction motor speed; and n_s is the motor synchronous torque.

The traction characteristic curve of the urban rail train is shown in Figure 4a. The urban rail train will be affected by the traction force and basic resistance of the motor under traction conditions. According to the literature [9], the speed curve of the urban rail train under traction conditions can be obtained, as shown in Figure 4b. The train performs a uniform acceleration motion in 0–15.7 s, the working range of the traction motor is the constant torque range and the speed reaches 42.6 km/h; the train performs variable acceleration motion in 15.7–70 s and the working range of the traction motor is the constant power range.

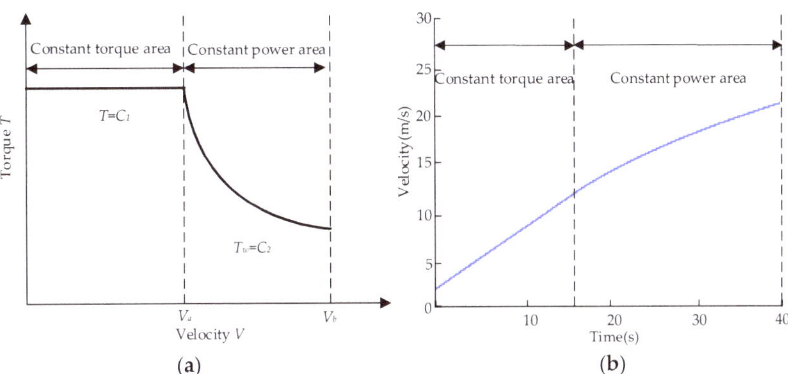

Figure 4. (**a**) Traction characteristic curve; and (**b**) speed curve under traction condition.

The evaluation index of the train's dynamic performance is an important basis for evaluating the dynamic characteristics of the train. When evaluating the safety of train operation, the derailment coefficient, wheel load reduction rate and other indicators are generally comprehensively evaluated. In this section, based on the dynamic simulation results of urban rail trains under traction conditions, evaluation indicators such as wheel–rail vertical force, wheel–rail lateral force, the derailment coefficient and the rate of wheel load reduction in six carriages are compared and analyzed, and the influence of train dynamics is analyzed.

4.1. Analysis of Safety and Critical Velocity

4.1.1. Security Analysis

Figure 5a,b show the wheel–rail vertical force and wheel–rail lateral force of each car of the urban rail train under traction conditions, respectively. The simulation data from the position of the axle box divides the 40 s traction condition into eight segments every 5 s and takes the peak value of each segment for comparative analysis. This method is adopted in the subsequent analysis in this section. The wheel–rail vertical force and the wheel–rail lateral force of the six carriages are monotonically increasing. The wheel–rail vertical force of M1 is the largest, and the variation range is 69.91–77.17 KN. The wheel–rail lateral force of M2 is the largest, and the variation range is 2.27–5.96 KN. The wheel–rail vertical force of T1 is smaller than that of T2. The wheel–rail vertical force of the four-section locomotive is significantly greater than that of the two-section trailer because the total mass of the locomotive is greater than that of the trailer and because of the effect of the traction transmission system on the train bogie.

Figure 6 shows the variation in the derailment coefficient of urban rail trains. The derailment coefficients of all cars are monotonically increasing. The derailment coefficient of M1 is the largest, followed by that of M4. The first car and the last car of the train are only connected to one car, which is greatly affected by the longitudinal impact of the train and has poor dynamic performance. The derailment coefficients of the four-section EMUs are significantly greater than those of two-section trailers, indicating that the trailers have better dynamic performance. T1 is smaller than T2, and the variation ranges are 0.016–0.048 and 0.019–0.051, respectively, indicating that the dynamic performance of the trailer in the front is better than that in the rear.

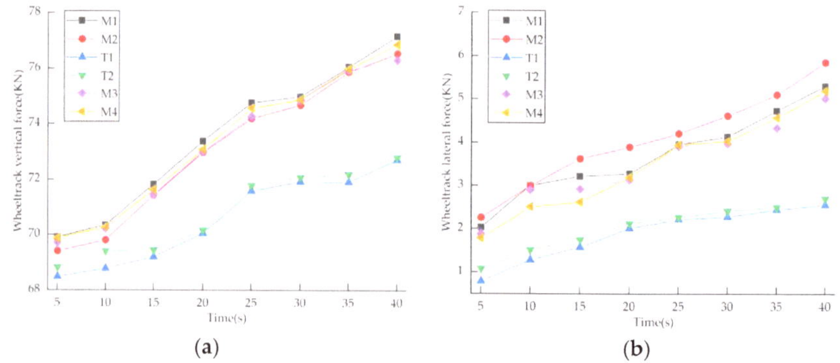

Figure 5. Wheel–rail force of urban rail train in operation: (**a**) wheel–rail vertical force; and (**b**) wheel–rail lateral force.

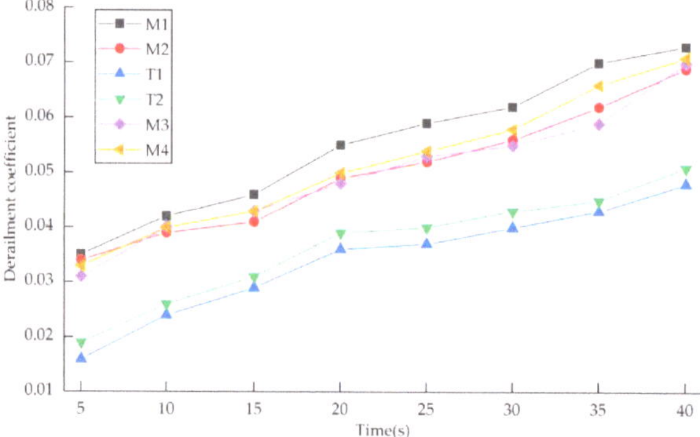

Figure 6. Derailment coefficient.

Figure 7 shows the change in the wheel load reduction rate of each car of the urban rail train. The wheel load reduction rate of all cars is monotonically increasing. The wheel load reduction rate of M1 is the largest, followed by M4, and their variation ranges are 0.050~0.090 and 0.049~0.089, respectively, whilst the wheel load reduction rate of the four-segment trains is significantly larger than those of the two-segment trailers, T1 is smaller than T2 and their variation ranges are 0.033~0.070 and 0.035~0.073, respectively. The variation characteristics of the wheel load reduction rate of urban rail trains are similar to the derailment coefficient, which verifies the previous analysis.

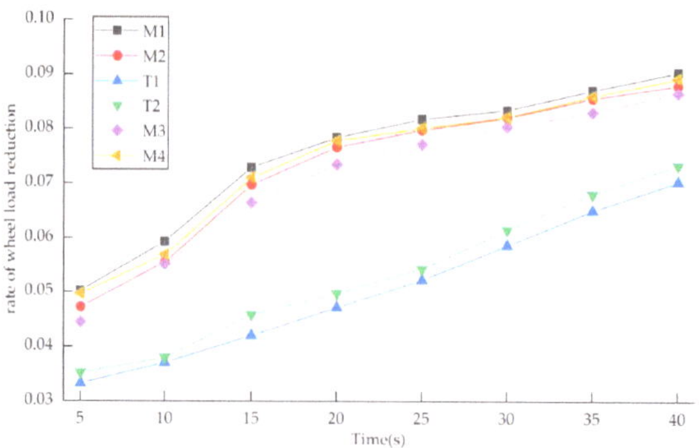

Figure 7. Rate of wheel load reduction.

4.1.2. Nonlinear Critical Speed Analysis

The critical speed is divided into a linear critical speed and a nonlinear critical speed. In general, the linear critical speed is 10~28% faster than the nonlinear critical speed. Therefore, calculating the critical speed of the train only needs to calculate its nonlinear critical speed. The nonlinear critical speed needs to give the train an initial speed that is much larger than the actual running speed and at the same time apply a force opposite to the running direction of the train body to make the train perform a uniform deceleration motion, let the train pass through a small section of lateral track excitation and then cancel so that the train continues to run on a smooth track. The nonlinear critical speed of the train is judged by observing whether the traverse amount of the wheelset converges in the postprocessing, and the minimum value of the convergence of the traverse amount of the wheelset is the nonlinear critical speed of the train.

The nonlinear critical speed simulation selects 300 km/h as the initial speed of the train, which is significantly greater than the actual running speed of the train. The simulation results of the single-section trailer are shown in Figure 8. The lateral offset of the four wheelsets of the single-section trailer model is within ±10 mm, and the convergence speed of the lateral offset of the four wheelsets is basically the same. Therefore, it can be concluded that the nonlinear critical speed of the model is 223 km/h.

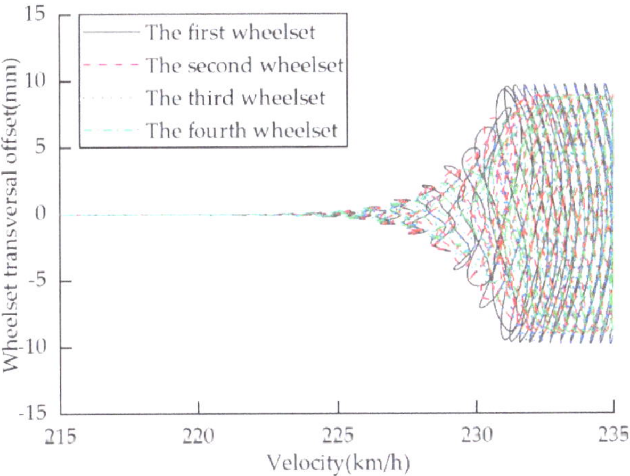

Figure 8. The nonlinear critical speed of the single trailer.

The simulation results of the marshaling train are shown in Figure 9. Figure 9 shows the "M1, M2, T1, T2, M3, M4" carriages of the six-car marshaling urban rail train. As a result, the nonlinear critical speed of each car relative to the single train model is different. The critical speeds of the T1 and T2 trailers are 217.5 km/h and 215.8 km/h, respectively, which are lower than the critical speed of the single-segment model. The critical speeds of the four motor cars M1, M2, M3 and M4 are 181.5 km/h, 183.3 km/h, 183.1 km/h and 181.2 km/h, respectively. Compared with the trailer, the motor vehicle has different parameters and more components, such as a gearbox and traction motor, so the critical speed of the motor vehicle is lower than that of the trailer. The critical speeds of M1 and M4 are lower than those of the other two, indicating that the dynamic performances of M1 and M4 are poor; the critical speed of T1 is greater than that of T2, indicating that the dynamic performance of T1 is better than that of T2, which verifies the previous analysis.

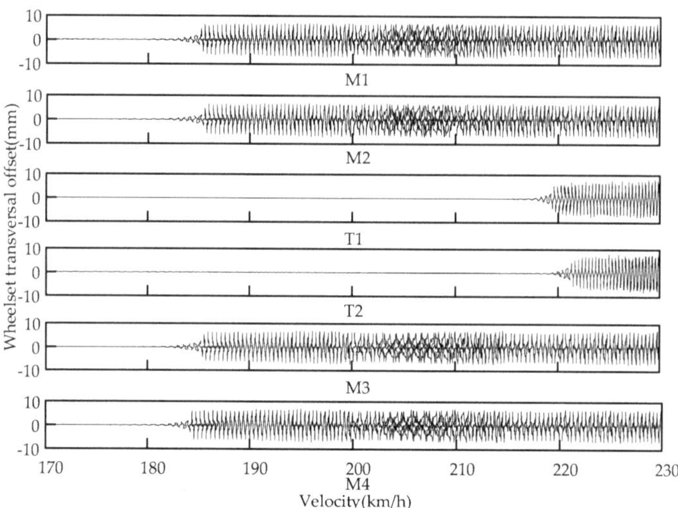

Figure 9. Nonlinear critical speed of each car of the marshaling urban rail train. Each curve represents the convergence rate of transverse movement of each carriage respectively.

4.2. Vibration and Comfort Analysis of Urban Rail Trains

4.2.1. Vibration Response Analysis

Figure 10a,b show the vertical acceleration and lateral acceleration of each carriage body, respectively. The vertical acceleration and lateral acceleration of the six-carriage car increase monotonically. In terms of vertical acceleration, the motor car is larger than the trailer, but the difference is not obvious. The variation range of M1 is the largest, in the range of $0.018{\sim}0.385$ m/s^2. T1 is the smallest, with a variation range of $0.011{\sim}0.365$ m/s^2. In terms of lateral acceleration, there is no obvious difference between the carriages, and the overall variation range is $0.022{\sim}0.472$ m/s^2.

Figure 10c,d show the vertical and lateral accelerations of the bogie frames of each car, respectively. The vertical and lateral accelerations of the six carriages monotonically increase. The vibration accelerations of M1 and M4 are relatively large, the vertical acceleration ranges are $0.262{\sim}1.523$ m/s^2 and $0.239{\sim}1.512$ m/s^2, respectively, and the lateral acceleration ranges are $0.083{\sim}1.062$ m/s^2 and $0.086{\sim}1.042$ m/s^2, respectively. The vibration acceleration of the four motor cars is greater than that of the two trailers because the traction motor will generate large vibration during operation, and the gear mechanism adopts a helical gear which will generate large vertical vibration and lateral vibration during operation. The vertical vibration acceleration has a significant effect.

Figure 10e,f show the vertical and lateral accelerations of the wheelsets of each carriage, respectively. The vertical and lateral accelerations of the six carriages monotonically increase. The vertical and lateral accelerations of the four motor cars are both greater than those of the two trailers, indicating that the traction transmission system has an obvious influence on the vibration acceleration of the train wheels. The ranges of the vertical and lateral accelerations of the motor car are $0.26{\sim}13.86$ m/s^2 and $0.15{\sim}11.84$ m/s^2, respectively, and the ranges of the vertical and lateral accelerations of the trailer are $0.13{\sim}3.47$ m/s^2 and $0.11{\sim}3.01$ m/s^2, respectively. The vibration acceleration of the axle box is significantly greater than that of the bogie frame, and the vibration acceleration of the bogie frame is significantly greater than that of the car body, indicating that the primary suspension and the secondary suspension suppress most of the vibration.

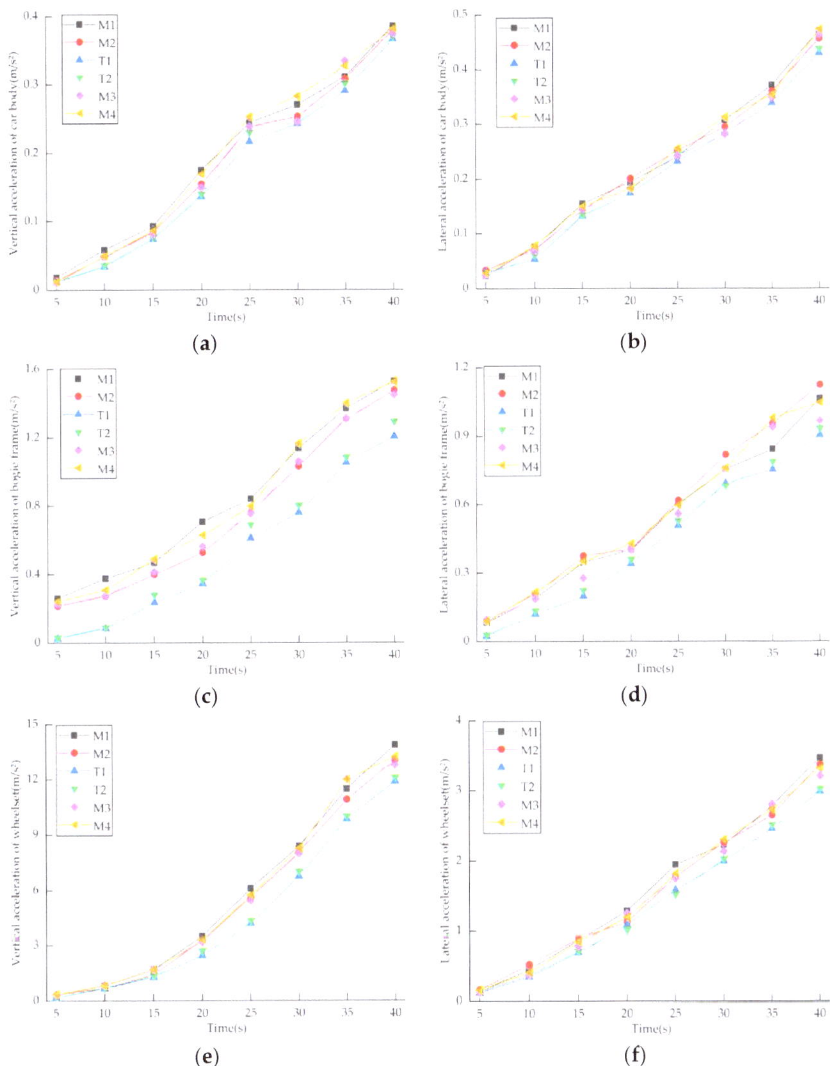

Figure 10. Vibration acceleration of each component of the marshaling urban rail train: (**a**) vertical acceleration of car body; (**b**) lateral acceleration of car body; (**c**) vertical acceleration of bogie frame; (**d**) lateral acceleration of bogie frame; (**e**) vertical acceleration of wheelset; and (**f**) lateral acceleration of wheelset.

The traction transmission system will generate a large vertical vibration during operation, which will affect the vibration response of each component of the train. Therefore, it is also important to compare and study the vibration frequency of each component of the motor car and the trailer. A short-time Fourier transform is performed on the vibration signals of the car body, bogie frame and axle box of the two adjacent carriages M2 and T1, and the time–frequency diagrams are obtained [36,37], as shown in Figures 11 and 12.

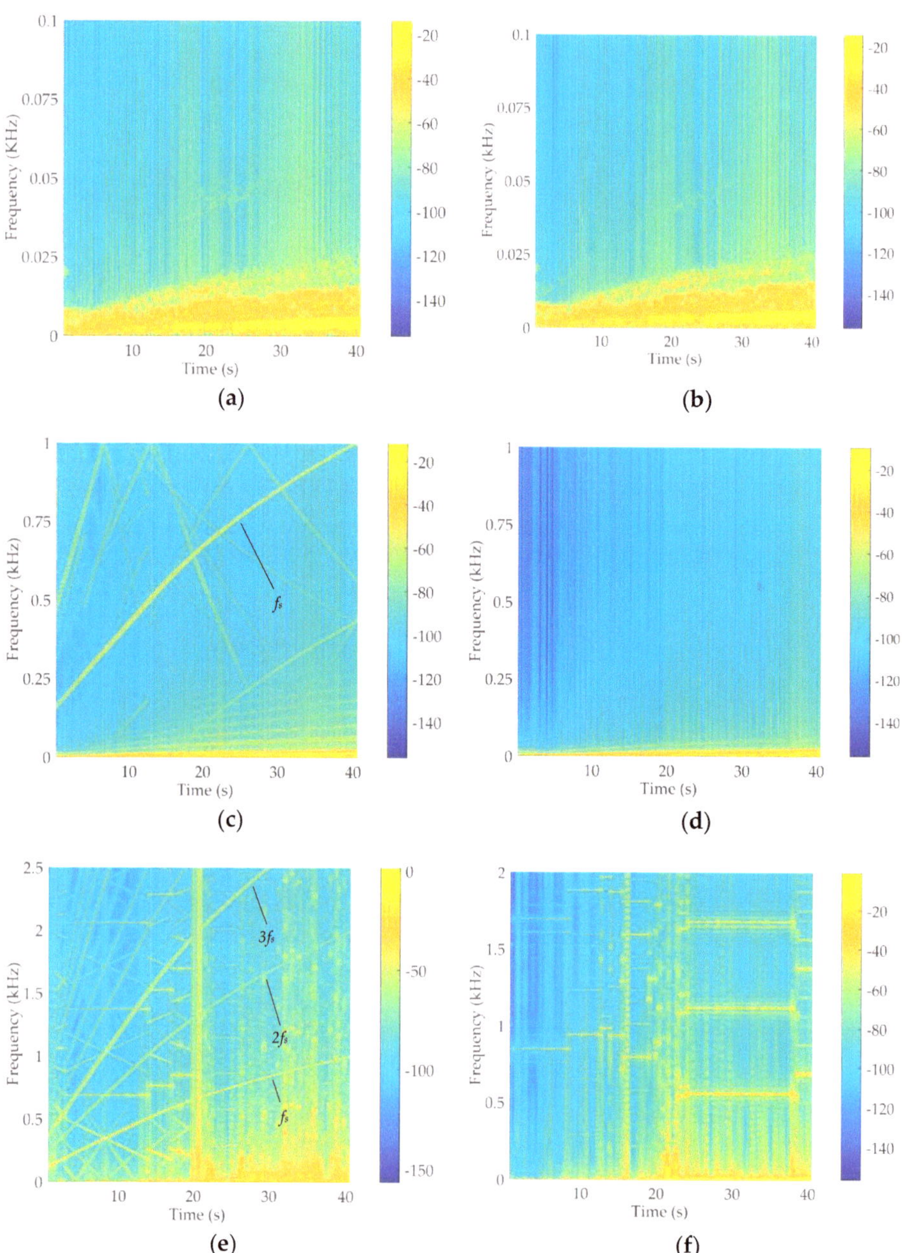

Figure 11. Time–frequency diagram of the vertical vibration acceleration of the motor car and trailer: (**a**) car body of the motor car; (**b**) car body of the trailer; (**c**) bogie frame of the motor car; (**d**) bogie frame of the trailer; (**e**) wheelset of the motor car; and (**f**) wheelset of the trailer.

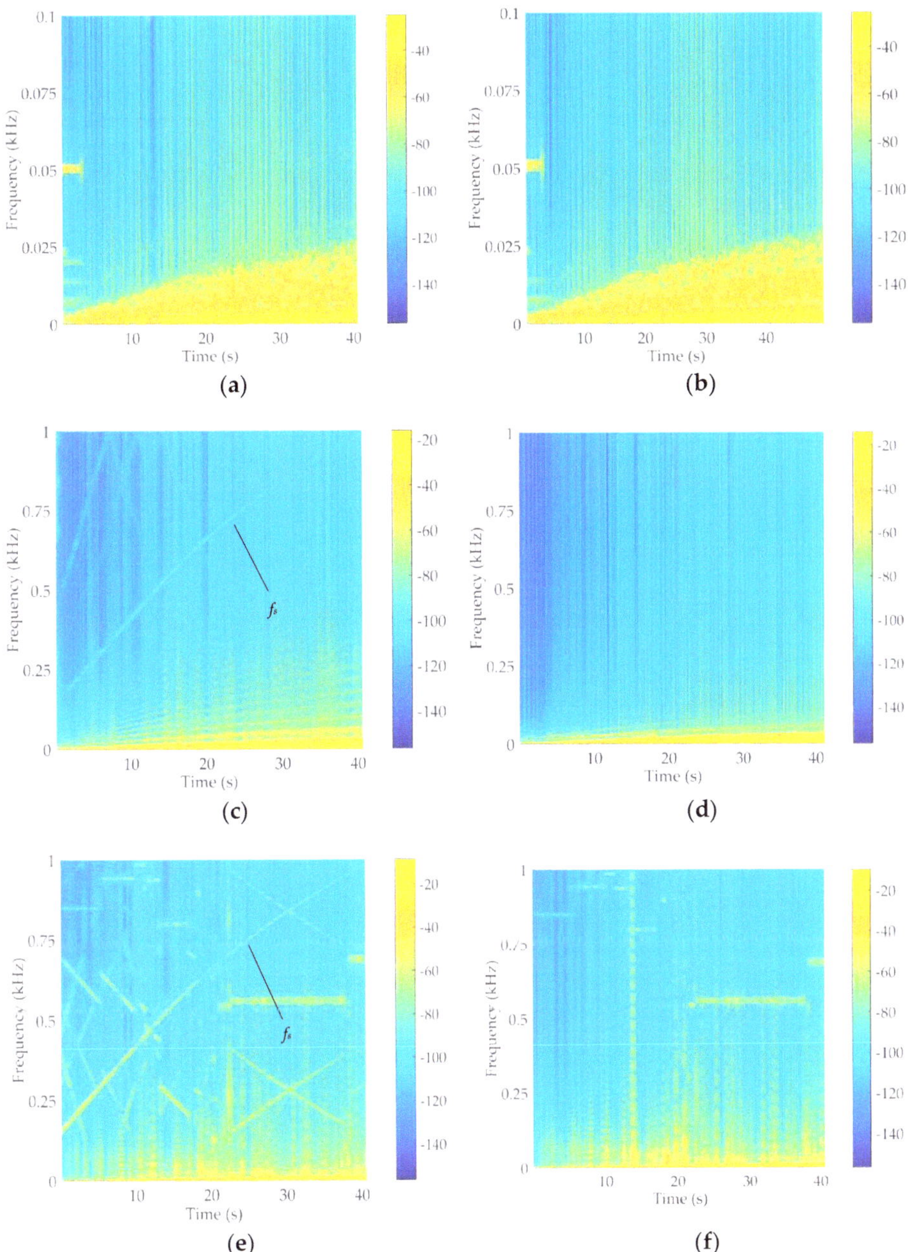

Figure 12. Time–frequency diagram of the lateral vibration acceleration of the motor car and trailer: (**a**) car body of the motor car; (**b**) car body of the trailer; (**c**) bogie frame of the motor car; (**d**) bogie frame of the trailer; (**e**) wheelset of the motor car; and (**f**) wheelset of the trailer.

Figures 11 and 12 show the time–frequency plots of the vertical and lateral vibration accelerations, respectively, of the M2 and T1 car bodies, bogie frames and wheelsets. There is an obvious energy bar for the vertical vibration and lateral vibration of the M2 car body, with a frequency band in the range of 0~25 Hz. This shows that the traction transmission system has little influence on the vibration frequency band of the motor car and trailer body. The vibration difference between the M2 and T1 bogie frames is obvious. The vertical vibration and lateral vibration energy bars of M2 are mainly concentrated in the ranges of 0~250 Hz and 0~300 Hz, and the meshing frequency of the gearbox can clearly be seen, indicating that the vibration of the gearbox will be obviously transmitted to the bogie frame. The energy bars of the vertical vibration and lateral vibration of the T1 bogie frame are mainly concentrated in 0~100 Hz. The vertical vibration and lateral vibration energy bars of the M2 and T1 wheel sets are mainly concentrated in the low-frequency range of 0~50 Hz. M2 clearly shows the meshing frequency and frequency multiplication of the gearbox, which is due to the hinge joint of the large gear and the axle. The vibration generated by the meshing of the gears will be directly passed to the wheelset. The meshing frequency of the gears is denoted by f_s, the double frequency is denoted by $2f_s$, and the triple frequency is denoted by $3f_s$. The frequency multiplication response wheel-to-body octave response gradually decreases because of the gradual reduction in the effect of the nonlinear term generated by the nonlinear forces during the Fourier transform. The traction transmission system has a significant impact on the low-frequency vibration of the bogie frame, and the meshing frequency of the gearbox will be significantly transmitted to the bogie frame and the wheelset.

4.2.2. Stability Analysis

The evaluation of the train stability mainly refers to the vibration acceleration, vibration amplitude, vibration frequency and vibration duration of the train body. The running quality is determined by the parameters of the vehicle itself, which is the most widely used dynamic performance of trains in the world. Sperling index is one of the indicators to evaluate the train stability, so this index is also the running stability evaluation Chinese standard specified in GB/T 5599-2019.

The sperling index W_s can be expressed as

$$W_s = 0.896 \sqrt[10]{\frac{a^3}{f} F(f)} \tag{2}$$

where a is the vibration acceleration, cm/s^2; f is the vibration frequency, Hz; and $F(f)$ is a weighting factor that considers the sensitivity of the human body to various frequencies of vibration. Here, $W_s \leq 2.5$ is excellent grade, $W_s \leq 2.75$ is good grade and $W_s \leq 3.0$ is a pass grade.

Figure 13 shows the vertical sperling index and the lateral sperling index of each car of the urban rail train during different time periods. In terms of the vertical sperling index, each carriage monotonically increases within 0~15 s and 20~40 s, and the value of the sperling index decreases within 15~20 s because the working condition of the traction motor changes from constant rotation during this time period. The torque area becomes a constant power area, and the torque output by the motor is reduced to optimize the sperling index of each carriage. With increasing vehicle speed, the influence caused by the traction motor gradually decreases, and the value of the stability index gradually increases. The change rate of the four motor cars within 15~20 s is greater than that of the two trailers, indicating that the traction motor has a greater impact on the EMU's stability index. The value of the vertical sperling index of M1 is the largest, followed by M2, with variation ranges of 1.11~2.25 and 1.09~2.23, respectively. In terms of the lateral sperling index, the values of the lateral sperling index of all carriages monotonically increase, and the difference in the lateral sperling index values of the six carriages is small. The vertical sperling index and lateral sperling index of each car of the urban rail train are both less than 2.5, so the stability level of the urban rail train is excellent.

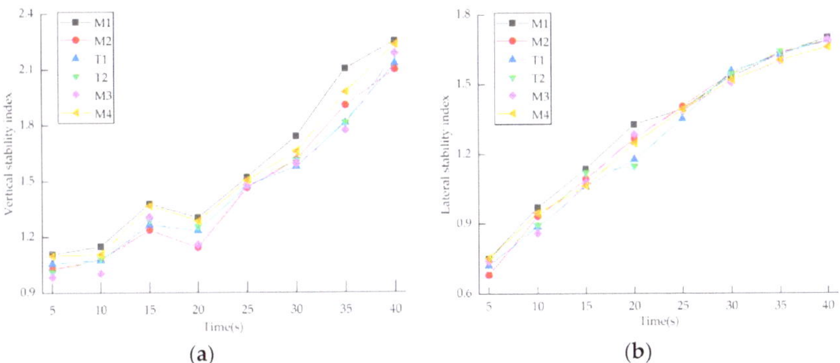

Figure 13. Sperling index of the marshaling urban rail train: (**a**) vertical sperling index; and (**b**) lateral sperling index.

5. Conclusions

(1) When calculating the nonlinear critical speed of urban rail trains, the calculation results of a single car body are quite different from the calculation results of marshaling trains, and the calculation results of marshaling trains are more in line with the actual situation. Therefore, the calculation of the marshaling train model for subsequent simulations is more realistic.

(2) In terms of safety indicators, dynamic indicators such as the wheel–rail force, wheel load reduction rate and derailment coefficient of the motor cars are significantly larger than those of the trailers, indicating that the safety of the motor car is significantly worse than that of the front trailer. In terms of running stability, the difference between the motor car and the trailer is small, and the traction motor will obviously improve the stability of the train when it transitions between the constant torque zone and the constant power zone. In the four-section motor car, the dynamic performance of the first and last sections is poor. In a two-section trailer, the dynamics of the rear car are poor.

(3) In terms of vibration response, the difference in the amplitude of the vibration acceleration between the motor car and the trailer body is small, the vibration acceleration of the motor car bogie frame and wheelset is significantly higher than that of the trailer and the low-frequency vibration frequency of the motor car is significantly larger than that of the trailer, indicating that the traction transmission system has a significant impact on the train bogie frame. The time domain and time–frequency response of vibration have a great influence.

The establishment of the marshaling urban rail train model with a traction transmission system in this paper is of great significance for more accurately determining the dynamic performance of each carriage of the marshaling train and the influence of the traction transmission system on each carriage. The dynamic performance of the train is optimized to improve its dynamic performance, and it can also simplify the maintenance of the marshaled urban rail train in the later stage.

Author Contributions: Conceptualization, Y.Z. (Yichao Zhang) and J.W.; methodology, Y.Z. (Yichao Zhang) and J.Y.; software, Y.Z. (Yichao Zhang) and Y.Z. (Yue Zhao); validation, Y.Z. (Yichao Zhang), J.Y., J.W. and Y.Z. (Yue Zhao); formal analysis, Y.Z. (Yichao Zhang), J.W. and Y.Z. (Yue Zhao); investigation, Y.Z. (Yichao Zhang) and J.W.; resources, J.Y. and J.W.; data curation, Y.Z. (Yue Zhao); writing—original draft preparation, Y.Z. (Yichao Zhang); writing—review and editing, Y.Z. (Yichao Zhang), J.W. and Y.Z. (Yue Zhao); supervision, J.Y.; project administration, J.Y. and J.W.; funding acquisition, J.Y. and J.W. All authors have read and agreed to the published version of the manuscript.

Funding: This research was funded by the National Natural Science Foundation of China (grant numbers 52272385, 52205083), the Beijing Natural Science Foundation (grant numbers L211008, 3214042) and the Pyramid Talent Training Project of Beijing University of Civil Engineering and Architecture (grant number JDYC20220827).

Conflicts of Interest: The authors declare no conflict of interest.

References

1. Peiffer, A.; Storm, S.; Röder, A.; Maier, R.; Frank, P.G. Active vibration control for high speed train bogies. *Smart Mater. Struct.* **2005**, *14*, 1. [CrossRef]
2. Simson, S.A.; Cole, C. Simulation of curving at low speed under high traction for passive steering hauling locomotives. *Veh. Syst. Dyn.* **2008**, *46*, 1107–1121. [CrossRef]
3. Ribeiro, D.; Cal Ada, R.; Delgado, R.; Brehm, M.; Zabel, V. Finite-element model calibration of a railway vehicle based on experimental modal parameters. *Veh. Syst. Dyn.* **2013**, *51*, 821–856. [CrossRef]
4. Wang, M.Y.; Sheng, X.Z.; Li, M.X.; Li, Y.G. Estimation of vibration powers flowing to and out of a high-speed train bogie frame assisted by time-domain response reconstruction. *Appl. Acoust.* **2022**, *185*, 108390. [CrossRef]
5. Ling, L.; Li, W.; Foo, E.; Wu, L.; Wen, Z.F.; Jin, X.S. Investigation into the vibration of metro bogies induced by rail corrugation. *Chin. J. Mech. Eng.* **2017**, *30*, 93–102. [CrossRef]
6. Yang, J.W.; Zhao, Y.; Wang, J.H.; Liu, C.D.; Bai, Y.L. Influence of wheel flat on railway vehicle helical gear system under Traction/Braking conditions. *Eng. Fail. Anal.* **2022**, *134*, 106022. [CrossRef]
7. Yang, J.W.; Sun, R.; Yao, D.C.; Wang, J.H.; Liu, C. Nonlinear Dynamic Analysis of high speed multiple units Gear Transmission System with Wear Fault. *Mech. Sci.* **2019**, *10*, 187–197. [CrossRef]
8. Wang, J.H.; Yang, J.W.; Lin, Y.L.; He, Y.P. Analytical investigation of profile shifts on the mesh stiffness and dynamic characteristics of spur gears. *Mech. Mach. Theory* **2022**, *167*, 104529. [CrossRef]
9. Wang, J.H.; Yang, J.W.; Li, Q. Quasi-static analysis of the nonlinear behavior of a railway vehicle gear system considering time-varying and stochastic excitation. *Nonlinear Dynam.* **2018**, *93*, 463–485. [CrossRef]
10. Wang, J.H.; Yang, J.W.; Bai, Y.L.; Zhao, Y.; He, Y.P.; Yao, D.C. A comparative study of the vibration characteristics of railway vehicle axlebox bearings with inner/outer race faults. *Proc. Inst. Mech. Eng. F-J. Rail Rapid Transit* **2021**, *235*, 1035–1047. [CrossRef]
11. Zhang, K.; Yang, J.W.; Liu, C.D.; Wang, J.H.; Yao, D.C. Dynamic Characteristics of a Traction Drive System in High-Speed Train Based on Electromechanical Coupling Modeling under Variable Conditions. *Energies* **2022**, *15*, 1202. [CrossRef]
12. Li, M.; Liu, Q.; Zhu, S.; Ai, S.; Chen, W.; Zhu, R. Contribution analysis of vibration transmission path of planetary reducer box based on velocity involvement loss. *J. Low Freq. Noise Vib. Act. Control* **2022**, *42*, 39–53. [CrossRef]
13. Zhang, T.; Jin, T.M.; Zhou, Z.W.; Chen, Z.G.; Wang, K.Y. Dynamic modeling of a metro vehicle considering the motor-gearbox transmission system under traction conditions. *Mech. Sci.* **2022**, *13*, 603–617. [CrossRef]
14. Wang, Z.W.; Cheng, Y.; Mei, G.M.; Zhang, W.H.; Huang, G.H.; Yin, Z.H. Torsional vibration analysis of the gear transmission system of high-speed trains with wheel defects. *Proc. Inst. Mech. Eng. F-J. Rail Rapid Transit* **2020**, *234*, 123–133. [CrossRef]
15. Chen, Z.G.; Zhai, W.M.; Wang, K.Y. A locomotive-track coupled vertical dynamics model with gear transmissions. *Veh. Syst. Dyn.* **2017**, *55*, 244–267. [CrossRef]
16. Chen, Z.G.; Zhai, W.M.; Wang, K.Y. Dynamic investigation of a locomotive with effect of gear transmissions under tractive conditions. *J. Sound Vib.* **2017**, *408*, 220–233. [CrossRef]
17. Chen, Z.G.; Zhai, W.M.; Wang, K.Y. Locomotive dynamic performance under traction/braking conditions considering effect of gear transmissions. *Veh. Syst. Dyn.* **2018**, *56*, 1097–1117. [CrossRef]
18. Zhang, T.; Chen, Z.G.; Zhai, W.M.; Wang, K.Y.; Wang, H. Effect of the drive system on locomotive dynamic characteristics using different dynamics models. *Sci. China Technol. Sc.* **2019**, *62*, 308–320. [CrossRef]
19. Wang, J.H.; Yang, J.W.; Zhao, Y.; Bai, Y.L.; He, Y.P. Nonsmooth Dynamics of a Gear-Wheelset System of Railway Vehicles Under Traction/Braking Conditions. *J. Comput. Nonlin. Dyn.* **2020**, *15*, 081003. [CrossRef]
20. Zhou, Z.W.; Chen, Z.G.; Spiryagin, M.; Wolfs, P.; Wu, Q.; Zhai, W.M.; Cole, C. Dynamic performance of locomotive electric drive system under excitation from gear transmission and wheel-rail interaction. *Veh. Syst. Dyn.* **2022**, *60*, 1806–1828. [CrossRef]
21. Zhou, Z.W.; Chen, Z.G.; Spiryagin, M.; Arango, E.B.; Wolfs, P.; Cole, C.; Zhai, W.M. Dynamic response feature of electromechanical coupled drive subsystem in a locomotive excited by wheel flat. *Eng. Fail. Anal.* **2021**, *122*, 105248. [CrossRef]
22. Wang, Z.W.; Mei, G.M.; Xiong, Q.; Yin, Z.H.; Zhang, W.H. Motor car-track spatial coupled dynamics model of a high-speed train with traction transmission systems. *Mech. Mach. Theory* **2019**, *137*, 386–403. [CrossRef]
23. Zhang, T.; Chen, Z.G.; Zhai, W.M.; Wang, K.Y. Establishment and validation of a locomotive-track coupled spatial dynamics model considering dynamic effect of gear transmissions. *Mech. Syst. Signal Process.* **2019**, *119*, 328–345. [CrossRef]
24. Wang, J.G.; He, G.Y.; Zhang, J.; Zhao, Y.X.; Yao, Y. Nonlinear dynamics analysis of the spur gear system for railway locomotive. *Mech. Syst. Signal Process.* **2017**, *85*, 41–55. [CrossRef]
25. Huang, G.H.; Xu, S.S.; Zhang, W.H.; Yang, C.J. Super-harmonic resonance of gear transmission system under stick-slip vibration in high-speed train. *J. Cent. South Univ.* **2017**, *24*, 726–735. [CrossRef]

26. Kia, S.H.; Henao, H.; Capolino, G.A. Mechanical health assessment of a railway traction system. In Proceedings of the MELECON 2008—The 14th IEEE Mediterranean Electrotechnical Conference, Ajaccio, France, 5–7 May 2008.

27. Huang, G.H.; Zhou, N.; Zhang, W.H. Effect of internal dynamic excitation of the traction system on the dynamic behavior of a high-speed train. *Proc. Inst. Mech. Eng. Part F J. Rail Rapid Transit* **2016**, *230*, 1899–1907. [CrossRef]

28. Chen, Z.W.; Zhu, G. Dynamic evaluation on ride comfort of metro vehicle considering structural flexibility. *Arch. Civ. Mech. Eng.* **2021**, *21*, 162. [CrossRef]

29. Pradhan, S.; Samantaray, A.K. Integrated modeling and simulation of vehicle and human multi-body dynamics for comfort assessment in railway vehicles. *J. Mech. Sci. Technol.* **2018**, *32*, 109–119. [CrossRef]

30. Belforte, P.; Cheli, F.; Diana, G.; Melzi, S. Numerical and experimental approach for the evaluation of severe longitudinal dynamics of heavy freight trains. *Veh. Syst. Dyn.* **2008**, *46*, 937–955. [CrossRef]

31. Tao, G.Q.; Liu, M.Q.; Xie, Q.L.; Wen, Z.F. Wheel-rail dynamic interaction caused by wheel out-of-roundness and its transmission between wheelsets. *Proc. Inst. Mech. Eng. Part F J. Rail Rapid Transit* **2022**, *236*, 247–261. [CrossRef]

32. Wu, Q.; Luo, S.; Wei, C.; Ma, W. Dynamics simulation models of coupler systems for freight locomotive. *J. Traffic Transp. Eng.* **2012**, *12*, 37–43.

33. Ding, H.; Zhu, T.; Xiao, S.; Yang, G.; Yang, B.; Lv, R. Study on the collision characteristics of a subway coupler. *Int. J. Crashworthines* **2022**, 1–15. [CrossRef]

34. Yadav, O.P.; Vyas, N.S. Influence of slack of automatic AAR couplers on longitudinal dynamics and jerk behaviour of rail vehicles. *Veh. Syst. Dyn.* **2022**, 1–21. [CrossRef]

35. Zou, R.M.; Luo, S.H.; Ma, W.H.; Wu, Q. Dynamic Characteristics of Metro Trains under Rescue Conditions. *Shock Vib.* **2020**, *2020*, 8869605. [CrossRef]

36. Lee, S.; Yu, H.; Yang, H.; Song, I.; Choi, J.; Yang, J.; Lim, G.; Kim, K.S.; Choi, B.; Kwon, J. A Study on Deep Learning Application of Vibration Data and Visualization of Defects for Predictive Maintenance of Gravity Acceleration Equipment. *Appl. Sci.* **2021**, *11*, 1564. [CrossRef]

37. Patil, S.; Pardeshi, S.; Patange, A. Health Monitoring of Milling Tool Inserts Using CNN Architectures Trained by Vibration Spectrograms. *Comput. Model. Eng. Sci.* **2023**, *136*, 177–199. [CrossRef]

Article

A Novel Strategy for Automatic Mode Pairing on the Model Updating of Railway Systems with Nonproportional Damping

Diogo Ribeiro [1,*], Cássio Bragança [2], Maik Brehm [3], Volkmar Zabel [4] and Rui Calçada [5]

[1]   CONSTRUCT-LESE, School of Engineering, Polytechnic of Porto, 4249-015 Porto, Portugal
[2]   Department of Structural Engineering, Federal University of Minas Gerais, Belo Horizonte 31270-901, Brazil
[3]   Merkle CAE Solutions GmbH, 89518 Heidenheim, Germany
[4]   Institute of Structural Mechanics, Bauhaus-University Weimar, 99423 Weimar, Germany
[5]   CONSTRUCT-LESE, Faculty of Engineering, University of Porto, 4200-465 Porto, Portugal
*   Correspondence: drr@isep.ipp.pt

Abstract: Mode pairing is a crucial step for the stability of any model-updating strategy based on experimental modal parameters. Automatically establishing a stable and assertive correspondence between numerical and experimental modes, in many cases, proves to be a very challenging task, especially in situations where complex mode shapes are present. This article presents a novel formulation for the automatic mode pairing between experimental and numerical complex modes based on an Energy-based Modal Assurance Criterion (EMAC). The efficiency of the proposed criterion was demonstrated on the basis of a case study involving the pairing between numerical and experimental modes of a passenger railway vehicle. A highly complex detailed FE numerical model of the vehicle was developed involving the modeling of the carbody, bogies and axles. A numerical damped modal analysis allowed obtaining the main global rigid-body and flexural modes of the vehicle's carbody, as well as several local modes associated to the vibration of specific components of the carbody. Due to the localized damping provided by the suspensions, these modes presented complex modal ordinates, especially for the rigid-body modes. The comparison between the results obtained from the application of the EMAC and the classical MAC criteria, on the pairing of five global mode shapes, proved that the EMAC criterion is much more assertive, avoiding mismatches between the experimental global modes and some of the local numerical modes with similar configurations, and, consequently, establishing the correct correspondences between experimental and numerical modes.

Keywords: model updating; automatic mode pairing; complex modal parameters; energy-based MAC

Citation: Ribeiro, D.; Bragança, C.; Brehm, M.; Zabel, V.; Calçada, R. A Novel Strategy for Automatic Mode Pairing on the Model Updating of Railway Systems with Nonproportional Damping. *Appl. Sci.* **2023**, *13*, 350. https://doi.org/10.3390/app13010350

Academic Editor: Suchao Xie

Received: 22 November 2022
Revised: 23 December 2022
Accepted: 23 December 2022
Published: 27 December 2022

1. Introduction

The overall use of numerical modeling techniques based on the finite element (FE) method, as well as experimental techniques for operational modal analysis (OMA), made the updating of numerical models based on modal parameters quite widespread [1]. These model-updating methodologies are widely used for: (i) developing highly accurate numerical models [2–6]; (ii) modeling structures under operational conditions with unknown levels of degradation and/or geometrical/mechanical parameters with very high levels of uncertainty [7,8]; (iii) monitoring the evolution of the structural behavior during retrofit operations [9]; and (iv) identifying structural damage [10,11], among others.

In most situations, these methodologies are based on the minimization of an objective function, composed by the residuals between numerical and experimental modal parameters, through the iterative variation of sensitive parameters of the numerical model [4]. During the optimization process, particularly due to variations on the numerical parameters' values, several changes in the order of the numerical mode shapes are frequently registered. To guarantee that these modifications do not affect the search for the optimal solution, it must be ensured that, in all iterations throughout the optimization process, the residuals of the objective function are calculated between an experimental mode and its correspondent numerical counterpart. Thus, the convergence of the optimization problem fundamentally depends on an efficient, automatic and stable mode-pairing technique to perform the correct assignment between numerical and experimental modes [12].

Most of the mode-pairing techniques are based on metrics for evaluating the correlation between two vectors (in this case, modal ordinate vectors), or between Frequency Response Functions (FRFs) [13,14]. Such metrics, applied to experimental and numerical results, allow to quantify the degree of correlation between the modes, and, consequently, to assign each experimental mode to its numerical counterpart [12]. The metrics based on FRFs are less widespread in the field of mode pairing compared to those based on modal ordinate vectors, mainly due to the difficulties associated with the estimation of the experimental FRFs in large-scale structures. Nevertheless, the Frequency Domain Assurance Criterion (FDAC), proposed by [15], can be used for mode pairing with FRFs. The FDAC computes the correlation between two FRFs for different frequency shifts, and provides, for each frequency, a scalar between 0 and 1, with 0 meaning no correspondence and 1 meaning full correspondence.

Among the modal-ordinate-vector-based criteria, one of the first to be applied is known as the Modal Scale Factor (MSF), proposed by [16]. The MSF is defined as the inner product of the two modal ordinate vectors to be compared, scaled by the inner product of either one of these vectors, making it significantly dependent on the normalization of the modes [17]. This dependence on the normalization factor sometimes causes a problem in practical applications, since it is often not possible to apply the same normalization to experimental and numerical modes, especially when using OMA techniques to gather experimental data. A very widespread criterion that is independent from normalization is the Modal Assurance Criterion (MAC), proposed by [16]. The MAC is defined as the square of the inner product of the two vectors, scaled by the product of the two inner products of the two vectors by themselves, resulting in a real scalar between 0 and 1 [18].

Based on the MAC, several other criteria for the correlation between modal ordinate vectors were proposed: the Partial MAC (PMAC) [19], in which only part of the modal ordinate vector is used in the calculations, focusing the analysis in a particular component or direction; the Extended MAC (MACX) [20], with an expanded assertiveness for complex modes, and its enhanced version (MACXP) [20], which, by a weighting with the poles of the corresponding mode shapes, improves the performance of the criterion under reduced spatial resolution; and the Weighted MAC (WMAC), also known as Normalized Cross Orthogonality (NCO), which incorporates the mass or stiffness matrices as a weighting matrix for the calculation of the MAC, remedying the shortcoming of the MAC not being a true orthogonality check [13]. However, due to the fact that usually not all degrees of freedom are instrumented, the application of the WMAC requires condensing the mass or stiffness matrices through model-reduction techniques [13,17]. Another issue regarding the WMAC is that, for complex mode shapes, a different formulation based on the state formulation of the orthogonality condition is required. This is due to the orthogonality conditions with respect to mass and stiffness matrices not being met for complex modes [21].

The mathematical formulation of some of these criteria will be briefly presented in Section 2 of the manuscript. A more detailed description of these and other criteria can be consulted, for example, in the References section [17,19,21,22].

A common issue associated with the application of these classic mode-pairing criteria arises when dealing with structures where local modes are associated to a particular structural component, as well as a set of components or parts of the structure that do not constitute the object of interest but also have global components, even if very small, with a shape similar to the true global modes. These modes, especially when they have close natural frequencies, can easily generate errors in the matching process and compromise the stability of the model-optimization algorithm [12,23].

Seeking to solve this issue, Brehm et al. [12] proposed an innovative criterion, the EMAC, which is based on the weighting of the MAC by the relative modal strain energy associated with different parts of the structure denominated as clusters. The authors demonstrated that, through the appropriate choice of clusters, it is possible to isolate modes from different parts of the structure quite efficiently. Applications of the EMAC for mode pairing can be consulted in the works of [3,5,23,24], in which its use proved to be essential in applications involving the model updating of FE models of railway bridges including the track.

Since it relies on the orthogonality conditions between the vibration modes and the stiffness matrix to calculate the modal strain energy, the EMAC, according to its original formulation, cannot be applied to problems involving complex modes. In many cases, especially in civil engineering structures where the modes are real or almost real, this does not represent a problem, and often, the pairing and model updating are performed considering the undamped modal problem. However, in the case of vehicles and structures with localized dampers, modes with a significant degree of complexity are present. In these situations, the EMAC ends up being at a disadvantage compared to criteria such as the FDAC, the MSF and the MAC, which can be applied to complex modes, and even more in relation to the MACX, which was specifically developed for these applications.

In this framework, the present work intends to share innovative contributions that, according to the authors' knowledge, are not sufficiently detailed in the existing literature, namely:

- The development of a mode-pairing formulation dedicated to complex modes based on an energy-based criterion and relying on a state-space formulation. The existing criteria for complex mode shapes reveal weaknesses and tend to fail in several situations;
- The evaluation of the performance of the developed mode-pairing criterion based on a case study involving a highly complex FE model of a railway vehicle and experimental modal parameters. In the existing mode pairing criteria, the validation is usually performed based on simple numerical or analytical examples. Additionally, the experimental restrictions associated with the positioning and number of sensors, noise and environmental interference create more challenging conditions to evaluate the performance of the pairing criteria.

This article presents the main existing mode-pairing criteria for complex modes, with a special emphasis on the criteria relying on the Energy-based Modal Assurance Criteria (EMAC). Regarding the EMAC, an innovative mathematical formulation based on a state-space model is detailed. Then, this criterion was applied to a case study involving the automatic mode pairing between the experimental and numerical modes of a passenger railway vehicle. In railway vehicles, the localized damping introduced by the suspension systems is responsible for the existence of complex mode shapes. In addition, the numerical model considers the flexibility of the carbody's elements, and consequently, the local modes of the panels that form the floor, walls and roof are present. These two aspects make the application of a criterion such as the EMAC particularly relevant to obtain a stable and robust automatic mode pairing.

2. Review of Existing Mode-Pairing Criteria

In this section, some of the most used criteria for mode pairing, which can be applied for systems with complex modes, are briefly presented. In addition to their mathematical formulation, the advantages and disadvantages of using each of these criteria are also presented.

2.1. Modal Assurance Criterion (MAC)

The Modal Assurance Criterion (MAC) is the most used criterion for pairing numerical and experimental modes of vibration [17,21,25]. It is defined as:

$$\mathrm{MAC}_{ij} = \frac{\left|\mathbf{\Phi}_i^H \mathbf{\Phi}_j\right|^2}{\mathbf{\Phi}_i^H \mathbf{\Phi}_i \mathbf{\Phi}_j^H \mathbf{\Phi}_j} \tag{1}$$

in which $\mathbf{\Phi}_i$ and $\mathbf{\Phi}_j$ are the vectors containing the modal ordinates of modes i and j, respectively, and ^{H} is the Hermitian transpose (conjugate transpose) operator.

According to [12], the main advantages of applying the MAC parameter are: (i) simple implementation; (ii) experimental information is not required on all degrees of freedom of the structure; and (iii) it does not depend on the normalization of the modal vectors. However, according to [21], the value of the MAC parameter strongly depends on the dimension of the modal vectors and is also particularly sensitive to the change in the higher amplitude ordinates. Due to these issues, pairing by the MAC value may be not sufficient in cases of complex structures, as well as in continuous structures or structures with partial continuity, such as bridges with several spans, as demonstrated by [23]. Examples of the application of the MAC criterion in mode pairing can be found in references [1,2,26,27].

2.2. Extended Modal Assurance Criterion (MACX)

As stated by Sternharz et al. [22], the MAC criteria might lead to inconclusive results in the case of modes with a significant level of complexity, especially in the presence of close or repeated modes. According to Vacher et al. [20], these inconclusive results are due to the fact that the MAC provides different results depending on the combinations made with the pairs of complex and complex-conjugate of the two mode shapes being compared. In order to address those issues, Vacher et al. [20] proposed the Extended Modal Assurance Criterion (MACX), defined as:

$$\text{MACX}_{ij} = \frac{\left(\left|\mathbf{\Phi}_i^H \mathbf{\Phi}_j\right| + \left|\mathbf{\Phi}_i^T \mathbf{\Phi}_j\right|\right)^2}{\left(\mathbf{\Phi}_i^H \mathbf{\Phi}_i + \left|\mathbf{\Phi}_i^T \mathbf{\Phi}_i\right|\right)\left(\mathbf{\Phi}_j^H \mathbf{\Phi}_j + \left|\mathbf{\Phi}_j^T \mathbf{\Phi}_j\right|\right)} \tag{2}$$

in which ^{T} is the transpose operator.

Compared to the MAC, the use of the MACX provides more consistent results in the case of complex mode shapes, but it is also influenced by the dimension of the modal ordinate vector and more sensible to variations in components of greater amplitude. Examples of its application can be found in references [28,29].

Aiming to improve the MACX performance in situations where the modal ordinate vectors contain information from only a few points, Vacher et al. [20] proposed an enhancement of the criterion called Pole-Weighted MACX (MACXP). This criterion incorporates information regarding the natural frequencies and damping ratios of the structure by weighting the MACX with the poles of the dynamic system.

The MACXP is defined as:

$$\text{MACXP}_{ij} = \frac{\left(\frac{\left|\mathbf{\Phi}_i^H \mathbf{\Phi}_j\right|}{\left|\lambda_i^* + \lambda_j\right|} + \frac{\left|\mathbf{\Phi}_i^T \mathbf{\Phi}_j\right|}{\left|\lambda_i + \lambda_j\right|}\right)^2}{\left(\frac{\mathbf{\Phi}_i^H \mathbf{\Phi}_i}{2\left|\text{Re}(\lambda_i)\right|} + \frac{\left|\mathbf{\Phi}_i^T \mathbf{\Phi}_i\right|}{2\left|\lambda_i\right|}\right)\left(\frac{\mathbf{\Phi}_j^H \mathbf{\Phi}_j}{2\left|\text{Re}(\lambda_j)\right|} + \frac{\left|\mathbf{\Phi}_j^T \mathbf{\Phi}_j\right|}{2\left|\lambda_j\right|}\right)} \tag{3}$$

in which λ_i and λ_j are the poles associated with modes i and j, and * is the complex-conjugate operator. It was demonstrated by Vacher et al. [20] that this weighting by the poles is capable of significantly improving the accuracy of the criterion in situations where there are few sample points. Examples of the application of this criterion can be consulted in [22].

2.3. Frequency Domain Assurance Criterion (FDAC)

The Frequency Domain Assurance Criterion (FDAC), proposed by Pascual et al. [15], is analogous to the MAC criteria, but it is calculated with the Frequency Response Functions (FRFs) with distinct frequency shifts. It is defined as:

$$\text{FDAC}\left(\omega_f, \omega_g\right) = \frac{\left|\sum_{p=1}^{N}\sum_{q=1}^{N} h_{pq}^{(x)}\left(\omega_f\right) h_{pq}^{*(a)}\left(\omega_g\right)\right|^2}{\left(\sum_{p=1}^{N}\sum_{q=1}^{N} h_{pq}^{(x)}\left(\omega_f\right) h_{pq}^{*(x)}\left(\omega_f\right)\right)\left(\sum_{p=1}^{N}\sum_{q=1}^{N} h_{pq}^{(a)}\left(\omega_g\right) h_{pq}^{*(a)}\left(\omega_g\right)\right)} \tag{4}$$

in which $h_{pq}^{(x)}\left(\omega_f\right)$ and $h_{pq}^{(a)}\left(\omega_g\right)$ are the values of the FRFs corresponding to an excitation at the Degree Of Freedom (DOF) p, measured at DOF q and at frequencies ω_f and ω_g, respectively [30]. The FDAC parameter allows for the analysis of the correspondence between two FRFs for all frequencies within a selected range. Such an operation results in something similar to a MAC matrix, although it is much denser given the large number of frequency values which can be used compared to the restricted number of modes used to compute the MAC matrix [13].

Several variants of this criterion are also found in the literature, such as the Response Vector Assurance Criterion (RVAC) [31], in which only one column of the FRF matrix is used to compute the FDAC; the improved FDAC [15], which takes into account the lags between the FRFs and prevents the pairing of FRFs with a lag of $180°$; and the complex FDAC [30], which is calculated without the modulus and conjugated operator in the numerator to account for the real and imaginary parts of the criterion.

The major drawbacks associated with the application of this mode-pairing criteria are related to the necessity of estimating experimental FRFs, which are not obtained when applying OMA techniques. Furthermore, the FDAC calculation implies a higher computational cost compared to the correlation criteria between modal order vectors, especially when there is a wide range and high resolution of frequencies. Examples of application of the FDAC criteria and its variants are found in references [32–34].

3. Mode Pairing Using the Energy-Based Modal Assurance Criterion (EMAC)

An efficient and robust mode-pairing criterion is a key aspect to assure the stability of any model-updating methodology based on experimental modal data. Undoubtedly, the MAC is the most used criterion for this task. However, due to its drawbacks (presented in Section 2), it might be unsuitable in some situations. Some of these issues may be solved by a proper weighting of the MAC values. The criterion based on strain energy gathers the information from the mathematical correlation between the modal vectors with the physical information of the degrees of freedom observed in the dynamic test and related to the stiffness or mass distribution.

In this criterion, the correspondence of the numerical modes with the experimental ones is carried out through the EMAC parameter (Energy-based Modal Assurance Criterion) which is given by:

$$\text{EMAC}_{ijk} = \prod_{jk}\text{MAC}_{ij} \tag{5}$$

This parameter results from the weighting of the MAC parameter by the relative modal strain energy ($\prod_{jk}$) of one or several groups of degrees of freedom of the numerical model, called clusters. Each experimental mode is paired with the numerical mode corresponding to the highest value of the EMAC parameter.

The success of the criterion based on the EMAC parameter largely depends on the selection of the degrees of freedom of the numerical model that form the various clusters. Clusters must allow for the separation of measured degrees of freedom from unmeasured degrees of freedom in the test. In complex structures, the clusters must also consider the different substructures constituted by groups of elements with a dynamic behavior different from the global structure.

This criterion was developed and successfully applied by Brehm et al. [12] for the case of real vibration modes. Its application to complex vibration modes based on a state-space formulation is innovative and is not reported in the bibliography in detail.

In this section, the mathematical formulation of the EMAC criterion is presented. First, the formulation proposed by Brehm et al. [12] for real modes is briefly presented, as it represents the basis for the understanding of the formulation for complex modes. Then, the mathematical formulation of the newly innovative approach to dealing with complex modes is presented.

3.1. Real Modes

In the case of real vibration modes, the relative modal strain energy uses the physical information of the stiffness matrix. Its calculation involves rearranging the modal vectors and dividing the stiffness matrix into submatrices that relate the different clusters [12].

Assuming that the modal matrix ($\mathbf{\Phi}$) is normalized in relation to the mass matrix, the modal stiffness matrix, which constitutes the orthogonality condition in relation to the stiffness matrix, is as follows:

$$\mathbf{\Phi}^T \mathbf{K} \mathbf{\Phi} = \begin{bmatrix} \ddots & & \\ & \omega_j^2 & \\ & & \ddots \end{bmatrix} \tag{6}$$

in which ω_j corresponds to the angular natural frequency of mode j. The total strain energy (Modal Strain Energy) associated with each vibration mode j (MSE_j) is equal to $1/2 \times \omega_j^2$.

The vector that contains the modal information of the numerical mode j can be rearranged by separating the degrees of freedom of the numerical model into n clusters:

$$\mathbf{\Phi}_j^T = \begin{bmatrix} \mathbf{\Phi}_{j1}^T \mathbf{\Phi}_{j2}^T \cdots \mathbf{\Phi}_{jn}^T \end{bmatrix}^T \tag{7}$$

In turn, the stiffness matrix is also divided into submatrices ($\mathbf{K}_{kl}$) that relate the degrees of freedom of clusters k and l, that is:

$$\mathbf{K} = \begin{bmatrix} \mathbf{K}_{11} & \mathbf{K}_{12} & \cdots & \mathbf{K}_{1n} \\ \mathbf{K}_{21} & \mathbf{K}_{22} & \cdots & \mathbf{K}_{2n} \\ \vdots & \vdots & \ddots & \vdots \\ \mathbf{K}_{n1} & \mathbf{K}_{n2} & \cdots & \mathbf{K}_{nn} \end{bmatrix} \tag{8}$$

where k and l take values equal to 1, 2, ..., n, and n is the total number of clusters.

The modal strain energy of vibration mode j with respect to cluster k (MSE_{jk}) is calculated based on the following expression:

$$\text{MSE}_{jk} = \frac{1}{2} \sum_{l=1}^{n} \mathbf{\Phi}_{jk}^T \mathbf{K}_{kl} \mathbf{\Phi}_{jl} \tag{9}$$

where $\mathbf{\Phi}_{jk}$ is the matrix that contains the modal information of the numerical mode j, corresponding to the degrees of freedom of cluster k; $\mathbf{K}_{kl}$ is the stiffness submatrix that relates the degrees of freedom of clusters k and l; and $\mathbf{\Phi}_{jl}$ is the matrix that contains the modal information of numerical mode j, corresponding to the degrees of freedom of cluster l.

The total strain energy of vibration mode j is given by:

$$\text{MSE}_j = \frac{1}{2} \sum_{k=1}^{n} \sum_{l=1}^{n} \mathbf{\Phi}_{jk}^T \mathbf{K}_{kl} \mathbf{\Phi}_{jl} = \frac{1}{2} \mathbf{\Phi}_j^T \mathbf{K} \mathbf{\Phi}_j = \frac{1}{2} \omega_j^2 \tag{10}$$

The relative strain energy (Π_{jk}) represents the portion of the total energy mobilized by the vibration mode j considering only the degrees of freedom of cluster k. It can be calculated based on Equations (9) and (10), that is:

$$\Pi_{jk} = \frac{\text{MSE}_{jk}}{\text{MSE}_j} = \frac{\sum_{l=1}^{n} \Phi_{jk}^{T} K_{kl} \Phi_{jl}}{\Phi_j^{T} K \Phi_j} \tag{11}$$

with $\text{MSE}_j \neq 0$. This parameter is a scalar that takes values in the range between 0 and 1. Finally, the EMAC_{ijk} is calculated through Equation (5).

3.2. Complex Modes

In cases where the damping matrix is not proportional to the mass and stiffness matrices, the so-called complex modes of vibration are present. These are characterized by complex numbers, i.e., they encompass both magnitude and phase information. In a complex vibration mode, the movements of the points occur with a time delay proportional to the phase difference, situated between $0°$ and $180°$. The representation of the deformed structure is usually performed with animations that present the values of the amplitudes of the various components of the modes at different instants of time [17].

In the case of complex vibration modes, the previously presented formulation cannot be directly applied since the orthogonality conditions with respect to the stiffness matrix (Equation (6)) are not met. Therefore, to obtain an expression for the modal strain energy it is necessary to resort to a formulation based on state-space equations.

In a state-space formulation, the system of second-order differential equilibrium equations, with dimension z, is transformed into a system of $2z$ first-order differential equations. To this end, the state vector $\mathbf{x}(t)$ is defined with z lines, which are constituted by the displacements and velocities of the z degrees of freedom of the structure:

$$\mathbf{x}(t) = \begin{bmatrix} \mathbf{q}(t) \\ \dot{\mathbf{q}}(t) \end{bmatrix} \tag{12}$$

Based on the state-space formulation, the equation of motion ($\mathbf{M}\ddot{\mathbf{q}} + \mathbf{C}\dot{\mathbf{q}} + \mathbf{K}\mathbf{q} = \mathbf{p}(t)$) can be rewritten as:

$$\mathbf{P}\dot{\mathbf{x}}(t) + \mathbf{Q}\mathbf{x}(t) = \left\{ \begin{matrix} \mathbf{p}(t) \\ 0 \end{matrix} \right\} \tag{13}$$

$\mathbf{P}$ and $\mathbf{Q}$ are defined in Equations (14) and (15), respectively, in terms of the stiffness, damping and mass matrices [35,36]:

$$\mathbf{P} = \begin{bmatrix} \mathbf{C} & \mathbf{M} \\ \mathbf{M} & 0 \end{bmatrix} \tag{14}$$

$$\mathbf{Q} = \begin{bmatrix} \mathbf{K} & 0 \\ 0 & -\mathbf{M} \end{bmatrix} \tag{15}$$

Assuming $q(t) = \varphi_j e^{\lambda_j t}$ as a solution for homogeneous differential equations results in the following eigenvalue problem:

$$\mathbf{Q} \cdot \mathbf{\Psi} = -\mathbf{P} \cdot \mathbf{\Psi} \cdot \Lambda_C \tag{16}$$

The eigenvalues (Λ) and the eigenvectors ($\mathbf{\Psi}$) can be related to matrices that contain the vibration modes (Θ) and λ_j, which characterize the dynamic behavior of the structure through the following expressions:

$$\Lambda_C = \begin{bmatrix} \Lambda & 0 \\ 0 & \Lambda^* \end{bmatrix} \tag{17}$$

$$\boldsymbol{\Psi} = \begin{bmatrix} \boldsymbol{\Theta} & \boldsymbol{\Theta}^* \\ \boldsymbol{\Theta}\boldsymbol{\Lambda} & \boldsymbol{\Theta}^*\boldsymbol{\Lambda}^* \end{bmatrix} \tag{18}$$

where

$$\boldsymbol{\Lambda} = \begin{bmatrix} \ddots & & \\ & \lambda_j & \\ & & \ddots \end{bmatrix} \qquad j = 1, \ldots, n \tag{19}$$

$$\boldsymbol{\Theta} = \begin{bmatrix} \ldots & \boldsymbol{\Phi}_j & \ldots \end{bmatrix} \qquad j = 1, \ldots, n \tag{20}$$

in which $\boldsymbol{\Phi}_j$ is a vector containing the modal ordinates of mode i, and λ_i is the associated eigenvalue [35,36].

Based on the previously presented formulation, the following orthogonality conditions can be derived:

$$\boldsymbol{\Psi}^T \mathbf{P} \boldsymbol{\Psi} = \begin{bmatrix} \ddots & & \\ & a_j & \\ & & \ddots \end{bmatrix} \tag{21}$$

$$\boldsymbol{\Psi}^T \mathbf{Q} \boldsymbol{\Psi} = \begin{bmatrix} \ddots & & \\ & b_j & \\ & & \ddots \end{bmatrix} \tag{22}$$

in which a_j and b_j play a role similar to the modal mass and modal stiffness for undamped vibration systems but are usually complex numbers [36]. Therefore, in the case of complex modes, the modal strain energy can be calculated based on the second orthogonality condition, and the total strain energy associated with mode j (MSE_j) is equal to $1/2 \times b_j$.

The matrix $\boldsymbol{\Psi}$ has a dimension of $2z \times 2n'$, where z is the number of degrees of freedom of the numerical model and n' is the total number of vibration modes. This matrix can be rearranged by separating the degrees of freedom of the numerical model into n clusters, resulting in:

$$\boldsymbol{\Psi} = \begin{bmatrix}
\Phi_{11} & \Phi_{12} & \cdots & \Phi_{1n'} & \Phi_{11}^* & \Phi_{12}^* & \cdots & \Phi_{1n'}^* \\
\Phi_{21} & \Phi_{22} & \cdots & \Phi_{2n'} & \Phi_{21}^* & \Phi_{22}^* & \cdots & \Phi_{2n'}^* \\
\vdots & \vdots & \ddots & \vdots & \vdots & \vdots & \ddots & \vdots \\
\Phi_{n1} & \Phi_{n2} & \cdots & \Phi_{nn'} & \Phi_{n1}^* & \Phi_{n2}^* & \cdots & \Phi_{nn'}^* \\
\Phi_{11}\lambda_1 & \Phi_{12}\lambda_2 & \cdots & \Phi_{1n'}\lambda_{n'} & \Phi_{11}^*\lambda_1^* & \Phi_{12}^*\lambda_2^* & \cdots & \Phi_{1n'}^*\lambda_{n'}^* \\
\Phi_{21}\lambda_1 & \Phi_{22}\lambda_2 & \cdots & \Phi_{2n'}\lambda_{n'} & \Phi_{21}^*\lambda_1^* & \Phi_{22}^*\lambda_2^* & \cdots & \Phi_{2n'}^*\lambda_{n'}^* \\
\vdots & \vdots & \ddots & \vdots & \vdots & \vdots & \ddots & \vdots \\
\Phi_{n1}\lambda_1 & \Phi_{n2}\lambda_2 & \cdots & \Phi_{nn'}\lambda_{n'} & \Phi_{n1}^*\lambda_1^* & \Phi_{n2}^*\lambda_2^* & \cdots & \Phi_{nn'}^*\lambda_{n'}^*
\end{bmatrix} \tag{23}$$

The vector that contains the modal information of the numerical mode j can be rearranged by separating the degrees of freedom of the n clusters. It has the following format:

$$\boldsymbol{\Psi}_j^T = \begin{bmatrix} \boldsymbol{\Phi}_{1j}^T \boldsymbol{\Phi}_{2j}^T \cdots \boldsymbol{\Phi}_{nj}^T (\boldsymbol{\Phi}_{1j}\lambda_j)^T (\boldsymbol{\Phi}_{2j}\lambda_j)^T \cdots (\boldsymbol{\Phi}_{nj}\lambda_j)^T \end{bmatrix}^T \tag{24}$$

In turn, the matrix $\mathbf{Q}$, with a dimension of $2z \times 2z$, is also divided into submatrices ($\mathbf{K}_{kl}$ and $\mathbf{M}_{kl}$) that relate the degrees of freedom of clusters k and l, that is:

$$\mathbf{Q} = \begin{bmatrix} \mathbf{K}_{11} & \mathbf{K}_{12} & \cdots & \mathbf{K}_{1n} & 0 & 0 & \cdots & 0 \\ \mathbf{K}_{21} & \mathbf{K}_{22} & \cdots & \mathbf{K}_{2n} & 0 & 0 & \cdots & 0 \\ \vdots & \vdots & \ddots & \vdots & \vdots & \vdots & \ddots & \vdots \\ \mathbf{K}_{n1} & \mathbf{K}_{n2} & \cdots & \mathbf{K}_{nn} & 0 & 0 & \cdots & 0 \\ 0 & 0 & \cdots & 0 & -\mathbf{M}_{11} & -\mathbf{M}_{12} & \cdots & -\mathbf{M}_{1n} \\ 0 & 0 & \cdots & 0 & -\mathbf{M}_{21} & -\mathbf{M}_{12} & \cdots & -\mathbf{M}_{2n} \\ \vdots & \vdots & \ddots & \vdots & \vdots & \vdots & \ddots & \vdots \\ 0 & 0 & \cdots & 0 & -\mathbf{M}_{n1} & -\mathbf{M}_{n2} & \cdots & -\mathbf{M}_{nn} \end{bmatrix} \tag{25}$$

in which k and l can assume values equal to 1, 2, $\ldots$, n, where n is the total number of clusters.

Based on this submatrix division of $\mathbf{Q}$, it is possible to form the matrices $\mathbf{Q}_{kl}$, which relate the degrees of freedom from clusters k and l, that is:

$$\mathbf{Q}_{kl} = \begin{bmatrix} \mathbf{K}_{kl} & 0 \\ 0 & -\mathbf{M}_{kl} \end{bmatrix} \tag{26}$$

Therefore, the modal strain energy from mode j with respect to cluster k (MSE_{jk}) can be calculated by:

$$\mathrm{MSE}_{jk} = \frac{1}{2} \sum_{l=1}^{n} \mathbf{\Psi}_{jk}^{T} \mathbf{Q}_{kl} \mathbf{\Psi}_{jl} \tag{27}$$

where $\mathbf{\Psi}_{jk}$ is the matrix that contains the information of numerical mode j, corresponding to the degrees of freedom of cluster k; $\mathbf{Q}_{kl}$ is the submatrix from $\mathbf{Q}$ relating the degrees of freedom from clusters k and l; and $\mathbf{\Psi}_{jl}$ is the matrix containing the modal information of numerical mode j, corresponding to the degrees of freedom from cluster l.

The total modal strain energy of mode j is given by:

$$\mathrm{MSE}_{j} = \frac{1}{2} \sum_{k=1}^{n} \sum_{l=1}^{n} \mathbf{\Psi}_{jk}^{T} \mathbf{Q}_{kl} \mathbf{\Psi}_{jl} = \frac{1}{2} \mathbf{\Psi}_{j}^{T} \mathbf{Q} \mathbf{\Psi}_{j} = \frac{1}{2} b_{j} \tag{28}$$

Therefore, the relative strain energy (Π_{jk}) represents the portion of the total energy mobilized by vibration mode j considering only the degrees of freedom of cluster k. It is given by:

$$\Pi_{jk} = \frac{\left| \mathrm{MSE}_{jk} \right|}{\left| \mathrm{MSE}_{j} \right|} = \frac{\left| \sum_{l=1}^{n} \mathbf{\Psi}_{jk}^{T} \mathbf{Q}_{kl} \mathbf{\Psi}_{jl} \right|}{\left| \mathbf{\Psi}_{j}^{T} \mathbf{Q} \mathbf{\Psi}_{j} \right|} \tag{29}$$

with $\mathrm{MSE}_{j} \neq 0$. Similarly to the case involving real modes, the relative strain energy (Π_{jk}) varies from 0 to 1.

4. Case Study

In this section, the presented EMAC formulation for complex modes is applied to a case study involving the mode pairing between numerical and experimental modes of a BBN tourist-class passenger railway vehicle. The BBN vehicle (Figure 1) is a 25.9 m and 55-ton car with two motor bogies and capacity for 62 passengers. This vehicle is part of the CPA 4000 series ("Alfa Pendular") train which operates in the line connecting the cities of Porto and Lisbon in Portugal.

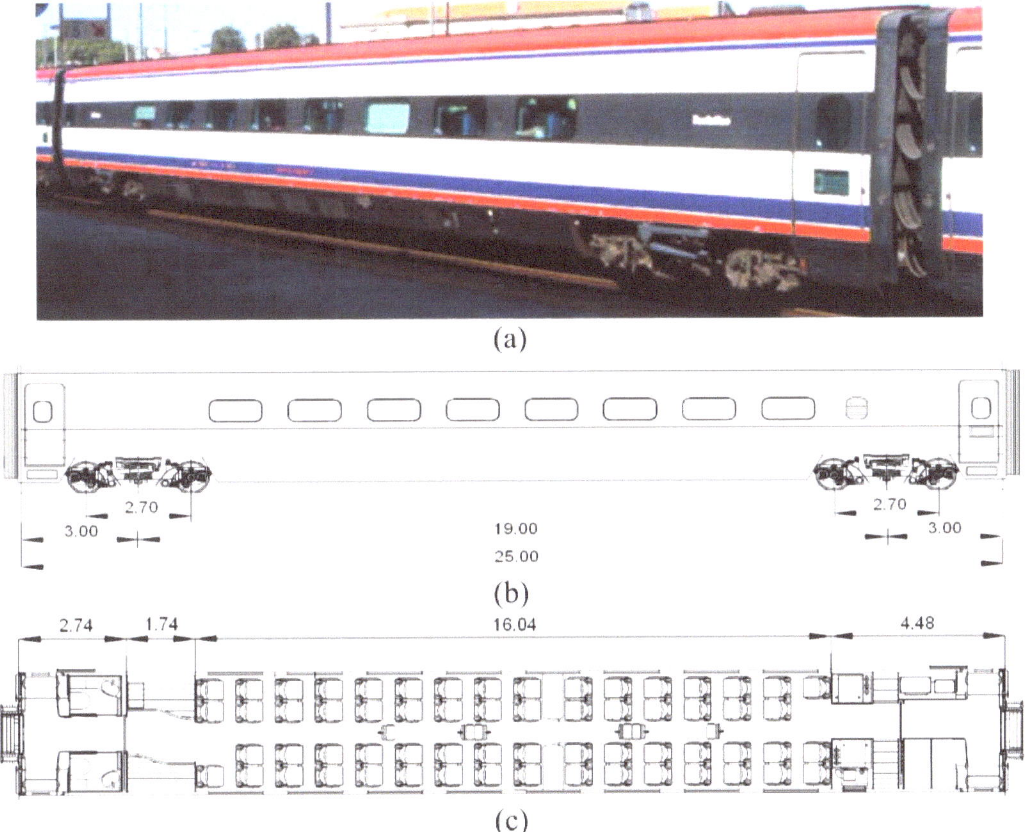

(a)

(b)

(c)

Figure 1. BBN vehicle: (**a**) perspective; (**b**) elevation; (**c**) floor plan.

4.1. Numerical Model

The numerical model of the vehicle (Figure 2) was developed based on shell, beam and spring-dashpot assemblies. Particularly, the beam elements were used for modeling the bogies, and the spring-dashpot assemblies were used to simulate the suspensions, the connecting rods and the tilting system. The shell elements were used to model the floor, the roof and the wall panels. The thickness of the shell elements was considered to match the cross-sectional area with one of the real panels. The real panels, however, are formed by an upper and lower plate connected by diagonal plates, thus presenting an orthotropic behavior to bending. To adequately represent this behavior, the inertia of the shell elements was corrected by the RMI (Ratio of the Bending Moment of Inertia) [37] parameter given by the ratio between the real inertia of the panel and that given by the shell element.

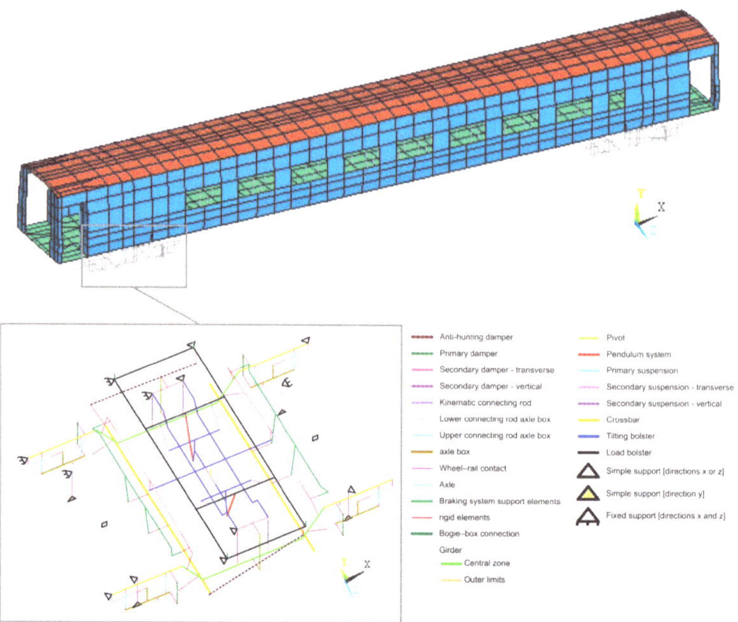

Figure 2. Overview of the BBN vehicle's numerical model.

The main parameters of the numerical model are depicted in Table 1, according to the information provided by the manufacturer of the BBN vehicle.

Table 1. Main parameters of the BBN vehicle's numerical model.

Parameter	Designation		Adopted Value	Unit
	Carbody			
K_{S1}	Vertical secondary suspension stiffness	Front bogie	256.4	kN/m
K_{S2}		Rear bogie		
c_S	Vertical secondary suspension damping		35	kNm/s
K_{ST}	Transverse secondary suspension stiffness		2500	kN/m
c_{ST}	Transverse secondary suspension damping		17.5	kNm/s
K_{Pend}	Rigidity of the pendulum system		0 (at rest)	kN/m
c_{AL}	Anti-hunting suspension damping		400	kNm/s
K_b	Stiffness of the tilting bolster–load bolster connection rod		20,000	kN/m
Δ_{alum}	Aluminum density		2700	kg/m^3
E_{alum}	Aluminum deformability module	Dir x	70	GPa
		Dir z	54.2	GPa
RMI_b	Corrective factor of the moment of inertia	Floor	90	-
RMI_p		Walls	114	-
RMI_c		Roof	386	-
ΔM_b	Additional mass	Floor	70	%
ΔM_p		Walls	20	%
ΔM_c		Roof	10	%

Table 1. *Cont.*

Parameter	Designation		Adopted Value	Unit
e_{bas}	Equivalent thickness	Floor	10.2	mm
e_{par}		Walls	10.3	mm
e_{cob}		Roof	8.8	mm
Bogies				
K_P	Primary suspension stiffness		564	kN/m
c_P	Primary suspension damping		18	kNm/s
K_{bls}	Axle-box connecting rod stiffness	Top	6.5	MN/m
K_{bli}		Bottom	25	MN/m
K_{rc}	Stiffness of the wheel–rail contact		1.5674×10^9	N/m
ΔM_{lc}	Additional mass	Girder (central zone)	42	kg/m
ΔM_{le}		Girder (extremities)	38	kg/m
ΔM_t		Crossmember	92	kg/m
ΔM_e		Axles	271	kg/m

Additionally, concentrated mass elements were used to incorporate the mass of some non-structural components and equipment at the wagon's floor and at specific locations distributed along the bogies. The adequate positioning of these elements is essential for accurately representing the vehicle's modal behavior. The positioning and corresponding values of these masses are depicted in Figure 3.

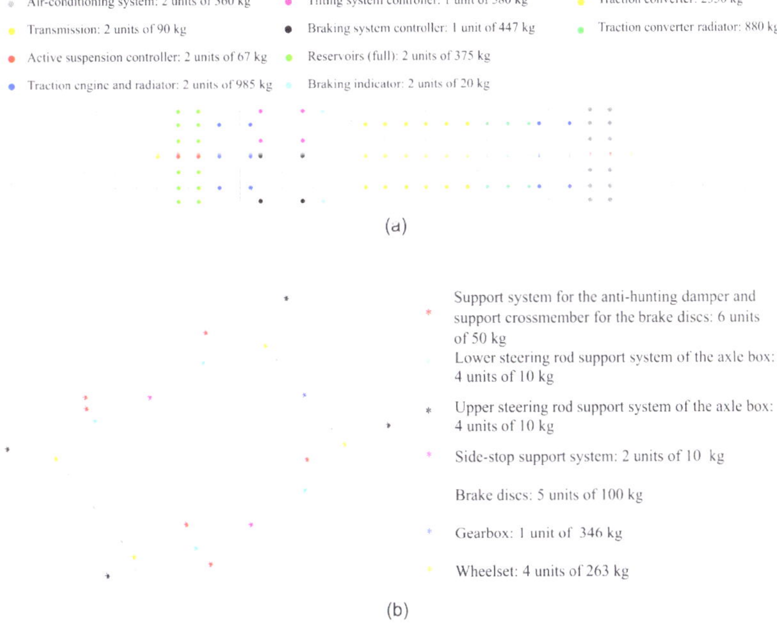

Figure 3. Mass elements in the finite element model: (**a**) carbody; (**b**) bogie.

Figure 4 depicts the modal configurations and natural frequencies of the vehicle's carbody. Among the rigid-body modes, 1C is a rotation about the x axis, 2C is a translation along the y axis, and 3C is a rotation about the z axis. Due to the relevant contribution of the localized damping provided by the suspension elements, these rigid-body modes presented a high degree of complexity. Among the deformation modes, 4C and 6C are, respectively, the first and second torsional modes, and 5C is the first bending mode.

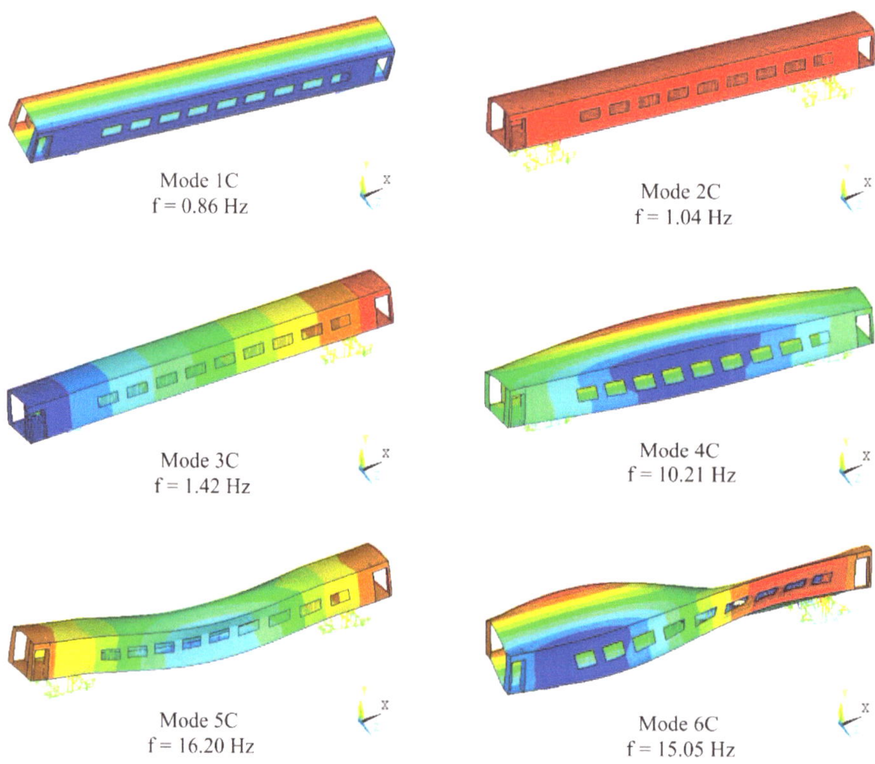

Figure 4. Numerical modes of the BBN vehicle's carbody.

Several local modes involving the bending of elements of the box, in particular the base, walls and roof, were also identified. As an example, Figure 5 illustrates the first two local modes (1L and 2L) that involve bending movements of the base of the carbody, with natural frequencies of the damped system equal to 8.85 Hz and 9.63 Hz, respectively.

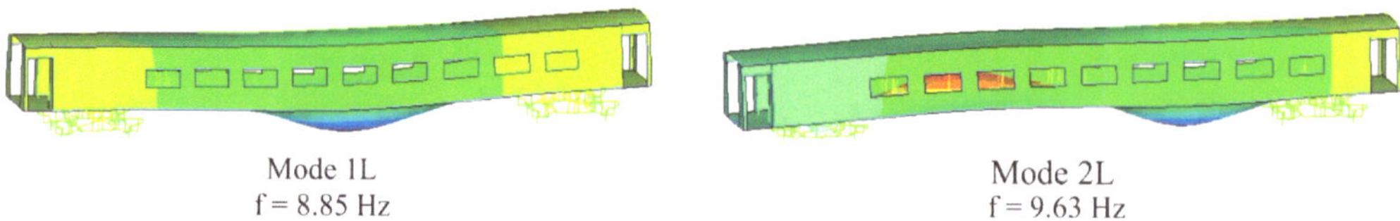

Figure 5. Numerically obtained local vibration modes of the housing.

As can be seen in Figure 5, despite being clearly local, modes 1L and 2L contain global displacement components, which might be confused with global modes and create extra difficulties for an automatic mode-pairing algorithm. Particularly, in Mode 1L, these global movements are very similar, in a smaller scale, to the first global bending mode (5C).

The vehicle damping matrix was constructed as the sum of a Rayleigh damping matrix and the matrix resulting from the scattering of the matrices of the elements with localized damping, particularly the primary, secondary and anti-hunting dampers. The addition of the matrices from the localized dampers to the Rayleigh damping matrix makes it nonproportional to the stiffness and mass matrices, which results in complex modes.

The Rayleigh constants were calculated setting damping coefficients equal to 2% for the 4C and 6C vibration modes. The evolution of the damping coefficient as a function of frequency for the Rayleigh damping portion is graphically represented in Figure 6.

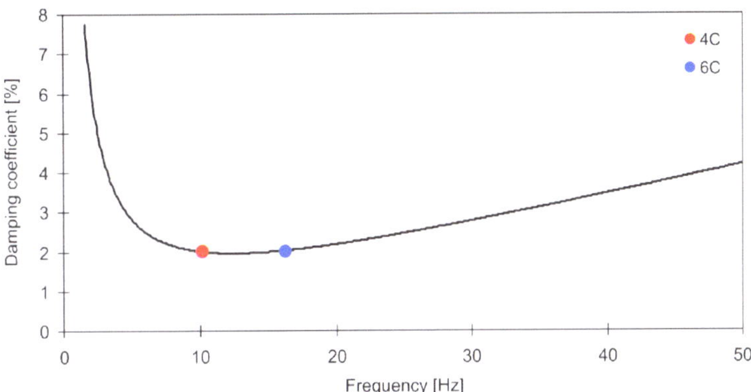

Figure 6. Rayleigh damping curve.

4.2. Mode Pairing

A dynamic test, described in detail in [4], was performed based on a set of 14 accelerometers distributed along the floor of the carbody. The tests allowed for the identification of 5 experimental vibration modes, depicted in Figure 7, which clearly correspond to the numerical modes 1C to 5C previously presented in Figure 4. In Figure 7, for each vibration mode, the respective natural frequency and Mode Complexity Factor (MCF) [38] are also indicated.

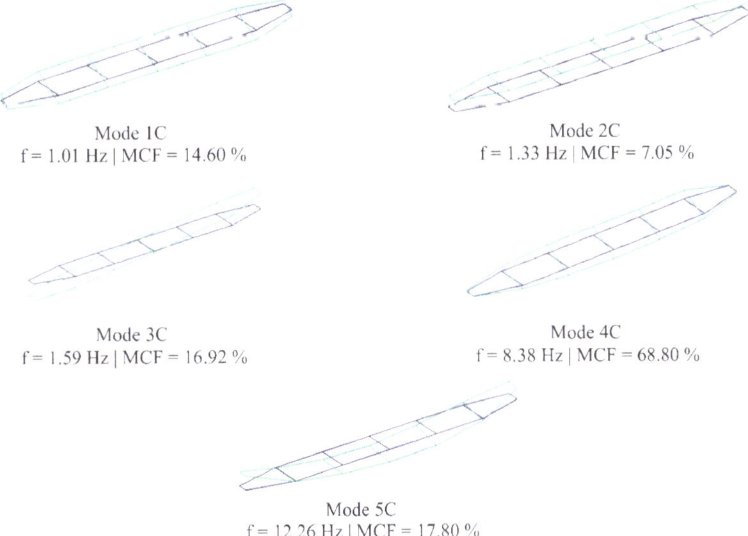

Figure 7. Experimental carbody modes.

To demonstrate the performance of the EMAC in automatically pairing these experimental modes to their numerical correspondents, the vehicle's FE model was divided into the four clusters presented in Figure 8. The carbody clusters, namely the floor, walls and roof, where each split into sub-clusters containing only the degrees of freedom associated with the vertical (y) and transverse (z) directions, respectively. Accordingly, seven separate clusters were obtained in total. In the "other elements" cluster, the remaining degrees of freedom of translation and rotation of the numerical model were included, encompassing, among others, the bogies and the seats.

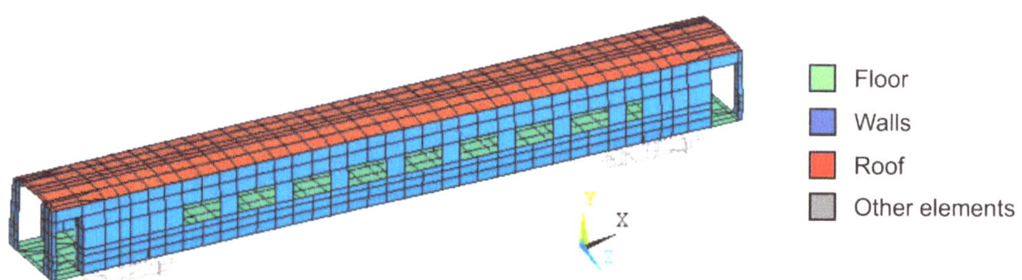

Figure 8. Identification of the clusters used in the numerical model of the BBN vehicle.

For each of these clusters, the relative modal strain energy (see Figure 9) was calculated through Equation (29) considering 80 vibration modes, obtained through a numerical modal analysis taking the damping into account.

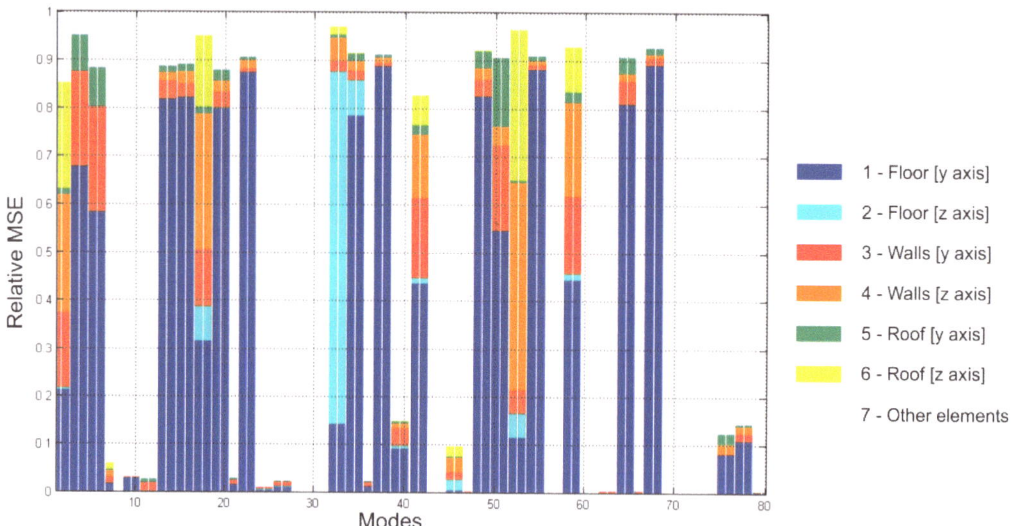

Figure 9. Relative MSE values for the different clusters and vibration modes, obtained based on the numerical model of the BBN vehicle.

Figure 10 presents the MAC and EMAC correlation matrices between the 80 numerical modes and the 5 experimentally identified modes (see Figure 7). In this Figure, the 80 numerical modes correspond to 40 complex-conjugate pairs. The EMAC values for the 2C, 3C and 5C modes were obtained by weighting the MAC by the modal strain energy of the clusters 1, 3 and 5 (see Figures 8 and 9). The EMAC values for the 1C and 4C modes resulted of the weighting of the MAC values by the modal strain energy of the clusters 1, 4 and 6 (see Figures 8 and 9). Cluster 1 is representative of the positioning of the sensors and the measurement direction used in the dynamic test of the carbody, since the sensors were installed on the vehicle's floor. The use of other clusters in the weighting of the MAC values, particularly in the rigid-body modes, allowed to highlight the interrelationship between the degrees of freedom of the base, walls and roof in the y and z directions.

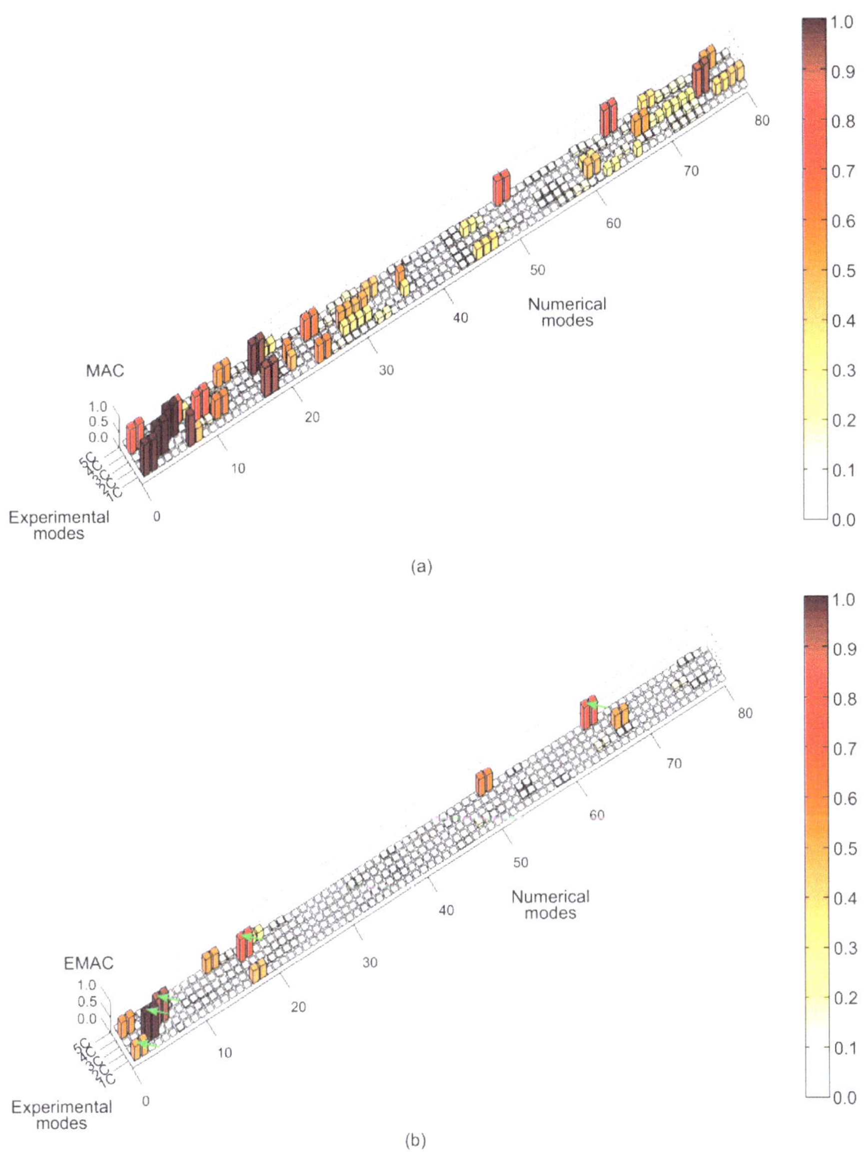

Figure 10. Pairing of experimental and numerical modes of the BBN vehicle based on the (**a**) MAC and (**b**) EMAC parameters.

As can be clearly seen, the EMAC matrix is significantly cleaner compared to the MAC matrix. The application of the EMAC allowed for the proper pairing the between experimental and the numerical modes, as indicated by the little green arrows in Figure 10b. On the other hand, the use of the MAC parameter resulted in a high level of correlation between the experimental modes and several numerical modes, which made it impossible to establish a proper pairing between experimental and numerical modes.

The EMAC parameter facilitated the pairing of the 1C and 5C experimental modes, especially. Figure 11 shows two numerical vibration modes, modes 7 and 50, which can be paired with experimental modes 1C and 5C, respectively, and which the use of the EMAC parameter allowed to exclude. Mode 7 is a transverse rotation mode of both bogies. It causes small-amplitude transverse-rotation movements of the box. Mode 50 is a global bending mode with local-box base movements.

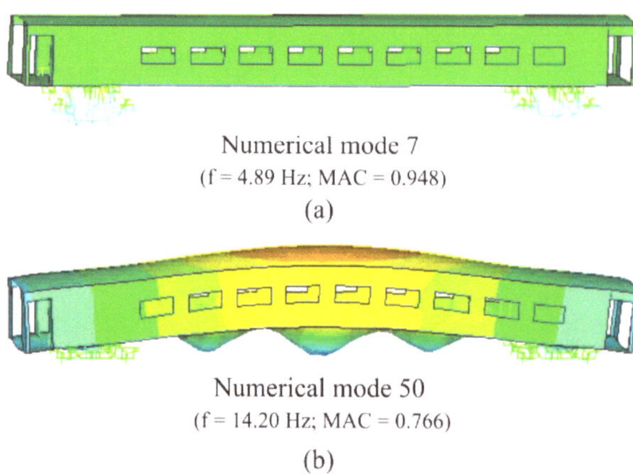

Numerical mode 7
(f = 4.89 Hz; MAC = 0.948)
(a)

Numerical mode 50
(f = 14.20 Hz; MAC = 0.766)
(b)

Figure 11. Numerical vibration modes likely to pair with (**a**) experimental mode 1C, and (**b**) experimental mode 5C.

It is also important to highlight the significant correlation between vibration modes 1C and 4C. As can be seen from Figure 7, due to the limited number of points where the experimental information was available, the deformed modal configurations of these two modes, from the perspective of the instrumented points, are very similar. The distinction between these two modes became more evident using the EMAC parameter.

5. Conclusions

This paper presents a novel approach to the problem of automatic pairing complex vibration modes through the expansion of the MAC criterion weighted by the modal strain energy (EMAC), for its application in problems involving complex mode shapes.

To enable the application of the EMAC to problems with complex modes, an expression for the modal strain energy was derived on the basis of the orthogonality conditions of a state-space formulation. The derived formulation allows to quantify the relative modal strain energy, used for the weighting of the MAC, in cases where the orthogonality condition between the modes and the stiffness matrix is not satisfied. Consequently, this made it possible to expand the criterion to more general applications.

Subsequently, the effectiveness of the proposed criterion was demonstrated through a case study involving the pairing of modes of a BBN-type passenger vehicle. A detailed numerical model of the vehicle was developed in ANSYS® based on shell, beam, spring-dashpot assemblies and concentrated mass elements. Particularly, the shell elements were used to represent the panels of the carbody, the beam elements were used to model the bogies, the spring-dashpot assemblies were used to model the suspension components, and the mass elements were used to represent the non-structural equipment and components of the vehicle. The localized damping effect introduced by the dampers led to a non-proportional damping matrix and, consequently, to complex modes.

A numerical modal analysis was performed, in which it was possible to identify the main rigid-body modes of the carbody in addition to the first structural modes associated with the bending and torsional movements of the carbody. In addition to these global modes, it was possible to identify several local modes, mainly for bending the carbody panels. These modes presented global components of small amplitude with a format very similar to some of the global modes, which imposes extra difficulties in the pairing process.

The numerical modes were paired with five experimental modes, from which information was available on the modal ordinate amplitudes of the 14 points that were instrumented on the vehicle floor. To demonstrate the efficiency of the proposed criterion, the pairing process was performed by the MAC and EMAC, with the aim of comparing the results obtained. After applying both criteria, the EMAC proved to be much more assertive in establishing the correct correspondences between numerical and experimental modes, even in challenging situations, due to the reduced number and positioning constraints of the sensors and the complexity of the numerical model. The weighting by the modal deformation energy, used in the EMAC, was able to significantly reduce the erroneous correspondences between the experimental global modes and the purely local modes verified when applying the MAC. In addition, even more assertive results are expected in situations with a higher number of sensors and larger spatial distribution.

In conclusion, the new formulation proposed for the application of the EMAC criterion to complex modes has proved to be very promising and represents an advance for future applications involving the updating of numerical models in the presence of non-proportional damping conditions.

Author Contributions: Conceptualization, D.R.; Methodology, M.B. and V.Z.; Software, D.R. and C.B.; Validation, D.R.; Writing—original draft, D.R.; Writing—review & editing, C.B., M.B. and V.Z.; Supervision, R.C. All authors have read and agreed to the published version of the manuscript.

Funding: This work was financially supported by Base Funding—UIDB/04708/2020 and Programmatic Funding—UIDP/04708/2020 of the CONSTRUCT—Instituto de I&D em Estruturas e Construções, funded by national funds through the FCT/MCTES (PIDDAC). The authors also wish to express their gratitude to Eng. João Pereira, from CP, and Engineers Rui Pereira, Nuno Freitas, Pedro Conceição and Carlos Touret, from EMEF, for their cooperation and provided information on the Alfa Pendular train and for the advantages granted on the performance of the dynamic tests.

Institutional Review Board Statement: Not applicable.

Informed Consent Statement: Not applicable.

Data Availability Statement: Data is unavailable due to privacy restrictions.

Conflicts of Interest: The authors declare no conflict of interest.

References

1. Bragança, C.; Neto, J.; Pinto, N.; Montenegro, P.A.; Ribeiro, D.; Carvalho, H.; Calçada, R. Calibration and Validation of a Freight Wagon Dynamic Model in Operating Conditions Based on Limited Experimental Data. *Veh. Syst. Dyn.* **2022**, *60*, 3024–3050. [CrossRef]
2. Silva, R.; Ribeiro, D.; Bragança, C.; Costa, C.; Arêde, A.; Calçada, R. Model Updating of a Freight Wagon Based on Dynamic Tests under Different Loading Scenarios. *Appl. Sci.* **2021**, *11*, 10691. [CrossRef]
3. Ribeiro, D.; Calçada, R.; Delgado, R.; Brehm, M.; Zabel, V. Finite Element Model Updating of a Bowstring-Arch Railway Bridge Based on Experimental Modal Parameters. *Eng. Struct.* **2012**, *40*, 413–435. [CrossRef]
4. Ribeiro, D.; Calçada, R.; Delgado, R.; Brehm, M.; Zabel, V. Finite-Element Model Calibration of a Railway Vehicle Based on Experimental Modal Parameters. *Veh. Syst. Dyn.* **2013**, *51*, 821–856. [CrossRef]
5. Meixedo, A.; Ribeiro, D.; Calçada, R.; Delgado, R. Global and Local Dynamic Effects on a Railway Viaduct with Precast Deck. In Proceedings of the 2nd International Conference on Railway Technology: Research, Development and Maintenance, Ajaccio, France, 8–11 April 2014. [CrossRef]
6. Ribeiro, D.; Bragança, C.; Costa, C.; Jorge, P.; Silva, R.; Arêde, A.; Calçada, R. Calibration of the Numerical Model of a Freight Railway Vehicle Based on Experimental Modal Parameters. *Structures* **2022**, *38*, 108–122. [CrossRef]
7. Clementi, F.; Pierdicca, A.; Formisano, A.; Catinari, F.; Lenci, S. Numerical Model Upgrading of a Historical Masonry Building Damaged during the 2016 Italian Earthquakes: The Case Study of the Podestà Palace in Montelupone (Italy). *J. Civ. Struct. Health Monit.* **2017**, *7*, 703–717. [CrossRef]

8. Costa, C.; Ribeiro, D.; Jorge, P.; Silva, R.; Arêde, A.; Calçada, R. Calibration of the Numerical Model of a Stone Masonry Railway Bridge Based on Experimentally Identified Modal Parameters. *Eng. Struct.* **2016**, *123*, 354–371. [CrossRef]
9. Pierdicca, A.; Clementi, F.; Fortunati, A.; Lenci, S. Tracking Modal Parameters Evolution of a School Building during Retrofitting Works. *Bull. Earthq. Eng.* **2019**, *17*, 1029–1052. [CrossRef]
10. Alves, V.N.; de Oliveira, M.M.; Ribeiro, D.; Calçada, R.; Cury, A. Model-Based Damage Identification of Railway Bridges Using Genetic Algorithms. *Eng. Fail. Anal.* **2020**, *118*, 104845. [CrossRef]
11. Huang, M.-S.; Gül, M.; Zhu, H.-P. Vibration-Based Structural Damage Identification under Varying Temperature Effects. *J. Aerosp. Eng.* **2018**, *31*, 04018014. [CrossRef]
12. Brehm, M.; Zabel, V.; Bucher, C. An Automatic Mode Pairing Strategy Using an Enhanced Modal Assurance Criterion Based on Modal Strain Energies. *J. Sound Vib.* **2010**, *329*, 5375–5392. [CrossRef]
13. Ewins, D.J. Model Validation: Correlation for Updating. *Sadhana Acad. Proc. Eng. Sci.* **2000**, *25*, 221–234. [CrossRef]
14. Lein, C.; Beitelschmidt, M. Comparative Study of Model Correlation Methods with Application to Model Order Reduction. In Proceedings of the ISMA 2014—International Conference on Noise and Vibration Engineering and USD 2014–International Conference on Uncertainty in Structural Dynamics, Leuven, Belgium, 15–17 September 2014; pp. 2683–2700.
15. Pascual, R.; Golinval, J.C.; Razeto, M. A Frequency Domain Correlation Techinique for Model Correlation and Updating. In Proceedings of the 15th International Modal Analysis Conference (IMAC XV), Orlando, FL, USA, 3–7 February 1997.
16. Allemang, R.J.; Brown, D.L. Correlation Coefficient for Modal Vector Analysis. In Proceedings of the International Modal Analysis Conference & Exhibit, Orlando, FL, USA, 6–9 February 1982; pp. 110–116.
17. Ewins, D.J. *Modal Testing: Theory, Practice and Application*, 2nd ed.; Wiley: Baldock, UK, 2009; ISBN 978-0863802188.
18. Pastor, M.; Binda, M.; Harčarik, T. Modal Assurance Criterion. *Procedia Eng.* **2012**, *48*, 543–548. [CrossRef]
19. Heylen, W.; Janter, T. Extensions of the Modal Assurance Criterion. *J. Vib. Acoust. Trans. ASME* **1990**, *112*, 468–472. [CrossRef]
20. Vacher, P.; Jacquier, B.; Bucharles, A. Extensions of the MAC Criterion to Complex Modes. In Proceedings of the ISMA 2010–International Conference on Noise and Vibration Engineering, including USD 2010, Leuven, Belgium, 20–22 September 2010; pp. 2713–2725.
21. Allemang, R.J. The Modal Assurance Criterion–Twenty Years of Use and Abuse. *Sound Vib.* **2003**, *37*, 14–21.
22. Sternharz, G.; Kalganova, T.; Mares, C.; Meyeringh, M. Comparative Performance Assessment of Methods for Operational Modal Analysis during Transient Order Excitation. *Mech. Syst. Signal Process.* **2022**, *169*, 108719. [CrossRef]
23. Ticona Melo, L.R.; Ribeiro, D.; Calçada, R.; Bittencourt, T.N. Validation of a Vertical Train–Track–Bridge Dynamic Interaction Model Based on Limited Experimental Data. *Struct. Infrastruct. Eng.* **2020**, *16*, 181–201. [CrossRef]
24. Ribeiro, D.; Calçada, R.; Brehm, M.; Zabel, V. Calibration of the Numerical Model of a Track Section over a Railway Bridge Based on Dynamic Tests. *Structures* **2021**, *34*, 4124–4141. [CrossRef]
25. Brehm, M. *Vibration-Based Model Updating: Reduction and Quantification of Uncertainties*; Bauhaus Universitat Weimar: Weimar, Germany, 2011.
26. Jaishi, B.; Ren, W.-X. Structural Finite Element Model Updating Using Ambient Vibration Test Results. *J. Struct. Eng.* **2005**, *131*, 617–628. [CrossRef]
27. Akiyama, Y.; Tomioka, T.; Takigami, T.; Aida, K.; Kamada, T. A Three-Dimensional Analytical Model and Parameter Determination Method of the Elastic Vibration of a Railway Vehicle Carbody. *Veh. Syst. Dyn.* **2020**, *58*, 545–568. [CrossRef]
28. Pereira, S.; Magalhães, F.; Gomes, J.P.; Cunha, Á. Modal Tracking under Large Environmental Influence. *J. Civ. Struct. Health Monit.* **2022**, *12*, 179–190. [CrossRef]
29. Bonisoli, E.; Lisitano, D.; Vigliani, A. Damping Identification and Localisation via Layer Method: Experimental Application to a Vehicle Chassis Focused on Shock Absorbers Effects. *Mech. Syst. Signal Process.* **2019**, *116*, 194–216. [CrossRef]
30. Pérez, M.A.; Serra-López, R. A Frequency Domain-Based Correlation Approach for Structural Assessment and Damage Identification. *Mech. Syst. Signal Process.* **2019**, *119*, 432–456. [CrossRef]
31. Van Der Auweraer, H.; Iadevaia, M.; Emborg, U.; Gustavsson, M.; Tengzelius, U.; Horlin, N. Linking Test and Analysis Results in the Medium Frequency Range Using Principal Field Shapes. In Proceedings of the the 23rd International Conference on Noise and Vibration Engineering, ISMA, Leuven, Belgium, 16–18 September 1998; pp. 129–136.
32. Cuadrado, M.; Pernas-Sánchez, J.; Artero-Guerrero, J.A.; Varas, D. Detection of Barely Visible Multi-Impact Damage on Carbon/Epoxy Composite Plates Using Frequency Response Function Correlation Analysis. *Measurement* **2022**, *196*, 111194. [CrossRef]
33. Pérez, M.A.; Manjón, A.; Ray, J.; Serra-López, R. Experimental Assessment of the Effect of an Eventual Non-Invasive Intervention on a Torres Guitar through Vibration Testing. *J. Cult. Herit.* **2017**, *27*, S103–S111. [CrossRef]
34. Pascual, R.; Razeto, M.; Golinval, J.C.; Schalchli, R. A Robust FRF-Based Technique for Model Updating. In Proceedings of the International Conference on Noise and Vibration Engineering ISMA, Leuven, Belgium, 16–18 September 2002; pp. 1037–1045.
35. Magalhães, F.M.R.L. *de Identificação Modal Estocástica Para Validação Experimental de Modelos Numéricos*; Universidade do Porto: Porto, Portugal, 2004.
36. Craig, R.R., Jr.; Kurdila, A.J. *Fundamentals of Structural Dynamics*, 2nd ed.; John Wiley & Sons: Hoboken, NJ, USA, 2006; ISBN 0471430447.

37. ANSYS Inc. *ANSYS®Theory Reference Manual*; ANSYS Inc: Canonsburg, PA, USA, 2007.
38. ARTeMIS. *ARTeMIS Extractor Pro–Academic License, User's Manual*; SVS: Aalborg, Denmark, 2009.

Article

Research on Geometric Parameters Optimization of Fixed Frog Based on Particle Swarm Optimization Algorithm

Rang Zhang *, Gang Shen and Xujiang Wang

Institute of Rail Transit, Tongji University, Shanghai 201804, China
* Correspondence: zhangrang@tongji.edu.cn

Abstract: In this paper, to improve the wheel/rail dynamic performance of the vehicle passing through the fixed frog area and improve the service life of the fixed frog, a geometric parameter optimization design method of the fixed frog area is proposed using particle swarm optimization (PSO). Based on the variable section rail profile interpolation algorithm and wheel/rail contact solution algorithm, the wheel/rail contact characteristics of the fixed frog area are analyzed. Then, the vehicle-fixed frog dynamic model is built using a MATLAB/Simulink platform to complete the dynamic calculation and analysis of the rail vehicle passing through the fixed frog area. Finally, based on the wheel/rail contact characteristics of the fixed frog area, and take wheel/rail forces as the optimization goal, the optimization design method for the wing rail lifting value and the nose rail height of the fixed frog area is proposed. The comparative analysis shows that the wheel/rail dynamic performance in the fixed frog area has been greatly improved after optimization, which verifies the feasibility of the optimization strategy.

Keywords: fixed frog; wheel/rail interaction; geometric parameter optimization; PSO

Citation: Zhang, R.; Shen, G.; Wang, X. Research on Geometric Parameters Optimization of Fixed Frog Based on Particle Swarm Optimization Algorithm. *Appl. Sci.* **2022**, *12*, 11549. https://doi.org/10.3390/app122211549

Academic Editor: Suchao Xie

Received: 23 October 2022
Accepted: 11 November 2022
Published: 14 November 2022

Publisher's Note: MDPI stays neutral with regard to jurisdictional claims in published maps and institutional affiliations.

1. Introduction

Turnout is the key part of the track structure of the railway system, and its role is to guide the rail vehicle from one track to another, so as to give full play to the transportation efficiency of the line. However, due to its complex structure, large number of applications, short service life, low traffic safety, limited train speed and high maintenance cost, it brings a great workload to the daily maintenance and overhaul of the railway public works department. Therefore, turnout, curve and joint are called the three weak links of track [1]. As the fixed frog turnout is cheaper and more reliable than the movable core rail turnout, it is widely used in a large number of the common railway, the heavy-haul line and the urban rail transit line [2].

The fixed frog turnout mainly includes switch parts, connecting parts and frog area parts. Among them, the gap from the throat of the frog to the actual point of nose rail is called the harmful space of a fixed frog. The wheel/rail relationship of the wheelset passing through a harmful space is very complex, which is also the main reason for limiting the vehicle crossing speed [3].

Due to the harmful space, the wheel/rail force in the frog is much greater than that in the general section line. This violent dynamic interaction becomes more and more serious with the speed increase of the train. This can not only cause strong wheel/rail impact and even derailment of vehicles, but also cause serious wear, fracture and other turnout diseases of the nose rail in the frog section. Therefore, it is particularly essential to research the wheel/rail interaction relationship of vehicles passing through the fixed frog area.

Many scholars have carried out extensive research on reducing wheel/rail impact in a fixed frog area. Ren [4,5] established the vehicle-frog vertical vibration model, calculated the wheel/rail impact force in the fixed frog through a self-made program, carrying out the wheel/rail dynamics research first in the fixed frog area in China. Wang [6,7] regarded the fixed frog as a finite length variable cross-section Euler beam, established a vehicle-turnout spatial coupling vibration model including the vibration of the fixed frog, the connecting part and the switch, and comprehensively research the influence of vehicle and track parameters on the wheel/rail relationship in the turnout area. Lagos [8] used four different treads and two different turnouts to conduct a dynamic simulation analysis of vehicle crossing, and results show the geometric design for turnouts play a very important role on the dynamic performance of vehicle crossing. Markine [9,10] established a complete dynamic model of the fixed frog to research the fatigue failure of the nose rail under the high-frequency wheel/rail impact load, analyzed the influence of the stiffness of the lower support structure in the fixed frog area on the contact fatigue failure of the nose rail, and found that the use of the under-rail pad with small stiffness is helpful to decrease the impact force on the rail. Grossoni [11] iteratively optimized the stiffness under the track in the fixed frog area based on Genetic Algorithm to reduce the common failure forms such as fatigue failure of the fixed frog. Anderson [12,13] established the wheel/turnout contact finite element model considering the characteristics of track variable stiffness, rail variable section and wheel/rail multi-point contact in detail, and analyzed the dynamic performance of the wheel/turnout in a wide frequency range. Blanco-Saura [14] built a detailed 3D finite element model for turnout by using ANSYS software, and combined with the vehicle dynamics model established by VAMPIRE Pro, analyzed the vertical dynamic response of turnout under dynamic load, especially the vertical dynamic force characteristics of center rail and switch rail area.

In the research on the optimization method for the wing rail lifting value and the nose rail height, Palsson [15] took the wheel/rail contact stress as the optimal goal, and proposes a cross geometric design strategy for the crossing to optimize the rail shape in the frog area, so as to reduce the rail damage. Wan [16] presented an optimization strategy for deceasing wheel/rail wear through changing nose rail profiles. Cao [17,18] obtained a more reasonable reduction value of a key section of nose rail through researching the wheel/rail static parameters and wheel/rail dynamic action performance in a fixed frog area. Xu [19] obtained a fixed frog optimization design method based on the ratio of the required height difference of wheel profile to the actual height difference of rail top profile in the fixed frog. Zhang [20] focused on the effects of the wing rail height value in fixed frog area on driving stability and wheel/rail dynamic action when vehicles pass through the turnout center.

Although the existing research has done a lot of in-depth and continuous work, there are still shortcomings. When the number of control sections of fixed frogs is large, the arrangement and combination schemes increase rapidly, which leads to the extremely low efficiency of trial and error method and is basically not feasible. Therefore, it is a difficult problem to propose a complete and standardized design process for the optimization design of key geometric parameters of the fixed frog area. This paper will focus on the wheel/rail interaction performance of rail vehicles passing through the fixed frog area, and decrease the wheel/rail impact by optimizing and improving geometric structure in the frog area, especially the wing rail lifting value and the nose rail height, so as to provide some theoretical basis for turnout design, application and maintenance. Additionally, to change the traditional design method of repeated trial and error and then dynamic verification, based on the comprehensive analysis of the wheel/rail contact relationship in the fixed frog area, a closed-loop design method of the geometric parameter optimization in the fixed frog area is proposed by using the particle swarm optimization algorithm, which has great engineering application value.

2. Wheel/Rail Contact Analysis in the Fixed Frog Area

To comprehensively explore the wheel/rail contact characteristics in a fixed frog area, variation laws of key wheel/rail contact parameters under different wheelset lateral displacement are analyzed; see Figure 1 for the calculation results. Figure 1a,b show that the wheelset passing the fixed frog area, the lateral coordinate of the wheel/rail contact points gradually increases from the throat with the wing rail continuously deviating from the gauge center line, and the vertical coordinate gradually decreases with the wing rail continuously height. This means that during the contact between the wing rail and wheel, wheel/rail contact points transfer from near the wheel profile rolling circle to the side away from the flange. When the wheelset is about 240 mm away from theoretical point of frog, the wheelset load begins to transfer to the nose rail, and coordinates of the wheel/rail contact point suddenly change. Additionally, wheel/rail contact point lateral coordinates instantaneously decrease and the vertical coordinates instantaneously increase. Then, with the continuous widening and lifting of the nose rail, lateral coordinates of wheel/rail contact points continue to increase, and the vertical coordinates continue to decrease and gradually return to the initial value.

From the perspective of the wheelset lateral shift, with the wheelset lateral shift increasing and moving towards the nose rail, the smaller the lateral coordinate jump amplitude of wheel/rail contact coordinate points and the larger vertical coordinate jump amplitude. At the same time, it also means that the shorter the contact time between the wing rail and wheel, the position of the section where the nose rail begins to bear load is closer to the theoretical point. Figure 1c shows that the rolling circle radius increases to more than 440 mm instantaneously, which means that the wheel flange contact occurred, resulting in a sharp increase in the vertical contact coordinate points and a risk that the wheelset will climb to the nose rail surface.

The above analysis shows that when the wheelset load is moved from the wing rail to the nose rail, the wheel will jump. If the wing rail height value is set unreasonably, it will directly lead to a large jump of the wheel, thus deteriorating the wheel/rail contact performance in the frog. In addition, to strictly guarantee the safety of the vehicle and prevent the wheelset from climbing the track in the wheel load transfer stage, the geometric structure design of the fixed frog should ensure that the wheelset passes through the turn-out center with a small lateral displacement.

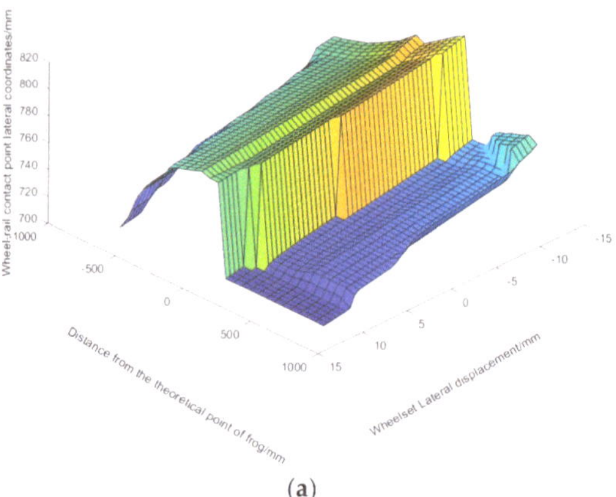

(a)

Figure 1. *Cont.*

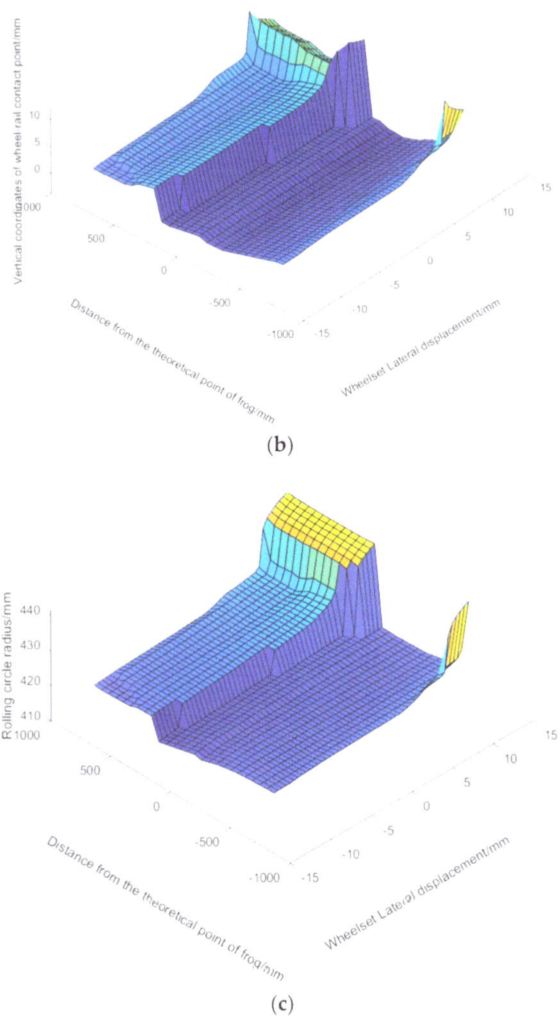

(b)

(c)

Figure 1. Wheel/rail contact calculation results of key sections: (**a**) Lateral coordinates of wheel/rail contact point; (**b**) Vertical coordinates of wheel/rail contact point; (**c**) Rolling circle radius.

3. Dynamics Modeling of Vehicle/Fixed Frog Interaction

3.1. Vehicle Dynamics Modeling

A dynamic model for the four-axle rail vehicle is built to more accurately simulate the dynamic characteristics of the rail vehicle passing the fixed frog area. The frog bodies of the vehicle dynamics system include 1 car body, 2 frames, and 4 wheelsets. In addition, the suspension system includes primary suspension and secondary suspension as shown in Figure 2. Assuming the vehicle runs at a uniform speed, the telescopic vibration of the wheelset, frames, and car body are negligible. Each rigid body considers 5 degrees of freedom (DOF), a total of 35 DOFs as shown in Table 1.

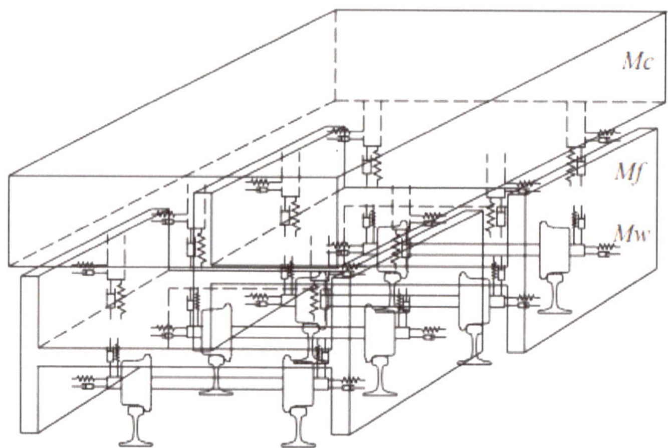

Figure 2. Calculation diagram of four-axle rail vehicle.

Table 1. Model DOFs.

Rigid Body	Lateral Motion	Vertical Motion	Rolling Motion	Pitch Motion	Yaw Motion	/
Car body	Y_c	Z_c	Φ_c	Θ_c	Ψ_c	/
Frames	Y_{bn}	Z_{bn}	Φ_{bn}	Θ_{bn}	Ψ_{bn}	$n = 1, 2$
Wheelsets	Y_{wi}	Z_{wi}	Φ_{wi}	Θ_{bi}	Ψ_{wi}	$i = 1, 2, 3, 4$

Based on the matrix assembly method [21], the general representation of the dynamic equation for the railway vehicle is:

$$\mathbf{M}\{\ddot{\mathbf{q}}\} + \mathbf{C}\{\dot{\mathbf{q}}\} + \mathbf{K}\{\mathbf{q}\} = \{\mathbf{F}\}$$

where, $\{\mathbf{q}\}$ is the DOF vector of the model;

$\mathbf{M}$ is the inertia matrix;

$\mathbf{C}$ is the damping matrix;

$\mathbf{K}$ is the stiffness matrix;

$\{\mathbf{F}\}$ is the external force matrix, ignoring the track irregularity, which is mainly composed of 3 parts:

$$\mathbf{F} = \mathbf{F}_{wr} + \mathbf{F}_{IG} + \mathbf{F}_c$$

The meaning of each force component is as follows:

$\mathbf{F}_{wr}$ is the force transmitted by the rail to the wheelset, including normal force and tangential force (creep force);

$\mathbf{F}_{IG}$ is gravity, inertia force, and Coriolis force;

$\mathbf{F}_c$ is the zero position non-equilibrium additional force due to curve line conditions.

This paper uses MATLAB/Simulink to build the railway vehicle dynamic model for the subsequent optimization solution, as shown in Figure 3.

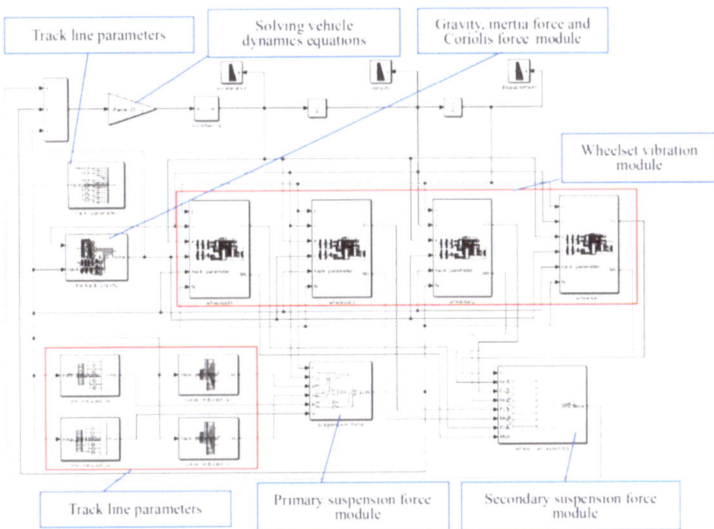

Figure 3. Vehicle dynamics model based on Simulink.

3.2. Rail Vibration System Model

According to the different problems studied, the rail dynamics models selected are different, mainly including the massless rail, the inertial rail, and the flexible track. Among them, the inertial rail model can be for simulation of complex wheel/rail contact: railway track dynamic performance in the switches and turnouts, contacts between wheel-back and rail, vehicle derailment, etc., so the inertial rail model is selected for dynamic calculation in the paper. Assuming wheel/rail contact of a certain section on one side, the rail is simplified as a mass block with 3 DOFs: lateral shift, vertical shift and torsion, and the mass block is connected with the foundation through equivalent spring and damping. Selecting right rail as an illustration, the dynamic modeling of the rail vibration system is carried out as shown in Figure 4.

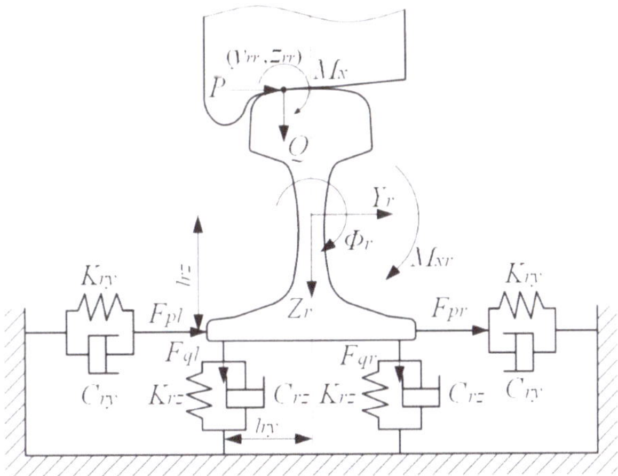

Figure 4. Equivalent model of track vibration model.

The coordinate of wheel/rail contact point is (y_{rr}, z_{rr}) in the rail coordinate system. P is the lateral component, Q is the vertical component, and M_x is the roll moment, respectively.

K_{ry} and C_{ry} are the lateral support stiffness and damping of sleeper, fastener, and other basic components to the rail. K_{rz} is the vertical foundation stiffness of sleeper, fastener, and other basic components to the rail, and C_{rz} is the damping. Therefore, the fulcrum reaction of the rail is shown in Equation (1).

$$\begin{cases} F_{ql} = -K_{rz}(z_r - \phi_r l_{ry}) - C_{rz}(\dot{z}_r - \dot{\phi}_r l_{ry}) \\ F_{qr} = -K_{rz}(z_r + \phi_r l_{ry}) - C_{rz}(\dot{z}_r + \dot{\phi}_r l_{ry}) \\ F_{pl} = -K_{ry}(y_r - \phi_r l_{rz}) - C_{ry}(\dot{z}_r - \dot{\phi}_r l_{rz}) \\ F_{pr} = -K_{ry}(y_r - \phi_r l_{rz}) - C_{ry}(\dot{z}_r - \dot{\phi}_r l_{rz}) \end{cases} \tag{1}$$

The torsional torque of the rail is shown in Equation (2):

$$M_{xr} = -K_{\phi r}\phi_r - C_{\phi r}\dot{\phi}_r \tag{2}$$

where, $K_{\phi r}$ is the rail torsional stiffness; $C_{\phi r}$ is the rail torsional damping.

Further, the vibration expression for the rail system can be obtained as shown in Equation (3):

$$\begin{cases} m_r \ddot{y}_r &= P + F_{pl} + F_{pr} \\ m_r \ddot{z}_r &= Q + F_{ql} + F_{qr} + m_r g \\ I_{rx}\ddot{\phi}_r &= M_x + P y_{rr} - Q z_{rr} + M_{xr} - F_{pl} l_{rz} - F_{pr} l_{rz} + F_{qr} l_{ry} - F_{ql} l_{ry} \end{cases} \tag{3}$$

where, m_r is the equivalent mass of rail, y_r, z_r, and ϕ_r are lateral, vertical shift, and torsional angle of the rail.

The inner back of the wheelset impacts the side of the guard rail, resulting in additional wheel/rail impact force. Due to the short action time and large amplitude of this wheel/rail impact force, it can be considered as the lateral impact force between wheel-back and guard rail [16].

To simulate the instantaneous impact load between wheel-back and guard rail, a contact condition is equivalent to the spring with great stiffness. Therefore, the wheel/rail elastic compression equation is:

$$\varepsilon = l_{wc} - l_{tc} + y_{rail} + 0.009 + d_a - \Delta_a(x) \tag{4}$$

where, k_a is the equivalent spring stiffness, l_{wc} is the lateral distance between wheel/rail contact point and the origin in wheelset translation coordinate system, l_{tc} is the lateral distance between wheel/rail contact point and the origin in the rail coordinate system, y_{rail} is the dynamic traverse of stock rail, d_a is the wheel flange thickness, $\Delta_a(x)$ is the wheel flange groove width at current position of the wheelset. Further, the wheel/rail impact force is:

$$F_a = \begin{cases} k_a \varepsilon, & \varepsilon > 0 \ \ (\text{Contact state}) \\ 0, & \varepsilon \leq 0 \ \ (\text{Non contact state}) \end{cases} \tag{5}$$

3.3. Real-Time Calculation for the Wheel/Rail Interaction

For the optimization problem in the paper, real-time calculation is indispensable. The transmission of wheel/rail creep force and normal forces between vehicle vibration system and rail vibration system are completed through the wheel/rail interaction module.

Figure 5 shows the calculation process of wheel/rail interaction. Firstly, calculate the wheel/rail real-time contact based on the wheelset and rail motion state at time t. Then, calculate wheel/rail creep rate, normal forces and contact forces between wheel-back and guard rail based on the wheel/rail real-time contact parameters and the wheel/rail real-time contact parameters, then calculate wheel/rail contact spots and creep coefficient based on the normal force and wheel/rail real-time contact parameters, and then calculate the creep force. Finally, the above creep forces, normal forces, and lateral impact forces are input into the vehicle model and the rail vibration system respectively, and motion states of the wheel and rail at $t + 1$ time is obtained, which starts again and again until the vehicle runs to the end of the frog. Based on the above analysis, Simulink is used to build a real-time calculation module for the wheel/rail interaction as shown in Figure 6.

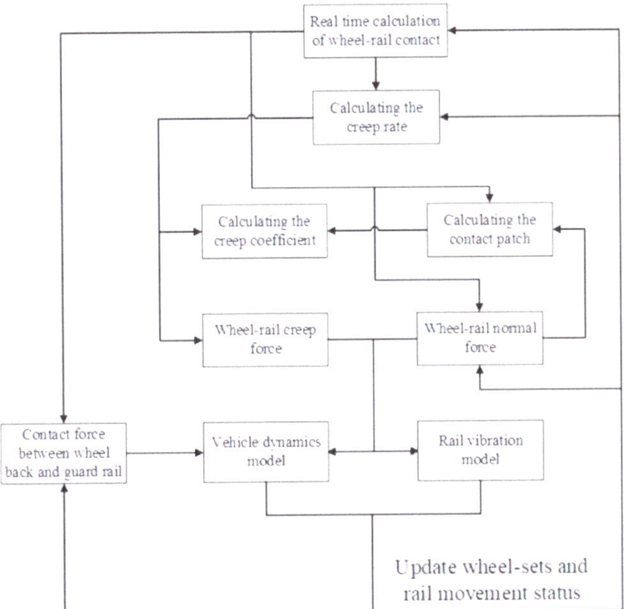

Figure 5. Calculation process of wheel/rail interaction.

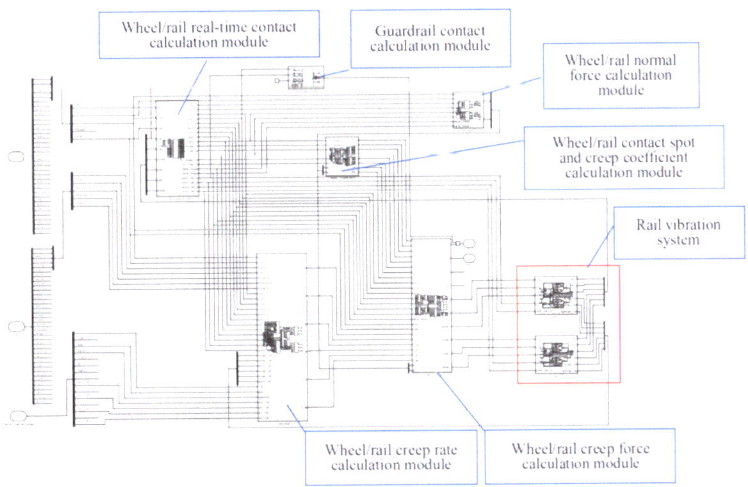

Figure 6. Wheel/rail interaction calculation module based on Simulink.

4. Geometric Parameter Optimization Design for the Fixed Frog Area

4.1. PSO Algorithm

PSO algorithm has faster calculation speed and better global search ability, and each particle will continuously follow its own position and flight speed according to the individual optimal solution and group optimal solution generated in the iterative process, and then track the present optimal particle to constantly find in the search space according to certain rules, so as to solve the complex optimization problem [22].

According to the particle swarm optimization algorithm, it is assumed that N particles are constantly searching for optimization in a D-dimensional search space, and the current position and flight speed of the i-th particle are recorded as Equations (6) and (7).

$$H_i = (h_{i1}, h_{i2}, \ldots, h_{iD}), \quad i = 1, 2, \ldots, N \tag{6}$$

$$V_i = (v_{i1}, v_{i2}, \ldots, v_{iD}), \quad i = 1, 2, \ldots, N \tag{7}$$

The individual optimal solution and group optimal solution of the i-th particle are recorded as:

$$P_{\text{best}} = (p_{i1}, p_{i2}, \ldots, p_{iD}) \tag{8}$$

$$G_{\text{best}} = (g_{i1}, g_{i2}, \ldots, g_{iD}) \tag{9}$$

Before finding the target solution, all particles update the flight speed and position at the next time according to Equation (10):

$$\begin{cases} v_{ij}(t+1) = \omega v_{ij}(t) + c_1 r_1 \left[p_{ij}(t) - h_{ij}(t) \right] + c_2 r_2 \left[g_j(t) - h_{ij}(t) \right] \\ h_{ij}(t+1) = h_{ij}(t) + v_{ij}(t+1) \end{cases} \tag{10}$$

where, h_{ij} is the current position of each particle, v_{ij} is the flight speed of each particle, and $v_{ij} \in [-v_{\max}, v_{\max}]$, c_1 is the individual learning factor, c_2 is the social learning factor, p_{ij} is the individual optimal solution, p_j is the overall optimal solution, r_1 and r_2 are the random number between [0, 1], and ω is the inertia weight.

In order to ensure that the PSO algorithm has a high probability of convergence to the global optimal position, the inertia weight can be adjusted dynamically ω to balance global search ability and local search ability, as shown in Equation (11):

$$\omega = \omega_{\max} - \frac{(\omega_{\max} - \omega_{\min})}{T_{\max}} t \tag{11}$$

where, $\omega_{\max}$ is the maximum inertia weight, $\omega_{\min}$ is the minimum inertia weight, $T_{\max}$ is the maximum iterations number, and t is the current iterations number.

4.2. Optimization Design Method for the Wing Rail Lifting Values

Figure 7 shows the plan view and side view of the wing rail, which deviates to both sides of gauge line with the longitudinal extension of frog. Section A is the throat of the frog, and section K is the section with the top width of the nose rail of 50 mm. Nine key sections are taken between section A and section K, then the wing rail can be interpolated from a total of 11 sections above. Section A is the starting point of deviation and lifting, so the lifting value of the wing rail at section A is set to zero ($h_A = 0$). Based on wheel/rail contact characteristics of the fixed frog area mentioned above, the wheel load transfer has been completed at section K, so the lifting value of the wing rail at section K is also set to zero ($h_k = 0$). The lifting value of the nine sections from section B to section J is set as a series of variable values h_k (k = 1~9), a series of complete wing rail profiles can be obtained by optimizing the wing rail lifting values of the nine sections.

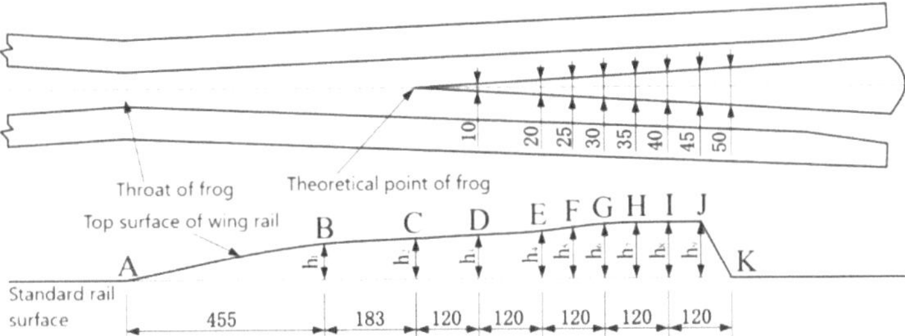

Figure 7. Plan and side view of the wing rail.

As the object of optimization are wheel/rail forces at the stock rail and the frog area, the objective function of the optimization model can be determined by Equation (12):

$$\begin{cases} f_1(h_k) = \max(|F_{yli}|) \\ f_2(h_k) = \max(|F_{zli}|) \\ f_3(h_k) = \max(|F_{yri}|) \\ f_4(h_k) = \max(|F_{zri}|) \end{cases} \quad i = 1, 2, 3, 4, \ k = 1, 2 \ldots 9 \tag{12}$$

where, F_{yli} and F_{yri} are wheel/rail lateral forces at the stock rail and the frog area, and F_{zli} and F_{zri} are wheel/rail vertical forces at the stock rail and the frog area.

According to the order of magnitude and importance, the above four objective functions are weighted in turn, and the total optimization objective function is shown in Equation (13):

$$f(h_k) = \omega_1 f_1(h_k) + \omega_2 f_2(h_k) + \omega_3 f_3(h_k) + \omega_4 f_4(h_k) \tag{13}$$

where, $\omega_1 \sim \omega_4$ are the weight coefficient corresponding to each objective function.

In order to ensure the smooth transfer of wheelset load from the wing rail to nose rail, the height of the wing rail top surface should be gradually increased until the maxi-mum value is reached. Therefore, the constraint conditions are set as Equation (14).

$$0 < h_k < h_{k+1} < h_{\max} \tag{14}$$

Wheel/rail contact points jump near the section with the width of the nose rail top surface of 20 mm. When wheelset rolls from the wing rail to the nose rail from the throat of the frog, if the wing rail height value at section E is set unreasonably, the nose rail will bear severe wheel/rail impact load at the weak section with narrow top width, so the minimum height value $h_{\min}$ shall be set, as shown in Equation (15).

$$h_4 > h_{\min} \tag{15}$$

The optimized wing rail shall ensure the driving safety of the vehicle passing through the fixed frog area. Therefore, the derailment coefficient and wheel load reduction rate at the stock rail side and the frog side shall meet the following constraints, as shown in Equations (16) and (17):

$$\max\left(\left|\frac{F_{yli}}{F_{zli}}\right|, \left|\frac{F_{yri}}{F_{zri}}\right|\right) < 1 \tag{16}$$

$$\max\left(\left|\frac{\overline{P} - F_{zli}}{\overline{P}}\right|, \left|\frac{\overline{P} - F_{zri}}{\overline{P}}\right|\right) < 0.6 \tag{17}$$

where, $\overline{P}$ is the average wheel load of the left and right wheels of the wheelset.

Therefore, the established optimization model of wing rail lifting values is as shown in Equations (18) and (19).

The objective function:

$$\min : f(h_1, h_2, \ldots, h_9) = f(h_k) \tag{18}$$

The constraint condition:

$$\begin{cases} 0 < h_k < h_{k+1} < h_{max} \\ h_4 > h_{min} \\ \max \left(\left| \frac{F_{yli}}{F_{zli}} \right|, \left| \frac{F_{yri}}{F_{zri}} \right| \right) < 1 \\ \max \left(\left| \frac{\overline{P} - F_{zli}}{\overline{P}} \right|, \left| \frac{\overline{P} - F_{zri}}{\overline{P}} \right| \right) < 0.6 \end{cases} \tag{19}$$

Figure 8 is the flow chart of the wing lifting values optimization algorithm based on PSO. The optimization process is as follows:

(1) Set the basic parameters of particle swarm optimization algorithm and initialize the population, and randomly generate N groups of wing rail lifting values according to the constraints.
(2) A complete set of frog profiles is generated according to each lifting value and the profile of each key section, and the vehicle dynamics are calculated.
(3) If the dynamic calculation results conform to safety conditions, the individual optimal solution and group optimal solution shall be updated according to the quality of the objective function value, and then judge whether the current iteration times have reached the maximum iterations. If the maximum iterations have been met, the optimization results shall be output.
(4) If the dynamic calculation results do not meet the safety requirements or the current iterations number does not reach the maximum iterations, obtain a new group N lifting values according to Formula (14) and (15), and repeat the above calculation steps until the optimal lifting values are obtained.

To verify the effectiveness of the wing rail lifting optimization algorithm, the dynamic performance of a metro vehicle passing through No. 12 fixed frog in the main direction at the speed of 80 km/h is calculated, and the effects of the non-optimized wing rail and the optimized wing rail on the dynamic performance are compared. See Table 2 for the main calculation parameters. According to the technical manual of railway public works design [23], the maximum height value of wing rail h_{max} is set to 5 mm, the minimum height value of key section E is set to $h_{min} = 2$ mm. The basic parameters of the PSO algorithm are set as follows according to the recommended values in literature [24]: population dimension $D = 9$, population size $N = 10$ and maximum flight speed $v_{max} = 1$, individual learning factor $c_1 = 1.5$, social learning factor $c_2 = 1.5$, maximum iterations $T_{max} = 100$, maximum inertia weight $\omega_{max} = 0.8$, and minimum inertia weight $\omega_{min} = 0.4$.

Table 2. Principal parameters.

Parameter	Value
Car body mass (kg)	41,910
Frame mass (kg)	4060
Wheelset mass (kg)	1670
Gauge (m)	1.435
Nominal rolling circle radius (m)	0.42

Figure 9 shows the evolution curve of the objective function value of the wing rail lifting optimization model solved by particle swarm optimization algorithm. It can be seen that when the number of iterations is about 40, the objective function value has tended to be stable. The optimization results for the wing rail lifting values are shown in Table 3.

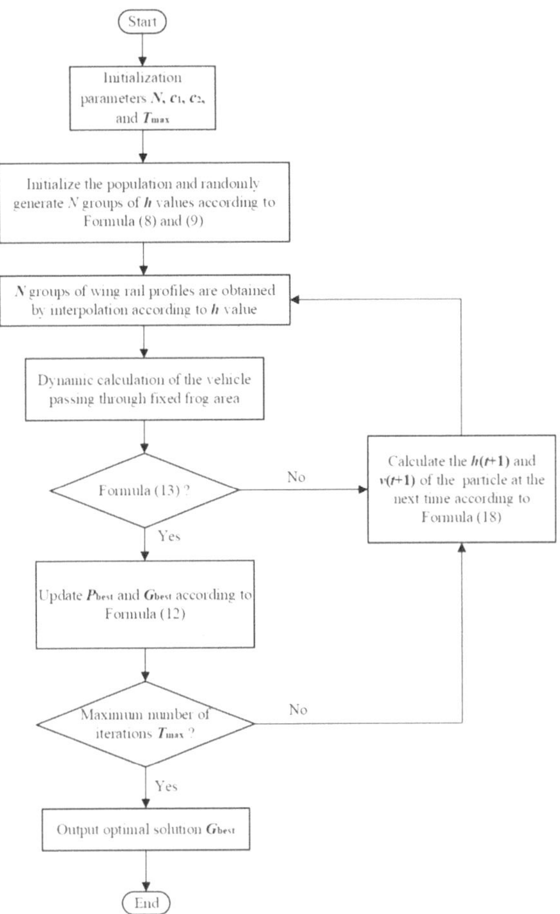

Figure 8. Flow chart of the wing rail optimization algorithm based on the PSO.

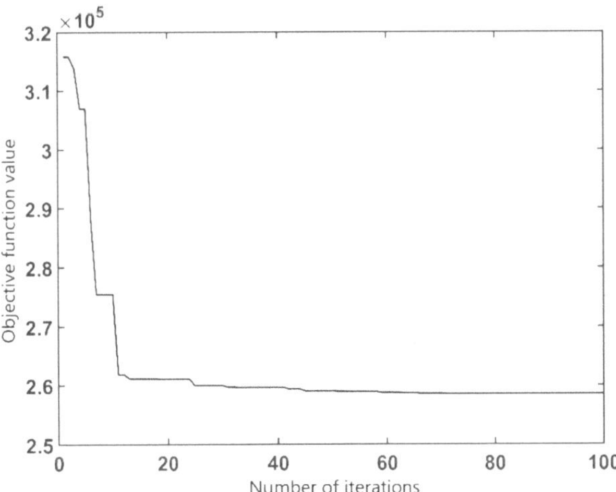

Figure 9. Evolution curve of the objective function value.

Table 3. Optimization results of the wing rail lifting value of each section.

Section Position	Optimized Value/mm
B	1.07
C	2.39
D	3.81
E	4.95
F	4.98
G	4.99
H	5
I	5
J	5

Figure 10 shows the change curve of wheelset vertical displacement of the vehicle passing through the fixed frog before and after optimization. Before wing rail optimization, the maximum vertical displacement of wheelset is 0.78 mm. After the wing rail optimization, the maximum vertical displacement of the wheelset is 0.51 mm, which is reduced by 34.6%, indicating that the optimized wing rail could greatly decrease the vertical vibration amplitude of the wheelset when passing through the frog and improve the running stability of the vehicle.

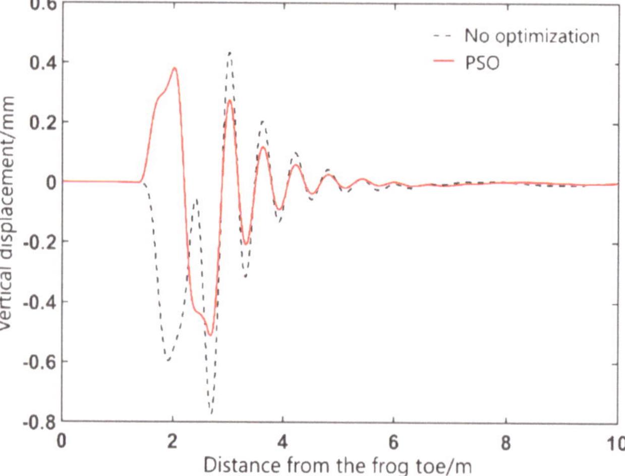

Figure 10. Wheelset vertical displacement.

Figures 11 and 12 show the wheel/rail force curve of the vehicle passing through the fixed frog area in main directions. On the stock rail side, after wing rail optimization, the maximum value of wheel/rail lateral force is reduced from 7.8 kN to 6.8 kN, with a decrease range of 12.8%, and the maximum value of wheel/rail vertical force decreases from 83.4 Kn to 75.1 kN, with a decrease of about 10%. On the frog side, after wing rail optimization, the maximum value of wheel/rail lateral force is reduced from 18.4 Kn to 11.4 kN, with a decrease range of 38%, and the maximum value of wheel/rail vertical force is reduced from 104 kN to 91 kN, with a decrease range of 12.5%. To sum up, the optimized lifting value of wing rail can greatly reduce the impact force between the wheel and rail. In addition, based on the wheel/rail force, the maximum value of the wheel load reduction rate is 0.53 and the maximum value of derailment coefficient is 0.2, indicating that the optimized wing rail can ensure the safety of vehicles passing through the turnout in the main direction.

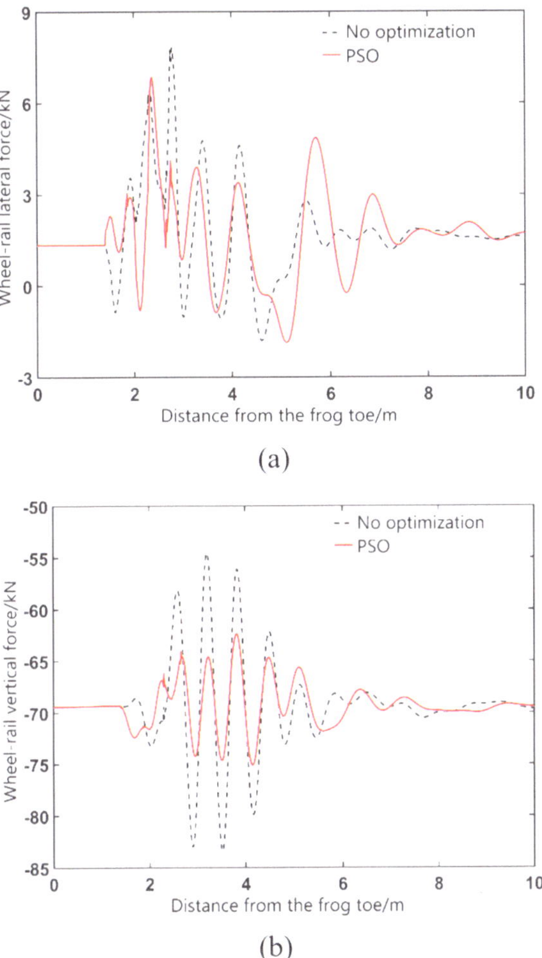

Figure 11. On the stock rail side: (**a**) Wheel/rail lateral force; (**b**) Wheel/rail vertical force.

4.3. Optimization Design Method for the Nose Rail Height

The plan and side schematic diagram of the nose rail in the fixed frog area is shown in Figure 13. The nose rail is located in the center of the two wing rails, symmetrically distributed relative to the gauge line, and continuously widened and heightened with the longitudinal direction of the frog from the theoretical point of frog, so as to meet the needs of the smooth crossing of wheels.

Section A is the theoretical point of frog, the top width of the nose rail at Section B is 20 mm, and the top width of the nose rail at section H is 50 mm. One section is set every 5 mm from section B to section H, and the profile of the nose rail can be interpolated from the above 8 key sections. Set the height of a total of 6 sections from section B to section G to the variable value h_k, $k = 1, 2 \ldots 6$, and set the height values of section A and section H to the fixed value h, respectively, h_A and h_H. A series of nose rail profiles can be generated by continuously adjusting the nose rail height values of the 6 sections.

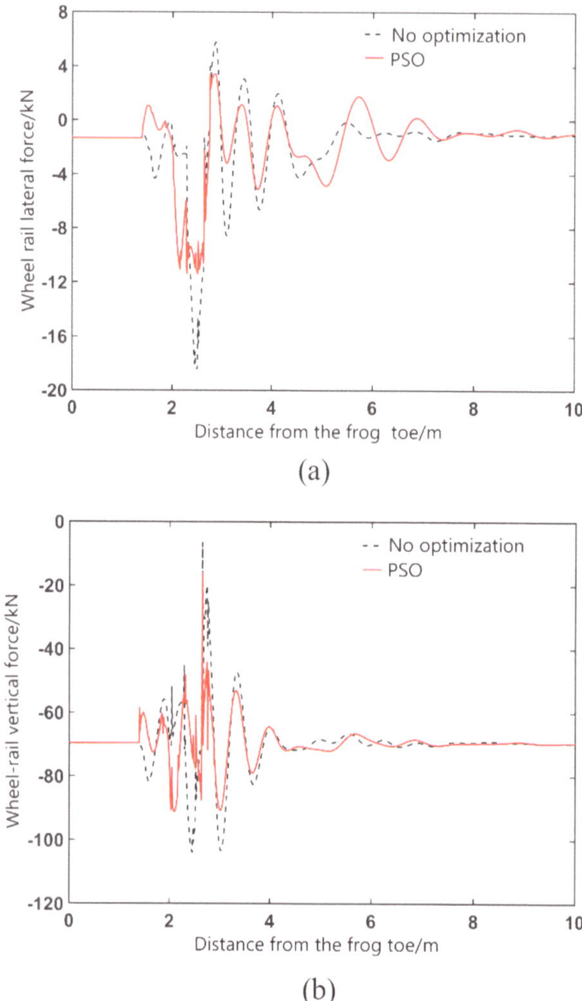

Figure 12. On the frog side: (**a**) Wheel/rail lateral force; (**b**) Wheel/rail vertical force.

In the same way, the wheel/rail force at the stock rail and the frog area is selected as the evaluation index of nose rail height. Therefore, the objective function of the optimization model is determined by Equation (12), and the total objective function is determined by Equation (13). To reduce the wheel/rail impact force generated when the wheel load transits from the wing rail to the nose rail, the height of the top surface of the nose rail should be gradually increased until it reaches the standard rail surface height. In order to make the optimized top surface of the nose rail smooth enough, the height value of each key section should be kept decreasing along the longitudinal direction of the frog, with the following constraints, as shown in Equation (20).

$$h_A > h_k > h_{k+1} > h_H \tag{20}$$

Similarly, the minimum height value h_{min} shall be set according to the bearing capacity of the nose rail, and the height value at section B shall meet the following constraints:

$$h_1 > h_{min} \tag{21}$$

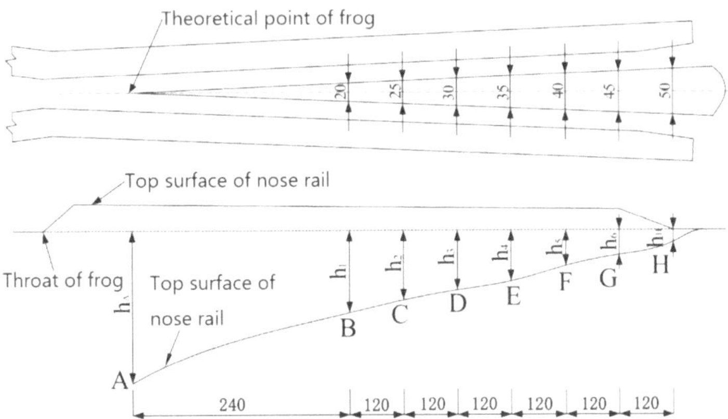

Figure 13. Plan and side view of the nose rail.

According to the technical manual of railway public works design, the nose rail height value of each fixed section is set as: the reduction value at Section A is $h_A = 6$ mm, the minimum height value at Section B is $h_{min} = 2$ mm, the height value at section H is $h_H = 0$ [23]. At the same time, the population dimension in particle swarm optimization algorithm is set to $D = 6$, and other parameters are the same as the wing rail lifting optimization design example. The calculation process is similar to the wing rail optimization method and will not be repeated here.

Therefore, the established optimization model of the nose rail height values is as follows.

The objective function:

$$\min : f(h_1, h_2, \ldots, h_6) = f(h_k) \tag{22}$$

The constraint condition:

$$\begin{cases} h_A > h_k > h_{k+1} > h_H \\ h_1 > h_{min} \\ \max\left(\left|\frac{F_{yli}}{F_{zli}}\right|, \left|\frac{F_{yri}}{F_{zri}}\right|\right) < 1 \\ \max\left(\left|\frac{\overline{P}-F_{zli}}{\overline{P}}\right|, \left|\frac{\overline{P}-F_{zri}}{\overline{P}}\right|\right) < 0.6 \end{cases} \tag{23}$$

Figure 14 shows the evolution curve of the objective function value of the nose rail height optimization model solved by particle swarm optimization algorithm. It can be seen that when the number of iterations is about 85, the objective function value has tended to be stable. The optimization results of the wing rail heightening value are shown in Table 4.

Table 4. Optimization results of the nose rail height value of each section.

Section Position	Optimized Value/mm
B	3.22
C	1.98
D	1.65
E	1.49
F	1.23
G	0.72

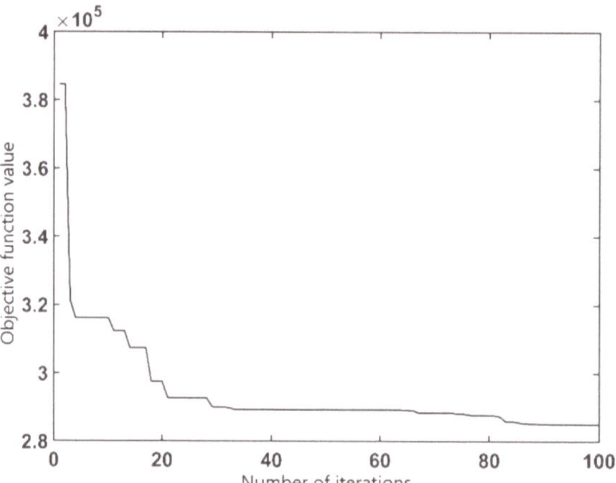

Figure 14. Evolution curve of the objective function value.

Figure 15 shows the change curve of wheel vertical displacement of the vehicle passing through the fixed frog before and after optimization. Before the nose rail optimization, the maximum vertical displacement of wheelset is 0.77 mm. After optimization, the maximum vertical displacement of the wheelset is 0.62 mm, which is reduced by 19.5%.

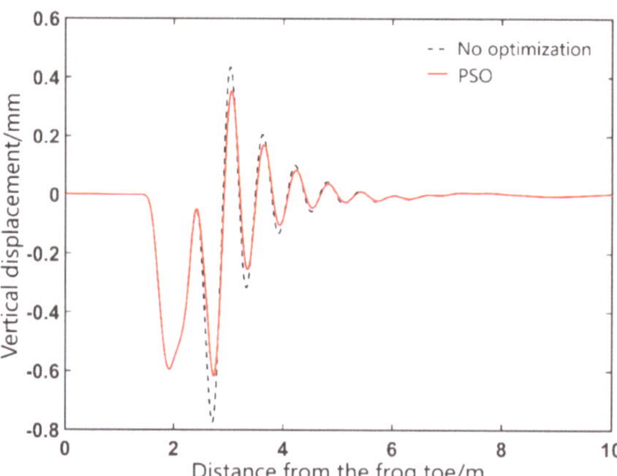

Figure 15. Wheelset vertical displacement.

Figures 16 and 17 show the wheel/rail force curve of the vehicle passing through the fixed frog area in main directions. On side of the stock rail side, after the nose rail optimization, the maximum value of wheel/rail lateral force is reduced from 7.8 kN to 6.6 kN, with a decrease range of 15.4%, and the maximum value of wheel/rail vertical force decreases from 83.4 kN to 80.5 kN, with a decrease of about 3.5%. On the side of the frog, after wing rail optimization, the maximum value of wheel/rail lateral force is reduced from 18.4 kN to 15 kN, with a decrease range of 18.5%, and the maximum value of wheel/rail vertical force is reduced from 106 kN to 96 kN, with a decrease range of 9.4%. To sum up, the optimized height values of nose rail can also reduce the impact force between the wheel and rail. In addition, according to the change curve of wheel/rail force, it can be calculated that the derailment coefficient and wheel load reduction rate are within the safe range, that is, the optimization algorithm does not affect the safety of the vehicle.

The effectiveness of the optimization method is verified by optimizing the wing rail lifting value and nose rail height of the fixed frog area. By analyzing the wheel/rail dynamic response in a fixed frog area before and after optimization, it is found that the optimized fixed frog can significantly improve the wheel/rail relationship in the fixed frog area, reduce the vertical impact force and lateral force of the wheel/rail, improve the running stability and driving safety of the vehicle, and thus extend the service life of the fixed frog. At the same time, by comparing and analyzing the optimization results of the wing rail lifting value and nose rail height, it is found that the optimization of wing rail lifting value improves the wheel/rail interaction performance in the frog area more.

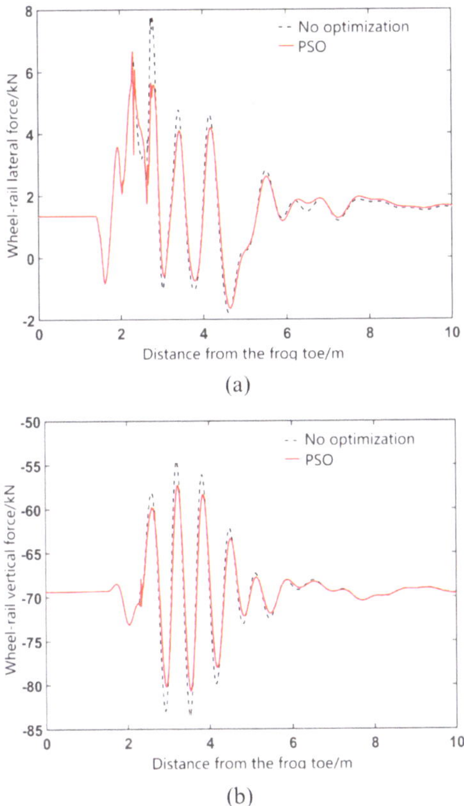

Figure 16. On the stock rail side: (**a**) Wheel/rail lateral force; (**b**) Wheel/rail vertical force.

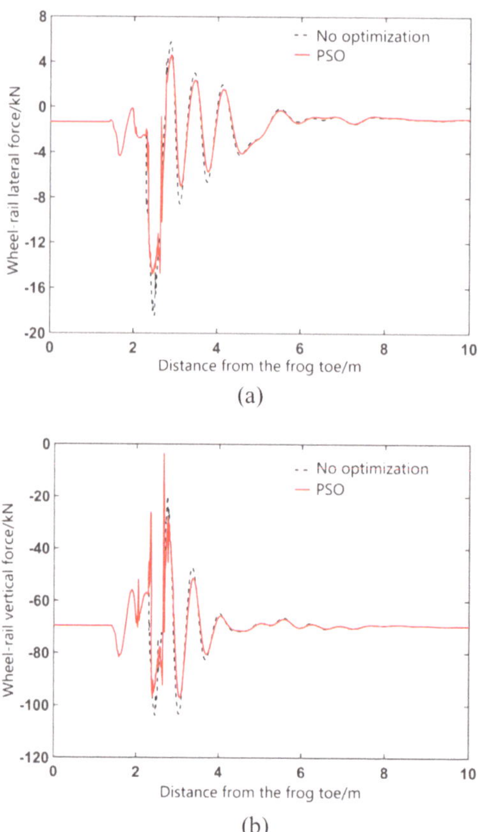

Figure 17. On the frog side: (**a**) Wheel/rail lateral force; (**b**) Wheel/rail vertical force.

5. Conclusions

In this paper, the wheel/rail contact characteristics in the fixed frog area are comprehensively analyzed, the optimization models of wing rail lifting values and nose rail height values in the fixed frog area are established, and the specific optimization design process based on particle swarm optimization algorithm is proposed. The main conclusions are as follows:

(1) If the wheelset is close to the nose rail with a large lateral displacement, there is a risk that the wheel will climb onto the nose rail. In order to reduce the vertical impact of wheelsets passing through the fixed frog area and improve the safety and stability of vehicles passing through the turnout center, the geometric parameters of the fixed frog should be set reasonably.

(2) After wing rail lifting values optimization, the maximum vertical displacement of the wheelset is reduced by 34.6%. On the stock rail side, the maximum wheel/rail lateral force is reduced by 12.8%, and the maximum wheel/rail vertical force is reduced by 10%. On the frog side, the maximum wheel/rail lateral force is reduced by 38%, and the maximum wheel/rail lateral force is reduced by 12.5%.

(3) After nose rail height values optimization, the maximum vertical displacement of the wheelset is reduced by 19.5%. On the stock rail side, the maximum wheel/rail lateral force is reduced by 15.4%, and the maximum wheel/rail vertical force is reduced by 3.5%. On the frog side, the maximum wheel/rail lateral force is reduced by 18.5%, and the maximum wheel/rail lateral force is reduced by 9.4%.

6. Future Work

The optimization algorithm proposed in this paper by using original PSO has a good effect on the optimization of geometric parameters of the fixed frog. However, with the continuous improvement and development of the random search algorithm based on group cooperation, there will be more options to solve the optimization problem of the fixed frog geometric parameters in the future, such as the Grey Wolf Optimizer (GWO), Sine Cosine Algorithm (SCA), Harris-Hawk Optimizer (HHO), Whale Optimization Algorithm (WOA), Arithmetic Optimization Algorithm (AOA), and their hybrid algorithm. In subsequent research, on the basis of the above algorithms, by comparing the characteristics of different algorithms, we can explore better optimization algorithms or hybrid algorithms to solve engineering problems.

Author Contributions: Methodology, validation, writing—original draft preparation, R.Z.; supervision, writing—review and editing, G.S.; investigation, software, X.W. All authors have read and agreed to the published version of the manuscript.

Funding: This research was funded by the Science and Technology Research and Development Plan of China Railway Corporation (Grant No. 2017G003-A).

Institutional Review Board Statement: Not applicable.

Informed Consent Statement: Not applicable.

Data Availability Statement: The authors confirm that the data supporting the findings of this study are available within the article.

Conflicts of Interest: The authors declare no conflict of interest.

References

1. Wang, P. *High Speed Railway Turnout Design Theory and Practice*, 1st ed.; Southwest Jiaotong University Press: Chengdu, China, 2011; pp. 12–13.
2. Ma, H.; Niu, Y.; Zou, X.; Zhang, J.; Yu, M. Dynamic and static contact performance analysis of wagon and fixed frog in heavy haul railway. *Sci. Technol. Eng.* **2020**, *20*, 14229–14233.
3. Chen, D. *Research on Optimization Design of Rail Grinding Profile in Turnout Switch Area*; Tongji University: Shanghai, China, 2019.
4. Ren, Z.; Zhai, W. Vertical dynamic simulation calculation of railway vehicle passing frog. *J. Southwest Jiaotong Univ.* **1997**, *1997*, 46–52.
5. Zhai, W.; Ren, Z. Study on vertical interaction between speed increasing train and turnout. *J. China Railw. Soc.* **1998**, *1998*, 34–39.
6. Wang, P. Spatial coupling vibration model of wheel rail system in turnout area and its application. *J. Southwest Jiaotong Univ.* **1998**, *1998*, 52–57.
7. Wang, P.; Liu, X.; Kou, Z. Discussion on the longitudinal distribution law of turnout vertical stiffness along the line. *J. Southwest Jiaotong Univ.* **1999**, *1999*, 18–22.
8. Lagos, R.F.; Alonso, A.; Vinolas, J.; Pérez, X. Rail vehicle passing through a turnout: Analysis of different turnout designs and wheel profiles. *Proc. Inst. Mech. Eng. Part F J. Rail Rapid Transit* **2012**, *226*, 587–602. [CrossRef]
9. Markine, V.L.; Shevitsov, I.Y. An Experimental study on crossing nose damage of railway turnouts in the Netherlands. In Proceedings of the Fourteenth International Conference on Civil, Structural and Environmental Engineering Computing, Stirlingshire, UK, 3–6 September 2013.
10. Markine, V.L.; Steenbergen, M.J.M.M.; Shevtsov, I.Y. Combatting RCF on switch points by tuning elastic track properties. *Wear* **2011**, *271*, 158–167. [CrossRef]
11. Grossoni, I.; Bezin, Y.; Neves, S. Optimisation of support stiffness at railway crossings. *Veh. Syst. Dyn.* **2017**, *56*, 1072–1096. [CrossRef]
12. Anderson, C.; Dahlberg, T. Wheel/rail impacts at a railway turnout crossing. *Proc. Inst. Mech. Eng. Part F J. Rail Rapid Transit* **1998**, *212*, 123–134. [CrossRef]
13. Anderson, C. *Modelling and Simulation of Train-Track Interaction Including Wear Prediction*; Chalmers University of Technology: Gothenburg, Sweden, 2003.
14. Blanco-Saura, A.E.; Velarte-Gonzalez, J.L.; Ribes-Llario, F.; Real-Herráiz, J.I. Study of the dynamic vehicle-track interaction in a railway turnout. *Multibody Syst. Dyn.* **2018**, *43*, 21–36. [CrossRef]
15. Palsson, B.A. Optimisation of railway crossing geometry considering a representative set of wheel profiles. *Veh. Syst. Dyn.* **2015**, *53*, 274–301. [CrossRef]
16. Wan, C.; Markine, V.L.; Shevitsov, I.Y. Improvement of vehicle-turnout interaction by optimizing the shape of crossing nose. *Veh. Syst. Dyn.* **2014**, *52*, 1517–1540. [CrossRef]

17. Cao, Y.; Wang, P.; Zhao, W. Design method for rigid frog based on wheel/rail contact parameters. *J. Southwest Jiaotong Univ.* **2012**, *47*, 605–610.
18. Cao, Y.; Wang, P. Optimization of nose depth for rigid frog. *J. Southwest Jiaotong Univ.* **2015**, *50*, 1067–1073.
19. Xu, J.; Wang, P. Optimization design method for rigid frog based on wheel/rail profile type. *China Railw. Sci.* **2014**, *35*, 1–6.
20. Zhang, P.; Zhu, X.; Lei, X.; Xiao, J. Influence of wing rail lifting value on dynamic characteristics of high-speed train crossing the turnout. *J. Railw. Sci. Eng.* **2019**, *16*, 2903–2912.
21. Shen, G. *Railway Vehicle System Dynamics*, 1st ed.; China Railway Press: Beijing, China, 2014; pp. 45–49.
22. Zhu, Q.; Dai, W.; Tan, X.; Li, C.; Xie, D. Multi-objective optimization control strategy of traction inverter based on particle swarm algorithm. *J. Tongji Univ. Nat. Sci.* **2020**, *48*, 287–295.
23. Compilation Group of the Public Works Bureau of the Ministry of Railways. *Railway Public Works Technical Manual Turnout*; China Railway Press: Beijing, China, 2012.
24. Wang, W. *Research on Particle Swarm Optimization Algorithm and Its Application*; Southwest Jiaotong University: Chengdu, China, 2012.

Article

The Limit of the Lateral Fundamental Frequency and Comfort Analysis of a Straddle-Type Monorail Tour Transit System

Fengqi Guo [1], Yanqiang Ji [1,*], Qiaoyun Liao [2], Bo Liu [2], Chenjia Li [1], Shiqi Wei [1] and Ping Xiang [1]

[1] School of Civil Engineering, Central South University, Changsha 410075, China
[2] Zhuzhou CRRC Special Equipment Technology Co., Ltd., Zhuzhou 412004, China
* Correspondence: 204812181@csu.edu.cn

Abstract: The straddle-type monorail tour transit system is a light overhead steel structure, and the lateral stiffness is generally low. However, the limit of the lateral natural vibration frequency is not clear in the current codes of China, and designers may ignore it. Weak lateral stiffness will lead to a violent vibration during vehicle operation and crowds walking, affecting human comfort and structural safety. Based on a practical project, we tested the acceleration of a monorail vehicle under full load conditions, and its running stability and ride comfort were assessed. Then, the impact of pedestrians on lateral vibration under some working conditions was measured and analyzed. Furthermore, the influence of different structural parameters on the lateral fundamental frequency was investigated. The results showed the following: (i) The vehicle's running stability and riding comfort was good. However, human comfort was poor due to the weak lateral stiffness of the structure, which was affected by human-induced vibration. (ii) The comprehensive response of the structure increased with the increase in walking frequency, increased with the increase in the number of people working or weight, and the growth speed slowed down. (iii) The structural stiffness was most sensitive to the change in steel column diameter. (iv) The recommended value of the lateral fundamental frequency limit for different spans of the straddle-type monorail tour transit system was put forward. The recommended lower limit of fundamental frequency for a 15 m span is 5.0 Hz, for an 18 m span it is 3.5 Hz, and for a 25 m span it is 2.8 Hz.

Keywords: straddle-type monorail tour transit system; lateral vibration; human-induced vibration; comfort level

Citation: Guo, F.; Ji, Y.; Liao, Q.; Liu, B.; Li, C.; Wei, S.; Xiang, P. The Limit of the Lateral Fundamental Frequency and Comfort Analysis of a Straddle-Type Monorail Tour Transit System. *Appl. Sci.* **2022**, *12*, 10434. https://doi.org/10.3390/app122010434

Academic Editors: Junhong Park and Sakdirat Kaewunruen

Received: 4 September 2022
Accepted: 13 October 2022
Published: 16 October 2022

Publisher's Note: MDPI stays neutral with regard to jurisdictional claims in published maps and institutional affiliations.

1. Introduction

In recent years, with the rapid development of tourism and the improvement in the awareness of natural landscape protection, more and more people have begun to think about how to develop the tourism industry while protecting the original style of the scenic spot and adapting to the complex and changeable terrain conditions. Therefore, the straddle-type monorail tour transit system (MTTS) has gradually been popularized. As a new type of rail transit, it has the advantages of a small land area, short construction period, strong climbing ability, and strong terrain adaptability [1–4]. More than 30 projects have been completed or are under construction in China, and typical projects are shown in Figure 1.

Unlike the traditional monorail systems and railways, most MTTS have adapted elevated steel structures with lightweight beams and heavy-weight vehicles, showing the characteristics of small dead loads and large live loads. Therefore, the dynamic response of the track structure during vehicle operation may be notable [5–8]. The tracking subsystem of MTTS is composed of a track structure and a maintenance and evacuation platform for relevant personnel. The track structure bears the vehicle load and acts as a rail. Different from the traditional wheel–rail contact, the wheel–rail contact of the MTTS is a rubber wheel contact, which also makes the research results of the traditional railway unable to be applied to the research of MTTS [9–11]. With the development of MTTS towards long-span use, the dynamic interaction problem of track structure is becoming highly prominent. While taking the vehicle or walking on the evacuation platform, a notable vibration response may be present, which may cause human discomfort or even reduce the safety and service life of the structure [12–14]. Thus, reducing the system's dynamic response [15–23] and improving the comfort of humans [24–27] have been researched topics of great concern.

Figure 1. Single track monorail project (**left**) and double track monorail project (**right**).

For the MTTS, the crowd belongs to an eccentric excitation force on the structure, and the periodic lateral load will generate with walking, resulting in the lateral vibration of the system. This problem has not received enough attention from engineers. In recent years, some large-span structures have also experienced evident lateral vibrations caused by pedestrians. The Millennium Bridge in London, UK, has experienced a severe lateral vibration due to pedestrians, which has seriously affected pedestrian comfort and safety, resulting in a significant phenomenon of "collective synchronization" of walking [28]. He [29] et al. considered people as a time-varying dynamic model. They established the coupling dynamic equation between the beam and the human to study the dynamic response of the people and the structure. The results showed that a person's natural frequency, walking frequency, and speed were important factors affecting the structure's dynamic response and the person's comfort. Many studies have also analyzed structural vibration and its impact on comfort and studied methods to reduce vibration and improve comfort [30–32]. The technical code for urban pedestrian overpasses and pedestrian underpasses [33] is the main code for designing pedestrian bridges in China. The frequency-adjustment method is adopted to consider the vertical vibration of the structure, and the vertical frequency is required to be no less than 3 Hz. However, the code does not provide further suggestions or methods for evaluating vibration serviceability that cannot meet the code's requirements, and the problem of lateral vibration is still unresolved. Although there are differences between MTTS and pedestrian bridges, the research results of pedestrian bridge vibration have significance for the research on pedestrian structure vibration of the evacuation platform of the MTTS.

Based on a practical project, firstly, the vehicle full load operation dynamic test was carried out by counterweight simulating passenger load, the single vibration "Sperling" index was analyzed to evaluate the vehicle's running stability, and the ride comfort was studied. Furthermore, in the human-induced vibration test, the different numbers of people and frequencies had different effects on the lateral vibration response of the structure. In this paper, a deterministic pedestrian load model was introduced to explore the influence of the number of people and step frequency on the structural response through the control variable method. The sensitivity of the parameters, such as the thickness of the track wall plate, the wall thickness of the column, the height, and the diameter, was analyzed using the finite element method. Finally, based on pedestrian comfort, the comfort standard specified in the German EN03 (2007) specification was taken as the basis, and in combination with some provisions of the domestic GB50458-2008 Code for Design of Straddle Monorail

Transit, the recommended limit values for the lateral fundamental frequency of different spans of the steel structure monorail system were proposed.

2. Test Overview

2.1. Project Introduction

The project included a circular line with a total length of 341.6 m. The experimental research was carried out only for three spans. The spans were 14.125 m, 17.875 m, and 20.768 m. The column diameters were 400 mm, 600 mm, and 700 mm. The section mainly included straddle-type monorail vehicles, tracks, columns, and evacuation platforms. The section form of the MTTS is shown in Figure 2.

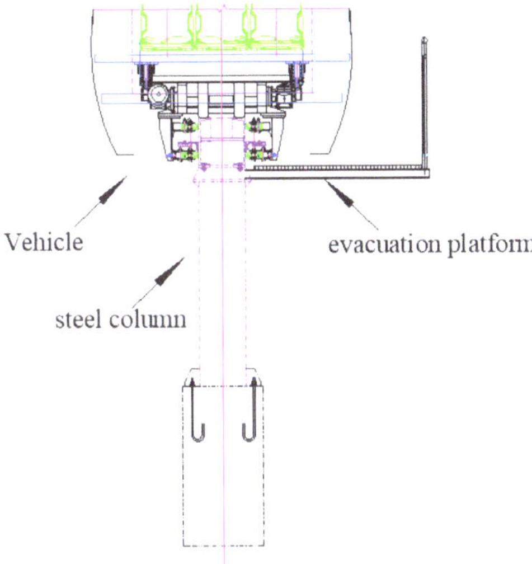

Figure 2. Section form of MTTS.

2.2. Test Content

The project was located in a construction area. According to the site conditions, we selected the period without construction, and the wind speed was less than 1.0 m/s (no wind or weak wind) to avoid the impact of the construction environment and wind-induced vibration on the test. The test was divided into three parts: (i) Dynamic characteristic test. The residual vibration method was adopted. The vehicle ran freely through the track, and then the free attenuation signal in the structural response was extracted for modal parameter identification. (ii) Vehicle running performance test. When the vehicle ran typically, we measured the three-dimensional acceleration of the vehicle floor, then analyzed its running stability and ride comfort. (iii) Human-induced vibration test. When pedestrians walked on the evacuation platform, the acceleration response at the maximum point of the track and the pier top amplitude were tested.

2.3. Layout of Measuring Points

The test instrument included eight acceleration sensors with a sensitivity of 0.33 m/s^2, two displacement meters with a sensitivity of 0.1183 mv/mm, and a YSV8008 dynamic signal-acquisition instrument with eight channels. The vibration response of the track under random load excitation was measured, and the natural vibration characteristics of the structure were obtained by spectrum analysis.

In the test, ten lateral and vertical acceleration measuring points were located at the mid-span and side column of the test section, numbered H/S 1–5 from left to right. Two amplitude measuring points were located at the top of two columns in the middle of the test section. The vehicle ran on the whole line under full load, and five measuring points were located on both sides' bottom mounting surface and the floor surface of the cushion. The layout of the measuring points is shown in Figure 3.

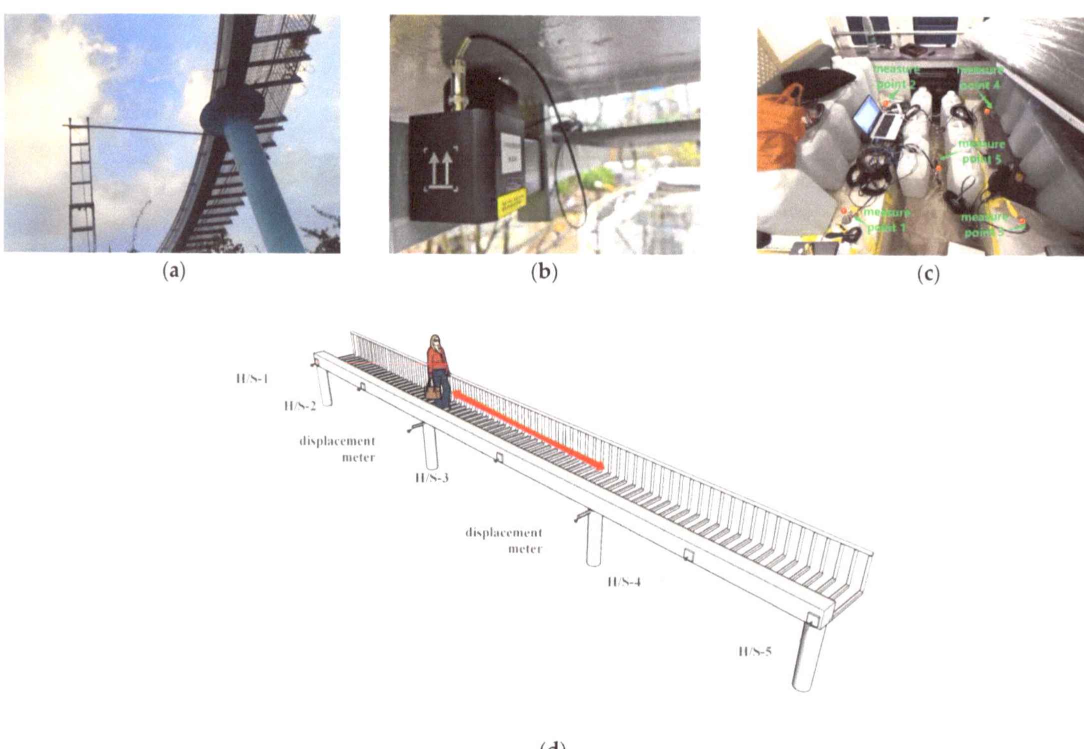

(a)　　　　(b)　　　　(c)

(d)

Figure 3. The layout of measuring points. (**a**) Amplitude measuring point; (**b**) vibration sensors; (**c**) vehicle measuring points; (**d**) overview of walking route and measuring points.

3. Dynamic Characteristics and Comfort Analysis

3.1. Dynamic Characteristics Analysis

The residual vibration method was used to study the structure's dynamic characteristics. After analysis, the results of the span acceleration spectrum analysis were similar. This paper extracted the signal of a mid-span acceleration sensor, analyzed the time-frequency domain characteristics of the attenuation signal, and obtained the structure's first four natural vibration frequencies, as shown in Table 1. At the same time, the finite element model was established by ANSYS for analysis. The theoretical analysis agreed with the experimental results, and the finite element model could be further applied to relevant research. We took lateral acceleration as an example. The test signal analysis and third-order vibration mode are shown in Figures 4 and 5.

Table 1. The first four frequencies of the structure.

Mode	Test		ANSYS	
	Lateral	Vertical	Lateral	Vertical
1	1.82	4.26	1.85	4.13
2	2.65	5.55	2.72	5.15
3	3.54	9.77	3.67	9.53
4	4.11	10.23	4.38	10.32

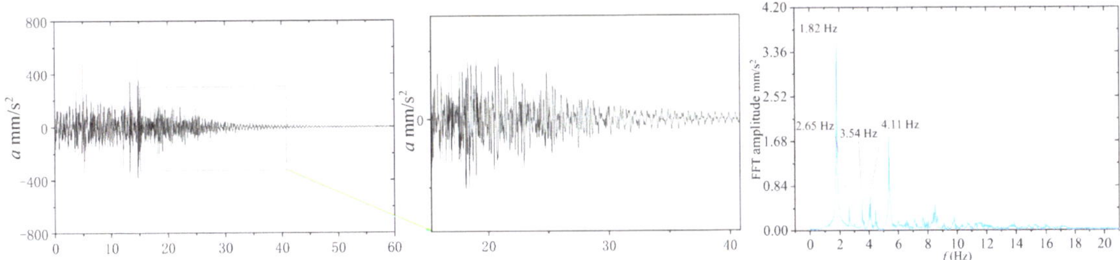

Figure 4. Time history and frequency domain signal of acceleration by the residual vibration method.

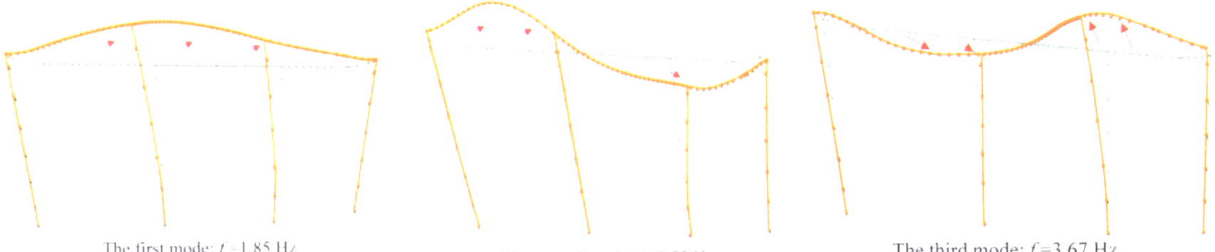

Figure 5. Mode and frequency.

The MTTS is a single-column structure. Combined with Figure 5 and Table 1, the first three-track mode was lateral bending, and the lateral fundamental frequency was 1.82 Hz. This indicates that its lateral stiffness was weak, and the lateral fundamental frequency was close to the walking frequency of tourists, which easily causes resonance. Therefore, in the design of MTTS, designers should focus on controlling the structure's lateral stiffness.

3.2. Vehicle Operation Performance Analysis

The MTTS belongs to amusement facilities with high running stability and ride comfort requirements. Ride comfort reflects passengers' subjective feelings about vibration frequency and magnitude. In addition to having the same effect as ride comfort, running stability reflects the train vibration's objective frequency and extent [34]. Various vibrations affect the train's running stability and ride comfort during vehicle operation.

The Sperling stability index w_t of a single vibration was analyzed and calculated according to the literature [35]. The vehicles running in the circle line at an average driving speed were evaluated according to the time history. After the 0.5–40 Hz band-pass filter filtered the vibration acceleration of each measuring point, the stability index was calculated every 5 m [36]. The method in UIC 513-1994 specification [37] of the International Union of Railways was used to study ride comfort. We took some measuring points as examples. The distribution of the stability and comfort index with time is shown in Figures 6 and 7.

The results showed that the lateral stability index of each measuring point was smaller than the vertical stability index at full load, indicating that the vehicle's lateral stability was better than the vertical stability. Only 1.4% of the vertical indicators exceeded the qualified line in the whole process. The comfort index met the "comfort" standard. Generally speaking, the vehicle operation performance under full load was good.

3.3. Human-Induced Vibration Analysis

The monorail system vehicle acts vertically on the track subsystem. With the track structure's refined fabrication and the curved sections' speed limit requirements, the track will not normally generate excessive vibration during vehicle driving. Relevant research [6,38] has shown that the monorail system vehicles' running stability and riding comfort are good. When pedestrians walk on the evacuation platform, the crowd load acts eccentrically on the main structure of the MTTS. The lateral stiffness of the MTTS is relatively weak, and the designers cannot ignore the lateral vibration caused by the crowd load. Therefore, this paper only studied the lateral response of human-induced vibration.

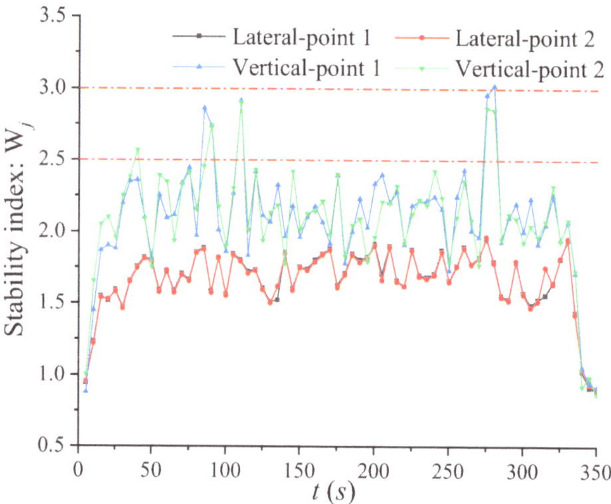

Figure 6. Stability index under full load operation.

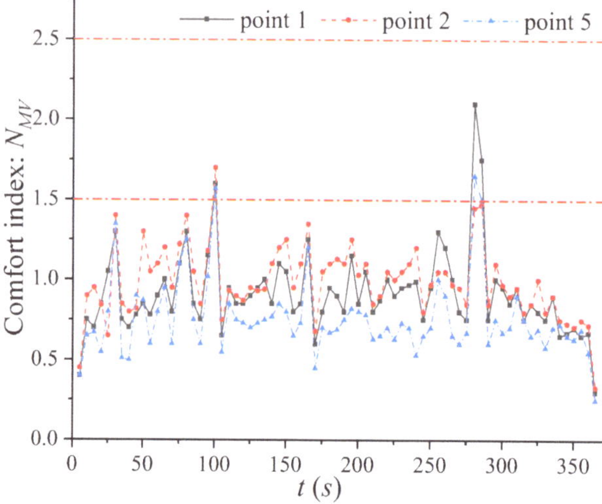

Figure 7. Comfort index under full load operation.

Three people (A, B, and C) participated in the human-induced vibration test. The test conditions were divided into multi-person and multi-frequency walking to study the impact of the number of excitations and frequency on the structural response. A stopwatch timer controlled the excitation frequency. The acquisition time of each condition was 10 s. At the same time, 10-person excitation conditions were simulated. Table 2 shows the specific condition information, and the walking route is shown in Figure 3d. The field test is shown in Figure 8.

When people generally walk on the evacuation platform, there is a slight lateral amplitude on the top of the column. The measured lateral amplitude on the top of a column is shown in Figure 9. It can be seen from the analysis that when the vibration was excited at the frequency of 2 Hz, the lateral amplitude of the column top was the measured maximum value in all cases, which is consistent with the results of the measured and finite element analysis that the fundamental frequency was about 2 Hz. The maximum amplitude of the pier top met the requirement that the lateral displacement shall not exceed 15 mm as specified in the reference [39].

Figure 8. Field test.

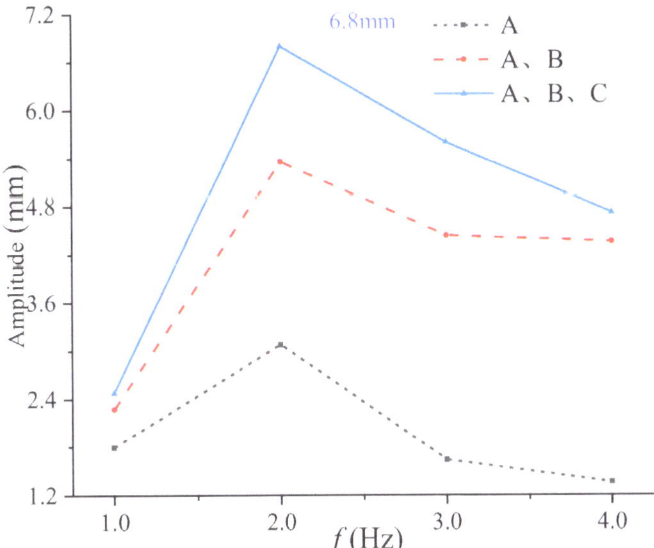

Figure 9. Lateral amplitude.

Table 2. Conditions information.

Test	Conditions	f/Hz	Participants	Excitation Source
Number of people test	1	2	A	Walk
	2	2	A, B	Walk in line
	3	2	A, B, C	Walk in line
	4	2	Ten people (70 kg/person)	Walk in line
Walk frequency test	5	1	A	Walk
	1	2		
	6	3		
	7	1	A, B	Walk in line
	2	2		
	8	3		
	9	1	A, B, C	Walk in line
	3	2		
	10	3		
	11	1	Ten people (70 kg/person)	Walk in line
	4	2		
	12	3		

Compared with the fundamental vertical frequency, the fundamental lateral frequency was smaller and closer to the human gait frequency, which more easily produces lateral vibration. The fundamental vertical frequency met the comfort requirements specified in the Chinese *CJJ69-95* code [33], so this paper only studied the lateral response of human-induced oscillation.

There are two methods to evaluate the comfort of human-induced vibration, avoiding the sensitive frequency method and limiting the dynamic response value method. Avoiding the sensitive frequency method requires that the natural vibration frequency of the structure deviates from the step frequency range to prevent man–bridge resonance. The limited dynamic response method requires that the maximum vibration acceleration of the system meet the comfort limit requirements [40–42]. The existing pedestrian bridge specifications at home and abroad have different provisions for human-induced vibration (Table 3). Among them, the Chinese specifications do not specify the lateral comfort standard. The German code considers the influence of the deviation between the structural frequency and the pedestrian step frequency on the structure and believes that the system will have a severe vibration response only when the fundamental structural frequency is close to the pedestrian step frequency [43,44]. We combined the resonance characteristics of modern steel structure bridges and the actual pedestrian crossing experience. This paper used the method of pedestrian acceleration limit specified in the design guide EN03 (2007) to evaluate comfort. The comfort distribution of human-induced vibration is shown in Figure 10.

Table 3. Current codes on human-induced vibration at home and abroad.

Current Codes	Evaluation Standard of Human-Induced Vibration Comfort	
	Vertical	**Lateral**
CJJ69-95 (China)	$f_{vertical} \geq 3\,\text{Hz}$	-
BS5400 (Englind)	$a_{max} \leq 0.5 f_{v1}{}^{0.5}\,\text{m/s}^2$	-
EN03(2007) (Germany)	$a_{max} \leq 1.0\,\text{m/s}^2$	$a_{max} < 1.0\,\text{m/s}^2$
Bro2004 (Sweden)	$a_{rms} \leq 0.5\,\text{m/s}^2$	-

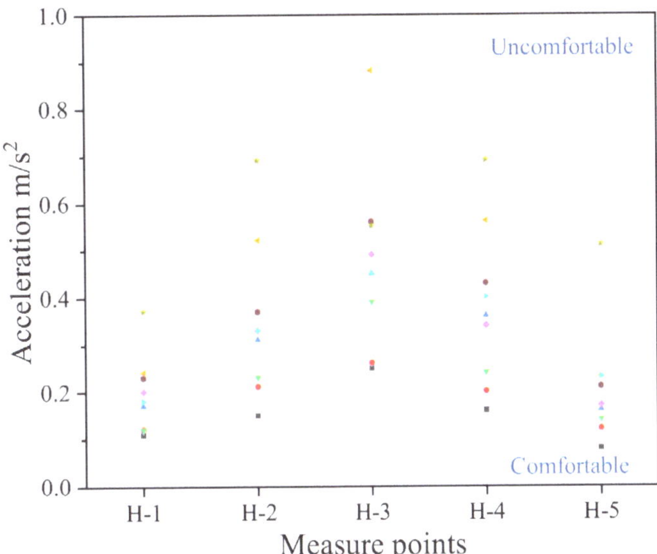

Figure 10. Comfort distribution of human-induced vibration.

Table 3 and Figure 10 show that the self-reported walking experience was relatively poor. For MTTS, even if the driving comfort is good, there is still a problem of poor pedestrian comfort. The lateral fundamental frequency of the structure should be controlled by considering the factors of pedestrian comfort. However, the current domestic specifications do not consider the impact of pedestrians on the lateral dynamic response of the structure, and there are hidden dangers in comfort and safety. Therefore, it is important to clarify the corresponding design parameters.

4. Analysis of Different Parameters

4.1. Pedestrian Load Model

During the complete gait cycle of pedestrian walking, the center of gravity fluctuates up and down, and the force acting on the evacuation platform changes constantly. The single-person continuous load model mainly includes the deterministic and random load models. The deterministic load model is commonly used in engineering to describe the walking load [45], as shown in Equation (1):

$$F_p(t) = G\left[1 + \sum_{i=1}^{n} \beta_i \sin(2i\pi f_p t - \Phi_i)\right] \tag{1}$$

where G is the pedestrian weight, n is the order of the function, t is the time, f_p is the walking frequency, β_i is the coefficient of the ith order Fourier series, and Φ_i is the phase angle of the ith order load. $\beta_1 = 0.2611, f_p - 0.2109, \beta_2 = 0.09, \beta_3 = 0.077, \Phi_1 = \Phi_2 = \Phi_3 = 0$.

According to relevant research results [29,46–48], this paper selected the single-person load step frequency of slow walking as 1.0 Hz, normal walking as 2.0 Hz, fast walking or running as 3.0 Hz, and the load walking curve is shown in Figure 11.

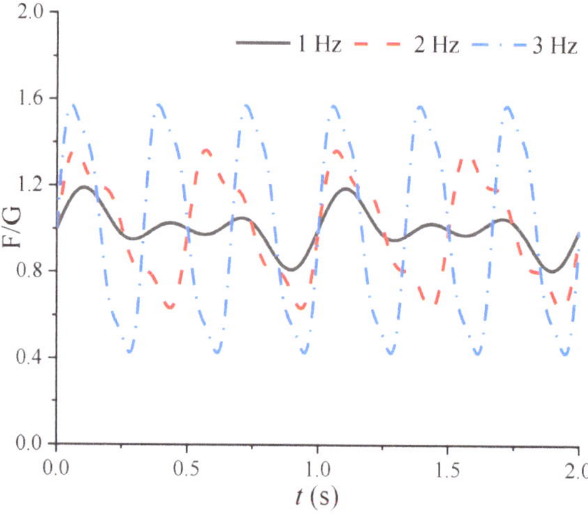

Figure 11. Time history curve of single person walking load.

4.2. Influence of the Number of People on the Structural Response

The root-mean-square (RMS) of acceleration is a function of time. Considering the influence of the overall acceleration response in the time history, it can better reflect the attribute of structural response than the peak acceleration [49,50]. This paper divided the acceleration time history data into calculation windows by 0.5 s, and the RMS of acceleration in each window was calculated by MATLAB (Equation (2)). For condition one, the acceleration time history and the RMS curve of measuring point H-3 are shown in Figure 12.

$$a_w = \left[\frac{1}{T} \int_0^T a_w^2(t)\,dt \right]^{\frac{1}{2}} \tag{2}$$

where $a_w(t)$ is the acceleration time history (m/s^2), T is the total time (s), and a_w is the RMS of acceleration within the calculation window.

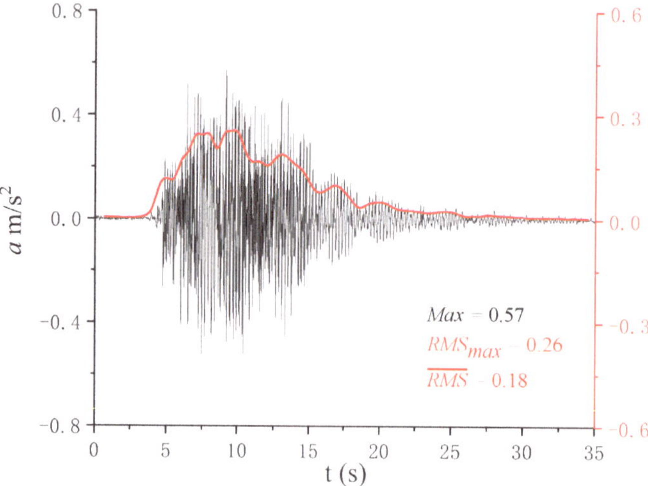

Figure 12. Acceleration time history and RMS.

To study the influence of the number of pedestrians on the structural response of MTTS, we arranged different numbers of pedestrians to walk back and forth for 10 s along the walking route shown in Figure 3d. The relationship between the number of pedestrians or weight and the acceleration response is shown in Figure 13, where a_{rms} represents the maximum *RMS* of acceleration. The results show that when the number of pedestrians increased from 1 to 2, the *RMS* of acceleration response rose significantly. When the number of pedestrians increased from 2 to 3, the increase in response was relatively small. The structural response increased with the number of people and body weight, and the growth rate slowed. In addition, the reaction of measuring points H-1 and H-5 under various working conditions was slight because the measuring points were far from the excitation position, and it had the same section as the column. The lateral restraint of the column on the structure made its vibration weak. The acceleration of measuring point H-3 was the maximum value under the same condition because the measuring point was located at the mid-span of the personnel walking section, where the excitation was the largest. Still, the constraint was relatively weak, making its vibration the most obvious.

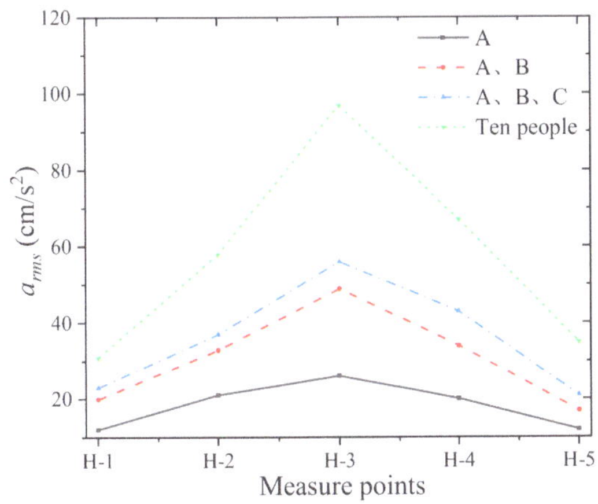

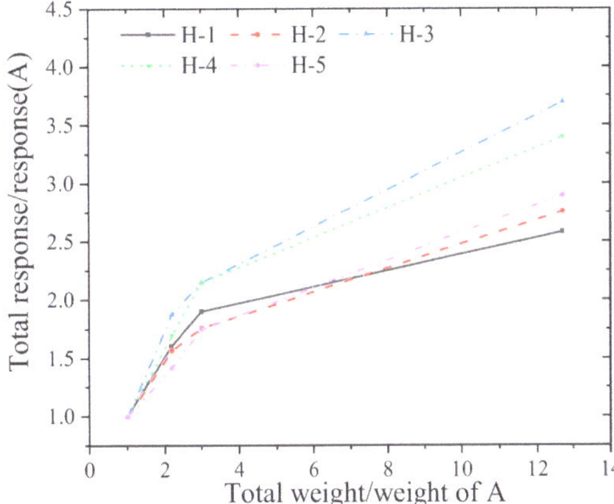

Figure 13. RMS_{max} at different numbers.

4.3. Influence of Walking Frequency on Structural Response

To study the influence of walking frequency on the structural response of MTTS, we arranged different numbers of pedestrians to step back and forth at 1, 2, and 3 Hz along the walking route shown in Figure 3d for 10 s. The relationship between walking frequency and the acceleration response of each measuring point is shown in Figure 14. The results show that the overall reaction of the track increased with the increase in walking frequency. For the same number of people, when the walking frequency increased from 1 Hz to 2 Hz, the structural acceleration response rose significantly. When the walking frequency increased from 2 Hz to 3 Hz, the ratio of excitation frequency to structural natural frequency (1.82 Hz) increased from 1.10 to 1.65, but the acceleration response of each measuring point decreased. When the walking frequency was 2 Hz, there was a possibility of man–bridge resonance. The measuring points H-1 and H-5 were in the same section with the column, and the lateral restraint of the column on the structure made its vibration weak.

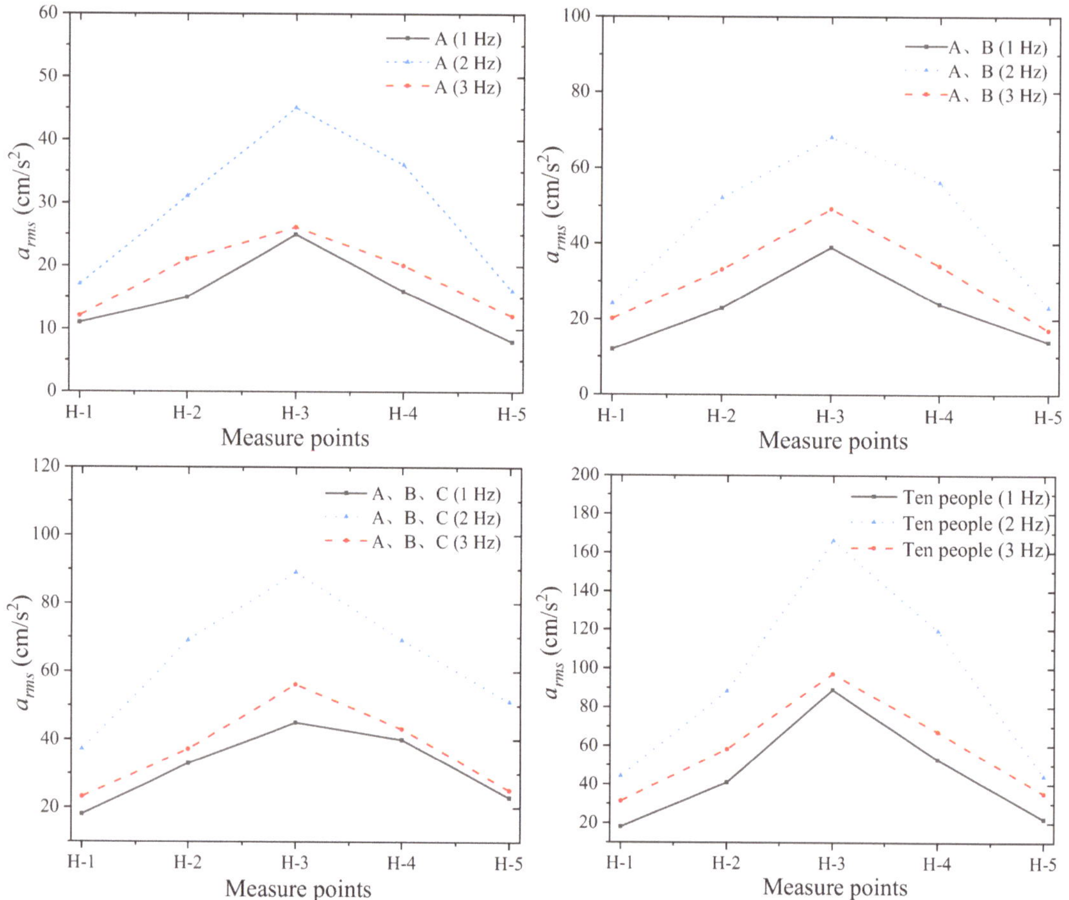

Figure 14. $RMS_{\max}$ at different frequency.

4.4. Methods of Increasing Structural Stiffness

It is necessary to analyze the fundamental frequency changes to evaluate the sensitivity of structural stiffness changes to different structural parameters. The structural parameters were consistent with the engineering examples analyzed in Section 2.1. The change of fundamental structural frequency under various conditions was analyzed by adjusting the track wall thickness and the column wall thickness, height, and diameter. The changing trend of fundamental structural frequency under different structural parameters is shown in Figure 15.

Through analysis, the fundamental frequency of the structure changed obviously with the structural parameters. If we adjusted the structural parameters slightly within a reasonable range, there were differences in the change range of the fundamental frequency of the structure. The fundamental frequency changed the fastest with the diameter of the column, the wall thickness of the column changed the second-fastest, and the change of speed with the height of the column was the slowest. That is, the structural stiffness was the most sensitive to the change in the diameter of the column, and the sensitivity to the height of the column was relatively weak. In design, when the structural rigidity does not meet the requirements, we suggest first considering adjusting the diameter of the column.

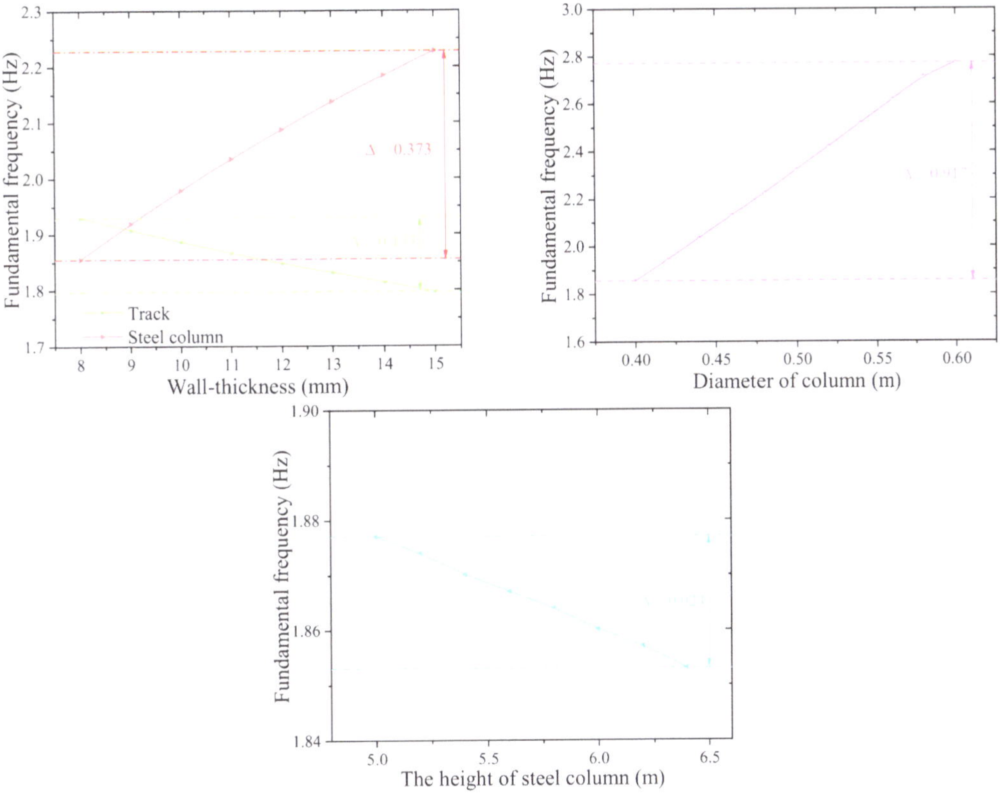

Figure 15. Change trend of fundamental frequency of structure.

5. Study on the Limit of the Fundamental Lateral Frequency

5.1. Analysis of Human-Induced Vibration

The ANSYS established the finite element model. According to the deterministic load model defined in Section 4.1, the time history analysis was carried out by walking in unison within the whole span. The acceleration response corresponding to the measuring point H-3 is shown in Figure 16. The simulation was in good agreement with the measured result.

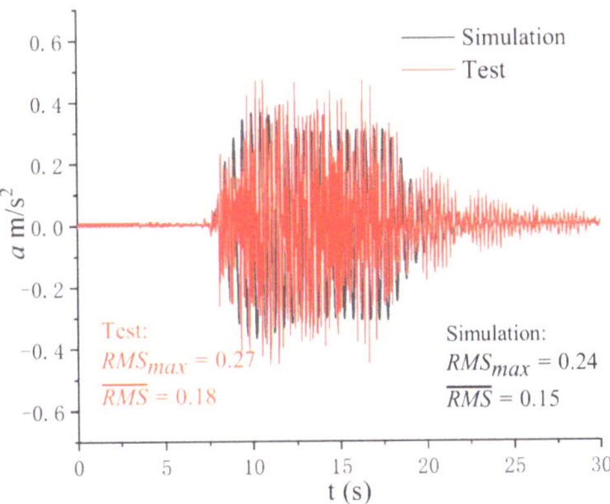

Figure 16. Comparison between simulation and measured.

Since the evacuation platform is the main structure for evacuating tourists, more than one person will likely be walking on it simultaneously. This paper used 15 people walking simultaneously, with an interval of 1 m between the front and back of the group, walking at a step frequency of 2 Hz. The lateral vibration peak acceleration did not exceed 0.3 m/s^2 according to the standards in the German EN03 (2007) code that analyzed the lateral fundamental frequency limit of domestic MTTS. We considered that the weight of a single person was 70 kg. The load walking curve is shown in Figure 17.

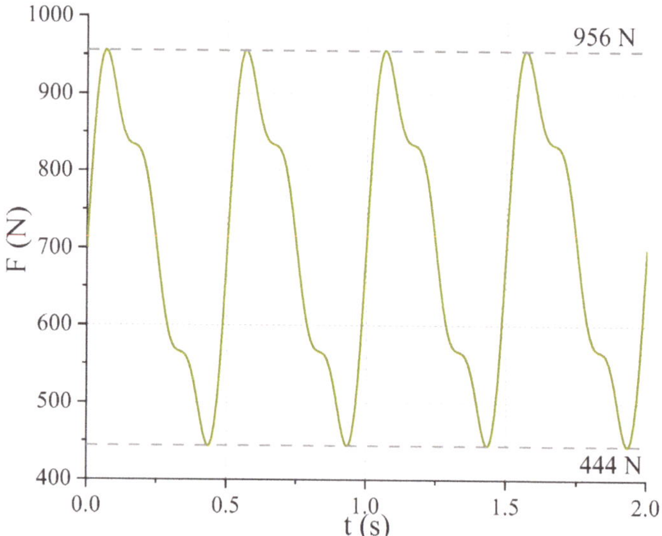

Figure 17. Time history of walking load.

5.2. Discussion on the Limit of the Fundamental Frequency

It is necessary to analyze their acceleration response to evaluate whether structures with different stiffnesses produce excessive vibration under pedestrian load. To study the influence of structural stiffness on the structural response of the MTTS, we used the walking load in Figure 17 as the excitation condition. The German EN03 (2007) code stipulates that the lateral acceleration of the structure is not more than 0.3 m/s^2 from the perspective of pedestrian comfort, and the lateral fundamental frequency limit of the straddle PC structure without pedestrians in the Code for Design of Straddle Monorail Traffic GB50458-2008 is 70/L (L represents the bridge span). This paper also considered avoiding the sensitive frequency and limiting the dynamic response. Through finite element analysis, it explored the value of the lateral fundamental frequency of the structure when the lateral acceleration was just below 0.3 m/s^2 to ensure that the pedestrian comfort reached the "comfort" standard. In combination with the horizontal fundamental frequency value specified in GB50458-2008 Code for Design of Straddle type Monorail Transit, the larger value of the two was taken as the lower limit of the fundamental lateral frequency of the span straddle type monorail tourism transportation system from the perspective of safety. For the span length of 18 m, the time history curve of the lateral acceleration response of the structure in the span with different lateral stiffnesses is shown in Figure 18. The lower limit of the fundamental lateral frequency of different spans is shown in Table 4.

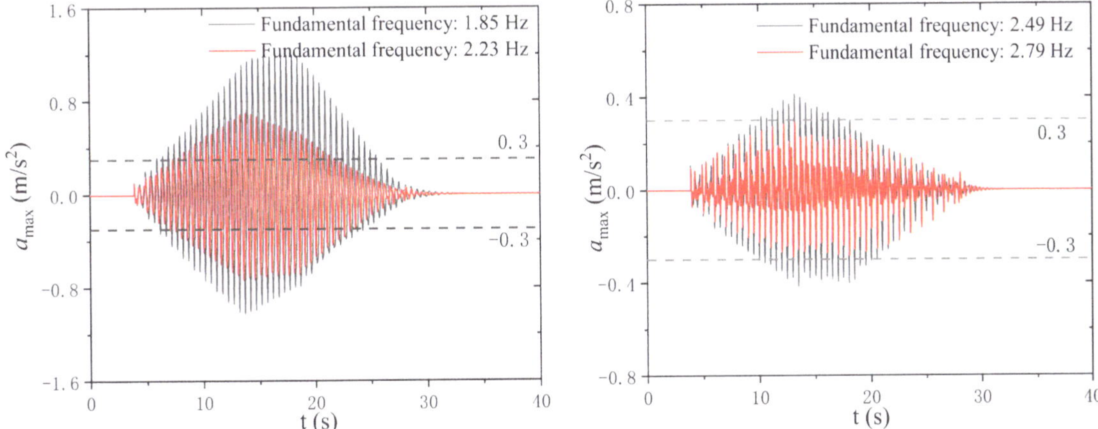

Figure 18. Time history of lateral acceleration in mid-span.

Table 4. Lower limit of horizontal fundamental frequency of different spans.

Span/m	Excitation Source	Lateral Acceleration: a_{max} (m/s^2)	Fundamental Lateral Frequency (Hz)	The Limit of Fundamental Frequency		
				Based on Acceleration	Standard Value (PC Structure)	Recommended Value
15		1.21	1.87	5.0	4.7	5.0
		0.29	4.93			
18	Full span crowd walking at 2 Hz	1.19	1.85	2.8	3.5	3.5
		0.29	2.79			
25		0.97	1.51	2.5	2.8	2.8
		0.28	2.42			

Note: In the projects under construction and operation, the 15 m, 18 m, and 25 m spans accounted for 87% of the total. The span selection is of universal significance.

It can be seen from Table 4 that, at the same span, with the increase in the lateral stiffness of the structure, that is, the increase in the lateral fundamental frequency, the lateral acceleration response in the middle of the span decreased. For MTTS, this paper suggests that the lower limit of the fundamental frequency for the span length of 15 m is 5.0 Hz, for 18 m it is 3.5 Hz, and for 25 m it is 2.8 Hz.

6. Conclusions

This paper used a straddle-type monorail tour transit system as the background for an experiment. We systematically studied vehicle running stability and pedestrian and riding comfort based on the tracks' dynamic characteristics. In addition, we used the finite element analysis method to control the lateral vibration of the MTTS and put forward the recommended value of the lateral fundamental frequency limit to optimize pedestrian comfort and further improve the evaluation index of comfort in the current codes of China. We can draw the following conclusions from our research:

(i) When the running stability and riding comfort of the MTTS are good, the comfort of human-induced vibration may also be poor. The lateral stiffness of the track is weak, and the fundamental lateral frequency is close to the activity frequency of tourists, which easily causes resonance. Thus, the designer should focus on controlling the structure's lateral stiffness;

(ii) When pedestrians walk on the evacuation platform, the comprehensive response of the structure increases with the increase in walking frequency. The increase in the number of pedestrians or total weight increases the structural response, and the growth speed decreases with the increase in number or weight;

(iii) The structural stiffness of the MTTS has a different sensitivity to different parameters and is most sensitive to the change in column diameter, followed by column wall thickness;

(iv) According to the test and finite element analysis, for the 15 m, 18 m, and 25 m span of MTTS, this paper suggests that the lower limit of the fundamental lateral frequency for the span length of 15 m is 5.0 Hz, for 18 m it is 3.5 Hz, and for 25 m it is 2.8 Hz.

Author Contributions: Methodology, validation, writing—original draft preparation, F.G. and Y.J.; investigation, software, Q.L. and B.L.; field test, conceptualization, C.L. and S.W.; supervision, writing—review and editing, P.X. All authors have read and agreed to the published version of the manuscript.

Funding: This research was funded by the Scientific research and innovation project of Central South University (1053320213475).

Institutional Review Board Statement: Not applicable.

Informed Consent Statement: Not applicable.

Data Availability Statement: The authors confirm that the data supporting the findings of this study are available within the article.

Conflicts of Interest: The authors declare no conflict of interest.

References

1. Ishikawa, K. Straddle-type monorail as a leading urban transport system for the 21st century. *Hitachi Rev* **2019**, *48*, 149–152.
2. Timan, P.E. Why Monorail Systems Provide a Great Solution for Metropolitan Areas. *Urban Rail Transit* **2015**, *1*, 13–25. [CrossRef]
3. Kikuchi, S.; Onaka, A. Monorail development and application in Japan. *J. Adv. Transp.* **1988**, *22*, 17–38. [CrossRef]
4. Kato, M.; Yamazaki, K. Straddle-type monorail systems with driverless train operation system. *Hitachi Rev.* **2019**, *53*, 25–29.
5. Guo, F.; Chen, K.; Gu, F.; Wang, H.; Wen, T. Reviews on current situation and development of straddle-type monorail tour transit system in China. *J. Cent. South Univ. Sci. Technol.* **2021**, *52*, 4540–4551.
6. Wang, P. *Study on Dynamic Response of Monorail Rapid-Transit Tour System Based on Measured Longitudinal Irregularity*; Central South University: Changsha, China, 2021.
7. Rahmatnezhad, K.; Zarastvand, M.R.; Talebitooti, R. Mechanism study and power transmission feature of acoustically stimulated and thermally loaded composite shell structures with double curvature. *Compos. Struct.* **2021**, *276*, 114557. [CrossRef]
8. Zarastvand, M.R.; Asadijafari, M.H.; Talebitooti, R. Improvement of the low-frequency sound insulation of the poroelastic aerospace constructions considering Pasternak elastic foundation. *Aerosp. Sci. Technol.* **2021**, *112*, 106620. [CrossRef]
9. Zarastvand, M.R.; Asadijafari, M.H.; Talebitooti, R. Acoustic wave transmission characteristics of stiffened composite shell systems with double curvature. *Compos. Struct.* **2022**, *292*, 115688. [CrossRef]
10. Zarastvand, M.R.; Ghassabi, M.; Talebitooti, R. Prediction of acoustic wave transmission features of the multilayered plate constructions: A review. *J. Sandw. Struct. Mater.* **2022**, *24*, 218–293. [CrossRef]
11. Gao, Q.; Cui, K.; Li, Z.; Li, Y. Numerical Investigation of the Dynamic Performance and Riding Comfort of a Straddle-Type Monorail Subjected to Moving Trains. *Appl. Sci.* **2020**, *10*, 5258. [CrossRef]
12. Nakagawa, C.; Suzuki, H. Effects of Train Vibrations on Passenger PC Use. *Q. Rep. RTRI* **2005**, *46*, 200–205. [CrossRef]
13. Suzuki, H. Effects of the range and frequency of vibrations on the momentary riding comfort evaluation of a railway vehicle. *Jpn. Psychol. Res.* **1998**, *40*, 156–165. [CrossRef]
14. Munawir, T.I.T.; Abu Samah, A.A.; Rosle, M.A.A.; Azlis-Sani, J.; Hasnan, K.; Sabri, S.; Ismail, S.; Yunos, M.N.A.M.; Bin, T.Y. A Comparison Study on the Assessment of Ride Comfort for LRT Passengers. *IOP Conf. Ser. Mater. Sci. Eng.* **2017**, *226*, 012039. [CrossRef]
15. Orvanäs, A. *Methods for Reducing Vertical Carbody Vibrations of a Rail Vehicle. Report in Railway Technology Stockholm*; KTH Engineering Sciences Department of Aeronautical and Vehicle Engineering, Division of Rail Vehicles: Stockholm, Sweden, 2010.
16. Takigami, T.; Tomioka, T. Investigation to Suppress Bending Vibration of Railway Vehicle Carbodies using Piezoelectric Elements. *Q. Rep. RTRI* **2005**, *46*, 225–230. [CrossRef]
17. Kamada, T.; Kiuchi, R.; Nagai, M. Suppression of railway vehicle vibration by shunt damping using stack type piezoelectric transducers. *Veh. Syst. Dyn.* **2008**, *46*, 561–570. [CrossRef]
18. Kamada, T.; Hiraizumi, K.; Nagai, M. Active vibration suppression of lightweight railway vehicle body by combined use of piezoelectric actuators and linear actuators. *Veh. Syst. Dyn.* **2010**, *48*, 73–87. [CrossRef]
19. Tomioka, T.; Takigami, T. Reduction of bending vibration in railway vehicle carbodies using carbody–bogie dynamic interaction. *Veh. Syst. Dyn.* **2010**, *48*, 467–486. [CrossRef]

20. Sugahara, Y.; Watanabe, N.; Takigami, T.; Koganei, R. Vertical vibration suppression system for railway vehicles based on primary suspension damping control—System development and vehicle running test results. *Q. Rep. RTRI* **2011**, *52*, 13–19. [CrossRef]
21. Tomioka, T. Reduction of car body elastic vibration using high-damping elastic supports for under-floor equipment. *Railw. Technol. Avalanche* **2012**, *41*, 245–270.
22. Aida, K.-I.; Tomioka, T.; Takigami, T.; Akiyama, Y.; Sato, H. Reduction of Carbody Flexural Vibration by the High-damping Elastic Support of Under-floor Equipment. *Q. Rep. RTRI* **2015**, *56*, 262–267. [CrossRef]
23. Dumitriu, M. A new passive approach to reducing the carbody vertical bending vibration of railway vehicles. *Veh. Syst. Dyn.* **2017**, *55*, 1787–1806. [CrossRef]
24. Schandl, G.; Lugner, P.; Benatzky, C.; Kozek, M.; Stribersky, A. Comfort enhancement by an active vibration reduction system for a flexible railway car body. *Veh. Syst. Dyn.* **2007**, *45*, 835–847. [CrossRef]
25. Dumitriu, M. Ride comfort enhancement in railway vehicle by the reduction of the carbody structural flexural vibration. *IOP Conf. Ser. Mater. Sci. Eng.* **2017**, *227*, 012042. [CrossRef]
26. Dumitriu, M. Study on Improving the Ride Comfort in Railway Vehicles Using Anti-Bending Dampers. *Appl. Mech. Mater.* **2018**, *880*, 207–212. [CrossRef]
27. Dumitriu, M.; Stănică, D.I. An approach to improving the ride comfort of the railway vehicles. *UPB Sci. Bull. Ser. D Mech. Eng.* **2020**, *82*, 81–98.
28. Zhang, G.; Ge, Y. Test and analysis of vibration characteristics of concrete continuous box girder pedestrian overpass. *J. Vib. Shock* **2009**, *28*, 102–106.
29. He, H.; Yan, W.; Zhang, A. Study on the coupling action between beam plate structure and human body under pedestrian excitation. *J. Vib. Shock* **2008**, *27*, 130–133.
30. Dumitriu, M.; I Stănică, D. Vertical bending vibration analysis of the car body of railway vehicle. *IOP Conf. Ser. Mater. Sci. Eng.* **2019**, *564*, 012104. [CrossRef]
31. Dumitriu, M.; Cruceanu, C. Influences of Carbody Vertical Flexibility on Ride Comfort of Railway Vehicles. *Arch. Mech. Eng.* **2017**, *64*, 219–238. [CrossRef]
32. Dumitriu, M.; Stănică, D.I. Influence of the Primary Suspension Damping on the Ride Comfort in the Railway Vehicles. *Mater. Sci. Forum* **2019**, *957*, 53–62. [CrossRef]
33. Ministry of Communications. *Technical Specification for Urban Pedestrian Overpass and Pedestrian Underpass. (CJJ69-95)*; China Communication Press: Beijing, China, 1995.
34. Zhou, J. *Study on Running Stability of Straddle Monorail Train in Chongqing*; Beijing Jiaotong University: Beijing, China, 2007.
35. State Railway Administration. *Specification for Dynamic Performance Evaluation and Test Evaluation of Rolling Stock: GB/T5599-2019*; China Standards Press: Beijing, China, 2019.
36. Cheng, Y.C.; Hsu, C.T. Running Safety and Comfort Analysis of Railway Vehicles Moving on Curved Tracks. *Int. J. Struct. Stab. Dyn.* **2014**, *14*, 1450004. [CrossRef]
37. International Union of Railways (UIC); European Committee for Standardization (CEN). *UIC Code 513 Guidelines for Evaluating Passenger Comfort in Relation to Vibration in Railway Vehicles*; International Union of Railways(UIC) & European Committee for Standardization(CEN): Paris, France, 1994.
38. Guo, F.; Wang, P. *Experimental Research Report on Mechanical Properties of Monorail Elevated Steel Structure Rapid Transit System*; School of Civil Engineering, Central South University: Changsha, China, 2021; pp. 1–41.
39. *GB/T 51234-2017*; Code for Design of Urban Rail Transit Bridges (with Description of Articles). China Architecture & Building Press: Beijing, China, 2017.
40. Zhu, Q.; Ma, F.; Zhang, Q.; Du, Y. Experimental study on vertical dynamic coupling effect of pedestrian-structure. *J. Build. Struct.* **2020**, *41*, 125–133.
41. Zhu, Q.; Li, H.; Du, Y.; Zhang, Q. Quantitative evaluation of vibration serviceability of pedestrian bridge under different walking speed. *J. Eng. Mech.* **2016**, *33*, 97–104.
42. Luo, X.; Zhang, J.; Shen, Z.; Zhang, Q.; Liu, S. Human-induced vibration control of curved beam footbridge with single inclined cable arch. *J. Vib. Shock.* **2020**, *39*, 83–92.
43. Dang, H.V. Influence of low-frequency vertical vibration on walking locomotion. *J. Struct. Eng.* **2016**, *142*, 04016120. [CrossRef]
44. Ma, B. Analysis of pedestrian induced lateral vibration and comfort of a steel structure pedestrian bridge. *China Munic. Eng.* **2017**, 8–11. [CrossRef]
45. Guan, J.; Tan, L.; Chen, X.; Zhang, Z.; Chang, H. Vibration comfort analysis of an indoor large-span steel corridor considering human-structure coupling. *Build. Struct.* **2021**, *51*, 43–49.
46. Cao, L.; Li, A.; Chen, X.; Zhang, Z. Vibration serviceability control of a long-span floor in large station room under crowd-induced excitation. *China Civ. Eng. J.* **2010**, *43*, 334–340.
47. Fu, X.; Qu, J.; Chen, X. Walking comfort analysis and control for the Expectation Bridge using combined time history and frequency spectra method. *China Civ. Eng. J.* **2011**, *44*, 73–80.
48. Xu, Q.; Li, A.; Zhang, Z.; Ding, Y. Research on vibration control of long-span suspension structure considering human comfort. *J. Vib. Shock* **2008**, *27*, 139–142.
49. Lu, Y.; Cheng, Y.; Cheng, Z.; Lv, Q.; Liu, Y. Experimental study on vibration serviceability of composite floor in a suspended structure. *J. Build. Struct.* **2020**, *41* (Suppl. 2), 263–269.

50. Ma, F.; Zhang, Z.; Xiao, X.; Li, A. Vibration response measurement and analysis of large-span steel floor structure at high-speed rail station under moving train and crowd excitations. *J. Build. Struct.* **2018**, *39*, 109–119.

applied sciences

Review

A Review of Recent Research into the Causes and Control of Noise during High-Speed Train Movement

Hongyu Yan [1,2,3], Suchao Xie [1,2,3,*], Kunkun Jing [1,2,3] and Zhejun Feng [1,2,3]

[1] Key Laboratory of Traffic Safety on Track, Ministry of Education, School of Traffic & Transportation Engineering, Central South University, Changsha 410075, China; 204201034@csu.edu.cn (H.Y.); jkk0711@csu.edu.cn (K.J.); 194201004@csu.edu.cn (Z.F.)
[2] Joint International Research Laboratory of Key Technology for Rail Traffic Safety, Changsha 410075, China
[3] National & Local Joint Engineering Research Center of Safety Technology for Rail Vehicle, Changsha 410075, China
* Correspondence: xsc0407@csu.edu.cn

Abstract: Since the invention of the train, the problem of train noise has been a constraint on the development of trains. With increases in train speed, the main noise from high-speed trains has changed from rolling noise to aerodynamic noise, and the noise level and noise frequency range have also changed significantly. This paper provides a comprehensive overview of recent advances in the development of high-speed train noise. Firstly, the train noise composition is summarized; next, the main research methods for train noise, which include real high-speed train noise tests, wind tunnel tests, and numerical simulations, are reviewed and discussed. We also discuss the current methods of noise reduction for trains and summarize the progress in current research and the limitations of train body panels and railroad sound barrier technology. Finally, the article introduces the development and potential future applications of acoustic metamaterials and proposes application scenarios of acoustic metamaterials for the specific needs of railroad sound barriers and train car bodies. This synopsis provides a useful platform for researchers and engineers to cope with problems of future high-speed rail noise in the future.

Keywords: high-speed train; noise control; acoustic metamaterial; sound barrier; rail vehicle

Citation: Yan, H.; Xie, S.; Jing, K.; Feng, Z. A Review of Recent Research into the Causes and Control of Noise during High-Speed Train Movement. *Appl. Sci.* **2022**, *12*, 7508. https://doi.org/10.3390/app12157508

Academic Editors: Edoardo Piana and Junhong Park

Received: 14 June 2022
Accepted: 19 July 2022
Published: 26 July 2022

Publisher's Note: MDPI stays neutral with regard to jurisdictional claims in published maps and institutional affiliations.

1. Introduction

The sound of trains is considered a disturbance to most residents around the railroad and train passengers, and the sound generated by trains can interfere with sleep, life, and work. As early as 1825, a letter from Leeds Intelligencer presents a record of train noise interfering with life [1].

Noise has always been a major threat to people's health, and several studies have shown that people exposed to noise for a long time have an increased risk of stroke [2], coronary heart disease [3], and many other cardiovascular and cerebrovascular diseases [4]. People who live in noisy environments for long periods also have significantly higher rates of endocrine disorders and breast cancer. Noise causes physical problems as well as sleep disturbances and mental problems [5].

Countries around the world have introduced various laws and regulations to protect people from noise hazards. The European Regional Environmental Noise Guidelines recommend that rail noise should be controlled to below 54 decibels (dB L_{den}). In 1974, the United States first enacted noise regulations to ensure the health of people. In China's railroad environmental noise management regulations, the railroad environmental noise emission standards implement Category 4b of the Sound Environmental Quality Standard (GB12525-90). This category requires no more than 70 decibels (dB L_{eq}) during the day and no more than 60 dB (dB L_{eq}) at night [6]. Japan promulgated noise standard values for the

Shinkansen in 1975. The values specified in decibels (L_{pA}.Smax) in these standards are 70 dB or less, mainly for residential areas, and 75 dB or less for other areas [7].

The rise in the industrial civilization of the nineteenth century brought the railroad, and people's pursuit of train speed drove the continuous development of train technology. Steam locomotives reached their peak in the 1930s when Gresley's streamlined "Mallard" locomotive reached a top speed of 202 km/h. In 1964, Japan's Shinkansen was the first to commercialize modern electric high-speed trains, operating at 210 km/h en route from Tokyo to Osaka.

Nowadays, the operating speed of high-speed trains has reached 350 km/h. In the future, the speed of high-speed trains will be further increased, and the noise problem will be further exacerbated, making the train noise problem a hot research topic among engineers with an interest in rail transportation. High-speed train noise is already a problem that cannot be ignored, and if this problem is not solved timeously, it will seriously affect the future development of high-speed trains.

To this end, this study presents a selective literature review focusing on:

(1) The causes of high-speed train noise and the distribution of the sound sources;
(2) The current main research methods for high-speed train noise;
(3) Traditional methods of high-speed train noise control;
(4) Potential uses of acoustic metamaterials in the area of high-speed train noise.

2. High-Speed Train Noise Composition

While the noise of early steam locomotives was generated by steam engines, the noise of modern electric high-speed trains consists of wheel-rail rolling noise and aerodynamic noise caused by train airflow [8].

2.1. High-Speed Train Aerodynamic Noise

The airborne noise generated by high-speed trains is divided into external airborne noise and internal airborne noise, of which the external airborne noise of high-speed trains is the main source of noise. When the driving speed of high-speed trains exceeds 300 km/h, the external airborne noise accounts for more than 50% of the total noise of the train [9], and as the speed of the train increases, the external aerodynamic noise of the train will increase in the ratio of six to eight times the speed of the train [10].

The aerodynamic noise sources are monopoles, dipoles, and quadrupoles. The aerodynamic noise generated by high-speed trains during the driving process is mainly caused by the dipole and quadrupole sound sources generated by the surface pressure fluctuations around high-speed trains. When running, the aerodynamic noise will change due to the speed and the surrounding environment [9].

The train's aerodynamic noise mainly comes from the pantograph area, bogie area, connection area, and the concave structural areas of the train's body surface. Smoothing the design of the train body [11], flow-field control, laying sound-absorbing material [12], and other measures are the main ways to reduce noise generation on the train body. Noise reduction in the pantograph region is currently achieved by optimizing the pantograph shape. Pantograph shape optimization can control the scale of eddy current shedding and thus reduce noise [13]. The addition of shields on both sides of the pantograph effectively diverts the airflow and prevents noise diffusion. The contribution of radiated noise from pantographs accounts for more than 10% of the total noise generated. Researchers mainly study the sound generation mechanism of pantographs and optimize them through experiments and numerical simulations. The bogie fairing can reduce the development of turbulence outside the fairing, thus reducing the noise from the bogie. Meanwhile, the use of a reasonable skirt design also reduces the influence of the train bottom spill on the car body and decreases noise propagation therefrom [14]. The inter-coach windshield region is considered one of the main sources of aerodynamic noise. At present, a fully enclosed outer windscreen is used to ensure that the external airflow and the inner windscreen do not come into contact, thus eliminating the vortex in the inner windscreen part, which can control the aerodynamic noise generated on the windscreen of the train [15].

The pantograph is the main noise source of aerodynamic noise in high-speed trains. Compared with the cavity and bogie, the pantograph is located on the top of the train, and the aerodynamic noise generated by it is difficult to be isolated by sound barriers. This paper counts some representative pantograph noise reduction measures in recent years, as shown in Table 1, which are found through comparison and analysis. For the noise reduction measures in the pantograph area, the current research can be summarized into two levels: one is to reduce the number of rods of the pantograph for the pantograph itself; second, passive noise reduction is applied to the shunt area.

Table 1. Comparison of noise reduction measures for pantographs.

Methods to Reduce Noise	Research Methods	Frequency Characteristics	Sound Pressure Level	Reference
Different strip spacing of the pantograph	CFD analysis	No effect	The sound pressure level of the standard noise measuring point is reduced by 2.8 dB with a spacing of 540 mm	[16]
Bionic pigeon feathers	CFD analysis	Around 1000 Hz (original model) Around 100 Hz (optimized model)	The total noise decreased by 10 dB	[17]
Pantograph insulators with elliptical section	CFD analysis	Tonal peaks are gradually reduced from 2 kHz	The peak sound pressure level of elliptical insulators decreased by 4 dB	[18]
Cylindrical rod and push rod applied to a layer of porous sound absorption material	CFD analysis, Wind tunnel test	No effect	The peak sound pressure level of the optimized pantograph decreased by 5 dB	[19]
Covering the fairings with a porous material	CFD analysis, Wind tunnel test	No effect	The optimized noise peak is 8 dB lower than the original noise peak	[20]
Covering the circular cylinder with metal foam	Wind tunnel test	Tonal peaks toward lower frequencies	The peak sound pressure level decreased by 5 dB at 216 km/h	[21]
Pantographs with and without the cavity	CFD analysis	No effect	The difference in OASPL between the pantographs with and without the cavity was approximately 4 dB at the side	[22]
The comparison of noise reduction effects for four types of pantographs covers	CFD analysis	No description	A pair of baffles with half of the height of the pantograph on both sides can lessen noise by about 3 dB	[23]
Noise contribution from high-speed train roof configuration of cavities, ramped cavities, flat roofs	CFD analysis	No effect	The flat roof with side insulation plates has the lowest overall noise levels.	[24]

In Table 1, we analyze the noise reduction measurements, research methods, frequency variation, and sound pressure level variation of each research. Through comparison, we find that researchers generally study the pantograph noise reduction problem by wind tunnel tests and CFD analysis, among which CFD analysis is the most widely used research method, which indicates that CFD technology has strong application value in the field of high-speed trains at present. In some of the studies, the results of wind tunnel tests also fully prove the reliability of CFD analysis. In the future, the flow simulation methods and computer performance algorithms will be the focus area of the researcher.

At present, optimizing the rod structure of the pantograph is a feasible measure to reduce the pantograph noise. By analyzing the noise frequencies of each study, we can find that optimizing the cylindrical rod and push rod of the pantograph will change the frequency of pantograph noise so that the peak frequency shifts to the low-frequency direction. Compared with high-frequency noise, low-frequency noise is more difficult to be absorbed by porous materials (polyurethane fiber, glass fiber wool, etc.). Using metamaterials to absorb low-frequency noise may be a feasible method. In Section 4, we further introduce the metamaterial for sound absorption. Bionic pigeon feathers shaped pantograph rods and covering the fairings with the porous material bring the most significant noise reduction effect, but the shape of the pigeon feathers will significantly increase the processing cost, and the porous material will change its pore structure under high wind conditions. Compared with other noise reduction measures, changing the strip spacing of the pantograph is the easiest to implement, but this has only been analyzed by CFD analysis, and the specific effect has yet to be experimented with.

An overall analysis of the studies in Table 1 shows that all pantograph noise reduction measures can improve the aerodynamic noise of high-speed trains to some extent, but the improvement is limited and difficult to be applied in actual high-speed train manufacturing. Perhaps there will be better measurements to improve the pantograph noise in the future, but at present, it is difficult to optimize the pantograph to reduce the noise. Therefore, in future research, improving the train body panels and sound barriers to stop noise propagation may be the best way to reduce aerodynamic noise.

2.2. High-Speed Train Mechanical Vibration Noise

2.2.1. Braking Noise

Brake squeal noise is loud and associated with high sound pressure and a sharp tone, which poses a hazard to people's physical and mental health. The friction of the brake disc generates brake squeal noise, friction generates wear, and the generation and development of brake noise are inseparable from frictional wear behavior.

In their study, Eriksson et al. [25] proposed that wear of the braking surface leads to significant randomness in braking noise spectra. Graf et al. [26] suggested that the uniform friction layer of the braking interface exerts an important influence on the occurrence of squeal noise. Renault et al. [27] estimated the effect between the surface morphological characteristics of the braking interface and the braking noise. Majcherczak et al. [28] discovered that material debris generated changes the friction coefficient and thus affects frictional noise. Massi et al. [29] found that when frictional squeal noise is generated during braking, a large amount of debris tends to accumulate on the material surface.

The aforementioned research found that high-speed train brake noise and tribology are inseparable, and now with the continuous improvement of materials and production processes of the train brake disc, frictional brake noise is further reduced.

2.2.2. Wheel-Rail Noise

Wheel-rail noise is mainly divided into three types: impact noise, rolling noise, and spike noise. When the train speed is less than 300 km/h, rolling noise is the main noise source [30]. Wheel-rail noise is generated by the relative motion between the wheels and the track due to the roughness of the rail [31]. The degree of wheel and rail roughness also determines the amplitude of vibration and other dynamic characteristics of each component during train travel. There is a mutual force between the wheels and the rails during motion of the train, and this force causes the rails and wheels to vibrate, thus radiating noise into the surrounding air. In 1976, to find the sound radiation between the wheel and rail, Remington et al. [32] proposed a simplified engineering method to predict wheel-rail noise by simplifying the wheel and rail as an infinitely long Eulerian beam and an Eulerian beam under continuous elastic support, respectively. Thompson considered the symmetry of the wheel based on Remington's work and calculated several typical wheels of that time by the finite element calculation method [33]. For the rail model, Thompson considered the influences of the rail sleeper [34] and wheel rotation [35], used the two-dimensional boundary element method to calculate the vibration sound radiation efficiency, and developed the wheel-rail noise prediction software "TWINS" according to the research results.

With the continuous improvement of the wheel-rail noise prediction model, the technique to eliminate wheel-rail noise is becoming increasingly mature. Reducing wheel and track surface roughness, improving the wheel-rail design, and using local shielding can better solve the train wheel-rail noise problem. Jones et al. [36] added shields to the bogies; Vincent et al. [37] eliminated the noise from the train wheels by changing the train track and sleepers; Bouvet et al. [38] and Cigada et al. [39] proposed improvements to the wheels to suppress wheel-rail noise by increasing the elasticity of the wheels; and Merideno et al. [40] changed the modalities of the wheels using sandwich-type dampers, thus reducing vibration and noise. For train wheels, Lee et al. [41] improved wheel wear by optimizing the curvature of the wheel webs so that the wheels maintain a stable noise level after long-term use. In terms of train rails, Chen et al. [42] believed that replacing rails on viaduct lines with damped rails can reduce noise while changing the peak frequency band of the noise.

3. High-Speed Train Noise Research Methods

At present, the noise of high-speed trains in motion is mainly studied by line tests, wind tunnel tests, and numerical simulations.

3.1. Real High-Speed Train Noise Tests

Real high-speed train noise tests can obtain the most realistic noise distribution during the train-driving process. Real high-speed train noise tests are generally divided into internal and external noise acquisition. Internal noise experiments on high-speed trains generally use different locations in the car for microphone arrays for acoustic signal acquisition [43], but the internal noise data alone cannot be used to assess the overall noise level in service and its effect on the surrounding environment: it is now generally used in sound-insulation research on high-speed train bodies [44].

Train exterior noise tests generally involve measurement of the noise distribution in space through single or multiple acoustic sensors and can also obtain the far-field field noise information of moving high-speed trains. The total sound pressure of the high-speed trains during the driving process and the contribution of different frequency bands of noise to the overall high-speed train interior and exterior noise can be measured using the acoustic signals collected by microphones at different measurement points. The arrangement of the acoustic sensors affects the collected noise data. As shown in Figure 1, the one-dimensional horizontal arrangement of the acoustic sensor array identifies the distribution of sound sources in the direction of motion of the train. The one-dimensional vertical arrangement can identify the sound source distribution at the height of the train [45]. The X-shaped sensor array identifies noise data in both horizontal and vertical directions relative to the high-speed train, and these data can be used to map the sound source distribution of the high-speed train [46]. By increasing the number of acoustic sensors, the microphone array is also arranged in spiral, star, or spherical forms.

The number of microphones in the microphone array further increases with the increase in the speed of the train. He et al. [47] used a star-shaped microphone array with 78 microphones to test the sound field of a high-speed train in a real-world experiment. Li et al. [48] employed a microphone array consisting of 78 microphones to measure the noise levels of high-speed trains traveling over viaducts and embankment sections, and Zhang et al. [49] used a microphone combination consisting of a 78-channel microphone array and microphones at a horizontal height of 3.5 m to study the sound field of high-speed trains travelling at a speed of 350 km/h. Noh [50] used a microphone array with 144 channels to conduct sound field tests on a high-speed train running at 390 km/h. As shown in Figure 2, these train noise experiments collected real high-speed train noise data and used the data to study the noise distribution and variations therein.

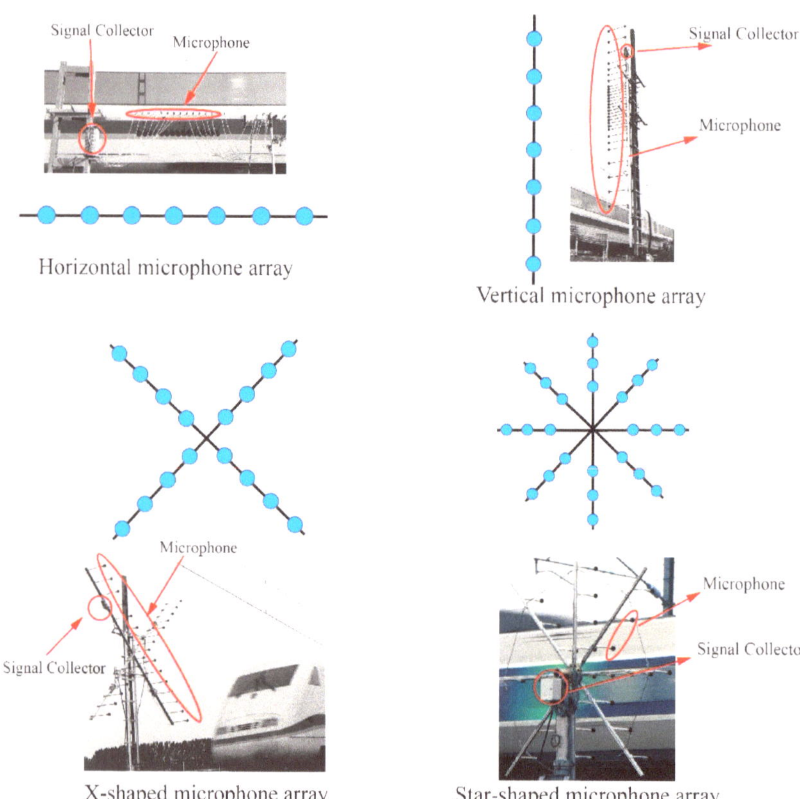

Figure 1. Microphone array arrangement form [45,46].

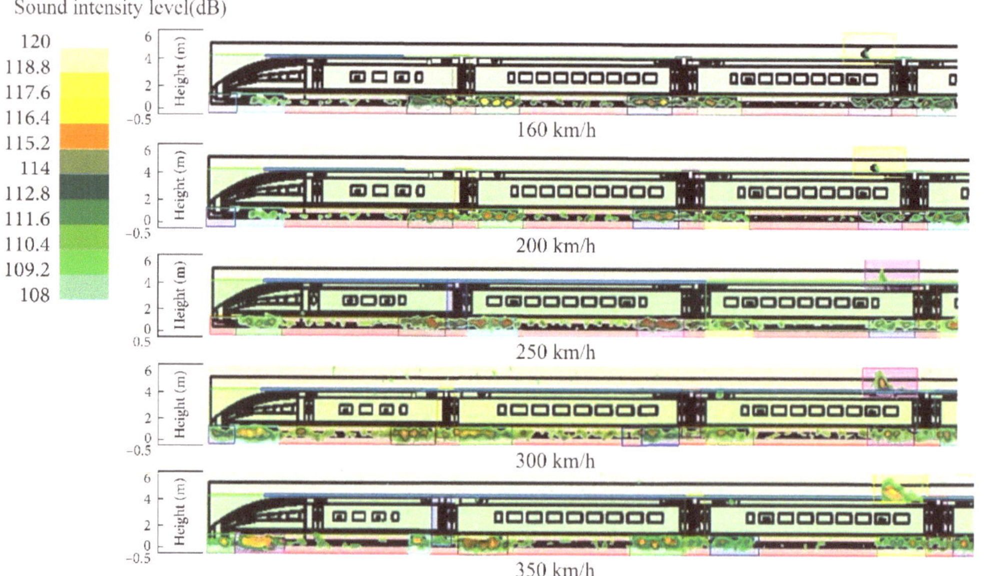

Figure 2. Noise distribution at different train speeds [48].

3.2. Wind-Tunnel Test

Compared to train noise experiments, wind tunnel experiments can be conducted in an indoor environment, independent of external climatic conditions, and can simulate external wind and other climatic conditions according to the hardware facilities of the wind tunnel, thus realizing the controllability of environmental factors. In aerodynamic research, wind tunnels are cutting-edge experimental testing tools. Both Ferrari and NASA have built wind tunnels to improve the aerodynamic performance of Formula 1 cars and space shuttles. As shown in Figure 3, wind tunnel facilities for high-speed trains typically employ large tunnels that use giant vanes at the entrance to generate strong airflow, with special grilles to reduce vortices in the airflow before it enters the laboratory. The noise collected in the wind tunnel test does not include mechanical noise, so it is possible to improve evaluations of the aerodynamic noise generated by the train.

Figure 3. Train isometric model.

As shown in Figure 3, to reduce the cost of wind tunnel experiments, scaled-down models are often used for experiments. In 1991, Baker et al. [51] used a 1/76 scale HST train model and a 1/20 scale TGV001 train model to evaluate the effects of different types of ground simulations on train drag. Willemsen et al. [52] experimentally investigated three different types of trains at a 1/10 scale using a German-Dutch wind tunnel, and the results of the study showed that high Reynolds numbers on the surface significantly improve the reliability of wind tunnel experiments on trains. Nagakura et al. [53] conducted wind tunnel experiments using a 1/5 scale model of the Shinkansen train, and they estimated the contribution of each source of noise generated by a Shinkansen train to the roadside noise level based on the experimental data. Zhu et al. [54] proposed a numerical simulation method for wheel-track noise of the high-speed train and verified the accuracy of this method through wind tunnel experiments. Zhang et al. [55] performed a large eddy simulation to research the unsteady flow near the pantograph of the DSA380 high-speed train and predicted the aerodynamic noise caused by the pantograph. Although wind tunnel experiments are cheaper and have a higher safety factor, current wind tunnel experiments are more demanding in terms of experimental environment and equipment for non-full-size flow field simulations due to the similarity criterion.

3.3. Numerical Simulation and Theoretical Research

To conduct train noise tests or wind tunnel tests relative to the various complex environments through which high-speed trains pass is costly, but the full range of flow field information can be determined quickly and easily through computational fluid dynamics (CFD). Computational fluid dynamics has various advantages, such as fewer restrictions and lower costs. Using the method of computational fluid dynamics, it is possible to use mathematical methods to discretize the flow field control equations in the grid of the computational region, solve the discrete numerical solutions, and obtain the aerodynamic noise of the train during travel according to the Navier–Stokes equations. As shown in Figure 4, the high-speed train model is huge and has a complex structure, so the mesh must contain many elements, often too onerous given current computational resources, to satisfy the need for a direct solution for the whole vehicle model. In CFD commercial software, the most common method is to ascertain the computational flow field characteristics by using Lighthill's acoustic approximation model and then use the acoustic analogy theory to simulate the noise propagation, thus improving the computational efficiency and saving computational resources.

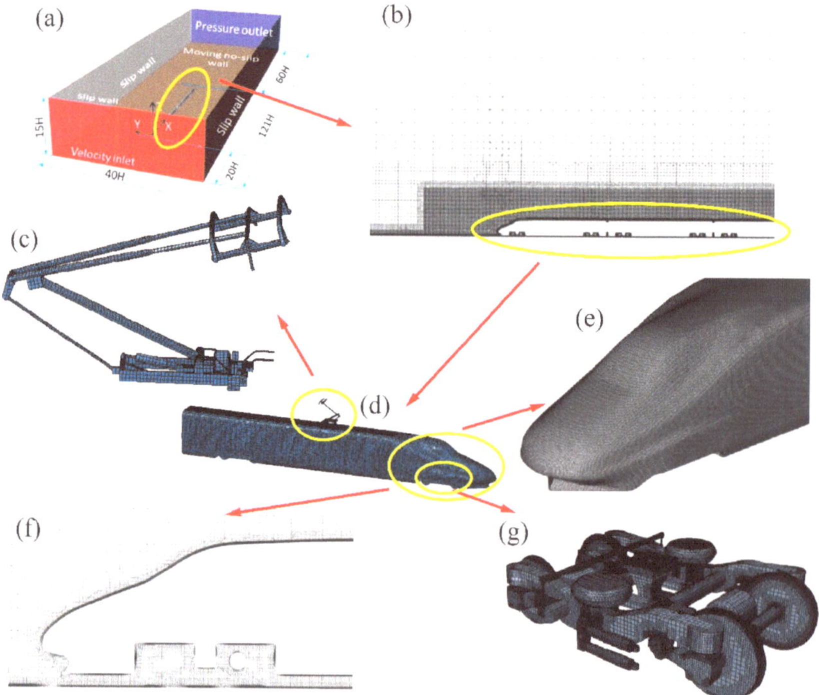

Figure 4. CFD meshing for high-speed trains. (**a**) Schematic of flow field computational domain [56]; (**b**) grid density variation in the computational domain [56]; (**c**) pantograph meshing [57]; (**d**) high-speed train complete grid [57]; (**e**) high-speed train head meshing [56]; (**f**) high-speed train head grid density variation [56]; (**g**) bogie meshing [58].

Many scholars have studied high-speed train noise using computational fluid dynamics. Sassa et al. [59] combined acoustic and fluid analyses to calculate the aerodynamic noise radiation from the surface vestibule side doors of high-speed trains. Takaishi et al. [60] derived the far-field integral equation for aerodynamic noise and calculated the aerodynamic noise conditions at the train pantograph. Masson et al. [61] developed a numerical model of the French TGV train based on the lattice Boltzmann method and obtained the noise distribution around a TGV high-speed train. Wang et al. [57] and Wu et al. [62] used large eddy current simulation and boundary element method to simulate the pantographs of high-speed trains and found that the pantograph noise is concentrated in the low-frequency range, and the pantograph aerodynamic noise is gradually dispersed in the high-frequency range. Sun et al. [63] conducted a simulation-based analysis of the pantograph and found that the slide plate, pantograph head, balance rod, insulators, bottom frame, and pull rod are the main sources of aerodynamic noise thereon. By comparing six models, Li et al. [64] found that the SST k-ω method is the most suitable for numerical simulation of train aerodynamic behavior in crosswinds. To cope with the problem of insufficient computational resources during the numerical computation of high-speed trains, Liu et al. [65] proposed a model for pantograph noise prediction based on the Reynolds number provided by a single component.

As shown in Figure 5, CFD software can give you the calculation results of running train vortex shedding. The rapid development of computers in the future will further promote the development of numerical simulation and theoretical research into high-speed trains to provide more abundant solutions to the noise problems associated with high-speed trains.

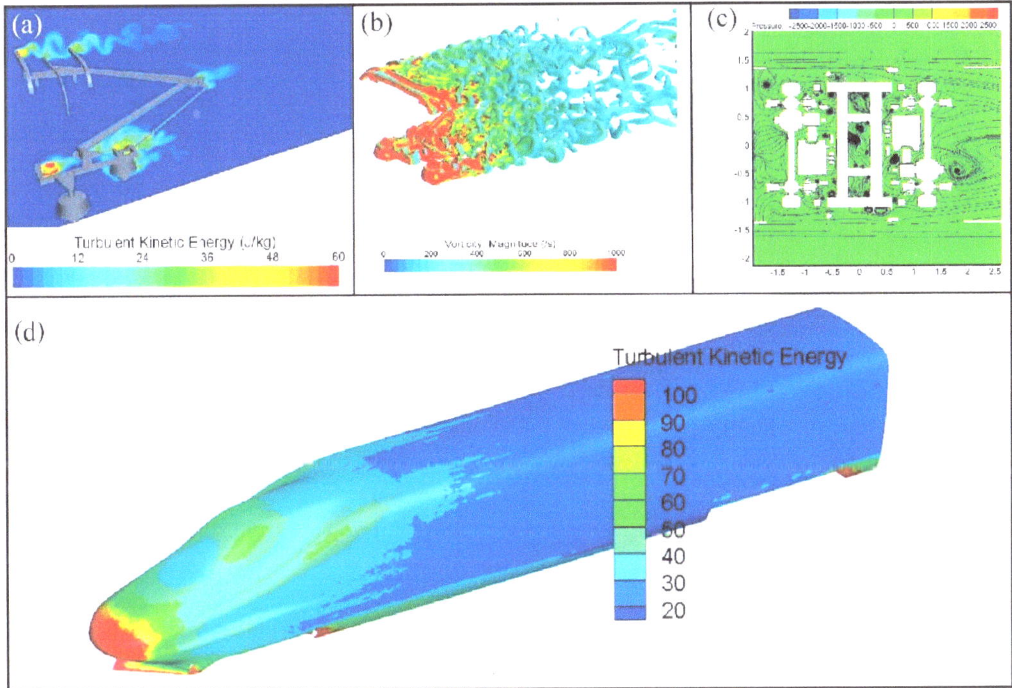

Figure 5. (**a**) Turbulent kinetic energy distribution of the pantograph [55]; (**b**) instantaneous iso-surface plots of the Q-criterion, colored by vorticity magnitude (Q = 10,000) [55]; (**c**) bogie speed flow chart [58]; (**d**) turbulent kinetic energy distribution of high-speed trains.

4. High-Speed Train Noise Control Methods

4.1. Traditional Noise Control Methods for High-Speed Trains

4.1.1. Train Vibration Reduction

High-speed trains will inevitably impose a greater impulse on railroad infrastructure, causing an increase in the vibration intensity of the infrastructure and the environment along the railroad line [66]. Under the action of the cyclic force caused by high-speed trains, excessive vibration amplitudes can damage the infrastructure, while this vibration also generates much noise, affecting the safety, comfort, and stability of train operation [67].

At present, the main active isolation measures and passive isolation are adopted to reduce high-speed train vibration and thus control high-speed train noise. Active isolation measures include floating slabs [68], highly resilient rail pads [69], and high-performance wheels [70]. All these measures can directly reduce the vibration of the train to a large extent. Passive isolation measures include open trenches and soft-filled barriers [71]. Although the above measures show a good damping effect, damping to reduce the actual engineering problems related to noise still faces certain difficulties due to the huge mass and fast speed of high-speed trains.

4.1.2. Train Body Sound Insulation

The noise inside the high-speed train is mainly transmitted from outside. The sound source and vibration excitation outside the train can be transmitted to the inside of the train through two main paths: airborne sound and structural sound transmission, thus forming the noise inside the train [72]. Therefore, improving the sound insulation performance of the high-speed train carriage panel structure is conducive to improving the acoustic environment inside the car. The train carriage panels are mainly extruded aluminum profile structures (floor, roof, and sidewalls) and double-plate cavity structures (windows). At present, extruded aluminum profiles are the main sheet structure of high-speed train bodies, and their acoustic performance is one of the important factors influencing the acoustic environment inside the vehicle.

Due to the special structural form of extruded aluminum profiles, their sound insulation performance is poor. Improving the sound insulation performance of aluminum profiles is a key focus of much research in academe. Xin et al. [73] studied the sound insulation characteristics of orthogonally rib-stiffened sandwich structures and corrugated core sandwich structures and the mechanism of sound radiation during the vibration of these structures. Xie et al. [74] developed an acoustic model of the statistical energy method for the amount of sound insulation of extruded profiles, and the model was able to favorably predict vibration excitation. Kim et al. [75] improved the sound insulation performance of aluminum profiles by filling. Qin et al. [76] employed an optimized finite element energy statistics method (FE-SEA) and verified its efficacy. Li et al. [77] used a waveguide finite element (2.5-d FE) model combined with the energy statistics method to investigate the sound insulation model of extruded aluminum profiles. Zhang et al. estimated the coupling loss factors between the structure and cavity and established a statistical energy analysis model to make predictions of bogie noise and bottom plate sound transmission. Based on the model, they developed a program that can quickly design the internal panel structure of a train and proposed a model that can predict the sound absorption and insulation characteristics of the composite bottom panel of a train [78,79]. Thereafter, they proposed to improve the noise insulation of the train by adding high-damping rubber and water-based damping as a coating to the train floor [80,81].

For the double-plate cavity structure of the window part of the train, Xin et al. [82] systematically studied the sound insulation characteristics of the finite and infinite double-plate cavity structure using the wave propagation method. Zhang et al. [83] established and validated their sound transmission loss model of the window by measuring the vibration and the interior noise level of a window of a high-speed train.

4.1.3. Railroad Sound Barriers

A sound barrier featuring easy installation and obvious noise reduction is an important method of traffic-noise management, which has been widely used and developed throughout the world [84]. Sound barriers mainly include wide top types, semi-enclosed types, fully enclosed types, etc. The diversity of sound barriers makes it possible to have different classification methods, which are based on appearance, line form, material, acoustic performance, structural form, etc. According to the differences in line form, the sound barriers can be classified into bridge sound barriers and road sound barriers; according to the differences in unit plate material, the sound barriers are classified into metal and non-metal sound barriers. As shown in Figure 6, the common structures of sound barriers are upright insert types, T types, inverted L types, Y types, multiple edge types, etc. [85]. In practical application, insert-type sound barriers are most widely used. May et al. [86] used proportional model experiments to compare the noise reduction effects of upright, T-shaped, and Y-shaped sound barriers and proposed, for the first time, that T-shaped sound barriers provide the best noise reduction effects. Defrance et al. [87] simulated the noise reduction effect of T-shaped sound barriers with sound-absorbing materials using the boundary element method and undertook experimental verification thereof. Baulac et al. [88] improved the acoustic performance by providing slots at the top of T-shaped sound barriers and optimizing the form and disposition of the slots using genetic algorithms. Oldham et al. [89] investigated the factors affecting the noise reduction effect of T-shaped sound barriers using numerical calculations and found that the additional noise reduction of the top structure is related to the locations of the source and receptor of the sound and the location and height of the sound barrier. Venckus et al. [90] studied the roof-inclination angle of upright-type sound barriers and found that the sound barrier was most effective in reducing high-frequency sound waves when the roof inclination angle was 120°. Zhang et al. [91] proposed a semi-enclosed sound barrier with slits and verified its sound insulation effect. At present, sound barriers provide good control of environmental noise along the railroad, but the noise frequency band they control is singular, and because most of them are reflective sound barriers, they will aggravate the noise inside the train instead.

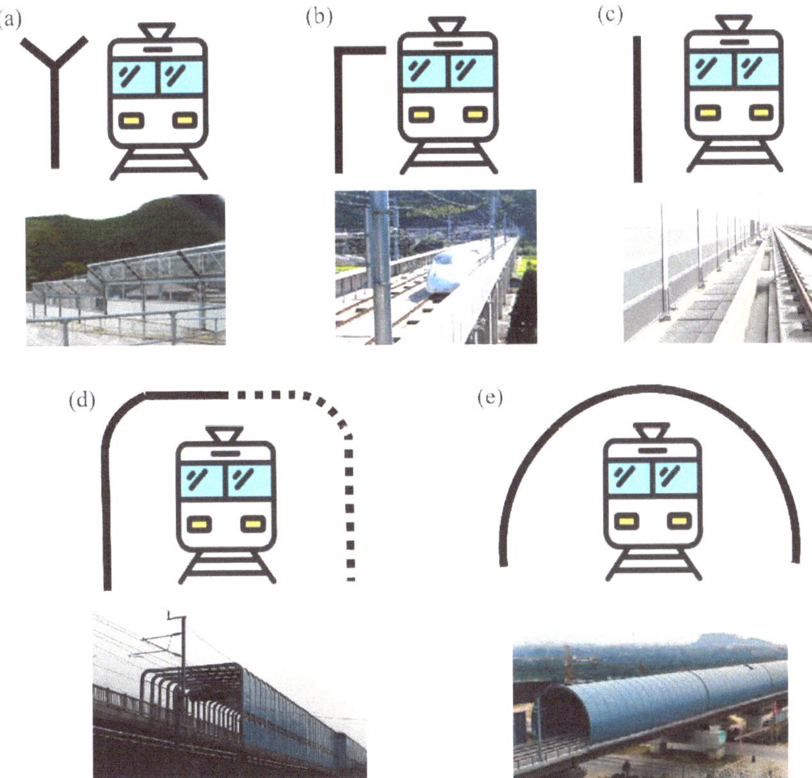

Figure 6. (**a**) Y-shaped sound barrier; (**b**) inverted L-shaped sound barrier; (**c**) upright type sound barriers; (**d**) semi-enclosed sound barrier; (**e**) fully enclosed sound barrier.

4.2. Acoustic Metamaterials Applied to High-Speed Train Noise Control

After the train speed is increased to 600 km/h, the aerodynamic noise and wheel-rail noise will be significantly increased, and the noise-reduction ability of the traditional extruded aluminum profiles used in the body of high-speed trains, the sound barriers along the railroad line, and the polyurethane foam materials laid inside the car body are all stretched to their operational limit. When traditional materials and structures fail to meet the needs of high-speed train development, acoustic metamaterials with exotic characteristics will become an important means by which to solve the noise problems associated with high-speed trains.

4.2.1. Definition and Development History of Acoustic Metamaterials

The term "metamaterial" is commonly used to describe artificial composites consisting of periodic or randomly arranged artificial subwavelength structures, a concept first introduced by Veselago in 1968 in the field of electromagnetism [92]. The emergence of electromagnetic metamaterials sharpens researchers' understanding of metamaterial theory, and the concept of metamaterials was also introduced in the fields of optics, mechanics, and heat transfer. The concept of acoustic metamaterials can be traced back to Narayanamurti [93], who first discovered that periodic structures could be used to control high-frequency phonon propagation but did not refer to such periodic structured materials as acoustic metamaterials. In 2000, Liu et al. [94] designed a small ball made of high-density lead wrapped in rubber and then proposed the theory of locally resonant phonon crystals using this model, opening the door to the study of acoustic metamaterials. After this, acoustic metamaterials have been further developed by artificially designing microstructures to allow materials to present limitations beyond their original natural laws. This can realize a series of idiosyncratic material functions, including acoustic stealth [95], acoustic directional transmission [96], acoustic negative refraction [97], acoustic focusing [98], low-frequency sound absorption [99], and so on. At present, acoustic metamaterials have been applied to solve engineering problems such as aircraft cabin noise reduction, automotive NVH, and building facades and have achieved better vibration and noise suppression effects [100].

Metamaterials often have different properties from those of traditional materials. For natural materials, material parameters such as mass density, Young's modulus, and Poisson's ratio are positive in the natural case, while for artificial structures, effective material parameters may become negative within a specific frequency range. Some mechanical metamaterials have been applied in the field of energy-absorbing structures and body structures of high-speed trains [101,102], and the application of metamaterials to reduce railroad noise will be one of the important means of dealing with the noise problem associated with future high-speed trains.

4.2.2. Metamaterials Applied to Railroad Sound Barriers

At present, railroad sound barriers are mainly upright insert-type sound barriers. The wind pressure fluctuations, when passing through such sound barriers, can lead to the loosening and breaking of bolts and the destruction of sound barrier panels [103]. When designing high-speed railroad sound barriers, not only should the sound insulation and sound absorption characteristics of the barrier be considered, but also the dynamic response of the sound barrier structure. Conventional sound barriers produce sound reflections that lead to increased noise levels within high-speed trains. For future high-speed railroad sound barriers, they inevitably need to meet codified sound insulation criteria and, at the same time, have both ventilation and noise absorption capabilities.

There is always a balance between the thickness of a sound insulation device and its ventilation capacity. Zhang et al. [104] designed a binary structure consisting of a coiled unit and a hollow tube with a thickness of less than one-fifth of the wavelength, which can block low-frequency sound from different directions while allowing 63% of the airflow to pass through. Wu et al. [105] designed a vented metamaterial absorber operating at low frequencies (<500 Hz) with only two absorption units, achieving high-efficiency absorption under vented conditions by using weak coupling of two identical split-tube resonators. Huang et al. [106] combined spiral channels and embedded apertures; this metamaterial structure can absorb low-frequency noise while maintaining the requisite thickness. Ghaffarivardavagh et al. [107] proposed a deep sub-acoustic wavelength metamaterial cell that includes nearly 60% of the open area allowing passage of air and also serves as a high-performance selective sound muffler. Wang et al. [108] proposed an acoustic metamaterial composed of many cells, and this open metamaterial contains a large hole in each cell to ensure airflow. Kumar et al. [109] integrated eight labyrinth cells of different configurations to form an acoustic metamaterial and introduced a herringbone channel to achieve ventilation. Xie et al. [110] proposed a metamaterial with a conchoidal cavity structure and embedded this metamaterial into a conventional concrete or metal sound barrier, which can significantly improve the sound absorption capacity of the sound barrier and prevent sound pollution of the surrounding high-rise buildings by high-speed railroads.

As shown in Figure 7, acoustic metamaterials can ensure better sound insulation and absorption effects under ventilation, but they have not been widely used in sound barriers due to their complicated production process and high cost of production. To increase the potential for wider engineering applications of acoustic metamaterials, researchers should further simplify their structure while considering the production cost in the design process and, furthermore, fit the specific use scenario for optimization.

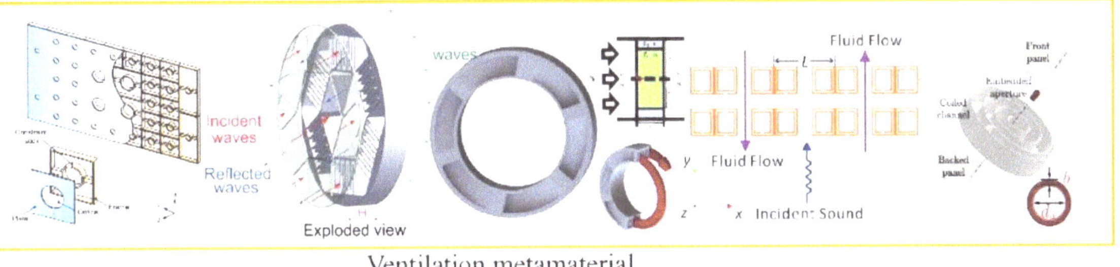

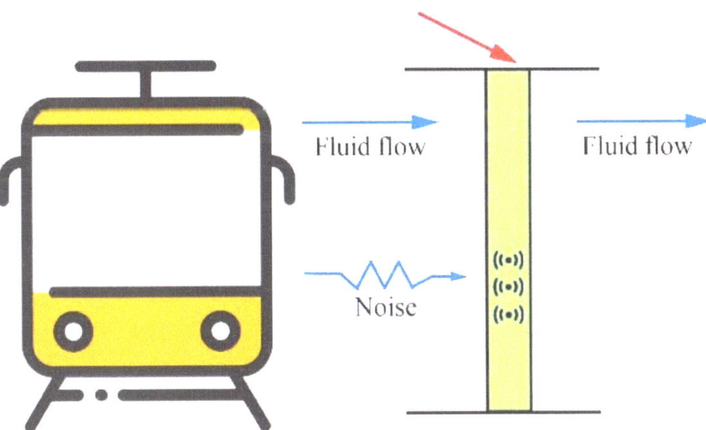

Figure 7. Acoustic metamaterials with ventilation properties applied in sound barrier.

4.2.3. Application of Metamaterials in Train Car Bodies

At present, the car body of high-speed trains is mainly fitted with a mixture of damping pulp and fibrous materials as sound insulation and vibration damping materials. At present, the train body shows poor sound insulation and absorption ability, especially for the long wavelengths of low-frequency noise, which passes more easily through the body structure and body gaps into the train interior. This impairs the quality of the acoustics within the train interior and adversely affects passenger comfort.

As shown in Figure 8, scholars have explored the application of acoustic metamaterials in engineering. Li et al. [111] combined microperforated plates and embedded partitions to design a controllable broadband acoustic unit with a thickness of only 70 mm. Liu et al. [112] added a separator plate with micro-perforations inside the cavity of a common Helmholtz resonator to demonstrate the absorber a multi-order absorption capability and expanded the absorption bandwidth of the absorber by combining eight absorbers. Long et al. [113] designed a multi-band quasi-perfect absorber constructed by a double-channel Mie resonator, which is more flexible, to achieve multi-band quasi-perfect absorption. The combination of labyrinth structure and micro-perforated plate structure often can produce better sound absorption. The labyrinth structure can partition the cavities, while the combination of different cavities and perforated plates can yield acoustic metamaterials with a wider frequency range. Zhang et al. [114] designed an acoustic metamaterial consisting of a single-hole plate and a labyrinth cavity combination, which achieved good sound absorption in the low-frequency range. Liu et al. [115] investigated an acoustic metamaterial that achieved nearly perfect sound absorption in the range of 380 to 3600 Hz by combining a variety of different micro-hole plates and labyrinth cavities. The honeycomb structure is the best topology covering two-dimensional planes with good mechanical properties and is often used as the core structure of high-speed train underlayment. Tang et al. [116] modified the traditional honeycomb sandwich panel by introducing micro-pores based on the honeycomb-corrugated hybrid core to acquire an acoustic metamaterial with good sound absorption in the low-frequency range. Peng et al. [117] designed a composite honeycomb structure by combining different microporous and honeycomb cavities to fabricate an acoustic metamaterial with 90% sound absorption in the range of 600 to 1000 Hz. Wang et al. [118] proposed a NOMEX honeycomb metamaterial with acoustic absorption capability based on NOMEX honeycomb, which can achieve quasi-perfect absorption against noise in high-speed train motion. Wu et al. [119] designed a hybrid metamaterial absorber based on a microperforated plate and a coiled Fabry channel, which can achieve more than 99% sound absorption at the resonant frequency (<500 Hz) of acoustic absorption. Xu et al. [120] designed a metamaterial consisting of three holes and three cavities connected in parallel and investigated the effects of different temperatures on its acoustic absorption. Xie et al. [121] added polyurethane material to the acoustic metamaterial composed of microporous plates and cavities to achieve continuous ultra-broadband acoustic performance.

Figure 9 analyzes the main noise sources of high-speed trains, and it can be found that the wheel-rail area noise, pantograph noise, and inter-coach gaps noise all rise to a great extent as the speed of high-speed trains increases. Among the three main noise sources, the pantograph noise increases most significantly with speed, and when the speed reaches 386 km/h, its main noise level exceeds that of the wheel-rail area and inter-coach gaps. The frequency variation of the main noise sources also deserves attention; compared with the noise in the range of 2000 Hz–4000 Hz, the improvement is more obvious in the range of 500 Hz–2000 Hz. The pantograph noise changes more significantly with speed, and the noise level in the 500 Hz–1500 Hz range reaches nearly 100 dB(A) after the speed reaches 386 km/h. The significant increase in noise levels in the mid-frequency and low-frequency can seriously affect passenger comfort.

Traditional acoustic materials mainly use porous structures, such as acoustic sponges, felt, glass fiber, polyurethane foam, etc. When sound waves pass through the pores of various materials, their kinetic energy is converted into thermal energy, leading to the dissipation of sound wave energy to attenuate noise; however, porous materials have poor absorption capacity for low-frequency noise, and the material thickness is strictly limited due to the size of the train, which makes it difficult for the current acoustic insulation car body design to deal with low-frequency noise. In contrast, metamaterials, with their unique material properties, can absorb low-frequency noise while maintaining a small thickness.

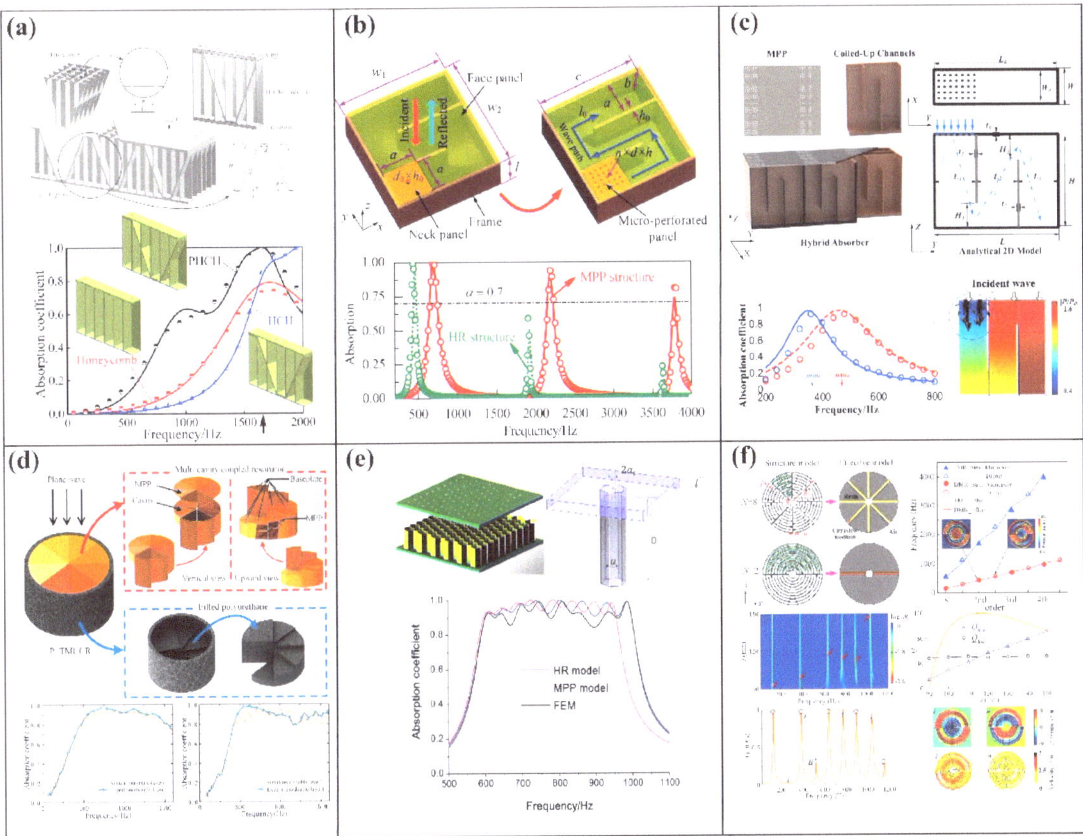

Figure 8. (**a**) Schematic of perforated honeycomb-corrugation hybrid (PHCH) metamaterial [116]. (**b**) Schematic of the broadband metamaterial unit [115]. (**c**) Schematic diagram of a hybrid absorber with microperforated plates and coiled channels [119]. (**d**) Schematic representation of a tunable multi-cavity coupled-resonator with polyurethane-filled slits [121]. (**e**) Schematics of the composite honeycomb sandwich panels [117]. (**f**) Multiband quasi-perfect low-frequency sound absorber based on double-channel Mie resonator [113].

According to the latest research on acoustic metamaterials shown in Figure 8, it can be found that acoustic metamaterials can cope with low-frequency and mid-frequency noise better and ensure a high noise absorption level in a thinner case. Figure 8d,e of the absorption coefficient curve can be found that in acoustic metamaterials in a wide range of frequencies, the sound absorption ability can reach more than 80%, and this acoustic characteristic can just meet the noise problems faced by high-speed trains. At present, acoustic metamaterials have been widely used in the fields of aerospace, ships, automobiles, and buildings. It is believed that acoustic metamaterials, with their excellent acoustic properties, will also become an important measure to solve the noise problem of high-speed trains.

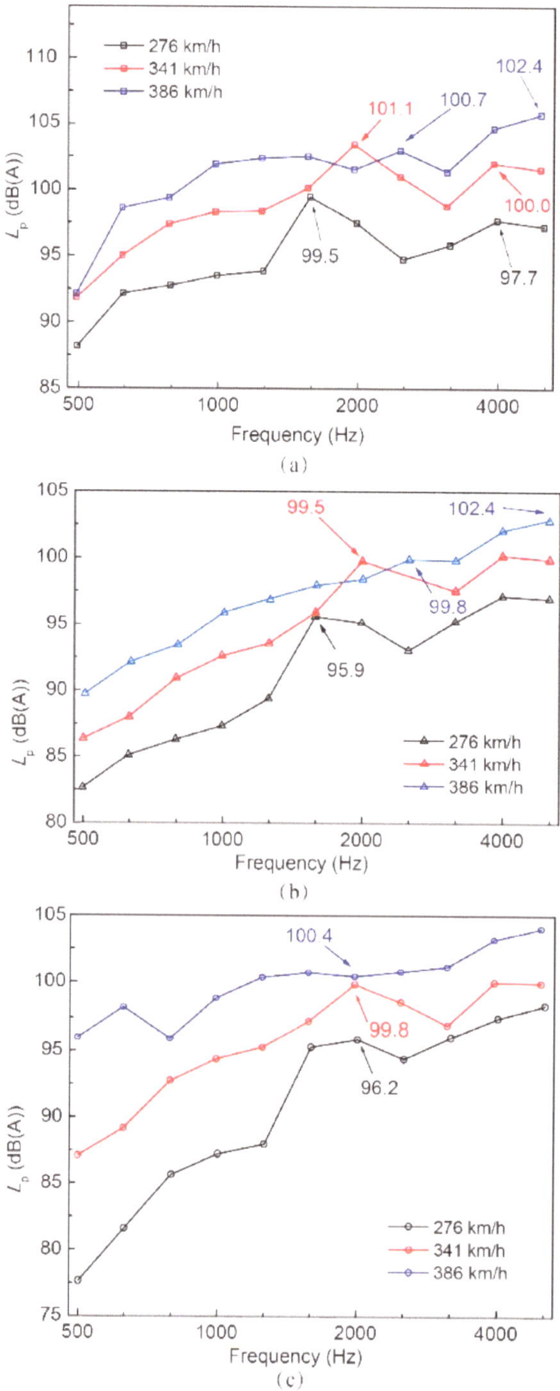

Figure 9. The main noise frequencies characteristics of high-speed trains: (**a**) wheel-rail area; (**b**) inter-coach spacing; (**c**) pantograph [47].

5. Summary and Prospect

With the increase in train speed and the development of modern train structures, the noise problem has become more pressing, and solutions thereto more complicated. The development of high-speed trains also brings new opportunities and challenges for future train noise research. To cope with the train noise problem, it is necessary to follow the development of science and adopt new materials, new research tools, and new theories. In this review, we have identified the following problems and made corresponding suggestions:

(1) The main component of high-speed train noise has gradually changed from mechanical vibration noise to aerodynamic noise, and with the further increase in running speeds in the future, the aerodynamic noise will be further intensified. The structure of the high-speed train pantograph, high-speed train skirt, and high-speed train connection should be further optimized, and the noise problem should be emphasized in the optimization process.

(2) The acoustic insulation level of traditional railroad sound barriers is limited and vulnerable to air pulsation stress. The introduction of acoustic metamaterials can largely improve their acoustic insulation performance, but at present, the structure of acoustic metamaterials is complicated, and their processing cost is high, so research on their manufacturing process and batch production is warranted.

(3) It is difficult for the current body structure of high-speed trains to cope with low-frequency noise, and the narrow body structure cannot be arranged with thicker materials to realize sound absorption and insulation. Thus, future acoustic metamaterials may be the best means of achieving sound absorption and insulation in the train body.

Author Contributions: Conceptualization, H.Y. and S.X.; methodology, H.Y.; software, H.Y.; formal analysis, H.Y., K.J. and Z.F.; writing—original draft preparation, H.Y., S.X., K.J. and Z.F.; writing—review and editing, H.Y., S.X., K.J. and Z.F.; supervision, S.X.; funding acquisition, S.X. All authors have read and agreed to the published version of the manuscript.

Funding: This research was funded by the National Natural Science Foundation of China (51775558). This paper was also supported by the Nature Science Foundation for Excellent Youth Scholars of Hunan Province (Grant No. 2019JJ30034) and the Shenghua Yu-ying Talents Program of the Central South University (Principle Investigator: Suchao Xie).

Institutional Review Board Statement: Not applicable.

Informed Consent Statement: Not applicable.

Data Availability Statement: Not applicable.

Acknowledgments: This research was undertaken at the Key Laboratory of Traffic Safety on Track (Central South University), Ministry of Education, China. The authors gratefully acknowledge the support from the National Natural Science Foundation of China (Grant No. 51775558). This paper was also supported by the Nature Science Foundation for Excellent Youth Scholars of Hunan Province (Grant No. 2019JJ30034) and the Shenghua Yu-ying Talents Program of the Central South University (Principle Investigator: Suchao Xie).

Conflicts of Interest: The authors declare no conflict of interest.

References

1. Siddall, W.R. *No Nook Secure: Transportation and Environmental Quality—Comparative Studies in Society and History*; Cambridge University Press: Cambridge, UK, 1974; Volume 16, pp. 2–23. [CrossRef]
2. Floud, S.; Blangiardo, M.; Clark, C.; de Hoogh, K.; Babisch, W.; Houthuijs, D.; Swart, W.; Pershagen, G.; Katsouyanni, K.; Velonakis, M.; et al. Exposure to Aircraft and Road Traffic Noise and Associations with Heart Disease and Stroke in Six European Countries: A Cross-Sectional Study. *Environ. Health* **2013**, *12*, 89. [CrossRef] [PubMed]
3. Banerjee, D.; Das, P.P.; Foujdar, A. Association between Road Traffic Noise and Prevalence of Coronary Heart Disease. *Environ. Monit. Assess.* **2014**, *186*, 2885–2893. [CrossRef] [PubMed]

4. Wothge, J.; Belke, C.; Möhler, U.; Guski, R.; Schreckenberg, D. The Combined Effects of Aircraft and Road Traffic Noise and Aircraft and Railway Noise on Noise Annoyance-An Analysis in the Context of the Joint Research Initiative NORAH. *Int. J. Environ. Res. Public Health* **2017**, *14*, 871. [CrossRef]

5. Beutel, M.; Brähler, E.; Ernst, M.; Klein, E.; Reiner, I.; Wiltink, J.; Michal, M.; Wild, P.; Schulz, A.; Münzel, T.; et al. Noise Annoyance Predicts Symptoms of Depression, Anxiety and Sleep Disturbance 5 Years Later. Findings from the Gutenberg Health Study. *Eur. J. Public Health* **2020**, *30*, 487–492. [CrossRef]

6. Gu, X.A. Railway Environmental Noise Control in China. *J. Sound Vib.* **2006**, *293*, 1078–1085. [CrossRef]

7. Segawa, T.; Fujimoto, M.; Saito, T.; Sakagoshi, O.; Tachibana, H. Assessment of Environmental Noise Immission in Japan. In Proceedings of the 2005 Congress and Exposition on Noise Control Engineering, Rio de Janeiro, Brazil, 7–10 August 2005; p. 6.

8. Li, T.; Qin, D.; Zhou, N.; Zhang, W. Step-by-Step Numerical Prediction of Aerodynamic Noise Generated by High Speed Trains. *Chin. J. Mech. Eng.* **2022**, *35*, 28. [CrossRef]

9. Thompson, D.J.; Latorre Iglesias, E.; Liu, X.; Zhu, J.; Hu, Z. Recent Developments in the Prediction and Control of Aerodynamic Noise from High-Speed Trains. *Int. J. Rail Transp.* **2015**, *3*, 119–150. [CrossRef]

10. Tan, X.; Yang, Z.; Tan, X.; Wu, X.; Zhang, J. Vortex Structures and Aeroacoustic Performance of the Flow Field of the Pantograph. *J. Sound Vib.* **2018**, *432*, 17–32. [CrossRef]

11. Zhu, H.; Wang, Y.; Huang, H.; Wang, J.; Xu, Q. Research on the Effect of Non-Smooth Convex Hull Structure of High-Speed EMU on Air Resistance and Friction Noise. In *Advances in Mechanical Design, Proceedings of the 2021 International Conference on Mechanical Design (ICMDE 2021), Sanya, China, 26–28 February 2021*; Tan, J., Ed.; Springer: Singapore, 2022; pp. 2109–2119.

12. Tang, R.; Zheng, W.; Li, S.F.; Liu, W.Y. Topology Optimization of Damping Material on Acoustic-Structural Systems for Minimizing Response Sensitivity. *Int. J. Perform. Eng.* **2021**, *17*, 26. [CrossRef]

13. Zhang, Y. Reduction of Aerodynamic Noise of High-Speed Train Pantograph. *J. Mech. Eng.* **2017**, *53*, 94. [CrossRef]

14. Liang, X.; Liu, H.; Dong, T.; Yang, Z.; Tan, X. Aerodynamic Noise Characteristics of High-Speed Train Foremost Bogie Section. *J. Cent. South Univ.* **2020**, *27*, 1802–1813. [CrossRef]

15. Dai, W.; Zheng, X.; Hao, Z.; Qiu, Y.; Li, H.; Luo, L. Aerodynamic Noise Radiating from the Inter-Coach Windshield Region of a High-Speed Train. *J. Low Freq. Noise Vib. Act. Control.* **2018**, *37*, 590–610. [CrossRef]

16. Li, T.; Qin, D.; Zhang, W.; Zhang, J. Study on the Aerodynamic Noise Characteristics of High-Speed Pantographs with Different Strip Spacings. *J. Wind. Eng. Ind. Aerodyn.* **2020**, *202*, 104191. [CrossRef]

17. Cao, Y.; Bai, Y.; Wang, Q. Complexity Simulation on Application of Asymmetric Bionic Cross-Section Rod in Pantographs of High-Speed Trains. *Complexity* **2018**, *2018*, e3087312. [CrossRef]

18. Yang, X.Y.; Xiao, Y.G.; Shi, Y. Shape Optimization of High-Speed Train Pantograph Insulators for Low Aerodynamic Noise. *Appl. Mech. Mater.* **2013**, *249*, 646–651. [CrossRef]

19. Wang, Y. Numerical Computation and Optimization Design of Pantograph Aerodynamic Noise. *J. Vibroeng.* **2016**, *18*, 1358–1369. [CrossRef]

20. Ikeda, M.; Mitsumoji, T.; Sueki, T.; Takaishi, T. Aerodynamic Noise Reduction of a Pantograph by Shape-Smoothing of Panhead and Its Support and by the Surface Covering with Porous Material. In *Noise and Vibration Mitigation for Rail Transportation Systems, Proceedings of the 10th International Workshop on Railway Noise, Nagahama, Japan, 18–22 October 2010*; Maeda, T., Gautier, P.-E., Hanson, C.E., Hemsworth, B., Nelson, J.T., Schulte-Werning, B., Thompson, D., de Vos, P., Eds.; Springer: Tokyo, Japan, 2012; pp. 419–426.

21. Liu, F.; Guo, H.; Hu, T.; Liu, P. Experimental Investigation on the Aeroacoustics of Circular Cylinders Covered with Metal Foam. In Proceedings of the 25th AIAA/CEAS Aeroacoustics Conference, Delft, The Netherlands, 20–23 May 2019.

22. Kim, H.; Hu, Z.; Thompson, D. Numerical Investigation of the Effect of Cavity Flow on High Speed Train Pantograph Aerodynamic. *Noise J. Wind Eng. Ind. Aerodyn.* **2020**, *201*, 104159. [CrossRef]

23. Yu, H.-H.; Li, J.-C.; Zhang, H.-Q. On Aerodynamic Noises Radiated by the Pantograph System of High-Speed Trains. *Acta Mech. Sin.* **2013**, *29*, 399–410. [CrossRef]

24. Kim, H.; Hu, Z.; Thompson, D. Effect of Different Typical High Speed Train Pantograph Recess Configurations on Aerodynamic Noise. *Proc. Inst. Mech. Eng. Part F J. Rail Rapid Transit* **2021**, *235*, 573–585. [CrossRef]

25. Eriksson, M.; Bergman, F.; Jacobson, S. On the Nature of Tribological Contact in Automotive Brakes. *Wear* **2002**, *252*, 26–36. [CrossRef]

26. Graf, M.; Ostermeyer, G.P. Instabilities in the Sliding of Continua with Surface Inertias: An Initiation Mechanism for Brake Noise. *J. Sound Vib.* **2011**, *330*, 5269–5279. [CrossRef]

27. Renault, A.; Massa, F.; Lallemand, B.; Tison, T. Experimental Investigations for Uncertainty Quantification in Brake Squeal Analysis. *J. Sound Vib.* **2016**, *367*, 37–55. [CrossRef]

28. Majcherczak, D.; Dufre'noy, P.; Nai¨t-Abdelaziz, M. Third Body Influence on Thermal Friction Contact Problems: Application to Braking. *J. Tribol.* **2005**, *127*, 89–95. [CrossRef]

29. Massi, F.; Berthier, Y.; Baillet, L. Contact Surface Topography and System Dynamics of Brake Squeal. *Wear* **2008**, *265*, 1784–1792. [CrossRef]

30. Chiacchiari, L.; Thompson, D.J.; Squicciarini, G.; Ntotsios, E.; Loprencipe, G. Rail Roughness and Rolling Noise in Tramways. *J. Phys. Conf. Ser.* **2016**, *744*, 012147. [CrossRef]

31. Thompson, D.J.; Hemsworth, B.; Vincent, N. Experimental Validation of the Twins Prediction Program for Rolling Noise, Part 1: Description of the Model and Method. *J. Sound Vib.* **1996**, *193*, 123–135. [CrossRef]

32. Remington, P.J. Wheel/Rail Rolling Noise, I: Theoretical Analysis. *J. Acoust. Soc. Am.* **1987**, *81*, 1805–1823. [CrossRef]

33. Thompson, D.J. Wheel-Rail Noise Generation, Part II: Wheel Vibration. *J. Sound Vib.* **1993**, *161*, 401–419. [CrossRef]

34. Thompson, D.J. Wheel-Rail Noise Generation, Part III: Rail Vibration. *J. Sound Vib.* **1993**, *161*, 421–446. [CrossRef]

35. Thompson, D.J. Wheel-Rail Noise Generation, Part IV: Contact Zone and Results. *J. Sound Vib.* **1993**, *161*, 447–466. [CrossRef]

36. Jones, C.J.C.; Thompson, D.J. Extended Validation of a Theoretical Model for Railway Rolling Noise Using Novel Wheel and Track Designs. *J. Sound Vib.* **2003**, *267*, 509–522. [CrossRef]

37. Vincent, N.; Bouvet, P.; Thompson, D.J.; Gautier, P.E. Theoretical Optimization of Track Components to Reduce Rolling Noise. *J. Sound Vib.* **1996**, *193*, 161–171. [CrossRef]

38. Bouvet, P.; Vincent, N.; Coblentz, A.; Demilly, F. Optimization of Resilient Wheels for Rolling Noise Control. *J. Sound Vib.* **2000**, *231*, 765–777. [CrossRef]

39. Cigada, A.; Manzoni, S.; Vanali, M. Vibro-Acoustic Characterization of Railway Wheels. *Appl. Acoust.* **2008**, *69*, 530–545. [CrossRef]

40. Merideno, I.; Nieto, J.; Gil-Negrete, N.; Giménez Ortiz, J.G.; Landaberea, A.; Iartza, J. Theoretical Prediction of the Damping of a Railway Wheel with Sandwich-Type Dampers. *J. Sound Vib.* **2014**, *333*, 4897–4911. [CrossRef]

41. Lee, S.; Lee, D.H.; Lee, J. Integrated Shape-Morphing and Metamodel-Based Optimization of Railway Wheel Web Considering Thermo-Mechanical Loads. *Struct. Multidisc. Optim.* **2019**, *60*, 315–330. [CrossRef]

42. Chen, Y.; Feng, Q.; Luo, K.; Xin, W.; Luo, X. Noise characteristics of urban rail transit viaduct installing damping rail. *J. Traffic Transp. Eng.* **2021**, *21*, 169–178. [CrossRef]

43. Zhang, J.; Xiao, X.; Sheng, X.; Zhang, C.; Wang, R.; Jin, X. SEA and Contribution Analysis for Interior Noise of a High Speed Train. *Appl. Acoust.* **2016**, *112*, 158–170. [CrossRef]

44. Zhang, J.; Xiao, X.; Sheng, X.; Fu, R.; Yao, D.; Jin, X. Characteristics of Interior Noise of a Chinese High-Speed Train under a Variety of Conditions. *J. Zhejiang Univ. Sci. A* **2017**, *18*, 617–630. [CrossRef]

45. Barsikow, B. Experiences with Various Configurations of Microphone Arrays Used to Locate Sound Sources on Railway Trains Operated by the DB AG. *J. Sound Vib.* **1995**, *193*, 283–293. [CrossRef]

46. Mellet, C.; Létourneaux, F.; Poisson, F.; Talotte, C. High Speed Train Noise Emission: Latest Investigation of the Aerodynamic/Rolling Noise Contribution. *J. Sound Vib.* **2006**, *293*, 535–546. [CrossRef]

47. He, B.; Xiao, X.; Zhou, Q.; Li, Z.; Jin, X. Investigation into External Noise of a High-Speed Train at Different Speeds. *J. Zhejiang Univ. Sci. A* **2014**, *15*, 1019–1033. [CrossRef]

48. Li, M.; Deng, T.; Wang, D.; Xu, F.; Xiao, X.; Sheng, X. An Experimental Investigation into the Difference in the External Noise Behavior of a High-Speed Train between Viaduct and Embankment Sections. *Shock. Vib.* **2022**, *2022*, 19. [CrossRef]

49. Zhang, J.; Xiao, X.; Wang, D.; Yang, Y.; Fan, J. Source Contribution Analysis for Exterior Noise of a High-Speed Train: Experiments and Simulations. *Shock. Vib.* **2018**, *2018*, 13. [CrossRef]

50. Noh, H.M. Noise-Source Identification of a High-Speed Train by Noise Source Level Analysis. *Proc. Inst. Mech. Eng. Part F J. Rail Rapid Transit* **2017**, *231*, 717–728. [CrossRef]

51. Baker, C.J.; Brockie, N.J. Wind Tunnel Tests to Obtain Train Aerodynamic Drag Coefficients: Reynolds Number and Ground Simulation Effects. *J. Wind. Eng. Ind. Aerodyn.* **1991**, *38*, 23–28. [CrossRef]

52. Willemsen, E. High Reynolds Number Wind Tunnel Experiments on Trains. *J. Wind. Eng. Ind. Aerodyn.* **1997**, *69*, 437–447. [CrossRef]

53. Nagakura, K. Localization of Aerodynamic Noise Sources of Shinkansen Trains. *J. Sound Vib.* **2006**, *293*, 547–556. [CrossRef]

54. Zhu, J.Y.; Hu, Z.W.; Thompson, D.J. Flow Simulation and Aerodynamic Noise Prediction for a High-Speed Train Wheelset. *Int. J. Aeroacoustics* **2014**, *13*, 533–552. [CrossRef]

55. Zhang, Y.; Zhang, J.; Sheng, X. Study on the Flow Behaviour and Aerodynamic Noise Characteristics of a High-Speed Pantograph under Crosswinds. *Sci. China Technol. Sci.* **2020**, *63*, 977–991. [CrossRef]

56. Guo, Z.; Liu, T.; Hemida, H.; Chen, Z.; Liu, H. Numerical Simulation of the Aerodynamic Characteristics of Double Unit Train. *Eng. Appl. Comput. Fluid Mech.* **2020**, *14*, 910–922. [CrossRef]

57. Wang, Y.; Wang, J.; Fu, L. Numerical Computation of Aerodynamic Noises of the High Speed Train with Considering Pantographs. *J. Vibroengineering* **2016**, *18*, 5588–5604. [CrossRef]

58. Yang, H.; Liu, D. Numerical Study on the Aerodynamic Noise Characteristics of CRH2 High-Speed Trains. *J. Vibroengineering* **2017**, *19*, 3953–3967. [CrossRef]

59. Sassa, T.; Sato, T.; Yatsui, S. Numerical Analysis of Aerodynamic Noise Radiation from a High-Speed Train Surface. *J. Sound Vib.* **2001**, *247*, 407–416. [CrossRef]

60. Takaishi, T.; Sagawa, A.; Nagakura, K.; Maeda, T. Numerical Analysis of Dipole Sound Source around High Speed Trains. *J. Acoust. Soc. Am.* **2002**, *111*, 2601–2608. [CrossRef] [PubMed]

61. Masson, E.; Paradot, N.; Allain, E. The Numerical Prediction of the Aerodynamic Noise of the TGV POS High-Speed Train Power Car. In *Notes on Numerical Fluid Mechanics and Multidisciplinary Design, Proceedings of the 10th International Workshop on Railway Noise, Nagahama, Japan, 18–22 October 2010*; Maeda, T., Gautier, P.-E., Hanson, C.E., Hemsworth, B., Nelson, J.T., Schulte-Werning, B., Thompson, D., de Vos, P., Eds.; Springer: Tokyo, Japan, 2012; pp. 437–444. [CrossRef]

62. Wu, J. Numerical Computation and Improvement of Aerodynamic Radiation Noises of Pantographs. *J. Vibroengineering* **2017**, *19*, 3939–3952. [CrossRef]
63. Sun, X.; Xiao, H. Numerical Modeling and Investigation on Aerodynamic Noise Characteristics of Pantographs in High-Speed Trains. *Complexity* **2018**, *2018*, 1–12. [CrossRef]
64. Li, T.; Qin, D.; Zhang, J. Effect of RANS Turbulence Model on Aerodynamic Behavior of Trains in Crosswind. *Chin. J. Mech. Eng.* **2019**, *32*, 85. [CrossRef]
65. Liu, X.; Zhang, J.; Thompson, D.; Iglesias, E.L.; Squicciarini, G.; Hu, Z.; Toward, M.; Lurcock, D. Aerodynamic Noise of High-Speed Train Pantographs: Comparisons between Field Measurements and an Updated Component-Based Prediction Model. *Appl. Acoust.* **2021**, *175*, 107791. [CrossRef]
66. Feng, S.; Zhang, X.; Wang, L.; Zheng, Q.; Du, F.; Wang, Z. In Situ Experimental Study on High Speed Train Induced Ground Vibrations with the Ballast-Less Track. *Soil Dyn. Earthq. Eng.* **2017**, *102*, 195–214. [CrossRef]
67. Li, T.; Su, Q.; Shao, K.; Liu, J. Numerical Analysis of Vibration Responses in High-Speed Railways Considering Mud Pumping Defect. *Shock. Vib.* **2019**, *2019*, 11. [CrossRef]
68. Liang, L.; Li, X.; Yin, J.; Wang, D.; Gao, W.; Guo, Z. Vibration Characteristics of Damping Pad Floating Slab on the Long-Span Steel Truss Cable-Stayed Bridge in Urban Rail Transit. *Eng. Struct.* **2019**, *191*, 92–103. [CrossRef]
69. Ge, X.; Ling, L.; Yuan, X.; Wang, K. Effect of Distributed Support of Rail Pad on Vertical Vehicle-Track Interactions. *Constr. Build. Mater.* **2020**, *262*, 120607. [CrossRef]
70. Liu, L.; Liu, H.; Lv, R. Research on Noise and Vibration Reduction of Damped Wheel-Rail. *Noise Vib. Worldw.* **2010**, *41*, 44–48. [CrossRef]
71. Connolly, D.; Giannopoulos, A.; Fan, W.; Woodward, P.K.; Forde, M.C. Optimising Low Acoustic Impedance Back-Fill Material Wave Barrier Dimensions to Shield Structures from Ground Borne High Speed Rail Vibrations. *Constr. Build. Mater.* **2013**, *44*, 557–564. [CrossRef]
72. Kim, T.; Kim, J. Comparison Study of Sound Transmission Loss in High Speed Train. *Int. J. Railw.* **2011**, *4*, 19–27. [CrossRef]
73. Xin, F.X.; Lu, T.J. Sound Radiation of Orthogonally Rib-Stiffened Sandwich Structures with Cavity Absorption. *Compos. Sci. Technol.* **2010**, *70*, 2198–2206. [CrossRef]
74. Xie, G.; Thompson, D.J.; Jones, C.J.C. A Modelling Approach for the Vibroacoustic Behaviour of Aluminium Extrusions Used in Railway Vehicles. *J. Sound Vib.* **2006**, *293*, 921–932. [CrossRef]
75. Kim, S.; Seo, T.; Kim, J.; Song, D. Sound-Insulation Design of Aluminum Extruded Panel in Next-Generation High-Speed Train. *Trans. Korean Soc. Mech. Eng. A* **2011**, *35*, 567–574. [CrossRef]
76. Qin, Z.H.; Zhang, Z.P.; Ren, F.; Li, H.B.; Liu, Z.H.; Zhang, Z.; Yuan, K.; Wang, Y.C. Application of FE-SEA Hybrid Method in Vibro-Acoustic Environment Prediction of Complex Structure. *J. Phys. Conf. Ser.* **2016**, *744*, 012238. [CrossRef]
77. Li, H.; Squicciarini, G.; Thompson, D.; Ryue, J.; Xiao, X.; Yao, D.; Chen, J. A Modelling Approach for Noise Transmission through Extruded Panels in Railway Vehicles. *J. Sound Vib.* **2021**, *502*, 116095. [CrossRef]
78. Zhang, J.; Xiao, X.; Sheng, X.; Yao, D.; Wang, R. An Acoustic Design Procedure for Controlling Interior Noise of High-Speed Trains. *Appl. Acoust.* **2020**, *168*, 107419. [CrossRef]
79. Zhang, J.; Yao, D.; Wang, R.; Xiao, X. Vibro-Acoustic Modelling of High-Speed Train Composite Floor and Contribution Analysis of Its Constituent Materials. *Compos. Struct.* **2021**, *256*, 113049. [CrossRef]
80. Zhang, J.; Yao, D.; Shen, M.; Wang, R.; Li, J.; Guo, S. Effect of Multi-Layered IIR/EP on Noise Reduction of Aluminium Extrusions for High-Speed Trains. *Compos. Struct.* **2021**, *262*, 113638. [CrossRef]
81. Yao, D.; Zhang, J.; Wang, R.; Xiao, X. Vibroacoustic Damping Optimisation of High-Speed Train Floor Panels in Low- and Mid-Frequency Range. *Appl. Acoust.* **2021**, *174*, 107788. [CrossRef]
82. Xin, F.X.; Lu, T.J.; Chen, C.Q. Vibroacoustic Behavior of Clamp Mounted Double-Panel Partition with Enclosure Air Cavity. *J. Acoust. Soc. Am.* **2009**, *124*, 3604–3612. [CrossRef]
83. Zhang, Y.; Xiao, X.; Thompson, D.; Squicciarini, G.; Wen, Z.; Li, Z.; Wu, Y. Sound Transmission Loss of Windows on High Speed Trains. *J. Phys. Conf. Ser.* **2016**, *744*, 012141. [CrossRef]
84. Wu, X.; Chen, X.; Liu, L.; Lu, W. Current Status and Prospects of Application of Sound Barrier for High-Speed Railway in China. *IOP Conf. Ser. Earth Environ. Sci.* **2019**, *300*, 032051. [CrossRef]
85. Ekici, I.; Bougdah, H. A Review of Research on Environmental Noise Barriers. *Build. Acoust.* **2003**, *10*, 289–323. [CrossRef]
86. May, D.N.; Osman, M.M. The Performance of Sound Absorptive, Reflective, and T-Profile Noise Barriers in Toronto. *J. Sound Vib.* **1980**, *71*, 65–71. [CrossRef]
87. Defrance, J.; Jean, P. Integration of the Efficiency of Noise Barrier Caps in a 3D Ray Tracing Method. Case of a T-Shaped Diffracting Device. *Appl. Acoust.* **2003**, *64*, 765–780. [CrossRef]
88. Baulac, M.; Defrance, J.; Jean, P. Optimisation with Genetic Algorithm of the Acoustic Performance of T-Shaped Noise Barriers with a Reactive Top Surface. *Appl. Acoust.* **2008**, *69*, 332–342. [CrossRef]
89. Oldham, D.J.; Egan, C.A. A Parametric Investigation of the Performance of T-Profiled Highway Noise Barriers and the Identification of a Potential Predictive Approach. *Appl. Acoust.* **2011**, *72*, 803–813. [CrossRef]
90. Venckus, Ž.; Grubliauskas, R.; Venslovas, A. The Research on the Effectiveness of the Inclined Top Type of a Noise Barrier. *J. Environ. Eng. Landsc. Manag.* **2012**, *20*, 155–162. [CrossRef]

91. Zhang, X.; Liu, R.; Cao, Z.; Wang, X.; Li, X. Acoustic Performance of a Semi-Closed Noise Barrier Installed on a High-Speed Railway Bridge: Measurement and Analysis Considering Actual Service Conditions. *Measurement* **2019**, *138*, 386–399. [CrossRef]
92. Veselago, V.G. The Electrodynamics of Substances with Simultaneously Negative Values of ε and μ. *Sov. Phys. Usp.* **1968**, *10*, 509–514. [CrossRef]
93. Narayanamurti, V.; Pohl, R.O. Tunneling States of Defects in Solids. *Rev. Mod. Phys.* **1970**, *42*, 201–236. [CrossRef]
94. Liu, Z.; Zhang, X.; Mao, Y.; Zhu, Y.Y.; Yang, Z.; Chan, C.T.; Sheng, P. Locally Resonant Sonic Materials. *Science* **2000**, *289*, 1734–1736. [CrossRef]
95. Zigoneanu, L.; Popa, B.-I.; Cummer, S.A. Three-Dimensional Broadband Omnidirectional Acoustic Ground Cloak. *Nat. Mater.* **2014**, *13*, 352–355. [CrossRef] [PubMed]
96. Lu, M.; Liu, X.-K.; Feng, L.; Li, J.; Huang, C.-P.; Chen, Y.-F.; Zhu, Y.-Y.; Zhu, S.-N.; Ming, N.-B. Extraordinary Acoustic Transmission through a 1D Grating with Very Narrow Apertures. *Phys. Rev. Lett.* **2007**, *99*, 174301. [CrossRef]
97. Lu, M.; Zhang, C.; Feng, L.; Zhao, J.; Chen, Y.-F.; Mao, Y.-W.; Zi, J.; Zhu, Y.-Y.; Zhu, S.-N.; Ming, N.-B. Negative Birefraction of Acoustic Waves in a Sonic Crystal. *Nat. Mater.* **2007**, *6*, 744–748. [CrossRef] [PubMed]
98. Zhang, S.; Yin, L.; Fang, N. Focusing Ultrasound with an Acoustic Metamaterial Network. *Phys. Rev. Lett.* **2009**, *102*, 194301. [CrossRef] [PubMed]
99. Romero-García, V.; Theocharis, G.; Richoux, O.; Merkel, A.; Tournat, V.; Pagneux, V. Perfect and Broadband Acoustic Absorption by Critically Coupled Sub-Wavelength Resonators. *Sci. Rep.* **2016**, *6*, 19519. [CrossRef] [PubMed]
100. Ang, L.Y.L.; Koh, Y.K.; Lee, H.P. Acoustic Metamaterials: A Potential for Cabin Noise Control in Automobiles and Armored Vehicles. *Int. J. Appl. Mech.* **2016**, *8*, 1650072. [CrossRef]
101. Xing, J.; Xu, P.; Zhao, H.; Yao, S.; Wang, Q.; Li, B. Crashworthiness Design and Experimental Validation of a Novel Collision Post Structure for Subway Cab Cars. *J. Cent. South Univ.* **2020**, *27*, 2763–2775. [CrossRef]
102. Gao, G.; Zhuo, T.; Guan, W. Recent Research Development of Energy-Absorption Structure and Application for Railway Vehicles. *J. Cent. South Univ.* **2020**, *27*, 1012–1038. [CrossRef]
103. Qiu, X.; Li, X.; Zheng, J.; Wang, M. Fluctuating Wind Pressure on Vertical Sound Barrier during Two High-Speed Trains Passing Each Other. *Int. J. Rail Transp.* **2022**, *10*, 1–18. [CrossRef]
104. Zhang, H.; Zhu, Y.; Liang, B.; Yang, J.; Yang, J.; Cheng, J. Omnidirectional Ventilated Acoustic Barrier. *Appl. Phys. Lett.* **2017**, *111*, 203502. [CrossRef]
105. Wu, X.; Au-Yeung, K.Y.; Li, X.; Roberts, R.C.; Tian, J.; Hu, C.; Huang, Y.; Wang, S.; Yang, Z.; Wen, W. High-Efficiency Ventilated Metamaterial Absorber at Low Frequency. *Appl. Phys. Lett.* **2018**, *112*, 103505. [CrossRef]
106. Huang, S.; Fang, X.; Wang, X.; Assouar, B.; Cheng, Q.; Li, Y. Acoustic Perfect Absorbers via Spiral Metasurfaces with Embedded Apertures. *Appl. Phys. Lett.* **2018**, *113*, 233501. [CrossRef]
107. Ghaffarivardavagh, R.; Nikolajczyk, J.; Anderson, S.; Zhang, X. Ultra-Open Acoustic Metamaterial Silencer Based on Fano-like Interference. *Phys. Rev. B* **2019**, *99*, 024302. [CrossRef]
108. Wang, X.; Luo, X.; Yang, B.; Huang, Z. Ultrathin and Durable Open Metamaterials for Simultaneous Ventilation and Sound Reduction. *Appl. Phys. Lett.* **2019**, *115*, 171902. [CrossRef]
109. Kumar, S.; Lee, H.P. Labyrinthine Acoustic Metastructures Enabling Broadband Sound Absorption and Ventilation. *Appl. Phys. Lett.* **2020**, *116*, 134103. [CrossRef]
110. Xie, S.; Yang, S.; Yan, H.; Li, Z. Sound Absorption Performance of a Conch-Imitating Cavity Structure. *Sci. Prog.* **2022**, *105*, 00368504221075167. [CrossRef] [PubMed]
111. Li, Y.; Assouar, B.M. Acoustic Metasurface Based Perfect Absorber with Deep Subwavelength Thickness. *Appl. Phys. Lett.* **2016**, *108*, 063502. [CrossRef]
112. Liu, C.R.; Wu, J.H.; Chen, X.; Ma, F. A Thin Low-Frequency Broadband Metasurface with Multi-Order Sound Absorption. *J. Phys. D Appl. Phys.* **2019**, *52*, 105302. [CrossRef]
113. Long, H.; Gao, S.; Cheng, Y.; Liu, X. Multiband Quasi-Perfect Low-Frequency Sound Absorber Based on Double-Channel Mie Resonator. *Appl. Phys. Lett.* **2018**, *112*, 033507. [CrossRef]
114. Zhang, C.; Hu, X. Three-Dimensional Single-Port Labyrinthine Acoustic Metamaterial: Perfect Absorption with Large Bandwidth and Tunability. *Phys. Rev. Appl.* **2016**, *6*, 064025. [CrossRef]
115. Liu, C.; Wu, J.; Yang, Z.; Ma, F. Ultra-Broadband Acoustic Absorption of a Thin Microperforated Panel Metamaterial with Multi-Order Resonance. *Compos. Struct.* **2020**, *246*, 112366. [CrossRef]
116. Tang, Y.; Ren, S.; Meng, H.; Xin, F.; Huang, L.; Chen, T.; Zhang, C.; Lu, T.J. Hybrid Acoustic Metamaterial as Super Absorber for Broadband Low-Frequency Sound. *Sci. Rep.* **2017**, *7*, 43340. [CrossRef]
117. Peng, X.; Ji, J.; Jing, Y. Composite Honeycomb Metasurface Panel for Broadband Sound Absorption. *J. Acoust. Soc. Am.* **2018**, *144*, EL255–EL261. [CrossRef] [PubMed]
118. Wang, D.; Xie, S.; Yang, S.; Li, Z. Sound Absorption Performance of Acoustic Metamaterials Composed of Double-Layer Honeycomb Structure. *J. Cent. South Univ.* **2021**, *28*, 2947–2960. [CrossRef]
119. Wu, F.; Xiao, Y.; Yu, D.; Zhao, H.; Wang, Y.; Wen, J. Low-Frequency Sound Absorption of Hybrid Absorber Based on Micro-Perforated Panel and Coiled-up Channels. *Appl. Phys. Lett.* **2019**, *114*, 151901. [CrossRef]

120. Xu, W.; Yu, D.; Wen, J. Simple Meta-Structure That Can Achieve the Quasi-Perfect Absorption throughout a Frequency Range of 200–500 Hz at 350 °C. *Appl. Phys. Express* **2020**, *13*, 047001. [CrossRef]
121. Xie, S.; Li, Z.; Yan, H.; Yang, S. Ultra-Broadband Sound Absorption Performance of a Multi-Cavity Composite Structure Filled with Polyurethane. *Appl. Acoust.* **2022**, *189*, 108612. [CrossRef]

applied sciences

Article

Analysis of Train Car-Body Comfort Zonal Distribution by Random Vibration Method

Zhaozhi Wu [1], Nan Zhang [1,*], Jinbao Yao [1] and Vladimir Poliakov [2]

[1] School of Civil Engineering, Beijing Jiaotong University, Beijing 100044, China; 19125893@bjtu.edu.cn (Z.W.); jbyao@bjtu.edu.cn (J.Y.)

[2] Bridge and Tunnels Department, Russian University of Transport, 127994 Moscow, Russia; pvy55@mail.ru

* Correspondence: nzhang@bjtu.edu.cn

Abstract: With the increase in train speeds on high-speed railways, the excitation frequency of track irregularities increases, which has a negative impact on train comfort and opposes the passengers' desire for high ride comfort. In addition, the uncertainty of train comfort results from the stationary randomness of track irregularities and the different zonal distribution in the car body. Therefore, the application of the stationary random vibration method to analyze the zonal distribution of train comfort and the relevant influencing factors is important to guarantee the passengers' comfortable experience in each ride and to provide a theoretical basis for comfort optimization. First, the train was modeled using eight independent vehicle elements. Second, the pseudo-excitation method was applied to obtain the theoretical zonal distribution of the Sperling index, an indicator of comfort, via the linearity of the power spectrum density of train acceleration. Third, considering various factors affecting train comfort, the results were compared with those calculated using the Monte Carlo method. It was found that the most comfortable area was located slightly in the front of the center of the car body. Improving track irregularities and reasonably controlling the speed of a train will increase the train's comfort, while it will deteriorate with a loss in car-body mass and damage to the secondary suspension system.

Keywords: train comfort zonal distribution; Sperling index; stationary random vibration method; pseudo–excitation method; Monte Carlo method; power spectrum density

Citation: Wu, Z.; Zhang, N.; Yao, J.; Poliakov, V. Analysis of Train Car-Body Comfort Zonal Distribution by Random Vibration Method. *Appl. Sci.* **2022**, *12*, 7442. https://doi.org/10.3390/app12157442

Academic Editor: Suchao Xie

Received: 17 June 2022
Accepted: 21 July 2022
Published: 25 July 2022

Publisher's Note: MDPI stays neutral with regard to jurisdictional claims in published maps and institutional affiliations.

1. Introduction

High-speed railways (HSRs) have been widely and rapidly developed in global transportation thanks to their advantages such as positive energy efficiency, safety, and reliability. The continuous acceleration of high–speed trains increases the excitation frequency of track irregularities and intensifies the interaction between the vehicle and the track, which has a negative impact on train comfort. Under these circumstances, people are increasingly dissatisfied with the inferior ride comfort of trains. In addition, the irregularity of the tracks, a type of stationary random excitation, and the different arrangement of seats in the car body are responsible for the uncertainty of ride comfort for passengers. Therefore, it is important to improve the ride comfort for passengers and apply the stationary random vibration method to analyze the zonal distribution of train comfort and the relevant influencing factors to provide a theoretical basis for comfort optimization.

Nowadays, research on train vibration comfort has attracted the attention of scientists all over the world. The main contribution of Albers et al. [1] was to develop the drive train and its assemblies to meet customers' vibration comfort requirements. Lee et al. [2] investigated the implementation of an e-Health train that provided passengers with optimized ride comfort as well as the opportunity to provide appropriate feedback on the changed ride comfort to improve passengers' feelings and the health service. The calculation and analysis results of Yu et al. [3] showed that: (1) the vibration of the waiting and store floor under the load of the standing crowd was small, and the comfort requirements could be met; (2) the vibration of the waiting and store floor under the load of the high-speed train was larger because the waiting and store floors were more flexible; and (3) the vibration comfort under the load of the high-speed train could meet the requirements of the corresponding code. Using frequency-domain analysis and time-domain analysis, the distribution of the main vibration frequency was given by Lu [4]. Ni et al. [5] presented an experimental study on the design of a tunable secondary suspension for high-speed trains using magnetorheological fluid dampers (hereafter referred to as MR dampers) to improve lateral ride comfort. Bao et al. [6] develop a fully coupled wind-vehicle-bridge (WVB) interaction model to evaluate the dynamic performance and ride comfort of the monorail-vehicle-bridge system in turbulent crosswinds. The TR08 car of the Shanghai Magnetic Train Demonstration Line was prototyped by Jingyu [7], and a simulation model of multibody dynamics was built. Zhu et al. [8] proposed a strategy to evaluate the comfort of the trestle in terms of the vibration of the train–bridge–trestle coupling. Alehashem et al. [9] showed that the designed MR dampers effectively reduced the rolling motion of the car body. Peng et al. [10] proposed a novel evaluation model to assess this, as the most commonly used international standard ISO 2631-1 is inappropriate.

The random vibration method is used to study the vibration response of the train system accurately and statistically significantly. Lu et al. [11] presented a detailed study on the effect of vibration modes on fatigue damage in a bogie frame of a high-speed train under random loading conditions while considering the extension of the excitation frequency range and the proportional increase of the high-frequency components. Li et al. [12,13] introduced the pseudo-excitation method (PEM) to solve the random oscillations induced by rail irregularities, and presented an efficient algorithm to analyze the random multipoint oscillations of train inverters. Tan et al. [14] established a refined 3D vibration model for a coupled system of train, rail, and beam, and solved it using the PEM. Jin et al. [15] developed a versatile semiactive suspension system with variable-stiffness magnetorheological elastomer isolators and variable-damping magnetorheological dampers for high-speed trains to improve ride comfort by avoiding car body resonance and dissipating vibration energy. Arnaud [16] proposed a new method to predict the random vibration of the train–track–bridge system during earthquakes based on the Hamiltonian Monte Carlo method (MCM). Yao et al. [17] used a combination of numerical simulation, field tests, and a random forest algorithm to predict the building vibration caused by a moving train. Guo et al. [18] proposed a method to monitor the deformation of the track slab using fiber optic sensing and an intelligent method to identify the deformation of the track slab using a random forest model.

Despite numerous research works, there is still a lack of a comprehensive analysis of train comfort from the point of view of zonal distribution and the random vibration method. In order to represent the zonal distribution of train comfort, flexible car bodies must be modeled with a larger number of degrees of freedom (DOFs) than those based on rigid-body dynamics theory, which significantly increases the complexity and inefficiency of the solution. In this case, we aimed to accurately determine the zonal distribution of the Sperling index by using stationary random excitation, including the PEM and MCM, and to analyze various influencing factors such as the degree of track irregularity, train speed, vehicle load, and damage to the secondary suspension system.

2. Establishment and Solution of the Motion Equation

2.1. Motion Equation of Train System Based on MCM

In the analysis of the train vibration, it was helpful to generate different series of random track irregularity excitations repeatedly. Later, each of them was solved using the Newmark-β method (NβM) to obtain the corresponding series of the random vibratory response of the train. When the solving process for one of series of random excitation is terminated, it is called once-sampling of MCM. The entire programming procedure was realized in MATLAB software (version R2022a). Before that, the relevant motion equation of train system needed to be established in accordance with the following assumptions [19]:

(1) The interaction is nonexistent among vehicle elements.
(2) Each vehicle element is composed of a rigid car-body, bogies, and wheel sets.
(3) The springs of a train are linearly elastic, while the dampers are linearly viscous.
(4) Each vehicle element is a linear stationary system; namely, the mass, damping, and stiffness matrices of train are constant.
(5) The train moves at a constant speed.
(6) The wheels and rails fit snugly. The relative displacement between the track and the bridge deck, as well as the elastic effect of the track–bridge system, are neglected.
(7) The track irregularity is a zero-mean stationary Gaussian random process.

Based on the assumptions above, a train subsystem is made up of a certain number of mutually independent 3D vehicle elements. Each of them consists of a car-body, two bogies, four wheel sets, and a two-layer spring-damper suspension system. Each vehicle element contains 15 independent DOFs for the car-body and bogies, and three dependent DOFs for each wheel set coupled respectively with the vertical, horizontal, and torsional motion states of 4 contact points. The DOFs of each vehicle element are listed in Table 1 and correspond to those in Figure 1.

Table 1. The DOFs of a vehicle element.

Vehicle Element i	DOF
Car body	5 DOFs: y_{ci}, z_{ci}, θ_{ci}, φ_{ci}, ψ_{ci} [1]
Bogies	10 DOFs: y_{fbi}, z_{fbi}, θ_{fbi}, φ_{fbi}, ψ_{fbi}, y_{rbi}, z_{rbi}, θ_{rbi}, φ_{rbi}, ψ_{rbi}
Wheel sets	12 DOFs: y_{w1i}, z_{w1i}, θ_{w1i}, y_{w2i}, z_{w2i}, θ_{w2i}, y_{w3i}, z_{w3i}, θ_{w3i}, y_{w4i}, z_{w4i}, θ_{w4i}

[1] y is the lateral sway, z is the vertical levitation, θ is the roll, φ is the pitch, and ψ is the yaw. Subscript "f" and "r" respectively denote the front and the rear bogie, while the numbers represent the orders of wheel sets. Subscript i denotes the number of a certain vehicle element and equals 1, 2, 3 … 8. Back-and-forth along the x-axis was neglected because the train ran at a constant speed according to the 7th assumption. Subscripts "c", "b", and "w", respectively refer to "car body", "bogies", and "wheel sets".

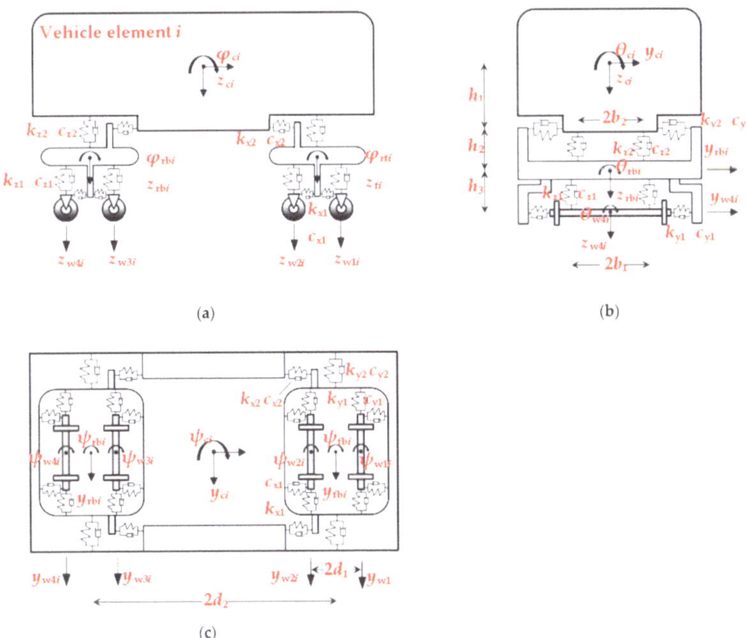

Figure 1. Vehicle element model: (**a**) front view; (**b**) lateral view; (**c**) top view.

According to the D'Alembert principle, the motion equation of a train can be expressed as:

$$M_t \ddot{u}_t + C_t \dot{u}_t + K_t u_t = T_s u_w + T_d \dot{u}_w \tag{1}$$

where M_t, C_t, and K_t are the mass, the damping, and the stiffness matrix, respectively. T_s and T_d are the projector matrices for the spring and the damper, respectively. All of them can be diagonally assembled from M_{ve}, C_{ve}, K_{ve}, T_{sve}, and T_{dve} of the vehicle elements, as given in Equation (2), where u_t is the vibratory displacement vector containing the DOFs of the car body and the bogies in Table 1, while u_w is the vibratory displacement vector containing the DOFs of the wheel-sets in Table 1. K_{ve} in Equation (3) contains the auto-coupling stiffness matrices of the car body and bogies, namely K_{cc} and K_{bb} expressed in Equations (5) and (6), as well as the cross-coupling stiffness matrix $K_{bc}(\eta)$, which represents the car–bogie interaction and differentiates the front and rear bogies using $\eta = 1$ and -1, respectively. M_{ve} is diagonally assembled using the mass of different components and their moments of inertia around different axes. T_{sve} in Equation (8) is a stiffness projector matrix multiplied by the displacement vector u_w of wheel–track contact points to form the force vector of the train. The expression of C_{ve} and T_{sve}, respectively similar to Equations (3) and (8), can be obtained simply by replacing the spring factor k with the corresponding damping factor c.

$$K_t = \mathrm{diag}\begin{bmatrix} K_{ve} & K_{ve} & \cdots & K_{ve} \end{bmatrix} \tag{2}$$

$$K_{ve} = \begin{bmatrix} K_{cc} & K_{bc}(1)^{\mathrm{T}} & K_{bc}(-1)^{\mathrm{T}} \\ K_{bc}(1) & K_{bb} & 0 \\ K_{bc}(-1) & 0 & K_{bb} \end{bmatrix} \tag{3}$$

$$M_{ve} = \mathrm{diag}(m_c, m_c, I_{xc}, I_{yc}, I_{zc}, m_b, m_b, I_{xb}, I_{yb}, I_{zb}, m_b, m_b, I_{xb}, I_{yb}, I_{zb}) \tag{4}$$

$$K_{cc} = 2 \begin{bmatrix} k_{y2} & 0 & -k_{y2}h_1 & 0 & 0 \\ 0 & k_{z2} & 0 & 0 & 0 \\ -k_{y2}h_1 & 0 & k_{y2}h_1^2 + k_{z2}b_2^2 & 0 & 0 \\ 0 & 0 & 0 & k_{x2}h_1^2 + k_{z2}d_2^2 & 0 \\ 0 & 0 & 0 & 0 & k_{x2}b_2^2 + k_{y2}d_2^2 \end{bmatrix} \tag{5}$$

$$K_{bb} = \begin{bmatrix} 2k_{y1} + k_{y2} & 0 & -2k_{y1}h_3 + k_{y2}h_2 & 0 & 0 \\ 0 & 2k_{z1} + k_{z2} & 0 & 0 & 0 \\ -2k_{y1}h_3 + k_{y2}h_2 & 0 & 2(k_{y1}h_3^2 + k_{z1}b_1^2) + k_{y2}h_2^2 + k_{z2}b_2^2 & 0 & 0 \\ 0 & 0 & 0 & 2(k_{x1}h_3^2 + k_{z1}d_1^2) + k_{x2}h_2^2 & 0 \\ 0 & 0 & 0 & 0 & 2(k_{x1}b_1^2 + k_{y1}d_1^2) + k_{x2}b_2^2 \end{bmatrix} \tag{6}$$

$$K_{bc}(\eta) = \begin{bmatrix} -k_{y2} & 0 & k_{y2}h_1 & 0 & -\eta k_{y2}d_2 \\ 0 & -k_{z2} & 0 & -\eta k_{z2}d_2 & 0 \\ -k_{y2}h_2 & 0 & k_{y2}h_1h_2 - k_{z2}b_2^2 & 0 & -\eta k_{y2}d_2h_2 \\ 0 & 0 & 0 & k_{x2}h_1h_2 & 0 \\ 0 & 0 & 0 & 0 & -k_{x2}b_2^2 \end{bmatrix} \tag{7}$$

$$T_{sve} = \begin{bmatrix} 0 & 0 \\ T_{sve}^b & 0 \\ 0 & T_{sve}^b \end{bmatrix} ; \quad T_{sve}^b = \begin{bmatrix} k_{y1} & 0 & 0 & k_{y1} & 0 & 0 \\ 0 & k_{z1} & 0 & 0 & k_{z1} & 0 \\ -k_{y1}h_3 & 0 & k_{z1}b_1^2 & -k_{y1}h_3 & 0 & k_{z1}b_1^2 \\ 0 & k_{z1}d_1 & 0 & 0 & -k_{z1}d_1 & 0 \\ k_{y1}d_1 & 0 & -k_{y1}d_1 & 0 & 0 \end{bmatrix} \tag{8}$$

In MCM, the trigonometric series superposition method is commonly used to generate the history samples of the track irregularity to form u_w. Therefore, u_w can be expressed in Equation (9) with the delay vector t_d, which reflects the phase difference among the wheel sets in Equation (10). In addition, the symbol "∘" in Equation (9) represents the Hadamard product, also known as the element-wise product between two matrices.

$$u_{\text{w}}(t) = \left[\sqrt{2} \sum_{k=1}^{N} \sqrt{S_{\text{ir}}(n_k)\text{d}n_k} \circ \cos(n_k v(t - t_{\text{d}}) + \theta_k) \right] \tag{9}$$

$$t_{\text{d}} = \left[t_{\text{d}}^{\text{ve}}(1) \quad t_{\text{d}}^{\text{ve}}(2) \quad \cdots \quad t_{\text{d}}^{\text{ve}}(n_v) \right]^{\text{T}} \tag{10}$$

$$t_{\text{d}}^{\text{ve}}(i) = \mathbf{1}_{1\times12}\text{diag}[(i-1)l_v I_3, \frac{2d_1 + (i-1)l_v}{V} I_3, \frac{2d_2 + (i-1)l_v}{V} I_3, \frac{2(d_1 + d_2) + (i-1)l_v}{V} I_3] \tag{11}$$

$$S_{\text{ir}}(n) = \left\{ \begin{bmatrix} S_{\text{a}}(n) & S_{\text{v}}(n) & S_{\text{c}}(n) \end{bmatrix} \begin{bmatrix} I_3 & I_3 & \cdots & I_3 \end{bmatrix}_{3\times12n_v} \right\}^{\text{T}} \tag{12}$$

where n_v is the total number of vehicle elements; S_{ir} is the given vertical track irregularity PSD matrix in Equation (12) composed of the alignment, vertical, and cross-level PSDs; n_k is an angular wave number sample; $\text{d}n$ is the bandwidth; and θ_k is a random phase angle obeying the uniform distribution U(0,2π). The integer order of k ranges from 1 to N.

2.2. Motion Equation of Train System Based on the PEM

The PEM proposed by LIN helps to establish a series of input pseudo simple harmonic excitation by the given random excitation PSD in order to obtain the structural response PSD and the variance [20]. In the PEM, the stationary pseudo solution can be calculated through the multiplication of the frequency response function matrix (FRF). Based on the same assumptions as those presented in Section 2.1, using the D'Alembert principle, the motion equation of the train can be expressed as:

$$\tilde{U}_{\text{t}}(\omega, t) = \begin{bmatrix} \tilde{u}_{\text{t}}^{\text{a}} & \tilde{u}_{\text{t}}^{\text{v}} & \tilde{u}_{\text{t}}^{\text{c}} \end{bmatrix} = (-\omega^2 M_{\text{t}} + j\omega C_{\text{t}} + K_{\text{t}})^{-1}(T_{\text{s}} + j\omega T_{\text{d}})\tilde{U}_{\text{w}}(\omega, t) \tag{13}$$

where $\tilde{U}t$ with the overhead tilde "~" is the displacement vibratory response of the train to represent the variables in the pseudo form as the function of angular frequency ω and time t. In addition, in Equation (13), it is composed of three types of displacement response subvectors respectively controlled by three independent directional types of track irregularity PSDs, namely the alignment S_{a}, the vertical S_{v}, and the cross-level S_{c}. This is because the PEM does not allow the linear superposition of different pseudo inputs defined by mutually independent PSDs, and only their PSDs obey the linear superposition principle.

Meanwhile, the pseudo velocity $\dot{\tilde{U}}t$ and acceleration $\ddot{\tilde{U}}t$ vibratory response vectors can be expressed as Equations (14) and (15). The pseudo displacement of wheel sets can be defined in Equation (16) by converting the angular wave number domain to the angular frequency domain; it contains three types of pseudo track irregularity motions, similar to the definition of Equation (13):

$$\dot{\tilde{U}}_{\text{t}}(\omega, t) = j\omega\tilde{U}_{\text{t}}(\omega, t) \tag{14}$$

$$\ddot{\tilde{U}}_{\text{t}}(\omega, t) = -\omega^2\tilde{U}_{\text{t}}(\omega, t) \tag{15}$$

$$\begin{aligned}
\tilde{U}_{\text{w}}(\omega, t) &= \begin{bmatrix} \tilde{u}_{\text{w}}^{\text{a}} & \tilde{u}_{\text{w}}^{\text{v}} & \tilde{u}_{\text{w}}^{\text{c}} \end{bmatrix} = \left\{ \begin{bmatrix} I_3 & I_3 & \cdots & I_3 \end{bmatrix}_{3\times12n_v} \right\}^{\text{T}} \circ \sqrt{\frac{S_{\text{ir}}(n)}{V}} \circ \exp[jnV(t - t_{\text{d}})] \\
&= \left\{ \begin{bmatrix} I_3 & I_3 & \cdots & I_3 \end{bmatrix}_{3\times12n_v} \right\}^{\text{T}} \circ \sqrt{S_{\text{ir}}(\omega)} \circ \exp[j\omega(t - t_{\text{d}})]
\end{aligned} \tag{16}$$

With the pseudo displacement vibratory response of the train in Equation (16), the PSD matrix of the displacement vibratory response can be calculated through the sum of the multiplication of the conjugate and the transpose of the corresponding pseudo vectors, as shown in Equation (17). Accordingly, the PSD matrix of acceleration vibratory response can be deduced through the multiplication of the PSD matrix of the displacement vibratory response and the fourth power of angular frequency ω. The exponential term with the variable of time t will be eliminated due to the multiplication of the conjugate and the transpose, which makes the PSD independent of time; namely, stationary. Accordingly, in Equation (18), the principal diagonal elements represents the auto-PSDs of the corresponding DOFs, while the remaining elements represent the cross-PSDs of two different DOFs.

$$S_U(\omega) = \tilde{U}_w{}^* \tilde{U}_w{}^T = \tilde{u}_t^{a^*} \tilde{u}_t^{a^T} + \tilde{u}_t^{v^*} \tilde{u}_t^{v^T} + \tilde{u}_t^{c^*} \tilde{u}_t^{c^T} \tag{17}$$

$$S_{\ddot{u}}(\omega) = \omega^4 S_U(\omega) = \omega^4 \begin{bmatrix} S_{y_{c1}y_{c1}}(\omega) & S_{y_{c1}z_{c1}}(\omega) & S_{y_{c1}\theta_{c1}}(\omega) & \cdots \\ S_{z_{c1}y_{c1}}(\omega) & S_{z_{c1}z_{c1}}(\omega) & S_{z_{c1}\theta_{c1}}(\omega) & \cdots \\ S_{\theta_{c1}y_{c1}}(\omega) & S_{\theta_{c1}y_{c1}}(\omega) & S_{\theta_{c1}\theta_{c1}}(\omega) & \cdots \\ \vdots & \vdots & \vdots & \ddots \end{bmatrix} \tag{18}$$

3. Sperling Index Based on Vibratory Acceleration History and PSD

3.1. Zonal Distribution Deduction of Sperling Index Based on Vibratory Acceleration History

The Sperling index, an indicator of stability, is a measurement method to determine the comfort of passengers and occupants on the rolling stock, as well as the status of transported goods. The evaluation is based on the measurement of the vehicle body's vibration acceleration. The less comfort the passengers experience, the higher the Sperling index will be [21].

To derive the Sperling index at a given location, the zonal acceleration should first be determined. In Figure 2, the vibration acceleration is linearly distributed in the space of the car body according to the principle of rigid dynamics. The vertical vibration acceleration at any zonal location in the horizontal plane of the car-body center can be determined by considering the horizontal coordinate (x,y), vertical, roll, and pitch acceleration of the car-body center. In this case, the coordinate origin was at the car-body center.

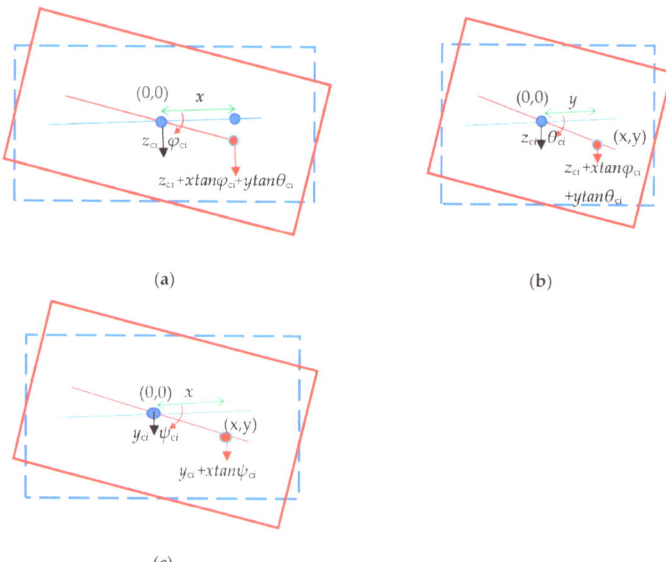

Figure 2. Linear zonal relation model of car-body motion between the center and a certain point. (**a**) front view of car-body motion of the vertical levitation and the pitch; (**b**) lateral view of car-body motion of the vertical levitation and the roll; (**c**) top view of car-body motion of the lateral sway and the yaw.

Therefore, the zonal distribution of the vertical acceleration can be derived using Equation (19). In the case of horizontal acceleration, the horizontal coordinate (x,y), lateral sway, and yaw acceleration of the car-body center are considered. Thus, the zonal distribution of the horizontal acceleration can be expressed as Equation (20).

$$\ddot{z}_{ci}(x,y,t) = \ddot{z}_{ci}(0,0,t) + x\ddot{\varphi}_{ci}(0,0,t) + y\ddot{\theta}_{ci}(0,0,t) \tag{19}$$

$$\ddot{y}_{ci}(x,y,t) = \ddot{y}_{ci}(0,0,t) + x\ddot{\psi}_{ci}(0,0,t) \tag{20}$$

Secondly, the corresponding acceleration amplitude spectrum within the prescribed interval of frequency must be calculated using fast Fourier transform (FFT). In this case, the single-frequency component for the Sperling index can be defined as Equation (21), while the 10th power root of the sum of the 10th power is described in Equation (22) for the Sperling index of multiple-frequency component vibrations for the car body of vehicle element i.

$$W_{ki}(x,y) = 3.57 \sqrt[10]{\frac{A_{ki}^3}{f_k} \circ F(f_k)} = 3.57 \sqrt[10]{\frac{\left|\text{FFT}_{[\ddot{y}_{ci}\,\ddot{z}_{ci}]}(f_k)\right|^3}{f_k} \circ [F_v(f_k) \quad F_h(f_k)]} \tag{21}$$

$$W_i = \sqrt[10]{\sum_{k=1}^{N} W_{ki}^{10}} \tag{22}$$

where A_k (unit: m/s^2) represents the amplitude vector of the vibration acceleration at the frequency point f_k (unit: Hz) after Fourier transform, $F(f_k)$ represents the frequency correction coefficient vector whose vertical and horizontal elements are sectionally expressed in Equations (23) and (24), and the number of frequency N is strongly related to the interval limit of frequency and the measurement duration. In accordance with the code GB/T5599-2019, the standard measurement duration is 5 s, the reciprocal of which is the bandwidth frequency df of the fast Fourier transform.

$$F_v(f_k) = \begin{cases} 0.325f_k^2 & (f_k = 0.5 \sim 5.9\text{Hz}) \\ 400/f_k^2 & (f_k = 5.9 \sim 20\text{Hz}) \\ 1 & (f_k = 20 \sim 40\text{Hz}) \end{cases} \tag{23}$$

$$F_h(f_k) = \begin{cases} 0.8f_k^2 & (f_k = 0.5 \sim 5.4\text{Hz}) \\ 650/f_k^2 & (f_k = 5.4 \sim 26\text{Hz}) \\ 1 & (f_k = 26 \sim 40\text{Hz}) \end{cases} \tag{24}$$

In general, the derivation of the zonal distribution of the Sperling index based on the course of the vibration acceleration is random, as the phase angles of the course of the track irregularities generated in each MCM vary. According to GB/T5599-2019, in order to roughly evaluate the train comfort, the average value of the Sperling index is often used.

3.2. Zonal Distribution Deduction of Sperling Index Based on Vibratory Acceleration PSD

To derive the zonal distribution of the Sperling index based on the PSD, the linearity of the PSD of the car-body acceleration must first be determined to form the auto-PSDs and cross-PSDs between the different DOFs.

The linearity of the PSD can be derived from the linearity of the correlation function of random signals. In signal processing, cross-correlation is a measure of the similarity of two series as a function of the displacement of one series relative to the other, while autocorrelation is the cross-correlation of a signal with a delayed copy of itself as a function of delay. Informally, it is the similarity between observations as a function of the time lag between them. Therefore, autocorrelation and cross-correlation can be defined as

Equations (25) and (26), where τ is the time lag, T is the period of the signal, and v and h are two signals [22].

$$R_{vv}(\tau) = \lim_{T \to \infty} \frac{1}{T} \int_{-\frac{T}{2}}^{\frac{T}{2}} v(t)v(t+\tau)dt \tag{25}$$

$$R_{vh}(\tau) = \lim_{T \to \infty} \frac{1}{T} \int_{-\frac{T}{2}}^{\frac{T}{2}} v(t)h(t+\tau)dt \tag{26}$$

There exists a signal p that is equal to the linear additional relation as $av + bh$, where a and b are two constants. Therefore, the auto-correlation of signal p can be deduced using Equation (27), from which it can be seen that the auto-correlation of the signal p is linked to the auto-correlation and cross-correlation of signals v and h.

$$\begin{aligned}
R_{pp}(\tau) &= R_{(av+bh)(av+bh)}(\tau) \\
&= \lim_{T \to \infty} \frac{1}{T} \int_{-\frac{T}{2}}^{\frac{T}{2}} a^2 v(t)v(t+\tau) + abv(t)h(t+\tau) + abh(t)v(t+\tau) + b^2 h(t)h(t+\tau)dt \\
&= a^2 R_{vv}(\tau) + b^2 R_{hh}(\tau) + ab[R_{hv}(\tau) + R_{vh}(\tau)]
\end{aligned} \tag{27}$$

The Wiener–Khinchin theorem illustrates that the Fourier transform of auto-correlation of a signal is equal to its PSD [23]. Therefore, the PSD of signal p can be deduced by the inverse Fourier transform given in Equation (28), namely the linearity of PSD.

$$S_{pp}(\tau) = \int_{-\infty}^{\infty} R_{pp}(\omega)e^{j\omega\tau}d\omega = a^2 S_{vv}(\tau) + b^2 S_{hh}(\tau) + ab[S_{hv}(\tau) + S_{vh}(\tau)] \tag{28}$$

Accordingly, the zonal relationship of the vertical and horizontal vibration acceleration between the center of the car body and a specific location on the same plane is given in Equations (19) and (20) in Section 3.1. Therefore, for the PSD of a vertical vibration acceleration at coordinate (x,y), not only the auto-PSDs for the vertical levitation, roll, and pitch of the car-body center, but also the cross-PSDs for the vertical levitation, roll, and pitch of the car-body center are multiplied by the corresponding coordinate and superimposed shortly thereafter in Equation (29). Similarly, the PSD of the horizontal vibration acceleration at coordinate (x,y) can be calculated using (30), which includes auto-PSDs for the lateral sway and yaw of the car-body center and cross-PSDs for the lateral sway and yaw of the car-body center.

$$S_{\ddot{z}[i]}(\omega, x, y) = y^2 S_{\ddot{\theta}_{ci}} + x^2 S_{\ddot{\varphi}_{ci}} + S_{\ddot{z}_{ci}} + x(S_{\ddot{z}_c \ddot{\varphi}_c} + S_{\ddot{\varphi}_c \ddot{z}_c}) + y(S_{\ddot{\theta}_{ci} \ddot{z}_{ci}} + S_{\ddot{z}_{ci} \ddot{\theta}_{ci}}) + xy(S_{\ddot{\theta}_{ci} \ddot{\varphi}_{ci}} + S_{\ddot{\varphi}_{ci} \ddot{\theta}_{ci}}) \tag{29}$$

$$S_{\ddot{y}[i]}(\omega, x, y) = x^2 S_{\ddot{\psi}_{ci}} + S_{\ddot{y}_{ci}} + x(S_{\ddot{y}_{ci} \ddot{\psi}_{ci}} + S_{\ddot{\psi}_{ci} \ddot{y}_{ci}}) \tag{30}$$

The auto-PSDs and the cross-PSDs above can be respectively withdrawn from the main and the counter diagonal elements of the vibratory acceleration PSD matrix in Equation (18). The PSD multiplied by the bandwidth $d\omega$ is equal to the square of amplitude spectrum. In this way, Equation (21), which describes the simple frequency component for the Sperling index, should be modified to be the form containing the quadratic term of amplitudes to be linked with the corresponding PSDs in Equation (31):

$$W_{ki}(x,y) = 3.57 \sqrt[\frac{20}{3}]{A_{ki}^2 \circ \left[\frac{F(f_k)}{f_k}\right]^{\frac{2}{3}}} = 3.57 \sqrt[\frac{20}{3}]{2[S_{\ddot{y}[i]}(2\pi f_k, x, y) \quad S_{\ddot{z}[i]}(2\pi f_k, x, y)] \circ \left[\frac{F(f_k)}{f_k}\right]^{\frac{2}{3}} df} \tag{31}$$

In the actual calculation, the Sperling index for multiple frequency components expressed in Equation (22) is equivalent to Equation (32):

$$W_i = \sqrt[10]{\sum_{k=1}^{N} W_{ki}^{10}} \approx \sqrt[\frac{20}{3}]{\sum_{k=1}^{N} W_{ki}^{\frac{20}{3}}} \tag{32}$$

Finally, through the combination of the equations above, the zonal distribution of the Sperling index in the vertical and horizontal can be expressed as Equation (33) in the form of a 20/3 root of the integration of PSDs.

$$W_i(x,y) = \sqrt[\frac{20}{3}]{4\pi \int_{0.5}^{40} 3.57^{\frac{20}{3}} \left[S_{\ddot{y}[i]}(2\pi f, x, y) \quad S_{\ddot{z}[i]}(2\pi f, x, y) \right] \circ \left\{ \frac{[F_v(f) \quad F_h(f)]}{f} \right\}^{\frac{2}{3}} df} \tag{33}$$

In general, the zonal distribution deduction of the Sperling index based on the vibratory acceleration PSD can comprehensively and precisely reflect the train comfort because the excitation is just formed of the PSD of track irregularities obtained through accurate statistics gathered by the relevant national railway department, and can be directly converted to the PSD of the response.

4. Case Study

4.1. Parameters of Track Irregularity and Train

The PSDs of track irregularities adopted for the case study were proposed by the Federal Railroad Administration (FRA) [24] based on a large amount of measured data fitted with the even functions expressed by the cutoff angle wave numbers (n_c and n_s with units rad/m) and roughness constants (A_a and A_v with units m/rad). The expressions of the PSDs for the alignment, vertical, and cross-level PSDs are given in Equation (34) through (36) and are shown in Figure 3. The PSDs were divided into Grades 1–6, and the corresponding parameters from Equations (34)–(36) are listed in Table 2.

$$S_a(n) = \frac{A_a n_c^2}{n^2(n^2 + n_c^2)} \tag{34}$$

$$S_v(n) = \frac{A_v n_c^2}{n^2(n^2 + n_c^2)} \tag{35}$$

$$S_c(n) = \frac{4 A_v n_c^2}{(n^2 + n_c^2)(n^2 + n_s^2)} \tag{36}$$

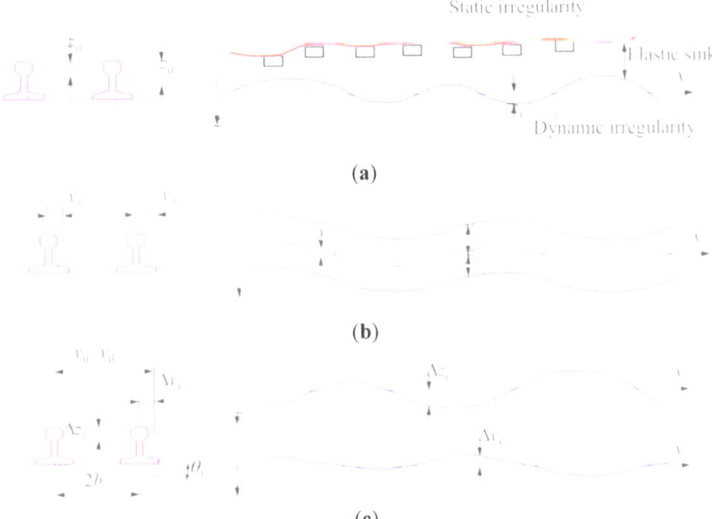

Figure 3. Illustrations of track irregularities: (**a**) vertical profile; (**b**) alignment; (**c**) cross-level.

Table 2. Parameters of the PSDs of track irregularities defined in FRA standard.

Items	Grade-1	Grade-2	Grade-3	Grade-4	Grade-5	Grade-6
Alignment roughness constant A_a	1.211×10^{-4}	1.018×10^{-4}	0.682×10^{-4}	0.538×10^{-4}	0.210×10^{-4}	0.034×10^{-4}
Vertical roughness constant A_v	3.363×10^{-4}	1.211×10^{-4}	0.413×10^{-4}	0.303×10^{-4}	0.076×10^{-4}	0.034×10^{-4}
Cutoff angular wave number n_c	0.6046	0.9308	0.8520	1.1312	0.8209	0.4380
Cutoff angular wave number n_s	0.8245	0.8245	0.8245	0.8245	0.8245	0.8245

The high-speed train CRH2 was used for analysis. To stabilize the response of the train to ensure correctness, the train began by running for an initial distance equivalent to a duration of 5 s. According to GB/T5599-2019 [25], the upper and lower limits of the angular wave number were determined. All the parameters of the train operation are listed in Table 3.

Table 3. Parameters of the train operation.

Items	Values
Half of longitudinal distance of wheel sets d_1	1.25 m
Half of longitudinal distance of bogies d_2	8.75 m
Half of lateral distance of springs in 1st suspension system b_1	1.00 m
Half of lateral distance of springs in 2nd suspension system b_2	1.00 m
Vertical distance, car body center to 2nd suspension system h_1	0.80 m
Vertical distance, 2nd suspension system to bogie center h_2	0.20 m
Vertical distance, bogie center to 1st suspension system h_3	0.10 m
Mass of wheel set m_w	2000 kg
Mass of bogie m_b	3000 kg
Moment of inertia of bogie in longitudinal direction I_{bx}	3000 kg·m^2
Moment of inertia of bogie in lateral direction I_{by}	3000 kg·m^2
Moment of inertia of bogie in vertical direction I_{bz}	3000 kg·m^2
Mass of car body m_b	40 t·m^2
Moment of inertia of car body in longitudinal direction I_{cx}	100 t·m^2
Moment of inertia of car body in lateral direction I_{cy}	2000 t·m^2
Moment of inertia of car body, about vertical direction I_{cz}	2000 t·m^2
Longitudinal damping of 1st suspension system/bogie side c_{x1}	1 kN·s/m
Lateral damping of 1st suspension system/bogie side c_{y1}	1 kN·s/m
Vertical damping of 1st suspension system/bogie side c_{z1}	20 kN·s/m
Longitudinal damping of 2nd suspension system/car-body side c_{x2}	60 kN·s/m
Lateral damping of 2nd suspension system/car-body side c_{y2}	60 kN·s/m
Vertical damping of 2nd suspension system/car-body side c_{z2}	10 kN·s/m
Longitudinal stiffness of 1st suspension system/bogie side k_{x1}	1000 kN/m
Lateral stiffness of 1st suspension system/bogie side k_{y1}	1000 kN/m
Vertical stiffness of 1st suspension system/bogie side k_{z1}	1000 kN/m
Longitudinal stiffness of 2nd suspension system/car-body side k_{x2}	200 kN/m
Lateral stiffness of 2nd suspension system/car-body side k_{y2}	200 kN/m
Vertical stiffness of 2nd suspension system/car-body side k_{z2}	200 kN/m
Lower limit of angular wave number n_{min}	π/V rad/m
Upper limit of angular wave number n_{max}	$80\pi/V$ rad/m
Sampling rate of angular wave number dn	$2\pi/5V$ rad/m
Lower limit of time t_{min}	0 s
Upper limit of time t_{max}	10 s
Time sampling rate dt	1/80 s

4.2. Demonstration of the Methodological Correctness

In this section, the methodological correctness is demonstrated by analyzing the probability characteristics of the random vibration acceleration of the car-body center. The expected value of the random process can be obtained by averaging the samples at each measurement time from the total sample. Due to ergodicity in all states of a stationary process, a sample history basically contains the properties reflecting all probability and statistical characteristics of the random process. The expected value of the family of random variables corresponds to the time average of a single random time sample, as shown in Equation (37), where v_j represents a particular random time sample of the family of random variables V. For the stationary Gaussian random process with a zero mean, the variance can be expressed as Equation (38). By combining Equation (25) with the time lag $\tau = 0$, the autocorrelation of the signal is equal to its variance.

$$\mu = \mathrm{E}[V(t)] = \mathrm{E}[v_j(t)] = \sum_{i=1}^{n} \frac{v_j(t)}{n} = \lim_{T \to \infty} \frac{1}{T} \int_{-\frac{T}{2}}^{\frac{T}{2}} v_j(t)dt \tag{37}$$

$$\sigma^2 = \mathrm{E}[V(t) - \mu]^2 = \mathrm{E}[v_j^2(t)] = \lim_{T \to \infty} \frac{1}{T} \int_{-\frac{T}{2}}^{\frac{T}{2}} v_j^2(t)dt = R_{vv}(0) \tag{38}$$

The track irregularities as input of the linear train system are a stationary Gaussian random process with a zero mean, which theoretically has the same properties of probability distribution as the train oscillation response. According to the Wiener–Khinchin theorem and the statistical control for a random signal with a zero mean value [26], the theoretical standard variance of the vertical vibration acceleration of the car body can be obtained via the square root of the integration of the PSD calculated using the PEM, as shown in Equation (31), similar to the horizontal acceleration. The preliminary history curve of the expected value and the standard variance of the history samples can be calculated by the statistical method at each time point.

$$\sigma_{\ddot{z}_{ci}} = \sqrt{\int_{-\infty}^{+\infty} S_{\ddot{z}_{ci}}(\omega)d\omega} \tag{39}$$

A vehicle element was simulated to run at a speed of 350 km/h for 5 s to cross a distance of a Grade-4 track irregularity. The PEM and MCM with 100 times of sampling were adopted to analyze the characteristics of the possibility distribution for the car-body center acceleration. In Figure 4, concerning the possibility distribution of the car-body center vibratory acceleration for only one vehicle element in the time history domain, it can be seen that:

1. In Figure 4a,c, the expected value μ and the limits of the 3σ normal distribution of the vibration response of the Car-Body center using MCM–NβM and PEM-FRF, respectively, are in substantial agreement. The response patterns are also well within the bounds of the 3σ principle for normal distributions. Compared to MCM, which required multiple solutions, the PEM was much more accurate and efficient.
2. In Figure 4b,d, the preliminary probability density curve calculated using MCM-NβM agrees with the theoretical normal distribution calculated using PEM-FRF.
3. In general, the train vibration acceleration under track irregularities obeyed the stationary zero-mean normal distribution, the statistical properties of which can be described by the 3σ-principle, and which is consistent with the original assumption regarding the track irregularities as a stationary zero-mean Gaussian random process. Therefore, it was demonstrated that the methodology was correct and suitable for the analysis outlined in the following sections.

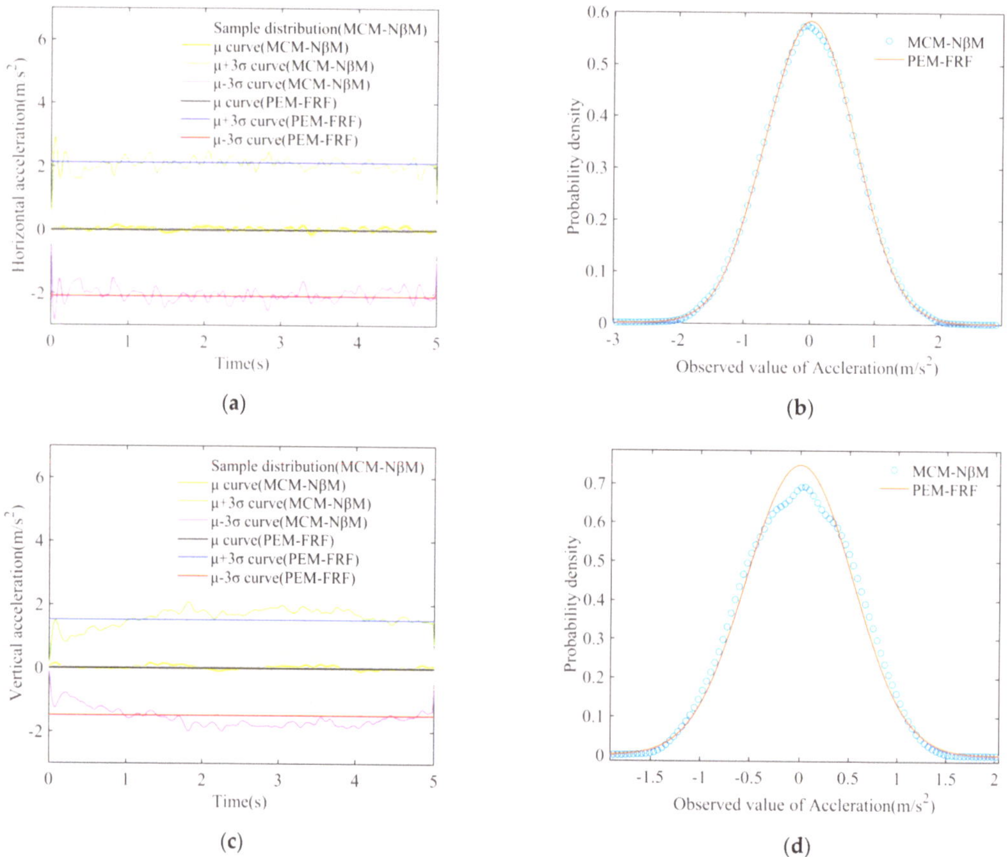

Figure 4. Car-Body center vertical and horizontal vibratory acceleration responses: (**a**) history samples distribution for the horizontal; (**b**) possibility density curve for the horizontal; (**c**) history samples distribution for the vertical; (**d**) possibility density curve for the horizontal.

4.3. Analysis of the Car-Body Center Comfort in Different Vehicle Elements

The train, which consisted of eight independent vehicle elements, was adopted for 5 s at a speed of 350 km/h over a track with Grade-4 track irregularities for the simulation. The MCM was applied 100 times to calculate the vibration acceleration of the car body at the center of each vehicle element and derive the Sperling indices, which are shown in Figure 5 in the form of box plots, and compared with the theoretical values calculated using the PEM. In Figure 5, it can be seen that:

1. The variations in the theoretical Sperling index of the car-body center according to the PEM between the different vehicle elements were insignificant for both the horizontal and vertical components. The theoretical Sperling index for the horizontal component was obviously higher than that for the vertical component;

2. In all sequences of vehicle elements, the difference between the upper and lower provisional range limits of the Sperling index sampling distribution according to MCM did not exceed 0.12 for both the horizontal and vertical components;

3. In general, the discomfort was strongly related to the horizontal vibration acceleration of the car body. In addition, the Sperling index was a stationary indicator of comfort that was independent of the order of vehicle elements when the train was subjected to track irregularities. In this way, the comfort of the center of the car body could be characterized by simply selecting a vehicle element to analyze.

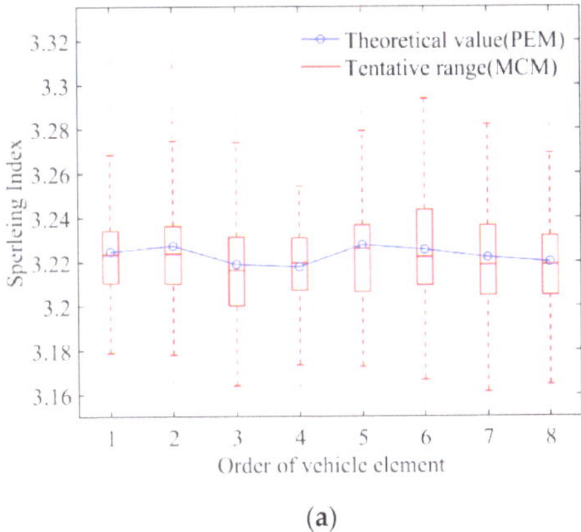
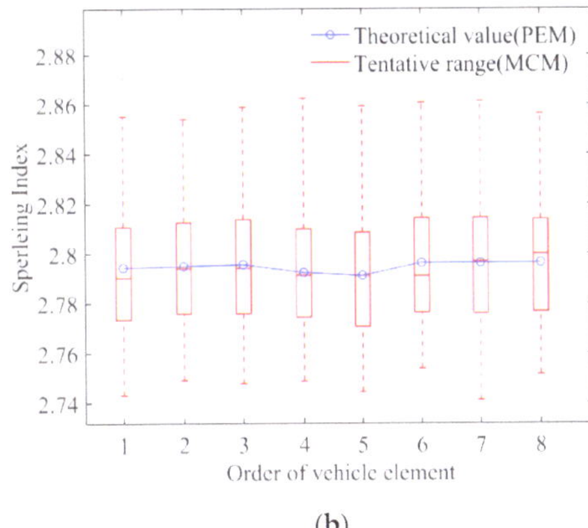

(a) **(b)**

Figure 5. Car-body center Sperling index as the function of the order of vehicle element using MCM and PEM: (**a**) horizontal Sperling index; (**b**) vertical Sperling index.

4.4. Analysis of the Zonal Distribution Characteristics of Train Comfort

In the simulation, a vehicle element was selected to traverse a track with Grade-6 track irregularities for 5 s at a speed of 350 km/h. MCM was applied 100 times to calculate the vibration acceleration of the car-body center of each vehicle element and then calculate the preliminary average of the zonal distribution of the Sperling index shown in Figure 6 according to Section 3.1 and compare it with the theoretical values calculated using the PEM according to the linearity of PSDs given in Section 3.2. To simplify the representation in the figure, the coordinate (x,y) was replaced by a length of the car body of 25 m and a width of the car body of 4 m. The mesh of the plane in which the car body was located was divided into 10 × 10. In Figure 6, it can be seen that:

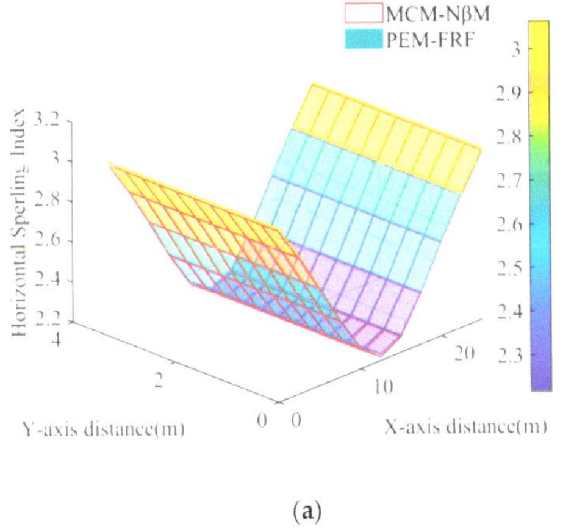
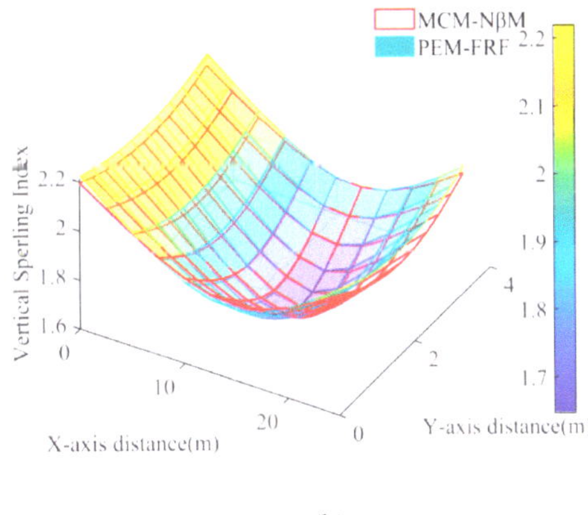

(a) **(b)**

Figure 6. Zonal distribution of car-body Sperling index using MCM and PEM: (**a**) horizontal Sperling index; (**b**) vertical Sperling index.

1. The mean network of the zonal distribution of the Sperling index calculated using MCM–NβM was almost identical to the theoretical network calculated using PEM-FRF for both the horizontal and vertical components;
2. In Figure 6a, the zonal distribution of the horizontal Sperling index is symmetric with respect to the pitch and roll axis of the car-body center. It has the cylindrical shape of the letter "V" and reaches the highest line at the rear and front edges of the car-body, respectively, while it reaches the lowest line on the pitch axis of the car-body;

3. In Figure 6b, the zonal distribution of the vertical Sperling index is a symmetrically curved surface with respect to the roll axis of the Car-Body center. It reaches the highest points at the four vertices of the $X-Y$ plane, where the center of the car body is located, while the lowest point is slightly in front of the center of the car body in the direction of travel;

4. Compared to Figure 5 in Section 4.3, the Sperling index of the center of the car body is smaller for both the horizontal and vertical components for Grade-6 track irregularities than for Grade-4 track irregularities;

5. In general, a small track irregularity had a negative effect on train comfort, and the zonal distribution of train comfort was not strictly symmetrical with respect to the center of the car body. The most comfortable area for the vertical component was near the front of the center of the car body, while the most comfortable area for the horizontal component was on the pitching axis of the center of the car body. The realistic evaluation of train comfort could be roughly characterized by the average value of the Sperling index during train operation, while the theoretical design of train comfort could be accurately determined by the PEM.

4.5. Analysis of Influence of the Quality of Track Irregularity on the Zonal Distribution Characteristics of Train Comfort

A vehicle element was simulated using the PEM to respectively cross a distance of track irregularity of Grade-1, Grade-3, and Grade-5 at a speed of 350 km/h to compare the different qualities of the track irregularities' influences on the zonal distribution of the Car-Body Sperling index, as shown in Figure 7. It can be seen in Figure 7 that:

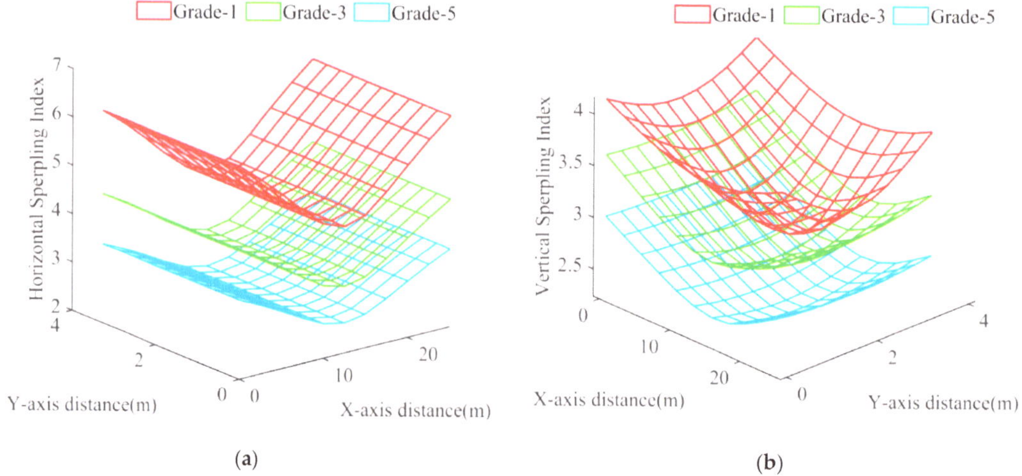

Figure 7. Zonal distribution of car-body Sperling index as influenced by the quality of the track irregularities using PEM: (**a**) horizontal Sperling index; (**b**) vertical Sperling index.

1. The symmetrical characteristic of the zonal distribution of the car-body Sperling index of the car body for the horizontal and vertical components was identical to that in Section 4.4;

2. With the deterioration in the quality of the track irregularities, the values of the Sperling index for the entire network increased significantly for both the horizontal and vertical components;

3. In general, train comfort deteriorated with the deterioration of the quality of track irregularities, so regular track maintenance is of great importance.

4.6. Analysis of Influence of the Train Speed on the Zonal Distribution Characteristics of Train Comfort

A vehicle element was simulated using the PEM to run a train with a track irregularity of Grade-2 at speeds of 150 km/h, 250 km/h, and 350 km/h to compare the influence of the different train speeds on the zonal distribution of the Car-Body Sperling index of the car body, as shown in Figure 8. In Figure 8, it can be seen that:

1. The symmetrical characteristic of the zonal distribution of the car-body Sperling index of the car body was identical to that shown in Section 4.4 for both the horizontal and vertical components;
2. As the train accelerated, the values of the car-body Sperling index for the entire network increased significantly for both the horizontal and vertical components;
3. In general, the train comfort deteriorated when the train traveled too fast. This was because the amplitude of the vibration velocity and the acceleration of the track irregularity contained the linear and quadratic terms of the train speed V, respectively. When the train accelerated, the amplitudes increased rapidly, which increased the input excitation and led to a significant increase in the vibration response of the car body. Therefore, appropriate control of a train's speed can help to improve passenger comfort.

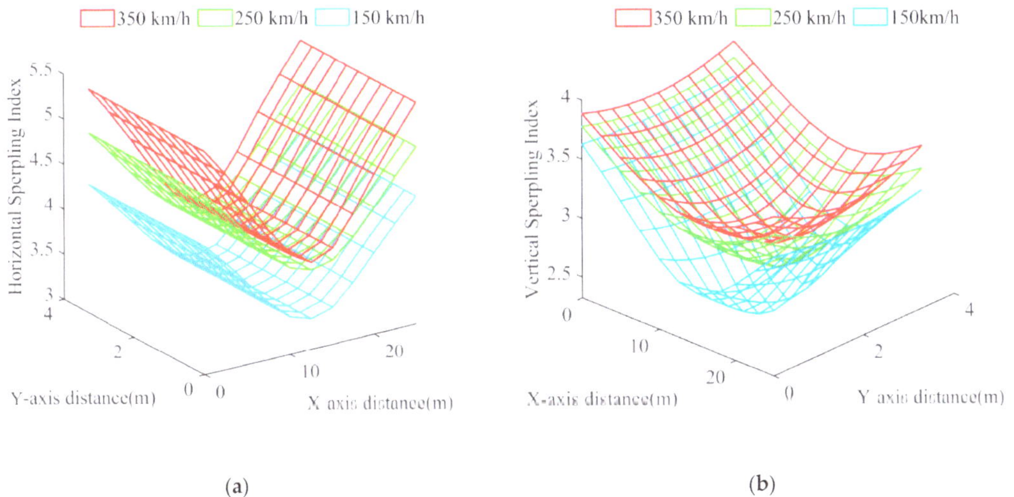

Figure 8. Zonal distribution of car-body Sperling index influenced by the train speed according to PEM: (**a**) horizontal Sperling index; (**b**) vertical Sperling index.

4.7. Analysis of Influence of the Car-Body Mass on the Zonal Distribution Characteristics of Train Comfort

In rigid dynamics, the moment of inertia of a car-body center is equivalently linear to the car-body mass. Considering the that center of mass is overlapped by the car-body center, the mass matrix of the vehicle element in Equation (4) can be respectively redefined as Equation (40) where α is the mass factor:

$$M_{ve} = \mathrm{diag}(\alpha m_c, \alpha m_c, \alpha I_{xc}, \alpha I_{yc}, \alpha I_{zc}, m_b, m_b, I_{xb}, I_{yb}, I_{zb}, m_b, m_b, I_{xb}, I_{yb}, I_{zb}) \qquad (40)$$

A vehicle element was simulated using the PEM to run on a track with a Grade-6 track irregularity for 5 s at a speed of 250 km/h to compare the influence of a full load (Full rated, $\alpha = 1.2$) and an empty load (Unladen, $\alpha = 1$) on the zonal distribution of the car body's Sperling index, as shown in Figure 9. In Figure 9, it can be seen that:

1. The symmetrical characteristics of the zonal distribution of the car body's Sperling index for the horizontal and vertical components were identical to those shown in Section 4.4;
2. With the additional mass of the car body, the values of the Sperling index for the entire mesh decreased significantly for both the horizontal and vertical components;
3. In general, the comfort of the train deteriorated with the loss in the car-body mass. Therefore, it is important to reasonably distribute the number of passengers during transfer and optimize the original mass design of the car body.

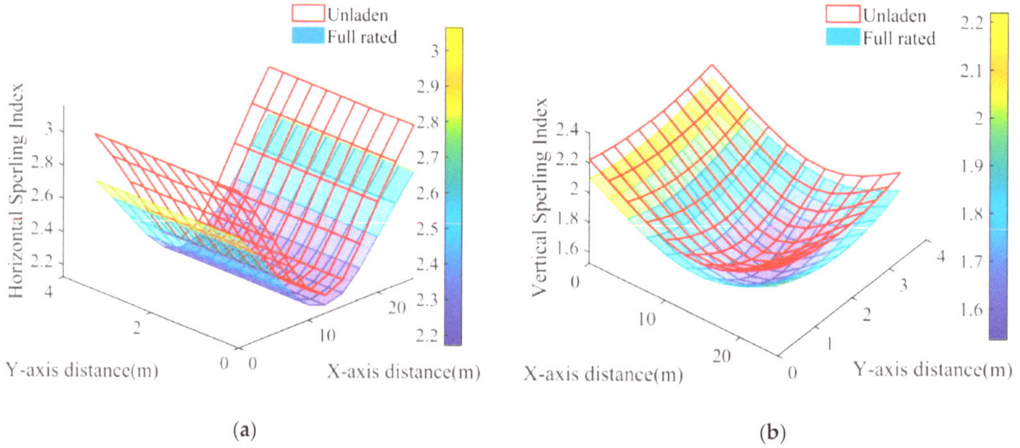

Figure 9. Zonal distribution of car-body Sperling index influenced by the addition of the car-body mass using PEM: (**a**) horizontal Sperling index; (**b**) vertical Sperling index.

4.8. Analysis of Influence of the Damage of Secondary Suspension System on the Zonal Distribution Characteristics of Train Comfort

The comfort of a train is highly dependent on the performance of the secondary suspension system. Reducing the stiffness and increasing the damping can help to simulate the damage of the secondary suspension system. In this way, one of the vertical springs and dampers in the secondary suspension system connecting the car body to the front bogie was multiplied by the damage factors d_{fk} and d_{fc}, respectively. Therefore, the corresponding stiffness submatrix of the vehicle element including Equations (3) and (5)–(7) was updated in Equations (41)–(44).

$$K_{\mathrm{ve}} = \begin{bmatrix} \overline{K}_{\mathrm{cc}} & \overline{K}_{\mathrm{bc}}(1)^{\mathrm{T}} & K_{\mathrm{bc}}(-1)^{\mathrm{T}} \\ \overline{K}_{\mathrm{bc}}(1) & \overline{K}_{\mathrm{bb}} & 0 \\ K_{\mathrm{bc}}(-1) & 0 & K_{\mathrm{bb}} \end{bmatrix} \tag{41}$$

$$\overline{K}_{\mathrm{cc}} = \begin{bmatrix} 2k_{y2} & 0 & -2k_{y2}h_1 & 0 & 0 \\ 0 & \frac{3+d_{\mathrm{fk}}}{2}k_{z2} & \frac{1-d_{\mathrm{fk}}}{2}k_{z2}b_2 & \frac{d_{\mathrm{fk}}-1}{2}k_{z2}d_2 & 0 \\ -2k_{y2}h_1 & \frac{1-d_{\mathrm{fk}}}{2}k_{z2}b_2 & 2k_{y2}h_1^2 + \frac{3+d_{\mathrm{fk}}}{2}k_{z2}b_2^2 & 0 & 0 \\ 0 & \frac{d_{\mathrm{fk}}-1}{2}k_{z2}d_2 & 0 & 2k_{x2}h_1^2 + \frac{3+d_{\mathrm{fk}}}{2}k_{z2}d_2^2 & 0 \\ 0 & 0 & 0 & 0 & 2k_{x2}b_2^2 + 2k_{y2}d_2^2 \end{bmatrix} \tag{42}$$

$$\overline{K}_{\mathrm{bb}} = \begin{bmatrix} 2k_{y1}+k_{y2} & 0 & -2k_{y1}h_3+k_{y2}h_2 & 0 & 0 \\ 0 & 2k_{z1}+\frac{3+d_{\mathrm{fk}}}{4}k_{z2} & \frac{1-d_{\mathrm{fk}}}{2}k_{z2}b_2 & 0 & 0 \\ -2k_{y1}h_3+k_{y2}h_2 & \frac{1-d_{\mathrm{fk}}}{2}k_{z2}b_2 & 2(k_{y1}h_3^2+k_{z1}b_1^2)+k_{y2}h_2^2+\frac{3+d_{\mathrm{fk}}}{4}k_{z2}b_2^2 & 0 & 0 \\ 0 & 0 & 0 & 2(k_{x1}h_3^2+k_{z1}d_1^2)+k_{x2}h_2^2 & 0 \\ 0 & 0 & 0 & 0 & 2(k_{x1}b_1^2+k_{y1}d_1^2)+k_{x2}b_2^2 \end{bmatrix} \tag{43}$$

$$\overline{K}_{bc}(1) = \begin{bmatrix} -k_{y2} & 0 & k_{y2}h_1 & 0 & -k_{y2}d_2 \\ 0 & -\frac{1+d_{fk}}{2}k_{z2} & 0 & -\frac{1+d_{fk}}{2}k_{z2}d_2 & 0 \\ -k_{y2}h_2 & 0 & k_{y2}h_1h_2 - \frac{1+d_{fk}}{2}k_{z2}b_2^2 & 0 & -k_{y2}d_2h_2 \\ 0 & 0 & 0 & k_{x2}h_1h_2 & 0 \\ 0 & 0 & 0 & 0 & -k_{x2}b_2^2 \end{bmatrix} \quad (44)$$

Similarly, the damping submatrix was formed by replacing the spring coefficient k and the spring damaged factor d_{fk} with the damping coefficient c and the damper damaged factor d_{fc}. The top row of the submatrix in the following equations means that damage was present.

In this case, a vehicle element was simulated using the PEM to traverse a track with a Grade-6 track irregularity for 5 s at a speed of 250 km/h and compare the effects of the original secondary suspension system (d_{fk} = 1 and d_{fc} = 1) and the damaged secondary suspension system (d_{fk} = 10 and d_{fc} = 0.1) on the zonal distribution of the car body's Sperling index, as shown in Figure 10. In Figure 10, it can be seen that:

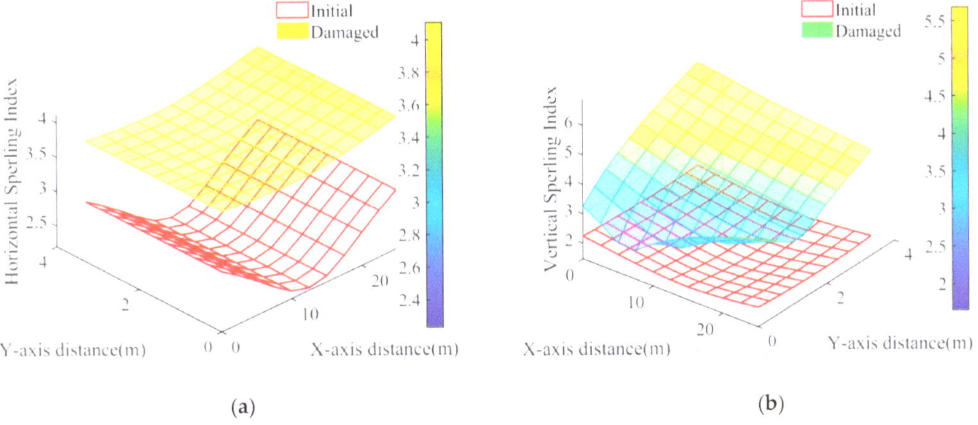

Figure 10. Zonal distribution of Car-Body Sperling index influenced by the damage of a vertical spring damper in secondary suspension system using PEM: (**a**) horizontal Sperling index; (**b**) vertical Sperling index

1. In Figure 10a, the symmetrical characteristic of the network of the horizontal Sperling index after damage to a vertical spring damper in the secondary suspension system was still consistent with the initial condition, but the overall magnitude of the horizontal Sperling index was much higher than the initial condition. This was because the damaged vertical spring damper affected the lateral car-body sway by influencing the car-body roll in conjunction with the vertical car-body sway;

2. It can be seen in Figure 10b that after the damage to the vertical spring damper in the secondary suspension system, the network of the vertical Sperling index could not maintain the uniformly curved surface as in the initial state, and reached the highest values when the damaged spring damper was located and gradually decreased toward the areas where the other healthy vertical spring dampers were located. The overall magnitude of the vertical Sperling index far exceeded that of the initial condition;

3. In general, the damage to the secondary suspension system led to an overall deterioration in the train comfort for both the horizontal and vertical components, even if only one local vertical spring damper in the secondary suspension system was damaged. The comfort of the train suffered more on the side where the damage occurred. The train comfort for the horizontal and vertical components were not independent of each other. Therefore, it is important to consider both components comprehensively when optimizing the train comfort.

5. Conclusions

This study established a spatial train model that adopted the random vibration method to study the zonal distribution of train comfort; the model could overcome the uncertainty caused by the randomness of track irregularities and the zonal difference in the vibratory acceleration in a car body. Our relevant conclusions are as follows:

1. Compared with the Monte Carlo method and depending on a large amount of samples, the pseudo-excitation Method, based on the linearity of the power spectrum density of the car-body vibratory acceleration, was more efficient to derive the accurate zonal distribution of the Sperling index.
2. The realistic evaluation of train comfort can be roughly characterized by the average value of the Sperling index during train operation, while the theoretical design of train comfort can be accurately determined using the pseudo–excitation method.
3. The vibration acceleration of the train during track irregularities is a stationary Gaussian random process with a zero mean value, the statistical properties of which can be described by the 3σ–principle.
4. The Sperling index is a stationary indicator of comfort that is independent of the order of the vehicle elements when the train is subjected to a track irregularity. In this way, train comfort can be characterized by simply selecting a vehicle element to analyze.
5. The zonal distribution of train comfort is not strictly symmetrical with respect to the center of the car body. The most comfortable area for the vertical component was located near the front of the center of the car body, while the most comfortable area for the horizontal component was located on the axis of the tilt axis of the center of the car-body center.
6. The comfort of the train deteriorated with a loss in mass of the Car-Body and with irregularities in the tracks, while reasonable control of the train speed, regular maintenance of the tracks, and reasonable distribution of the number of passengers when changing trains improved the comfort of the train.
7. The overall comfort of the train deteriorated even if only one local vertical spring damper in the secondary suspension system was damaged. It suffered more on the side where the damage was present. The comfort of the train was not independent of the horizontal and vertical components. To optimize the comfort of the train, it is therefore important to consider both components comprehensively.

Author Contributions: Z.W. came up with the concept, realized the simulation, analyzed the data, and edited the draft of manuscript. N.Z. conducted the literature review, checked the computations, wrote the draft of the manuscript, replied to reviewers' comments, and revised the final version. J.Y. verified the simulated data and polished the article. V.P. corrected the model of simulation and debugged the programming. All authors have read and agreed to the published version of the manuscript.

Funding: This project received funding from the National Natural Science Foundation of China (Grant No. 52178101).

Institutional Review Board Statement: Not applicable.

Informed Consent Statement: Not applicable.

Data Availability Statement: The data presented in this study are available upon request from the corresponding author.

Acknowledgments: We thank the National Natural Science Foundation of China for the support of this work.

Conflicts of Interest: The authors declare no conflict of interest.

References

1. Albers, A.; Lerspalungsanti, S.; Düser, T.; Ott, S.; Wang, J. A Systematic Approach to Support Drive Train Design Using Tools for Human Comfort Evaluation and Customer Classification. In Proceedings of the ASME 2008 International Design Engineering Technical Conferences and Computers and Information in Engineering Conference, Brooklyn, NY, USA, 3–6 August 2008; pp. 651–662. [CrossRef]
2. Lee, Y.; Shin, K.; Song, Y.; Han, S.; Lee, M. Research of Ride Comfort for Tilting Train Simulator Using ECG. *IFMBE Proc.* **2009**, *22*, 1906–1909. [CrossRef]
3. Yu, Y.L.; He, W. Comfortable Evaluation of Bridge-Station Combined Large Span High Speed Railway Station under High Speed Train Load. *Appl. Mech. Mater.* **2013**, *477–478*, 727–731. [CrossRef]
4. Lu, L. The effects forthe ride comfort of high-speed trains' operating conditions on the open and tunnel track. In Proceedings of the 2015 12th IEEE International Conference on Electronic Measurement & Instruments (ICEMI), Qingdao, China, 16–18 July 2015; Volume 3, pp. 1344–1349. [CrossRef]
5. Ni, Y.; Ye, S.; Song, S. An experimental study on constructing MR secondary suspension for high-speed trains to improve lateral ride comfort. *Smart Struct. Syst.* **2016**, *18*, 53–74. [CrossRef]
6. Bao, Y.; Zhai, W.; Cai, C.; Zhu, S.; Li, Y. Dynamic interaction analysis of suspended monorail vehicle and bridge subject to crosswinds. *Mech. Syst. Signal Process.* **2021**, *156*, 107707. [CrossRef]
7. Huang, J.; Zhou, X.; Shang, L.; Wu, Z.; Xu, W.; Wang, D. Influence analysis of track irregularity on running comfort of Maglev train. *Transp. Syst. Technol.* **2018**, *4*, 129–140. [CrossRef]
8. Zhu, Y.; Chen, G.; Li, X.; Lu, Y.; Liu, J.; Xu, T. Comfort assessment for rehabilitation scaffold in road-railway bridge subjected to train-bridge-scaffold coupling vibration. *Eng. Struct.* **2020**, *211*, 110426. [CrossRef]
9. Alehashem, S.M.S.; Ni, Y.Q.; Liu, X.Z. A Full-Scale Experimental Investigation on Ride Comfort and Rolling Motion of High-Speed Train Equipped With MR Dampers. *IEEE Access* **2021**, *9*, 118113–118123. [CrossRef]
10. Peng, Y.; Wu, Z.; Fan, C.; Zhou, J.; Yi, S.; Peng, Y.; Gao, K. Assessment of passenger long-term vibration discomfort: A field study in high-speed train environments. *Ergonomics* **2022**, *65*, 659–671. [CrossRef] [PubMed]
11. Lu, Y.; Xiang, P.; Dong, P.; Zhang, X.; Zeng, J. Analysis of the effects of vibration modes on fatigue damage in high-speed train bogie frames. *Eng. Fail. Anal.* **2018**, *89*, 222–241. [CrossRef]
12. Li, Y.-C.; Feng, S.-J.; Chen, H.-X.; Chen, Z.-L.; Zhang, D.-M. Random vibration of train-track-ground system with a poroelastic interlayer in the subsoil. *Soil Dyn. Earthq. Eng.* **2019**, *120*, 1–11. [CrossRef]
13. Li, Y.; Liu, Y.; He, G.; Chen, J.; Peng, X. Multipoint Random Vibration Analysis of Train Converter Equipment. In Proceedings of the 2019 IEEE Vehicle Power and Propulsion Conference (VPPC), Hanoi, Vietnam, 14–17 October 2019; pp. 1–3. [CrossRef]
14. Tan, S.; Yu, Z.; Shan, Z.; Mao, J. Influences of train speed and concrete Young's modulus on random responses of a 3D train-track-girder-pier coupled system investigated by using PEM-ScienceDirect. *Eur. J. Mech. A/Solids* **2018**, *74*, 297–316. [CrossRef]
15. Jin, T.; Liu, Z.; Sun, S.; Ren, Z.; Deng, L.; Yang, B.; Christie, M.D.; Li, W. Development and evaluation of a versatile semi-active suspension system for high-speed railway vehicles. *Mech. Syst. Signal Process.* **2019**, *135*, 106338. [CrossRef]
16. Arnaud, W. Hamiltonian Monte Carlo with application to train-track-bridge coupled interactions subjected to seismic excitation with uncertainties. 2021; *preprint*. [CrossRef]
17. Yao, J.; Fang, L. Building Vibration Prediction Induced by Moving Train with Random Forest. *J. Adv. Transp.* **2021**, *2021*, 6642071. [CrossRef]
18. Guo, G.; Cui, X.; Du, B. Random-Forest Machine Learning Approach for High-Speed Railway Track Slab Deformation Identification Using Track-Side Vibration Monitoring. *Appl. Sci.* **2021**, *11*, 4756. [CrossRef]
19. Xia, H.; Zhang, N.; Guo, W. *Dynamic Interaction of Train-Bridge Systems in High-Speed Railways: Theory and Applications*; Springer: Berlin/Heidelberg, Germany, 2018. [CrossRef]
20. Lin, J.; Zhang, Y.; Li, Q.; Williams, F. Seismic spatial effects for long-span bridges, using the pseudo excitation method. *Eng. Struct.* **2004**, *26*, 1207–1216. [CrossRef]
21. Feng, Z.; Xiao, H. Analysis of Stability and Vibration Transmission Law of Type A Metro Vehicles. *J. Phys. Conf. Ser.* **2021**, *1846*, 012029. [CrossRef]
22. Gerekos, C.; Bruzzone, L.; Imai, M. A Coherent Method for Simulating Active and Passive Radar Sounding of the Jovian Icy Moons. *IEEE Trans. Geosci. Remote Sens.* **2019**, *58*, 2250–2265. [CrossRef]
23. Koyama, A.; Nicholson, D.A.; Andreev, M.; Rutledge, G.C.; Fukao, K.; Yamamoto, T. Discretized Wiener-Khinchin theorem for Fourier-Laplace transformation: Application to molecular simulations. *arXiv* **2020**, arXiv:2004.097922020. [CrossRef]
24. Wang, T. Impact in a railway truss bridge. *Comput. Struct.* **1993**, *49*, 1045–1054. [CrossRef]
25. *GB/T-5599-2019*; Specification for Dynamic Performance Assessment and Testing Verification of Rolling Stock. Standardization Administration: Beijing, China, 2019.
26. Leibovich, N.; Barkai, E. Aging Wiener-Khinchin Theorem. *Phys. Rev. Lett.* **2015**, *115*, 080602. [CrossRef] [PubMed]

Article

Mobile Device-Based Train Ride Comfort Measuring System

Yuwei Hu [1], Lanxin Xu [1], Shuangbu Wang [2], Zhen Gu [1] and Zhao Tang [1,*]

[1] Traction Power National Key Laboratory, Southwest Jiaotong University, Chengdu 610031, China; huyuwei2tpl@163.com (Y.H.); xinlnitia@163.com (L.X.); guzheng16399@163.com (Z.G.)
[2] Institute of Smart City and Intelligent Transportation, Southwest Jiaotong University, Chengdu 611756, China; shuangbuwang@swjtu.edu.cn
[*] Correspondence: tangzhao@swjtu.edu.cn

Abstract: As an important train performance quality, comfort depends on vibration and noise data measured on a running train. Traditional vibration and noise measurement tools are facing challenges in terms of collecting big data, portability, and cost. With the continuous upgrade of mobile terminal hardware, the built-in sensors of mobile phones have the ability to undertake relatively complex data measurement and processing tasks. In this study, a new type of train comfort measurement system based on a mobile device is developed by using a built-in sensor to measure the vibration and noise. The functions of the developed system include the real-time display of three-way vibration acceleration, lateral and vertical Sperling indicators, sound pressure level, and train comfort-related data storage and processing. To verify the accuracy of the mobile device-based train ride comfort measuring system (DTRCMS), a comparison of test results from this system and from the traditional measuring system is conducted. The comparison results show that the DTRCMS is in good agreement with the traditional measuring system. The relative error in three-direction acceleration and Sperling values is 2~10%. The fluctuation range of the noise measured by DTRCMS is slightly lower than that of the professional noise meter, and the relative error is mainly between 1.5% and 4.5%. Overall, the study shows that using mobile devices to measure train comfort is feasible and practical and has great potential for big data-based train comfort evaluation in the future.

Keywords: rail vehicle; ride comfort; vibration measurement; noise measurement; Sperling index

Citation: Hu, Y.; Xu, L.; Wang, S.; Gu, Z.; Tang, Z. Mobile Device-Based Train Ride Comfort Measuring System. *Appl. Sci.* **2022**, *12*, 6904. https://doi.org/10.3390/app12146904

Academic Editor: Cesare Biserni

Received: 9 June 2022
Accepted: 4 July 2022
Published: 7 July 2022

Publisher's Note: MDPI stays neutral with regard to jurisdictional claims in published maps and institutional affiliations.

1. Introduction

Train ride comfort has a great impact on the competitive advantage of rail transit and other modes of transportation. To date, evaluation of train ride comfort has mainly used the vibration and noise data of the train [1–3]. In the 1950s and 1960s, a series of comfort evaluation standards and calculation criteria were initially formed. In 1941, Sperling and Helberg of the German Railway Vehicle Research Institute gave the famous Sperling index, and this empirical formula has been widely used in many countries [4]. Chinese specifications ("Test Appraisal Method and Evaluation Standard for Dynamic Performance of Railway Locomotives" and "Code for Dynamic Strength and Dynamic Performance of High-speed Test Trains") are also mainly evaluated based on this index. The International Organization for Standardization (ISO) proposed the 1/3 octave band method and the total weighted value to calculate the comfort evaluation index [5]. The International Union of Railways (UIC) promulgated and implemented the UIC513 standard in 1994 [6]. It can be seen that there are already a few standards and methods for evaluating train comfort. However, they need to be improved in the following aspects:

1. The data used in the existing evaluation standard or methods are not enough to cover full runtime and lifecycle. This is due to the fact that traditional vibration and noise meters cannot collect such large amounts of data at any given time due to cost and portability concerns;

2. Compared with train response data, human response data are more direct in evaluating train ride comfort, but traditional vibration and noise meters, such as train sensor systems and DSP-based systems, cannot directly measure the human response data [7], and the production of the comfort index requires second processing and calculation.

Accurate results of train comfort require a large number of subjects to perform a mass of tests under the actual operating environment. Currently, without the support of low-cost and easy-to-operate measurement tools, testers often have to make a compromise to reduce the sample, which greatly decreases the accuracy of the results [8]. It can be seen that the development of portable and accurate train ride comfort measurement tools that support mass data storage plays an important role in accurately reflecting passenger comfort and train operating status and improving comfort evaluation standards.

At present, the common measurement methods include inspection vehicle monitoring, precision sensor monitoring, and on-site manual measurement [9]. These traditional measurement methods have some shortcomings, such as high maintenance cost, poor portability, and nonsupport of real-time display and data processing. In order to improve measurement efficiency, Molly et al. [10] used virtual instrument technology and designed a portable train comfort and stability testing system integrating data acquisition, processing, storage, and analysis. Using the Qt development platform, Li et al. [11] designed a set of comfort measurement systems for high-speed railway automatic driving systems which achieves a quantitative detection of the impact of the high-speed railway automatic driving curve on passenger experience. Chang et al. [12] also tried to apply a mobile phone to evaluate train vibration comfort and initially verified its feasibility and measurement accuracy. Although these system instruments can complete the measurement of noise or train vibration and the evaluation of comfort, the sensor of most designs is separated from the control display interface. It is a single-function type with low portability.

In recent years, with the continuous progress in operating systems, core processors, and sensing devices, the functions of smartphones have become increasingly powerful and have been preliminarily applied in non-destructive testing, airport noise measurement, and other fields [13–18]. Shiferaw et al. [19] used smartphone sensors to measure traffic-induced ground vibration and grasp ground health in real time. Saurabh Garg et al. [20] used smartphones to record and analyze the car noise of various high-speed railway systems and to perform data analysis on the operating noise in a train passenger compartment. Partridge et al. [21] used a sinusoidal excitation filter to calibrate the acceleration sensor of a smartphone and analyzed the flattest road in an ambulance through the massive big data collected by the mobile phone. Liu et al. [11] explained the origin of vibration and noise in railway trains in more detail and summarized the technical standards and common methods for a series of studies on noise prediction, measurement analysis, and noise reduction in railway trains. Liu [22] used mobile phone sensors to measure the vibration acceleration data of vehicles passing through bridges and realized the functions of data saving, setting sampling frequency, and real-time data display.

In this paper, a new type of train ride comfort measurement system is proposed based on a mobile terminal with built-in sensors. Not only are the functions of real-time display of three-way vibration acceleration, lateral and vertical Sperling indicators, sound pressure level, and data storage and processing realized, but measurement accuracy is also ensured. The accuracy and usability of DTRCMS are verified by test comparisons with professional measurement equipment.

2. Train Ride Comfort Measurement System

The system function is divided into two modules, namely, the vibration measurement module and the noise measurement module. The built-in sensor and built-in microphone of a mobile phone are used as the system input; the mobile phone interface is used as the output of the system, and data correlation processing is performed, including algorithm processing of acceleration and noise, real-time data storage processing, etc. [23]. The overall use case diagram of the system is shown in Figure 1:

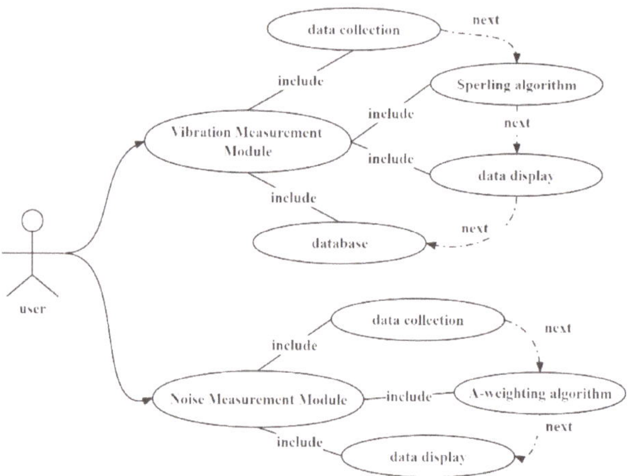

Figure 1. The overall use case diagram of the system.

The functions of the two modules are composed of four parts: data acquisition, data processing, data display, and data storage and evaluation. The overall implementation process and functions of the system are shown in Figure 2.

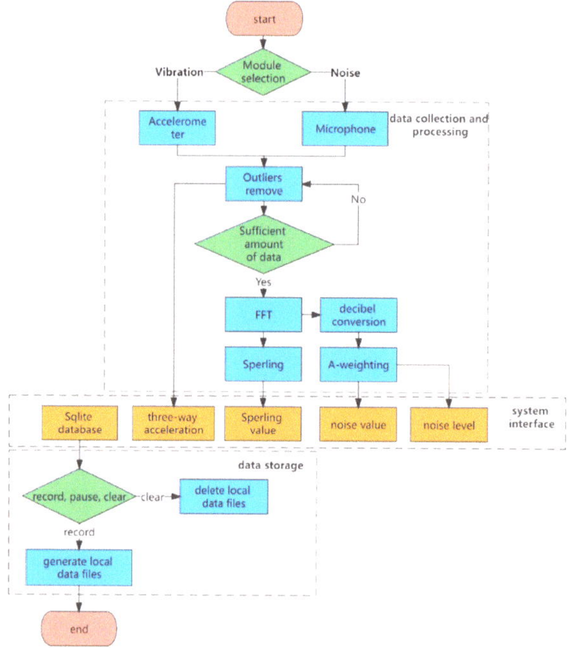

Figure 2. The overall flow of the system.

2.1. Data Acquisition

Data acquisition includes acceleration and noise. For collecting acceleration data, the built-in linear acceleration sensor was chosen to avoid the influence of gravity and to prepare for the calculation required by the Sperling index. At the same time, an appropriate sampling frequency should be adopted to prevent the frequency from being too fast to display real-time data unstably. This system selected the SENSOR_DELAY_GAME mode (50 Hz), which is one of the four sampling frequency modes of Android.

For collecting noise data, the Android system provides developers with the MediaRecorder class, which can realize an audio recording function. By creating an object of the MediaRecorder class and calling its corresponding method, the recording of audio and the acquisition of noise amplitude are realized.

2.2. Data Processing

Data processing includes acceleration processing and noise processing. The acceleration processing part includes the fast Fourier transform (FFT) and the Sperling algorithm. The noise processing part includes the conversion of the sound amplitude to the decibel value, the A-weighting algorithm, and the calculation of the final average decibel value of the entire measurement process.

2.2.1. FFT

FFT is an algorithm based on the characteristics of parity symmetry and opposite signs of virtual and real after sampling of the time-domain signal by discrete Fourier transform (DFT).

2.2.2. Equivalent Continuous A-Weighting

In the field of measuring noise, simply measuring decibels is no longer enough to describe the feeling that is closer to what the human ear hears. In order to reflect the high sensitivity of the human auditory system around the frequency of 3 kHz and the low sensitivity at the frequency of 60 Hz, the usual processing method is to use the A sound level weighting curve to weight the signals of different frequencies. Its formula is:

$$A(f) = 20lg\left[\frac{f_4^2 f^4}{(f^2 + f_1^2)(f^2 + f_2^2)^{\frac{1}{2}}(f^2 + f_3^2)^{\frac{1}{2}}(f^2 + f_4^2)}\right] - A_{1000}, \tag{1}$$

where f is the frequency of the input signal; $A_{1000} = -2.000$ dB is a normalization constant expressed in decibels; $f_1 = 20.6$ Hz; $f_2 = 107.7$ Hz; $f_3 = 737.9$ Hz; and $f_4 = 12194.0$ Hz.

The A-weighted frequency weighting coefficient $\alpha_A(f)$ can be obtained from the following formula:

$$\alpha(f) = \frac{f_4^2 f^4}{(f^2 + f_1^2)(f^2 + f_2^2)^{\frac{1}{2}}(f^2 + f_3^2)^{\frac{1}{2}}(f^2 + f_4^2)}, \tag{2}$$

The A-weighting coefficient in the frequency domain can be obtained as:

$$X_A(k) = \alpha_A(f_k)X(k), \tag{3}$$

For discontinuous and unstable noises such as urban vehicle traffic noise and train running noise, a single A-weighted sound level cannot be accurately reflected. Therefore, this system introduces the equivalent continuous A sound level, which uses the mean of the noise energy to evaluate the impact of noise on people according to the time-average method.

According to Parseval's theorem, the total energy of the same signal in the time domain and the total energy in the frequency domain are always the same, that is:

$$\sum_{n=0}^{N-1}|x(n)|^2 = \frac{1}{N}\sum_{k=0}^{N-1}|X(k)|^2, \tag{4}$$

where N is the number of sampling points; $x(n)$ is the discrete-time domain signal obtained by sampling the signal; and $X(k)$ is the Fourier transform corresponding to $x(n)$.

By using Equation (4), the average energy of the A-weighted signal can be obtained, and the corrected form is:

$$\hat{P} = \sqrt{\frac{1}{N} \sum_{n=0}^{N-1} |x(n)|^2} = \sqrt{\frac{1}{N^2} \sum_{k=1}^{N-1} |X_A(k)|^2}, \tag{5}$$

Then, according to the sound level calculation formula, the A-weighted sound level can be obtained as:

$$L_A = 20 \lg \left(\frac{\hat{P}}{2 \times 10^{-5}} \right) + 2, \tag{6}$$

2.2.3. Sperling Index Algorithm

The Sperling index is an evaluation index, summarized by Sperling et al. [24], in which vibration data are measured through a large number of experiments with human physiological sensation. It reflects the running quality of the vehicle itself and the riding comfort of passengers. Its formula is:

$$W = 0.896 \sqrt[10]{\frac{a^3}{f} F(f)}, \tag{7}$$

where a is the measured vehicle vibration acceleration (unit: cm/s^2); f is the frequency of the acceleration signal (unit: Hz); and $F(f)$ is the frequency weighting coefficient.

Studies have shown that the human body has very different sensitivities to vibration in different frequency bands [25]. In vertical vibration, the human body is most sensitive to vibrations at frequencies from 4 to 8 Hz; in longitudinal vibration, the human body is more sensitive to vibrations below 2 Hz. According to GB5599-2019 "Railway Vehicle Dynamic Performance Evaluation and Test Qualification Specification", the frequency weighting formula of the Sperling index is shown in Table 1:

Table 1. The frequency weighting formula of the Sperling index.

Lateral Vibration/Hz		Vertical Vibration/Hz	
0.5~5.4	$F(f) = 0.8f^2$	0.5~5.9	$F(f) = 0.325f^2$
5.4~26	$F(f) = 650/f^2$	5.9~20	$F(f) = 400/f^2$
>26	$F(f) = 1$	>20	$F(f) = 1$

W, obtained from formula (7), is the value at a single frequency f. In practice, the vibration contains acceleration values of different frequencies. After performing the Fourier transform, it is brought into formula (8) to obtain the value of each group of frequencies. After the W value, the final Sperling index can be calculated by the formula:

$$W = \sqrt[10]{W_1{}^{10} + W_2{}^{10} + W_3{}^{10} + \cdots W_n{}^{10}}, \tag{8}$$

where W_i is the Sperling index at the i-th frequency.

2.3. Data Display

Data presentation types should be as diverse as possible in order to enhance the visual performance of the user interface. The interface of the vibration measurement module mainly includes functions such as time display, curve display, real-time data display, calculation of Sperling index value, and database upload. The overall layout adopts a linear layout. The interface of the noise measurement module mainly includes a large disc display, real-time data value display, curve display, and related keys. The overall layout adopts a constraint layout, which is convenient. The user interface is shown in Figure 3.

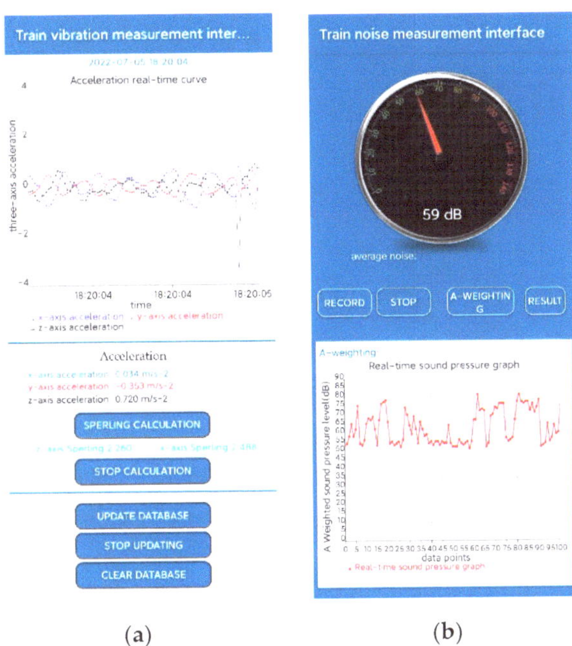

(a)   (b)

Figure 3. The user interface of the DTRCMS: (**a**) the vibration measurement interface; (**b**) the noise measurement interface.

2.4. Data Storage and Evaluation

In order to facilitate the acceleration data and Sperling comfort index data in the vibration module for further data analysis on the computer in the later stage, the data saving function was added to the vibration module area. SQLite was selected for data storage; it is a built-in lightweight database for mobile phones, convenient for operation, and it can record various real-time data at the same time.

In the noise measurement module, the result evaluation interface was introduced. The sound pressure level obtained during the measurement period is calculated to obtain the average decibel value (the average equivalent sound pressure value) during the measurement period, and the value is generated through the result button event and passed into the resulting interface. The users can select the running line (high-speed rail or subway) to get whether the noise value during this period exceeds the noise standard limit.

3. System Test and Application

In order to verify the feasibility of the DTRCMS, professional noise and vibration measuring instruments were used to compare with the DTRCMS. The Sperling index calculated by the mobile phone was also compared and analyzed to verify the accuracy and reliability of the Sperling train comfort evaluation.

3.1. Technical Standards of Measuring Equipment

3.1.1. Professional A Sound Level Noise Meter

1. Sampling frequency: 50 Hz, real-time measurement, fast response;
2. A-weighted processing of ambient noise;
3. Resolution: 0.1 dB, accuracy error: ±1.5 dB;
4. The measurement range is 30 dB~130 dB, and the frequency response range is 31.5 Hz~8.5 kHz;
5. The time constant is 125 ms, and the time weighting is fast.

3.1.2. Professional Vibration Accelerometer

6.   Integrated high-precision acceleration sensor and gyroscope;
7.   The acceleration measurement range is ± 16 g, and the accuracy is 0.0001 g;
8.   The data output frequency is 100 Hz; the interface adopts serial TTL communication, and the baud rate is 115,200.

3.1.3. Honor 9X Mobile Phone Measuring Instrument

9.   CPU Kirin 810 GPU Turbo3.0;
10.   Built-in sensors: acceleration sensor, mobile phone microphone, gravity sensor, pressure sensor, etc.

3.2. System Test

The noise test conducted experiments by playing an audio file in AAC format based on the ISO standard, with a sampling rate of 44.1 kHz and a bit rate of 192 Kbps. The two measuring instruments recorded two data every second, and the experiment time was 150 s. Among them, Figure 4 shows the data comparison curve diagram of the two measuring instruments in this test, and Figure 5 shows the relative error with interval distribution.

In Figure 4, the data trends of the two curves are roughly the same. The data fluctuation of the mobile phone measuring instrument is slightly lower than the standard value of the professional instrument, and the amplitude is generally slightly lower. The reasons for generating the lower fluctuation may include the following:

1.   Hardware: the limitations of the type and quality of built-in MEMS microphones in smartphones (small size, circuit board position, dynamic range and signal-to-noise ratio responsiveness, etc.);
2.   Software: the correction method of the application, the time delay of data processing, and the influence of software running in the background;
3.   Others: the influence of various obstacles (phone protection covers, microphone openings clogged by dust), different operating systems, different mobile phone brands, and environmental vibration.

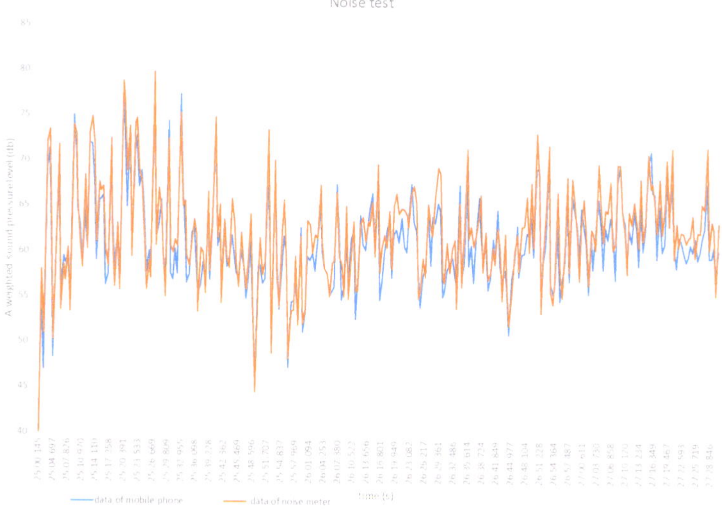

Figure 4. Comparison of noise data.

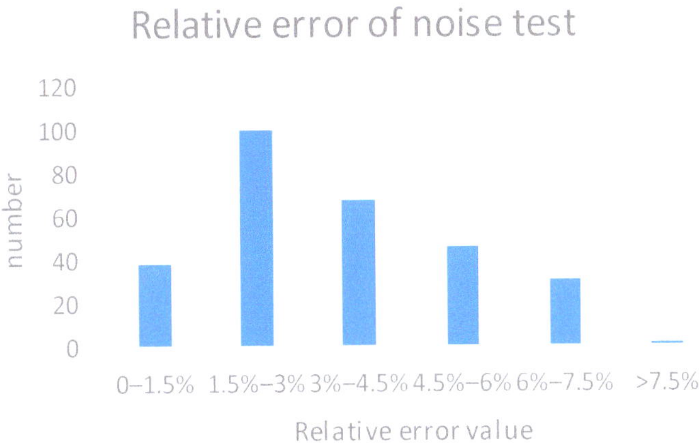

Figure 5. Distribution of relative error intervals.

It can be seen from Figure 5 that the relative errors of the two groups of data are mainly distributed between 1.5% and 4.5%; the maximum error is 7.83%, and the overall error is within a reasonable range.

In the acceleration test, the mobile phone measuring instrument and the professional measuring instrument were fixed on a wooden partition, and the vibration environment was created by artificial shaking. For the convenience of comparison, two data were recorded every second, and the experiment time is 120 s.

Figure 6 shows the relative error distribution interval diagram of the three-axis acceleration, and the experimental data include a total of 256 points. As shown in this figure, the relative error of the three-axis acceleration is generally distributed between 2% and 10%; the individual data error is greater than 14%, and the minimum error is less than 2%. In the three-axis data, the relative error data distributions of the y-axis and the x-axis are relatively concentrated, mainly distributed in 2~8%, and the z-axis data distribution is relatively uniform, mainly distributed in 0~10%.

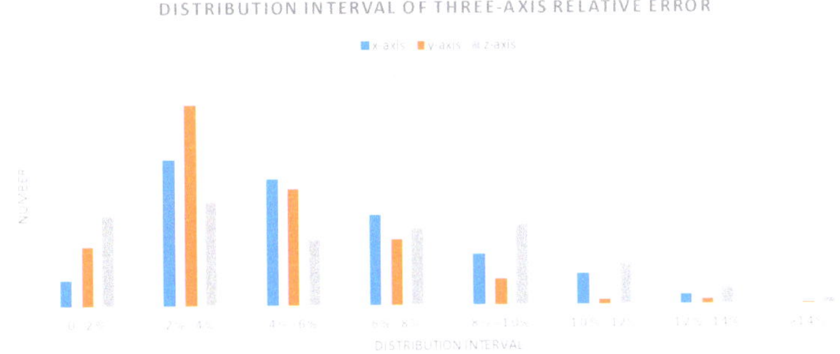

Figure 6. Distribution of acceleration relative error.

The Sperling index test adopted the same method as the acceleration test, and the vibration test environment was built by artificial shaking. The Sperling index of the mobile phone measuring instrument was extracted from the database, and the Sperling index of the professional acceleration instrument was calculated from its acceleration value through MATLAB. The sampling frequency of the two instruments was 50 Hz. A Sperling index value was calculated for every 1000 acceleration data, and the experiment time was set to be 55 s. There are 10 sets of data in total, as shown in Table 2.

Table 2. Comparison of Sperling index values of mobile phone/professional instrument.

No.	Horizontal Sperling		Relative Tolerance	No.	Vertical Sperling		Relative Tolerance
	Mobile Phone	Professional Equipment			Mobile Phone	Professional Equipment	
1	2.40	2.37	1.14%	1	2.56	2.58	0.66%
2	2.41	2.39	0.96%	2	2.45	2.49	1.67%
3	3.64	3.49	4.41%	3	3.28	3.35	2.12%
4	3.49	3.54	1.24%	4	3.23	3.23	0.03%
5	3.91	3.79	3.35%	5	3.26	3.35	2.66%
6	3.44	3.50	1.66%	6	2.97	2.93	1.19%
7	3.11	2.97	4.71%	7	2.94	2.86	2.62%
8	3.06	2.99	2.14%	8	3.08	3.10	0.49%
9	3.03	2.96	2.43%	9	3.08	3.10	0.87%
10	3.11	3.15	1.07%	10	3.05	2.99	1.97%

In Table 2, there is little difference between the Sperling index value calculated by the mobile phone and the Sperling index value calculated by MATLAB; i.e., the lateral error is less than 5%, and the vertical error is less than 3%.

The data comparison shows that there is still a certain error between the mobile phone and the professional tester. The reasons for generating the above errors may include the following:

1. On the whole, due to the limitations of mobile phone cost, sensor volume, and built-in location, the measurement accuracy and sampling performance of the built-in acceleration sensor and microphone in mobile phones are not as good as professional equipment, but the gap is not obvious;
2. The computing power of the mobile phone CPU is uneven, which leads to a certain delay in the A-weighting calculation and Sperling value calculation. Subsequent cloud computing can effectively solve this problem by decomposing computing tasks into the cloud;
3. During the experiment, the positions of the two devices cannot be completely consistent, resulting in different forces, relative jitter, or non-unique variables. In the follow-up, big data processing methods can be used to filter out human factors;
4. The test time is relatively short, and the number of tests is small (all tests are short-term tests of several minutes, and the number of measurements is small; differences between different mobile phones are not considered), which leads to limited test data and causes certain statistical errors.

The overall experimental results show that the relative errors are in a normal range, and it is feasible to use the mobile phone measuring instrument as a train comfort measurement tool in order to reflect train comfort.

3.3. System Application

Taking Chengdu Metro Line 6 as the test train line for the DTRCMS application, the noise measurement module was applied between Wangcongci Station and Shuxin Avenue Station; the vibration of standing posture was measured between Shuxin Avenue Station and Xipu Station, and the vibration of sitting posture was measured between Tianyu Road Station and Wangcongci station. The test roadmap is shown in Figure 7.

Figure 7. Test roadmap.

The sound pressure levels during the noise measurement process are shown in Figure 8. The data basically remained between 65 and 75 decibels. The decibel value of the running train was in the range of 70 to 75 decibels. When the train stops and starts, the noise can reach 80 decibels. According to the GB14892-2006 (Urban Rail Transit Vehicle Noise Limits and Measurement Methods) issued by China, the noise limit of subway passenger compartments is 83 decibels, and the noise measured by DTRCMS did not exceed this limit in the whole test process.

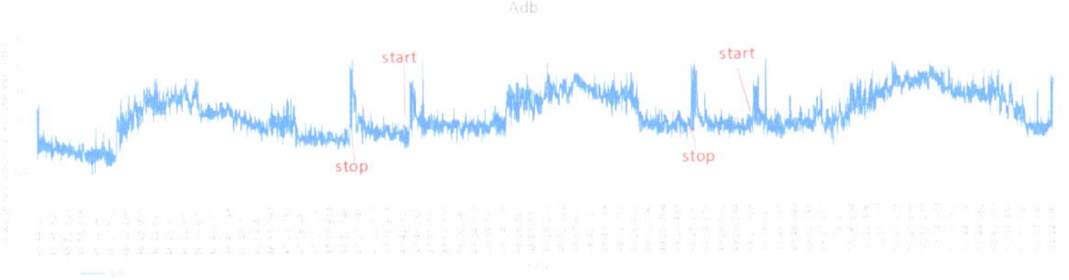

Figure 8. Decibels in the test process.

In the application of the vibration measurement module, the mobile phone was placed horizontally on the seat and the floor to simulate the two passenger states of sitting and standing, respectively. The x-axis direction of the mobile phone points towards the front of the train, as shown in Figure 9.

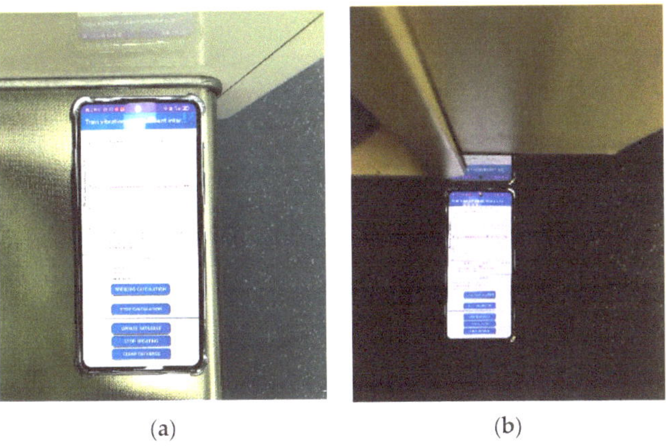

(**a**)   (**b**)

Figure 9. Application site of vibration measurement: (**a**) the mobile phone placed on the seat; (**b**) the mobile phone placed on the floor.

The measurement time of the standing posture vibration measurement process was 12 min; the measurement time of the sitting posture vibration measurement process was 15 min. Its three-axis acceleration curve is shown in Figure 10. The acceleration fluctuations of the two states were similar. In the subway operation stage, the acceleration fluctuations of the three axes were obvious, and they were all in the range of $-0.5\sim0.5$ m/s^2. When the subway stopped and started, the x-axis acceleration increased sharply, and the maximum value reached 1.5 m/s^2. During the stop phase of the subway, the three-direction acceleration changed smoothly.

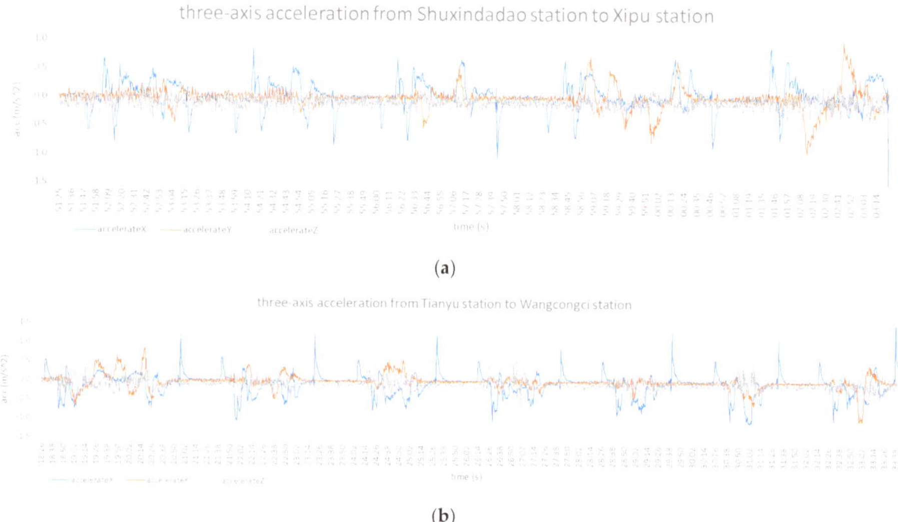

Figure 10. Three-way acceleration: (**a**) three-axis acceleration from Shuxin Avenue station to Xipu station; (**b**) three-axis acceleration from Tianyu station to Wangcongci station.

Horizontal and vertical Sperling indicators corresponding to the two states are shown in Figure 11, where the value of the vertical Sperling indicator fluctuated between 1.5 and 2, and the value of the horizontal Sperling index fluctuated around 3 (the value was around 3.5 during subway operation and around 1.5 during the stops). According to the Sperling rating table of locomotives and rolling stock in China, during the whole measurement process, for the horizontal Sperling index value, the subway operation stage is scored as "pass" or "good", due to the more obvious vibration in individual periods, and in the stop phase, the score is "excellent". For the vertical Sperling index value, the whole process is relatively stable, and the score is "excellent". The measured data of the two states are roughly consistent with actual human experience.

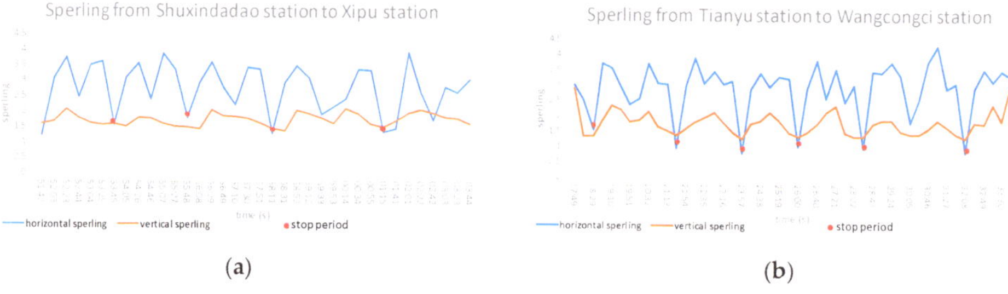

Figure 11. Sperling index value: (**a**) Sperling from Shuxin Avenue station to Xipu station; (**b**) Sperling from Tianyu station to Wangcongci station.

4. Conclusions

Based on smartphone mobile devices, this study develops a new portable train comfort measurement system (DTRCMS). The availability, accuracy, and reliability of the system are verified by comparison with a professional test system in an on-board test. The main conclusions are as follows:

1. Mobile devices can realize multi-module algorithm integration to meet real-time comfort assessment and visual presentation. The FFT, Sperling algorithm, A-weighting algorithm, and other calculation processes were integrated into the system, and noise and vibration data processing could be performed in real time.
2. The measurement accuracy of the built-in sensor in the mobile phone was verified by a comparative test. The relative error of the noise test data was mainly distributed in 1.5~5%. The overall trend of the three-way acceleration values was the same, and the relative error was 2~10%. The difference in the Sperling index was small, and the relative error was less than 5%.
3. The availability and reliability of the DTRCMS were verified by an on-board application on Chengdu Metro Line 6. During the application process, the decibel value basically remained between 65 and 75 db, which is consistent with the actual listening experience. The Sperling value was between 2 and 3, and the measured data of the two states were roughly consistent with actual human experience.

In summary, the DTRCMS not only provided almost all train ride comfort-related functions, such as noise and vibration measuring and Sperling index calculation, but it is also more convenient and faster than a traditional test meter. In addition, the challenges in the existing riding comfort evaluation standards and methods were discussed, and an optimization direction was introduced.

5. Future Works

The DTRCMS provides a more convenient and faster tool for train ride comfort measurement. In the error results of the comparison test, the DTRCMS is not far from professional equipment in data measurement and calculation. In practical applications, the DTRCMS's result are similar to passengers' feelings, which shows a certain feasibility. However, due to the limitations of mobile phone hardware compared with professional meters, there is still room for optimization in terms of improving accuracy. Currently, in the era of big data, follow-up research can further upgrade and improve the DTRCMS by uploading the measurement data to a big data cloud platform and evaluating comfort through massive amounts of data collected by the DTRCMS. In this way, current train comfort assessment data can be quickly obtained in real time; train running quality can be accurately reflected; and even train running tracking status can be monitored at the same time.

Author Contributions: Conceptualization, Z.T. and Y.H.; methodology, Y.H.; software, Y.H. and Z.G.; validation, Y.H.; formal analysis, Y.H.; investigation, Y.H.; resources, L.X.; data curation, L.X.; writing—original draft preparation, Y.H. and L.X.; writing—review and editing, S.W. and Z.T.; visualization, Y.H.; supervision, Y.H.; project administration, Z.T.; funding acquisition, Z.T. All authors have read and agreed to the published version of the manuscript.

Funding: This research was funded by "Network Collaborative Manufacturing and Smart Factory" of the National Key R&D Program, grant number 2020YFB1711402.

Institutional Review Board Statement: Not applicable.

Informed Consent Statement: Not applicable.

Data Availability Statement: Not applicable.

Conflicts of Interest: The authors declare no conflict of interest.

References

1. Born, A.C. Effects of rail vibrations on humans and mitigation solutions. *J. Acoust. Soc. Am.* **2019**, *146*, 2897. [CrossRef]
2. Li, Q.; Duhamel, D.; Luo, Y.; Yin, H.P. Improved Methods for In-situ Measurement Railway Noise Barrier Insertion Loss. *Trans. Nanjing Univ. Aeronaut. Astronaut.* **2018**, *35*, 58–68.
3. Kim, Y.G.; Kwon, H.B.; Kim, S.W.; Kim, C.K.; Kim, T.W. Correlation of ride comfort evaluation methods for railway vehicles. *Proc. Inst. Mech. Eng. Part F J. Rail Rapid Transit* **2003**, *217*, 73–88. [CrossRef]
4. Sperling, E. Verfahren zur beurteilung der laufeigenschaften von eisenbahnwesen. *Organ Für Die Fortschr. Des Eisenb.* **1941**, *12*, 176–187.
5. Oborne, D.J. Whole-Body Vibration and International Standard ISO 2631: A Critique. Human Factors. *J. Hum. Factors Ergon. Soc.* **1983**, *25*, 55–69. [CrossRef] [PubMed]
6. UIC. *513 R, Guidelines for Evaluating Passenger Comfort in Relation to Vibration in Railway Vehicle*; International Union of Railways: Paris, France, 1994.
7. Narayanamoorthy, R.; Khan, S.; Berg, M.; Goel, V.K.; Saran, V.H.; Harsha, S.P. Determination of Activity Comfort in Swedish Passenger Trains. In Proceedings of the 8th World Congress on Railway Research (WCRR 2008), Seoul, Korea, 18–22 May 2008.
8. Lauriks, G.; Rongsong, W. Evaluation of the comfort level of railway vehicles. *Smart Rail Transit* **2004**, *1*, 24–29.
9. Cong, J.-L.; Gao, M.-Y.; Wang, Y.; Chen, R.; Wang, P. Subway rail transit monitoring by built-in sensor platform of smartphone. *Front. Inf. Technol. Electron. Eng.* **2020**, *21*, 1226–1238. [CrossRef]
10. Li, M.; Hongping, Y.; Xin, H. Design and Research on the test system of train comfort and stability index based on DSP. Electronic and Automation Control Conference (IAEAC 2017). In Proceedings of the 2017 IEEE 2nd Advanced Information Technology, Chongqing, China, 25–26 March 2017; pp. 2174–2179.
11. Liu, X.B.; Zhang, J.R.; Li, P.; Le, V.Q. Energy Measurement of Bubble Bursting Based on Vibration Signals. *Chin. Phys. Lett.* **2012**, *29*, 64701. [CrossRef]
12. Chang, S.; Wang, H.; Ren, L. ISO2631 comfort meter for rail vehicle based on Android System. *Railw. Comput. Appl.* **2020**, *29*, 65–70.
13. Manikandan, K.G.; Pannirselvam, K.; Kenned, J.J.; Kumar, C.S. Investigations on suitability of MEMS based accelerometer for vibration measurements. *Mater. Today Proc.* **2020**, *45*, 6183–6192. [CrossRef]
14. Wu, T.; Shao, S.; Tang, L.; Zhou, Z. A Vision-Based Sensor for Structural Vibration Measurement Based on Laser Strip Center Extraction. In Proceedings of the 8th International Conference on Vibration Engineering, Shanghai, China, 23–26 July 2021; p. 423.
15. Zhong, J.; Jin, J.; Yang, Y.; Sun, Q. Implementation of Cloud- And Android- Based Indoor Illuminance Collection, Transmission and Visualization. In Proceedings of the 2022 3rd Asia Service Sciences and Software Engineering Conference, Macau, Macao, 24–26 February 2022; pp. 171–179.
16. Wenwen, T.; Feng, X.; Lu, L.; Liping, F. Estimating traffic flow states with smart phone sensor data. *Transp. Res. Part C* **2021**, *126*, 103062.
17. Kumar, L.; Tallam, T.; ChikkaKrishna, N.K.; Reddy, M.K.; Reddy, S.P. Response Type Road Roughness Measuring System from a Vehicle Mounted Android Smartphone. In Proceedings of the 2022 IEEE Delhi Section Conference (DELCON), Delhi, India, 11–13 February 2022; pp. 1–4.
18. Kumar, L.; Tallam, T.; Kumar, C.N. Assessment of Ride Quality and Road Roughness by Measuring the Response from a Vehicle Mounted Android Smartphone. *IOP Conf. Ser. Earth Environ. Sci.* **2022**, *982*, 012062. [CrossRef]
19. Marie, S.H. Measuring traffic induced ground vibration using smartphone sensors for a first hand structural health monitoring. *Sci. Afr.* **2021**, *11*, e00703.
20. Garg, S.; Lim, K.M.; Lee, H.P. Recording and Analyzing Carriage Noise of various High-Speed Rail Systems Using Smartphones. *Acoust. Aust.* **2020**, *48*, 121–130. [CrossRef]
21. Partridge, T.; Gherman, L.; Morris, D.; Light, R.; Leslie, A.; Sharkey, D.; McNally, D.; Crowe, J. Smartphone monitoring of in-ambulance vibration and noise. *Proc. Inst. Mech. Eng. Part H J. Eng. Med.* **2021**, *235*, 428–436. [CrossRef] [PubMed]
22. Liu, Z.; Zhu, D.; Yang, L.; Yang, D. Bridge vibration detection test with Android mobile phone sensor. *Sci. Technol. Rep.* **2017**, *35*, 80–86. [CrossRef]
23. González-Pérez, A.; Matey-Sanz, M.; Granell, C.; Casteleyn, S. Using mobile devices as scientific measurement instruments: Reliable android task scheduling. *Pervasive Mob. Comput.* **2022**, *81*, 101550. [CrossRef]
24. Dumitriu, M.; Stănică, D.I. Study on the Evaluation Methods of the Vertical Ride Comfort of Railway Vehicle—Mean Comfort Method and Sperling's Method. *Appl. Sci.* **2021**, *11*, 953. [CrossRef]
25. Kralik, J.; Kralik, J. Experimental and Sensitivity Analysis of the Vibration Impact to the Human Comfort. *Procedia Eng.* **2017**, *190*, 480–487. [CrossRef]

Article

Three-Dimensional Acoustic Analysis of a Rectangular Duct with Gradient Cross-Sections in High-Speed Trains: A Theoretical Derivation

Yanhong Sun, Yi Qiu, Lianyun Liu * and Xu Zheng *

College of Energy Engineering, Zhejiang University, Hangzhou 310027, China; sunyanhong@zju.edu.cn (Y.S.); yiqiu@zju.edu.cn (Y.Q.)
* Correspondence: lianyun.liu@zju.edu.cn (L.L.); zhengxu@zju.edu.cn (X.Z.)

Featured Application: The purpose of this work was to provide a detailed derivation process of the 3D analytical solution and TMs of the RDGCs based on the previous studies on the variable ducts and propose a certain reference for designing and improving the acoustic characteristics of the duct systems used in high-speed trains.

Abstract: Rectangular ducts used in the air-conditioning system of a high-speed train should be carefully designed to achieve optimal acoustic and flow performance. However, the theoretical analysis of the rectangular ducts with gradient cross-sections (RDGC) at frequencies higher than the one-dimensional cut-off frequency is rarely published. This paper has developed the three-dimensional analytical solutions to the wave equations of the expanding and shrinking RDGCs. Firstly, a homogeneous second-order variable coefficient differential equation is derived from the wave equations. Two coefficients of the solution to the differential equation are set to zero to ensure convergence. Secondly, the transfer matrices of the duct systems composed of multiple RDGCs are derived from the three-dimensional solutions. The transmission losses of the duct systems are then calculated from the transfer matrices and validated with the measurement. Finally, the acoustic performance and flow efficiency of the RDGCs with different geometries are discussed. The results show that the REC with double baffles distributed transversely has good performance in both acoustic attenuation and flow efficiency. This study shall provide a helpful guide for designing rectangular ducts used in high-speed trains.

Keywords: theoretical derivation; three-dimensional wave equation; rectangular duct with gradient cross-sections; transfer matrix

Citation: Sun, Y.; Qiu, Y.; Liu, L.; Zheng, X. Three-Dimensional Acoustic Analysis of a Rectangular Duct with Gradient Cross-Sections in High-Speed Trains: A Theoretical Derivation. *Appl. Sci.* **2022**, *12*, 5307. https://doi.org/10.3390/app12115307

Academic Editor: Suchao Xie

Received: 3 May 2022
Accepted: 23 May 2022
Published: 24 May 2022

Publisher's Note: MDPI stays neutral with regard to jurisdictional claims in published maps and institutional affiliations.

1. Introduction

Rectangular ducts are used in the air-conditioning systems of high-speed trains+ to guide airflow. Some rectangular ducts adopt the oblique baffles (Figure 1) to form the varying cross-sections which improve the performance of the ducts in noise attenuation [1].

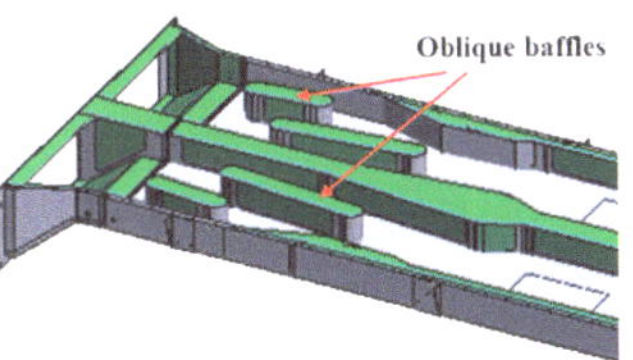

Figure 1. Rectangular ducts used in a Fuxing bullet train.

The current studies on the duct acoustics of high-speed train are practiced based on simulation, such as the hybrid method of finite element and statistical energy analysis

(FE-SEA) [1,2]. Most scholars have focused their research on ducts suitable for any vehicle based on a variety of methods. Both analytical and numerical methods have been widely used in the field of duct acoustics. Generally, the numerical methods, such as the finite element method (FEM) [3–6], boundary element method (BEM) [7] and computational fluid dynamics (CFD) method [8–10], are popular for analyzing duct systems with complex geometries. Assis et al. [4] proposed a spectral FEM approach to compute the transfer matrix (TM) of duct systems with arbitrary geometries. Liu et al. [8] proposed the time domain CFD approaches to predict the acoustic performance of the duct systems without and with mean flow.

On the other hand, the analytical method is more efficient in computation than the numerical methods. Two types of analytical methods have been used to investigate a duct system with varying cross–sections. The Wentzel–Kramers–Brillouin (WKB) method [11–15] utilizes the high frequency approximation that allows to neglect certain terms in the nonlinear governing equations of the media in a duct. Subrahmanyam et al. [12] used the WKB approximation to derive the exact solutions for one-dimensional (1D) ducts with area variations in the absence of mean flow. Rani et al. [14,15] derived a WKB-type solution to the generalized Helmholtz equation in 1D ducts with nonuniform cross-sectional areas and inhomogeneity in mean flow. The WKB approximation is less accurate than solving the full wave equations of the duct. The solutions to the wave equations of 1D ducts with varying cross-sections have been studied for decades [16–23]. Pillai et al. [24] developed the 1D solution to a horn-like rectangular duct at frequencies lower than 250 Hz. However, the three-dimensional (3D) solutions to the wave equations in rectangular ducts with gradient cross-sections (RDGCs), which are more accurate at higher frequencies than the 1D solutions, are rarely seen.

The objective of this paper is to develop the 3D analytical solutions to the wave equations of the expanding or shrinking RDGCs at frequencies up to 1600 Hz. The derived solutions are used to obtain the TMs and transmission losses (TLs) of the duct systems consisted of RDGCs, which have been validated with the measured results. Lastly, the effects of the RDGC geometries on the acoustic performance and flow efficiency of the duct systems are discussed.

The organization of this paper is as follows: Section 2 develops the 3D analytical solutions to the wave equations of RDGCs. In Section 3, the TMs of the RDGCs are derived. In Section 4, several duct systems consisting of RDGCs are modeled to obtain the TMs. The TLs and pressure losses of the duct systems with different RDGC geometries are obtained and discussed in Section 5. Finally, the conclusions are presented in Section 6.

2. 3D Analytical Solutions to the Wave Equations of a RDGC

2.1. 3D Solutions for a Straight Rectangular Duct

Figure 2 shows a uniform rectangular duct with a width of b and a height of h. The 3D wave equation of the rectangular duct is given by [25]

$$\frac{\partial^2 p}{\partial t^2} = c^2 \nabla^2 p,\tag{1}$$

where p, t and c are the sound pressure, advancing time and sound velocity, respectively. The Laplacian operator ∇^2 is given as follows:

$$\nabla^2 = \frac{\partial^2}{\partial x^2} + \frac{\partial^2}{\partial y^2} + \frac{\partial^2}{\partial z^2}\tag{2}$$

where x, y and z are the Cartesian coordinates shown in Figure 2.

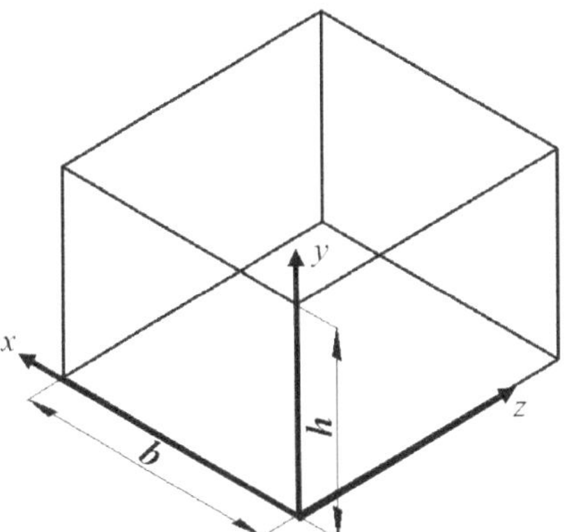

Figure 2. A straight rectangular duct.

The general solution of Equation (1) [25] is

$$p(x,y,z,t) = \left(C_1 e^{-jk_z z} + C_2 e^{+jk_z z}\right)\left(e^{-jk_x x} + C_3 e^{+jk_x x}\right)\left(e^{-jk_y y} + C_4 e^{+jk_y y}\right)e^{j\omega t} \tag{3}$$

where j is the imaginary unit. C_1, C_2, C_3 and C_4 are the coefficients to be determined with boundary conditions. k_x and k_y are the wave numbers in the x and y direction, respectively. k_z is defined as

$$k_z = \sqrt{k^2 - k_x{}^2 - k_y{}^2} \tag{4}$$

where $k = \omega/c$ and ω is the angular frequency. The solution of the rigid-walled duct with a width of b and a height of h is given by

$$p(x,y,z,t) = \sum_{m=0}^{\infty} \sum_{n=0}^{\infty} \cos\left(\frac{m\pi x}{b}\right)\cos\left(\frac{n\pi y}{h}\right)p(z,t) \tag{5}$$

where $p(z,t) = \left(C_{1,mn}e^{-jk_{z,mn}z} + C_{2,mn}e^{+jk_{z,mn}z}\right)e^{j\omega t}$ and $k_{z,mn} = \sqrt{k^2 - (m\pi/b)^2 - (n\pi/h)^2}$. Here, $\cos(m\pi x/b)\cos(n\pi y/h)$ is an eigenfunction representing the wave shape in the x-y plane at the (m, n) mode. $C_{1,mn}$ and $C_{2,mn}$ are the amplitudes of the waves at the (m, n) mode propagating in the positive and negative z directions.

2.2. 3D Solutions for a RDGC

A RDGC with either expanding (positive θ) or shrinking (negative θ) sections is shown in Figure 3. θ is the angle between the bevel edge and the z axis. b_i and b_o are the widths of the inlet and the outlet, respectively.

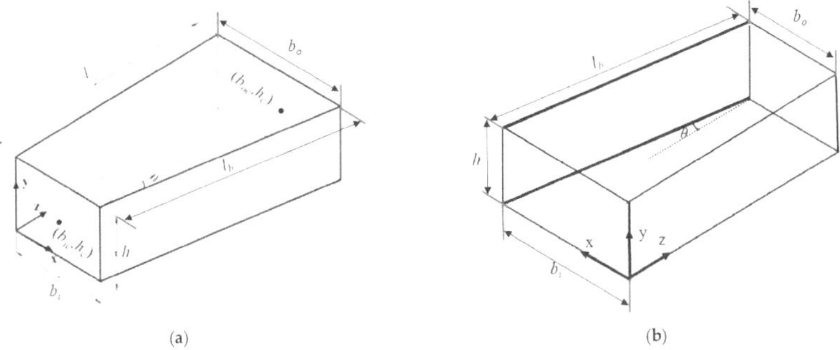

(a)

(b)

Figure 3. (**a**) A rectangular duct with expanding sections, (**b**) a rectangular duct with shrinking sections.

Since the cross-sectional area S changes along the z direction, the 1D sound wave equation in the z direction is obtained with modifying Equation (1) as

$$\frac{1}{S}\frac{\partial}{\partial z}\left(S\frac{\partial p}{\partial z}\right) = \frac{1}{c^2}\frac{\partial^2 p}{\partial t^2}.$$

(6)

For an expanding duct, S is given by

$$S = (z\tan\theta + b_i)h$$

(7)

Substituting Equation (7) into Equation (6) yields a homogeneous second-order variable coefficient differential equation as follows:

$$\frac{\partial^2 p}{\partial z^2} + \frac{\tan\theta}{z\tan\theta + b_i}\frac{\partial p}{\partial z} = \frac{1}{c^2}\frac{\partial^2 p}{\partial t^2}.$$

(8)

The solution of Equation (8) is given by [26]

$$p(z,t) = \{B_+ J_0(k_z\alpha) + B_- J_0(-k_z\alpha) + C_+ K_0(-jk_z\alpha) + C_- K_0(jk_z\alpha)\}e^{j\omega t}$$

(9)

where $\alpha = z + b_i/\tan\theta$. The quantities B_+, B_-, C_+ and C_- are the corresponding amplitudes. $J_0(.)$ is the zeroth order of the Bessel function of the first kind, and $K_0(.)$ is the zeroth order of the modified Bessel function of the second kind [27]. When $k_z\alpha$ becomes imaginary, the values of $J_0(k_z\alpha)$ and $J_0(-k_z\alpha)$ are possible to be infinite. To converge the solution, B_+ and B_- are set to zero. As a result, Equation (9) is simplified as

$$p(z,t) = \{C_+ K_0(-jk_z\alpha) + C_- K_0(jk_z\alpha)\}e^{j\omega t}.$$

(10)

Substituting Equation (10) into Equation (5), with b replaced by $b_i + z\tan\theta$, the 3D solution of an expanding RDGC is derived as

$$p(x,y,z,t) = \sum_{m=0}^{\infty}\sum_{n=0}^{\infty}\cos\left(\frac{m\pi x}{\alpha\tan\theta}\right)\cos\left(\frac{n\pi y}{h}\right)\{C_{+,mn}K_0(-jk_{z,mn}\alpha) + C_{-,mn}K_0(jk_{z,mn}\alpha)\}e^{j\omega t}$$

(11)

$$k_{z,mn} = \sqrt{k^2 - \left(\frac{m\pi}{\alpha\tan\theta}\right)^2 - (n\pi/h)^2}$$

(12)

where $C_{+,mn}$ and $C_{-,mn}$ are the coefficients to be determined with boundary conditions. Solve the momentum equation [25]

$$\frac{\partial v}{\partial t} = -\frac{1}{\rho_0}\nabla\cdot p$$

(13)

to give the particle velocity v of an expanding RDGC as follows:

$$v(x,y,z,t) = -\frac{1}{\rho_0\omega}\sum_{m=0}^{\infty}\sum_{n=0}^{\infty}k_{z,mn}\cos\left(\frac{m\pi x}{\alpha\tan\theta}\right)\cos\left(\frac{n\pi y}{h}\right)\{C_{+,mn}K_1(-jk_{z,mn}\alpha) - C_{-,mn}K_1(jk_{z,mn}\alpha)\}e^{j\omega t}$$

$$+ \left(-\frac{1}{j\rho_0\omega}\right)\underbrace{\sum_{m=0}^{\infty}\sum_{n=0}^{\infty}\frac{m\pi x}{\alpha^2\tan\theta}\sin\left(\frac{m\pi x}{\alpha\tan\theta}\right)\cos\left(\frac{n\pi y}{h}\right)\{C_{+,mn}K_0(-jk_{z,mn}\alpha) + C_{-,mn}K_0(jk_{z,mn}\alpha)\}e^{j\omega t}}_{X}$$

(14)

where ρ_0 is the ambient air density and $K_1(.)$ is the first order of the modified Bessel function of the second kind.

The derivation of the solutions for a shrinking RDGC is similar to that of an expanding RDGC and gives the same equations of Equations (11) and (14) with a negative θ.

3. Derivation of the TM for a RDGC

The transfer matrix **T** ($\begin{bmatrix} A & B \\ C & D \end{bmatrix}$) is used to describe the relationship as follows:

$$\begin{bmatrix} p_i \\ v_i \end{bmatrix} = \begin{bmatrix} A & B \\ C & D \end{bmatrix} \begin{bmatrix} p_o \\ v_o \end{bmatrix} \tag{15}$$

where v_i and v_o are the average particle velocities at the inlet and outlet of a duct system. The quantities p_i and p_o are the inlet and outlet average sound pressures defined with

$$p_i = \frac{1}{S_i} \iint_{S_i} p(x,y,0)\mathrm{d}x\mathrm{d}y \tag{16}$$

$$p_o = \frac{1}{S_o} \iint_{S_o} p(x,y,l)\mathrm{d}x\mathrm{d}y \tag{17}$$

where l is the length of the duct. The quantities S_i and S_o are the inlet and outlet cross-sectional areas, respectively. The four elements of the TM can be obtained as follows:

$$\begin{cases} A = (p_i/p_o)|_{v_o=0} \\ C = (v_i/p_o)|_{v_o=0} \end{cases} \tag{18}$$

$$\begin{cases} B = \{(p_i - Ap_o)/v_o\}|_{v_i=0} \\ D = (-Cp_o/v_o)|_{v_i=0} \end{cases}. \tag{19}$$

The elements A, B, C and D can be calculated from Appendix A.

The transfer matrix **T'** ($\begin{bmatrix} A' & B' \\ C' & D' \end{bmatrix}$) of the shrinking RDGC is derived from the **T** with a negative θ.

4. The TMs and TLs of Rectangular Expansion Chambers (RECs)

4.1. The TMs of the RECs with One or Double Baffles

To simplify the setup of a validation experiment, the circular ducts with a diameter of 50 mm are added at the inlet and outlet of the rectangular chamber with a height (h) of 150 mm. The dimensions of the REC with one baffle are shown in Figure 4. The dimensional parameters of REC are designed according to those of the branch rectangular ducts in high speed trains. The center **o** of the baffle coincides with that of the REC. All the RECs in the following sections have the same circular ducts and rectangular chamber as those in Figure 4.

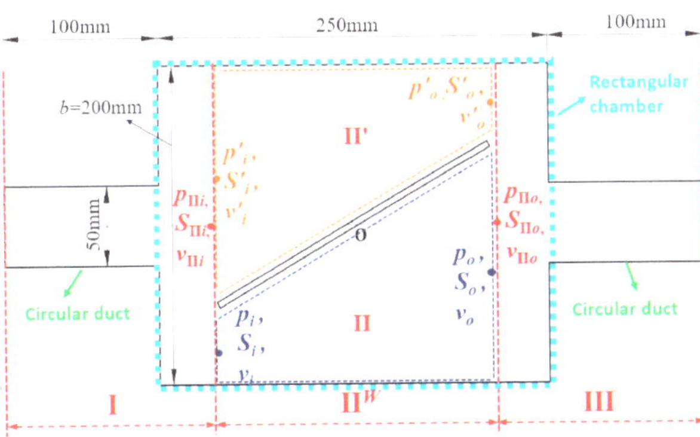

Figure 4. The gemometry of the REC with one baffle.

The components of the REC with one baffle are specified in Table 1.

Table 1. Description of each component of the REC with one baffle.

Unit	TM	Annotations
I	$\mathbf{T_I}$	—made of two uniform ducts and one sudden expansion section
II^W	$\mathbf{T}_{\mathrm{II}}^{W}$	—made of an expanding RDGC (II) and a shrinking RDGC (II')
III	$\mathbf{T_{III}}$	—made of two uniform ducts and one sudden contraction section

The transfer matrices ($\mathbf{T_{II}}$ and $\mathbf{T_{II}}'$) of the expanding RDGC and the shrinking RDGC are given by

$$\mathbf{T_{II}} = \begin{bmatrix} A_{II} & B_{II} \\ C_{II} & D_{II} \end{bmatrix}, \mathbf{T}'_{II} = \begin{bmatrix} A'_{II} & B'_{II} \\ C'_{II} & D'_{II} \end{bmatrix}. \tag{20}$$

According to the notation in Figure 4, the state variables at the two ends of each RDGC are related by

$$\begin{bmatrix} p_i \\ v_i \end{bmatrix} = \mathbf{T_{II}} \begin{bmatrix} p_o \\ v_o \end{bmatrix}, \begin{bmatrix} p'_i \\ v'_i \end{bmatrix} = \mathbf{T}'_{II} \begin{bmatrix} p'_o \\ v'_o \end{bmatrix}. \tag{21}$$

The continuity of pressure and mass velocity at the inlet and outlet of the component II^W gives

$$p_{II\,i} = p_i = p'_i, p_{II\,o} = p_o = p'_o \tag{22}$$

$$S_{II\,i} v_{II\,i} = S_i v_i + S'_i v'_i, S_{II\,o} v_{II\,o} = S_o v_o + S'_o v'_o \tag{23}$$

where $'$ denote the variables of the shrinking RDGC. $S_{II\,i}$ and $S_{II\,o}$ represent the cross-sectional areas at the inlet and outlet of the component II^W, respectively. Solving simultaneously Equations (21)–(23) yields [28]

$$\begin{bmatrix} p_{II\,i} \\ v_{II\,i} \end{bmatrix} = \mathbf{T}_{II}^{W} \begin{bmatrix} p_{II\,o} \\ v_{II\,o} \end{bmatrix} = \begin{bmatrix} A_{II}^{W} & B_{II}^{W} \\ C_{II}^{W} & D_{II}^{W} \end{bmatrix} \begin{bmatrix} p_{II\,o} \\ v_{II\,o} \end{bmatrix}. \tag{24}$$

The terms of the TM are given by

$$\begin{cases} A_{II}^{W} = \dfrac{A'_{II} B_{II} S_o + A_{II} B'_{II} S'_o}{B_{II} S_o + B'_{II} S'_o} \\[2ex] B_{II}^{W} = \dfrac{B_{II} B'_{II} S_{II\,o}}{B_{II} S_o + B'_{II} S'_o} \\[2ex] C_{II}^{W} = \dfrac{\left(D_{II} S_o S'_i - D'_{II} S'_o S_i\right)\left(A'_{II} - A_{II}\right)}{\left(B_{II} S_o + B'_{II} S'_o\right) S_{II\,i}} + \dfrac{S'_i C_{II} + S_i C'_{II}}{S_{II\,i}} \\[2ex] D_{II}^{W} = \dfrac{\left(D_{II} S_o S'_i - D'_{II} S'_o S_i\right) B'_{II} S_{II\,o}}{\left(B_{II} S_o + B'_{II} S'_o\right) S'_o S_{II\,i}} + \dfrac{S_{II\,o} S_i D'_{II}}{S_o S_{II\,i}} \end{cases} \tag{25}$$

The TM of the REC with one baffle is calculated with

$$\mathbf{T}_1 = \mathbf{T}_I \mathbf{T}_{II}^{W} \mathbf{T}_{III} = \begin{bmatrix} T_{11} & T_{12} \\ T_{21} & T_{22} \end{bmatrix}. \tag{26}$$

where $\mathbf{T}_I$ and $\mathbf{T}_{III}$ are calculated with the analytical solutions in the refs. [29,30].

The cut-off frequencies at the mode (m, n) of a duct with a rectangular section can be calculated with

$$f_{cut-off} = \frac{c}{2} \sqrt{\left(\frac{m}{b}\right)^2 + \left(\frac{n}{h}\right)^2}. \tag{27}$$

Table 2 shows the calculated cut-off frequencies with $m, n \leq 2$ of the rectangular chamber (Figure 4). The maximum cut-off frequency at $(2, 2)$ mode is 2858.3 Hz. As a result, the number of modes with $m, n \leq 2$ is enough to investigate the acoustic characteristics of the RECs at frequencies below 1600 Hz.

Table 2. Modal frequencies (Hz) of the rectangular chamber.

n \ m	0	1	2
0	0	857.5	1715.0
1	1143.3	1429.2	2061.2
2	2286.7	2442.2	2858.3

Figures 5 and 6 show the geometries of the RECs with double baffles distributed either axially or transversely. The distance between the centers (o_1 and o_2) of the baffles is 100 mm. The $\mathbf{T}_I$ and $\mathbf{T}_{III}$ in Figures 5 and 6 are exactly the same as those in Figure 4. $\mathbf{T}_{II}^W$ and $\mathbf{T}_V^W$ in Figure 5 are calculated with the Equation (25), while the $\mathbf{T}_{IV}$ of the straight rectangular duct is obtained with the analytical solutions in the ref. [29].

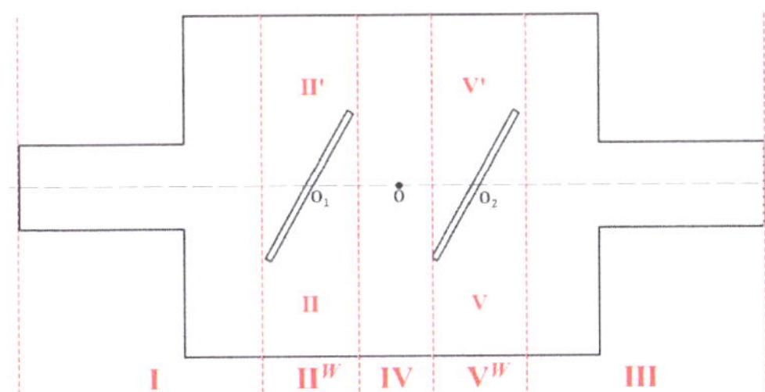

Figure 5. The REC with double baffles distributed axially.

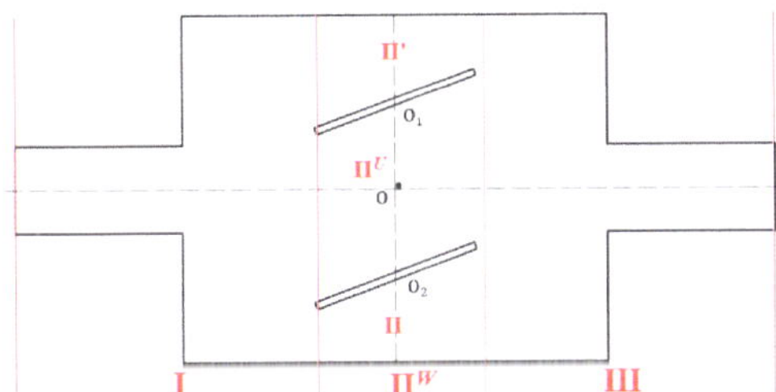

Figure 6. The REC with double baffles distributed transversely.

The transfer matrices ($\mathbf{T}_{II}$, $\mathbf{T}_{II}'$ and $\mathbf{T}_{II}^U$) of the expanding RDGC, the shrinking RDGC and the uniform duct (II^U) in Figure 6 are given by

$$\mathbf{T}_{II} = \begin{bmatrix} A_{II} & B_{II} \\ C_{II} & D_{II} \end{bmatrix}, \mathbf{T}_{II}' = \begin{bmatrix} A_{II}' & B_{II}' \\ C_{II}' & D_{II}' \end{bmatrix}, \mathbf{T}_{II}^U = \begin{bmatrix} A_{II}^U & B_{II}^U \\ C_{II}^U & D_{II}^U \end{bmatrix}. \tag{28}$$

The transfer matrix ($\mathbf{T}_{II}^W$) of the component II^W is calculated with a similar process presented in the Equations (21)–(24) and the four elements are given by

$$\begin{cases} A_{II}^W = \dfrac{A_{II}^U B_{II}' B_{II} S_o^U + A_{II}' B_{II}^U B_{II} S_i^U + A_{II} B_{II}' B_{II}^U S_o}{B_{II}' B_{II} S_o^U + B_{II}^U B_{II} S_o' + B_{II}' B_{II}^U S_o} \\[2ex] B_{II}^W = \dfrac{B_{II}' B_{II}^U B_{II} S_{II\,o}}{B_{II}' B_{II} S_o^U + B_{II}^U B_{II} S_o' + B_{II}' B_{II}^U S_o} \\[2ex] C_{II}^W = \dfrac{A_{II}^U B_{II}' B_{II} S_o^U + A_{II}' B_{II}^U B_{II} S_o' + A_{II} B_{II}' B_{II}^U S_o}{B_{II}' B_{II} S_o^U + B_{II}^U B_{II} S_o' + B_{II}' B_{II}^U S_o} \left(\dfrac{D_{II}' S_i'}{B_{II}' S_{II\,i}} + \dfrac{D_{II}^U S_i^U}{B_{II}^U S_{II\,i}} + \dfrac{D_{II} S_i}{B_{II} S_{II\,i}} \right) \\[2ex] \qquad - \left(\dfrac{A_{II}' D_{II}' S_i'}{B_{II}' S_{II\,i}} + \dfrac{A_{II}^U D_{II}^U S_i^U}{B_{II}^U S_{II\,i}} + \dfrac{A_{II} D_{II} S_i}{B_{II} S_{II\,i}} \right) + \dfrac{S_i' C_{II}' + S_i^U C_{II}^U + S_i C_{II}}{S_{II\,i}} \\[2ex] D_{II}^W = \dfrac{S_{II\,o}(D_{II}' B_{II}^U B_{II} S_i' + D_{II}^U B_{II}' B_{II} S_i^U + D_{II} B_{II}' B_{II}^U S_i)}{S_{II\,i}(B_{II}' B_{II} S_o^U + B_{II}^U B_{II} S_o' + B_{II}' B_{II}^U S_o)} \end{cases} \tag{29}$$

where S_i^U and S_o^U are the cross-sectional areas of the inlet and outlet of the component II^U. The TM ($\mathbf{T}_{2a}$) of the REC with double baffles distributed axially and the TM ($\mathbf{T}_{2t}$) with double baffles distributed transversely are given by

$$\mathbf{T}_{2a} = \mathbf{T}_I \mathbf{T}_{II}^W \mathbf{T}_{IV} \mathbf{T}_V^W \mathbf{T}_{III} \tag{30}$$

$$\mathbf{T}_{2t} = \mathbf{T}_I \mathbf{T}_{II}^W \mathbf{T}_{III}. \tag{31}$$

4.2. Geometries of the RECs

Table 3 shows the geometries of the RECs with different baffle configurations. All the baffles in the RECs have a thickness of 4 mm.

Table 3. The geometries of the RECs with different baffle configurations.

Type	Case		Parameters
Type 1	Case 1-0		$l_b = 0.50b, \theta = 40°$
	Case 1-1		$l_b = 0.50b, \theta = 20°$
	Case 1-2	The REC with one baffle	$l_b = 0.50b, \theta = 60°$
	Case 1-3		$l_b = 0.30b, \theta = 40°$
	Case 1-4		$l_b = 0.40b, \theta = 40°$
Type 2a	Case 2a-0		$l_b = 0.50b, \theta = 40°$
	Case 2a-1		$l_b = 0.50b, \theta = 20°$
	Case 2a-2	The REC with double baffles distributed axially	$l_b = 0.50b, \theta = 60°$
	Case 2a-3		$l_b = 0.30b, \theta = 40°$
	Case 2a-4		$l_b = 0.40b, \theta = 40°$
Type 2t	Case 2t-0		$l_b = 0.50b, \theta = 40°$
	Case 2t-1		$l_b = 0.50b, \theta = 20°$
	Case 2t-2	The REC with double baffles distributed transversely	$l_b = 0.50b, \theta = 60°$
	Case 2t-3		$l_b = 0.30b, \theta = 40°$
	Case 2t-4		$l_b = 0.40b, \theta = 40°$

4.3. Calculation and Measurement of the TLs for the RECs

The TL of a REC can be calculated from the derived TM as follows:

$$\mathrm{TL} = 20 \log_{10} \left(\frac{1}{2} \left| T_{11} + \frac{T_{12}}{\rho_0 c} + T_{21} \cdot \rho_0 c + T_{22} \right| \right). \tag{32}$$

To verify the accuracy of the analytical method, the TLs of the RECs (Case 1-1, Case 2a-1 and Case 2t-1) were measured with the two-load method [31] shown in Figure 7. The sound source was located at the outside of a semi-anechoic room where a REC with baffles was located. An acoustic stimulus was introduced into the REC through a metal duct, where two microphones were placed with a distance of 40 mm. The other two microphones were located with the same distance at the duct connected to the outlet of the REC. A Brüel & Kjær (B&K) 3560 C Module was adopted to acquire the data sampled with a frequency of 16,384 Hz. The experimental parameters are given in Table 4.

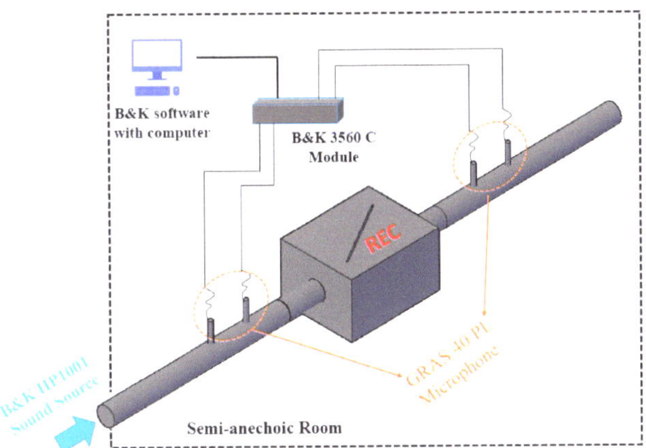

Figure 7. Experimental setup for measuring the TL of a REC.

Table 4. Experimental parameters.

Parameters	Values
Temperature	24.5 °C
Relative humidity	29.2%
Pressure	101,300 Pa

5. Results

5.1. Experimental Validation of the Calculated Results

In order to verify the 3D analytical method, the TLs obtained by experiment, 3D analytical method and FEM are shown in Figure 8. The FEM model and boundary conditions are shown in Appendix B. Generally, the calculated results are in good agreement with the measured results and the accuracy of the analytical method is validated to a certain degree. However, the frequencies of TL peaks from FEM agree less with the experimental results than those obtained by the 3D analytical method in this paper. The inaccuracy of the measured TLs below 200 Hz shown in the Case 1-1 of Figure 8 should be attributed to the insufficient energy of the sound source in this frequency range. As a result, the TLs below 200 Hz are not presented in the other cases of Figure 8. The discrepancies between the experiment and the 3D analytical method at higher frequencies may be caused by the following reasons. First, the analytical method regards the REC as rigid, while the prototypes under test are made of plastic with certain elasticity. Second, the damping in the air is ignored with the analytical method. Third, the ignorance of the $B_+J_0(k_z\alpha)$, $B_-J_0(-k_z\alpha)$ in the Equation (9) and the X in Equation (14) also causes errors.

5.2. TLs of the RECs

The TLs of the RECs with one or double baffles are shown in Figures 9 and 10. It can be seen that the peaks and troughs of the TL curves of all types move to a lower frequency with the decreasing θ and increasing l_b. Although the Type 2a has more TL peaks, it is worse in performance than the Type 2t at frequencies from 500 Hz to 1100 Hz. Generally, the Type 2t is better in acoustic performance than the other types, especially at frequencies from 600 Hz to 1100 Hz.

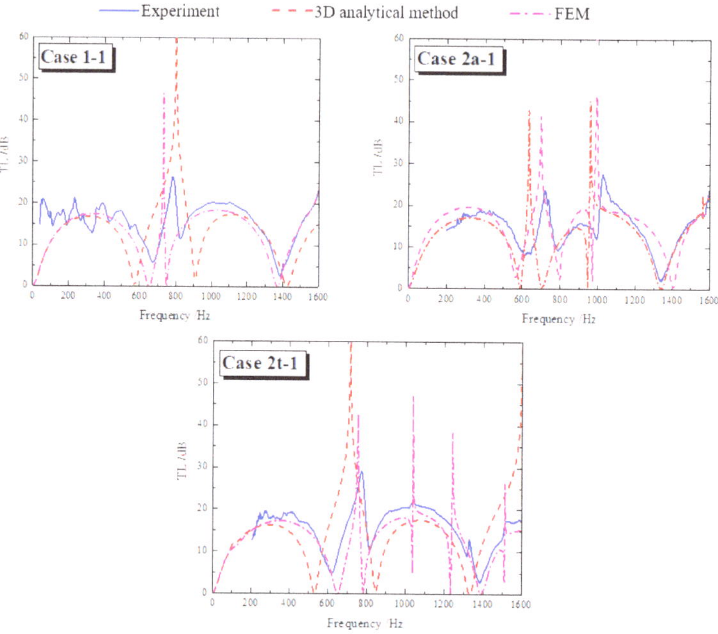

Figure 8. The comparison of TLs obtained by experiment, 3D analytical method and FEM.

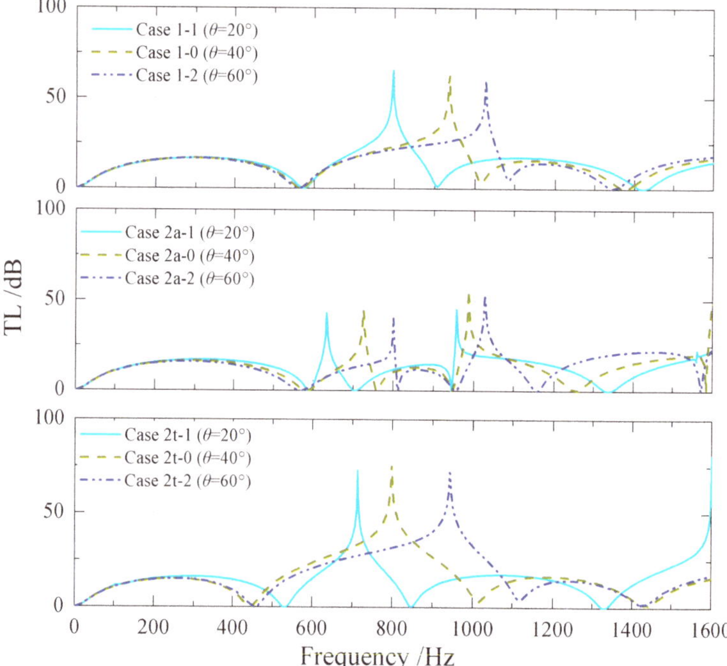

Figure 9. TLs of the RECs with different baffle angles.

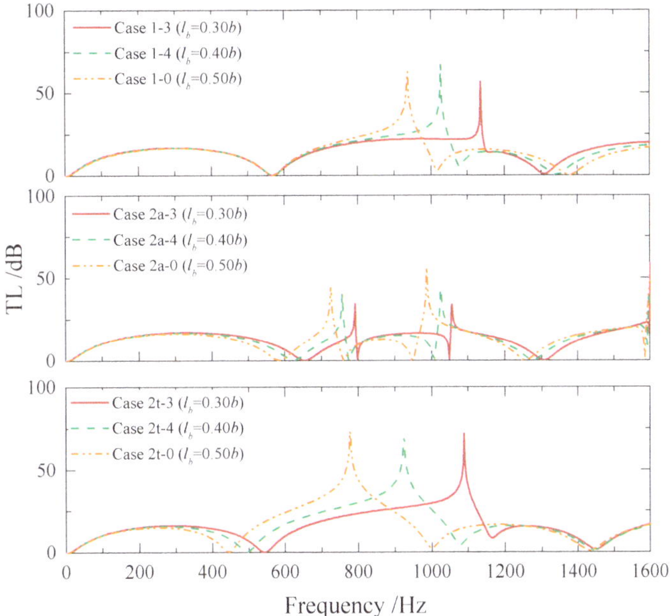

Figure 10. TLs of the RECs with different baffle lengths.

5.3. Pressure Losses of the RECs

The improvement of the acoustic performance of a duct system cannot be at the expense of flow efficiency. A CFD model [32] (Figure 11), using the standard k-ε turbulence model, is adopted to calculate the difference (pressure loss) between the area-weighted average pressures at the inlet and outlet of the RECs. The model is discretized by about one million unstructured tetrahedral meshes with a mesh size of 3 mm. The inlet has a velocity of 10 m/s and the outlet has a zero gauge pressure.

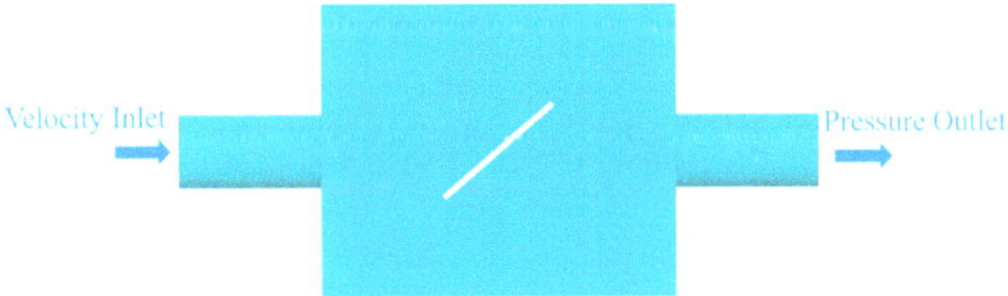

Figure 11. CFD model of the REC with one baffle.

Table 5 shows the pressure losses of the RECs with one baffle or double baffles. It can be seen that the pressure losses of the Case 1-0 and Case 2a-0 are higher than the Case 2t-0. Therefore, the REC with double baffles distributed transversely (Type 2t) has good performance in both flow efficiency and TL.

Table 5. The pressure losses of the RECs with different baffle configurations.

Case 1-0	Case 2a-0	Case 2t-0
141 Pa	140.2 Pa	117.7 Pa

The influence of the baffle angles and lengths on the pressure losses of the RECs are shown in Figures 12 and 13. Figure 12 shows that the pressure losses of Type 1 and Type 2t increase as θ increases, while the pressure loss of Type 2a decreases as θ increases from 40° to 60°. Figure 13 shows that Type 1 and Type 2a have much higher pressure losses than Type 2t, while the pressure loss of Type 2t increases faster than Type 1 and Type 2a as l_b increases. In general, Type 2t has better performance in flow efficiency than the other types, especially at small values of θ and l_b.

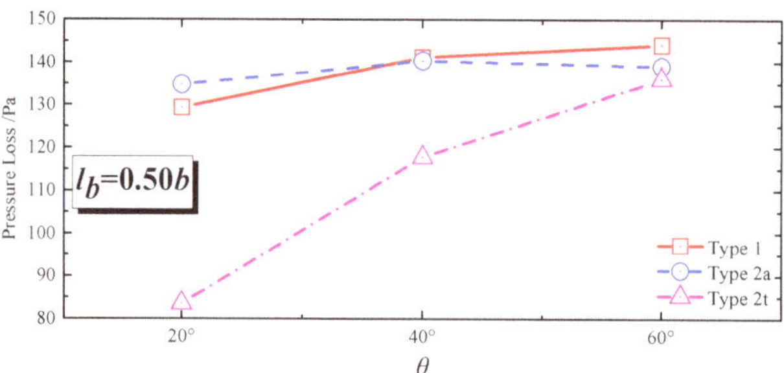

Figure 12. The pressure losses of the RECs with different baffle angles.

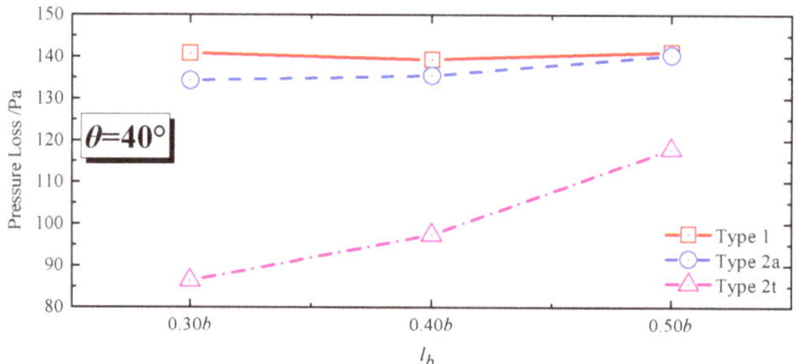

Figure 13. The pressure losses of the RECs with different baffle lengths.

6. Conclusions

Here, the 3D analytical solutions to the wave equations of the expanding and shrinking RDGCs were derived. The TMs of the RECs consisted of multiple expanding and shrinking RDGCs, which were then calculated from the 3D solutions. The TLs calculated from the TMs were validated with the measured results.

In the derivation of the 3D analytical solution and TMs of the RDGC, the ignorance of some infinite and complex terms is risky, but it can simplify the formulas and reduce the computation. These behaviors are proved to be practicable by experiments and the TLs obtained by the theories in this paper are accurate to a certain extent.

According to the TLs of the RECs with different baffle configurations, the peaks and troughs of the TL curves of all types move to a higher frequency with the increasing angle (θ) between the bevel edge and the axial direction and move to a lower frequency with the increasing length (l_b) of the baffle. The REC with double baffles distributed transversely (Type 2t) is better in acoustic performance than the other types at frequencies from 600 Hz to 1100 Hz. On the other hand, although the pressure losses of all types of RECs increase as θ or l_b increases, Type 2t always has a lower pressure loss than other types. In summary, Type 2t generally has good performance in both acoustic attenuation and flow efficiency.

This achievement of research shall provide a certain reference for designing and improving the acoustic characteristics of the duct systems used in high-speed trains.

Author Contributions: This article was prepared through the collective efforts of all the authors. Conceptualization, methodology, and writing—original draft preparation, Y.S.; validation, Y.S. and L.L.; writing—review and editing, Y.Q. and X.Z. All authors have read and agreed to the published version of the manuscript.

Funding: This research was supported by the National Natural Science Foundation of China (Grant No. 51975515 and No. 51905474).

Institutional Review Board Statement: The study did not involve humans or animals.

Informed Consent Statement: The study did not involve humans.

Data Availability Statement: The data that support the findings of this study are available from the corresponding author upon reasonable request.

Conflicts of Interest: The authors declared no potential conflict of interest with respect to the research, authorship and/or publication of this article.

Appendix A. Derivation of the A, B, C and D in the TM

We set $v(x, y, 0)$ and $v(x, y, l)$ as $\bar{v}$ and 0, respectivley, where $\bar{v}$ is the harmonic excitation with a constant amplitude. As a result, v_i and v_o are also equal to $\bar{v}$ and 0. To simplify the expression, $jk_{z,mn}\alpha$ is denoted as β_{mn}. Substituting $v(x, y, 0) = \bar{v}$ into Equation (14) and eliminating the $e^{j\omega t}$ yields

$$-\frac{1}{\rho_0\omega}\sum_{m=0}^{\infty}\sum_{n=0}^{\infty}k_{z,mn,i}\cos\left(\frac{m\pi x}{b_i}\right)\cos\left(\frac{n\pi y}{h}\right)\{C_{+,mn}K_1(-\beta_{i,mn}) - C_{-,mn}K_1(\beta_{i,mn})\} = \bar{v}. \tag{A1}$$

The X term in Equation (14) is ignored here, because it is too complicated to derive a concise solution. The rationality of this mathematical operation has been verified by the following experimental results.

Substituting $v(x, y, l) = 0$ into Equation (14) and ignoring the term X yields

$$C_{+,mn}K_1(-\beta_{o,mn}) - C_{-,mn}K_1(\beta_{o,mn}) = 0. \tag{A2}$$

The quantities $\beta_{i,mn}$ and $\beta_{o,mn}$ represent the β_{mn} at $z = 0$ and $z = l$, respectively. $k_{z,mn,i}$ and $k_{z,mn,o}$ represent the $k_{z,mn}$ at $z = 0$ and $z = l$, respectively.

In order to obtain the coefficients $C_{+,mn}$ and $C_{-,mn}$, operating the both sides of Equation (A1) with $\iint_{S_i}\cos(m'\pi x/b_i)\cos(n'\pi y/h)\mathrm{d}x\mathrm{d}y$ [28] yields

$$\begin{aligned}&\iint_{S_i}\bar{v}\cos\left(\frac{m'\pi x}{b_i}\right)\cos\left(\frac{n'\pi y}{h}\right)\mathrm{d}x\mathrm{d}y\\&= -\frac{1}{\rho_0\omega}\sum_{m,m'=0}^{\infty}\sum_{n,n'=0}^{\infty}k_{z,mn,i}\iint_{S_i}\cos(\frac{m'\pi x}{b_i})\cos(\frac{n'\pi y}{h})\cos(\frac{m\pi x}{b_i})\cos(\frac{n\pi y}{h})\mathrm{d}x\mathrm{d}y\{C_{+,mn}K_1(-\beta_{i,mn}) - C_{-,mn}K_1(\beta_{i,mn})\}.\end{aligned} \tag{A3}$$

According to the orthogonality property of eigenfunctions, Equation (A3) is transformed with $m = m'$ and $n = n'$ to the following equation:

$$\iint_{S_i}\bar{v}\cos\left(\frac{m\pi x}{b_i}\right)\cos\left(\frac{n\pi y}{h}\right)\mathrm{d}x\mathrm{d}y = -\frac{1}{\rho_0\omega}k_{z,mn,i}\iint_{S_i}\cos^2(\frac{m\pi x}{b_i})\cos^2(\frac{n\pi y}{h})\mathrm{d}x\mathrm{d}y\{C_{+,mn}K_1(-\beta_{i,mn}) - C_{-,mn}K_1(\beta_{i,mn})\}. \tag{A4}$$

The coefficients $C_{+,mn}$ and $C_{-,mn}$ can be calculated from Equations (A2) and (A4) as follows:

$$C_{+,mn} = -\rho_0c\bar{v}\frac{kK_1(\beta_{o,mn})}{k_{z,mn,i}W(k_{z,mn})}\frac{\iint_{S_i}\cos(\frac{m\pi x}{b_i})\cos(\frac{n\pi y}{h})\mathrm{d}x\mathrm{d}y}{\iint_{S_i}\cos^2(\frac{m\pi x}{b_i})\cos^2(\frac{n\pi y}{h})\mathrm{d}x\mathrm{d}y} \tag{A5}$$

$$C_{-,mn} = \frac{C_{+,mn}K_1(-\beta_{o,mn})}{K_1(\beta_{o,mn})} \tag{A6}$$

where

$$W(k_{z,mn}) = K_1(-\beta_{i,mn})K_1(\beta_{o,mn}) - K_1(\beta_{i,mn})K_1(-\beta_{o,mn}). \tag{A7}$$

Substituting Equations (A5) and (A6) into Equation (11), the 3D solution of the pressure is obtained as follows:

$$p(x, y, z) = -\rho_0c\bar{v}\sum_{m=0}^{\infty}\sum_{n=0}^{\infty}\frac{V(k_{z,mn}, z)}{W(k_{z,mn})}\frac{k}{k_{z,mn,i}}\frac{\cos(\frac{m\pi x}{b_i+z\tan\theta})\cos(\frac{n\pi y}{h})\iint_{S_i}\cos(\frac{m\pi x}{b_i})\cos(\frac{n\pi y}{h})\mathrm{d}x\mathrm{d}y}{\iint_{S_i}\cos^2(\frac{m\pi x}{b_i})\cos^2(\frac{n\pi y}{h})\mathrm{d}x\mathrm{d}y} \tag{A8}$$

where

$$V(k_{z,mn}, z) = \mathrm{K}_1(\beta_{o,mn})\mathrm{K}_0(-\beta_{mn}) + \mathrm{K}_0(\beta_{mn})\mathrm{K}_1(-\beta_{o,mn}). \tag{A9}$$

Substituting Equation (A8) with $z = 0$ into Equation (16) to obtain the average pressure at the inlet as follows:

$$
\begin{aligned}
p_i &= \frac{1}{S_i}\iint_{S_i}\left\{-\rho_0 c\bar{v}\sum_{m=0}^{\infty}\sum_{n=0}^{\infty}\frac{V(k_{z,mn},0)}{W(k_{z,mn})}\frac{k}{k_{z,mn,i}}\frac{\cos(\frac{m\pi x}{b_i})\cos(\frac{n\pi y}{h})\iint_{S_i}\cos(\frac{m\pi x}{b_i})\cos(\frac{n\pi y}{h})dxdy}{\iint_{S_i}\cos^2(\frac{m\pi x}{b_i})\cos^2(\frac{n\pi y}{h})dxdy}\right\}dxdy\\[2mm]
&= -\rho_0 c\bar{v}_1\left\{
\begin{array}{l}
\underbrace{\dfrac{V(k,0)}{W(k)}}_{E_{11}} + \underbrace{\sum_{m=1}^{\infty}\dfrac{V(k_{z,m0,i},0)}{W(k_{z,m0})}\dfrac{2k}{k_{z,m0,i}}\left(\dfrac{2}{m\pi}\Psi_{b1}\right)^2}_{E_{12}}\\[6mm]
+\underbrace{\sum_{n=1}^{\infty}\dfrac{V(k_{z,0n,i},0)}{W(k_{z,0n})}\dfrac{2k}{k_{z,0n,i}}\left(\dfrac{2}{n\pi}\Psi_{h1}\right)^2}_{E_{13}}\\[6mm]
+\underbrace{\sum_{m=1}^{\infty}\sum_{n=1}^{\infty}\dfrac{V(k_{z,mn,i},0)}{W(k_{z,mn})}\dfrac{4k}{k_{z,mn,i}}\left(\dfrac{4}{mn\pi^2}\Psi_{b1}\Psi_{h1}\right)^2}_{E_{14}}
\end{array}
\right\}\\[2mm]
&= -\rho_0 c\bar{v}_1 E_1
\end{aligned}
\tag{A10}
$$

where

$$E_1 = E_{11} + E_{12} + E_{13} + E_{14} \tag{A11}$$

$$\Psi_{b1} = \cos\left(\frac{m\pi b_{ic}}{b_i}\right)\sin\left(\frac{m\pi}{2}\right) \tag{A12}$$

$$\Psi_{h1} = \cos\left(\frac{n\pi h_c}{h}\right)\sin\left(\frac{n\pi}{2}\right). \tag{A13}$$

The quantity E_{11} is obtained with $m = 0$ and $n = 0$. E_{12} is obtained with $m \geq 1$ and $n = 0$. E_{13} is obtained with $m = 0$ and $n \geq 1$. Additionally, E_{14} is obtained with $m \geq 1$ and $n \geq 1$. (b_{ic}, h_c) are the coordinates of the center point at the inlet shown in Figure 3.

Substituting Equation (A8) with $z = l$ into Equation (17) to obtain the average pressure at the outlet as follows:

$$
\begin{aligned}
p_o &= \frac{1}{S_o}\iint_{S_o}\left\{-\rho_0 c\bar{v}_1\sum_{m=0}^{\infty}\sum_{n=0}^{\infty}\frac{V(k_{z,mn},l)}{W(k_{z,mn})}\frac{k}{k_{z,mn,i}}\frac{\cos(\frac{m\pi x}{b_i+l\tan\theta})\cos(\frac{n\pi y}{h})\iint_{S_i}\cos(\frac{m\pi x}{b_i})\cos(\frac{n\pi y}{h})dxdy}{\iint_{S_i}\cos^2(\frac{m\pi x}{b_i})\cos^2(\frac{n\pi y}{h})dxdy}\right\}dxdy\\[2mm]
&= -\rho_0 c\bar{v}\left\{
\begin{array}{l}
\underbrace{\dfrac{V(k,l)}{W(k)}}_{E_{21}} + \underbrace{\sum_{m=1}^{\infty}\dfrac{V(k_{z,m0,o},l)}{W(k_{z,m0})}\dfrac{2k}{k_{z,m0,i}}\left(\dfrac{2}{m\pi}\right)^2\Psi_{lb2}\Psi_{b1}}_{E_{22}}\\[6mm]
+\underbrace{\sum_{n=1}^{\infty}\dfrac{V(k_{z,0n,o},l)}{W(k_{z,0n})}\dfrac{2k}{k_{z,0n,i}}\left(\dfrac{2}{n\pi}\right)^2\Psi_{h1}\Psi_{h2}}_{E_{23}}\\[6mm]
+\underbrace{\sum_{m=1}^{\infty}\sum_{n=1}^{\infty}\dfrac{V(k_{z,mn,o},l)}{W(k_{z,mn})}\dfrac{4k}{k_{z,mn,i}}\left(\dfrac{4}{mn\pi^2}\right)^2\Psi_{b1}\Psi_{h1}\Psi_{lb2}\Psi_{h2}}_{E_{24}}
\end{array}
\right\}\\[2mm]
&= -\rho_0 c\bar{v}E_2
\end{aligned}
\tag{A14}
$$

where

$$E_2 = E_{21} + E_{22} + E_{23} + E_{24} \tag{A15}$$

$$\Psi_{h2} = \cos\left(\frac{n\pi h_c}{h}\right)\sin\left(\frac{n\pi}{2}\right) \tag{A16}$$

$$\Psi_{lb2} = \cos\left(\frac{m\pi b_{oc}}{b_o}\right)\sin\left(\frac{m\pi}{2}\right). \tag{A17}$$

The quantity E_{21} is obtained with $m = 0$ and $n = 0$. E_{22} is obtained with $m \geq 1$ and $n = 0$. E_{23} is obtained with $m = 0$ and $n \geq 1$. Additionally, E_{24} is obtained with $m \geq 1$ and $n \geq 1$. (b_{oc}, h_c) are the coordinates of the center point at the outlet in Figure 3.

The elements A and C can be calculated from Equations (18), (A10) and (A14) with the following equations

$$A = \frac{-\rho_0 c\bar{v} E_1}{-\rho_0 c\bar{v} E_2} = \frac{E_1}{E_2} \tag{A18}$$

$$C = \frac{v_i}{-\rho_0 c\bar{v}_1 E_{12}} = \frac{\bar{v}}{-\rho_0 c\bar{v} E_2} = \frac{-1}{\rho_0 c E_2}. \tag{A19}$$

We set $v(x, y, 0)$ and $v(x, y, l)$ as 0 and $\bar{v}$, respectively, and obtain $v_i = 0$ and $v_o = \bar{v}$. Substituting $v(x, y, 0) = 0$ into Equation (14) yields

$$C_{+,mn} K_1(-\beta_{i,mn}) - C_{-,mn} K_1(\beta_{i,mn}) = 0. \tag{A20}$$

Substituting $v(x, y, l) = \bar{v}$ into Equation (14) yields

$$-\frac{1}{\rho_0 \omega} \sum_{m=0}^{\infty} \sum_{n=0}^{\infty} k_{z,mn,o} \cos\left(\frac{m\pi x}{b_i + l\tan\theta}\right) \cos\left(\frac{n\pi y}{h}\right) \{C_{+,mn} K_1(-\beta_{o,mn}) - C_{-,mn} K_1(\beta_{o,mn})\} = \bar{v}. \tag{A21}$$

Operating the both sides of Equation (A21) with $\iint_{S_o} \cos\{m\pi x/(b_i + l\tan\theta)\} \cos(n\pi y/h)$ $dxdy$ and using the same transformation with $m = m'$ and $n = n'$ as in Equation (A3) yields

$$\begin{aligned}
&\iint_{S_o} \bar{v} \cos\left(\frac{m\pi x}{b_i + l\tan\theta}\right) \cos\left(\frac{n\pi y}{h}\right) dxdy \\
&= -\frac{1}{\rho_0 \omega} k_{z,mn,o} \iint_{S_o} \cos^2\left(\frac{m\pi x}{b_i + l\tan\theta}\right) \cos^2\left(\frac{n\pi y}{h}\right) dxdy \{C_{+,mn} K_1(-\beta_{o,mn}) - C_{-,mn} K_1(\beta_{o,mn})\}.
\end{aligned} \tag{A22}$$

The coefficients $C_{+,mn}$ and $C_{-,mn}$ can be obtained from Equations (A20) and (A22) as follows:

$$C_{+,mn} = \rho_0 c\bar{v} \frac{k K_1(\beta_{i,mn})}{k_{z,mn,o} W(k_{z,mn})} \frac{\iint_{S_o} \cos\left(\frac{m\pi x}{b_i + l\tan\theta}\right) \cos\left(\frac{n\pi y}{h}\right) dxdy}{\iint_{S_o} \cos^2\left(\frac{m\pi x}{b_i + l\tan\theta}\right) \cos^2\left(\frac{n\pi y}{h}\right) dxdy} \tag{A23}$$

$$C_{-,mn} = \frac{C_{+,mn} K_1(-\beta_{i,mn})}{K_1(\beta_{i,mn})}. \tag{A24}$$

Substituting Equations (A23) and (A24) into Equation (11), the 3D solution of the pressure is obtained as follows:

$$p(x, y, z) = \rho_0 c\bar{v} \sum_{m=0}^{\infty} \sum_{n=0}^{\infty} \frac{U(k_{z,mn}, z)}{W(k_{z,mn})} \frac{k}{k_{z,mn,o}} \frac{\cos\left(\frac{m\pi x}{b_i + z\tan\theta}\right) \cos\left(\frac{n\pi y}{h}\right) \iint_{S_o} \cos\left(\frac{m\pi x}{b_i + l\tan\theta}\right) \cos\left(\frac{n\pi y}{h}\right) dxdy}{\iint_{S_o} \cos^2\left(\frac{m\pi x}{b_i + l\tan\theta}\right) \cos^2\left(\frac{n\pi y}{h}\right) dxdy} \tag{A25}$$

where

$$U(k_{z,mn}, z) = K_1(\beta_{i,mn}) K_0(-\beta_{mn}) + K_0(\beta_{mn}) K_1(-\beta_{i,mn}). \tag{A26}$$

Substituting Equation (A25) with $z = 0$ into Equation (16) to obtain the average pressure at the inlet as follows:

$$p_i = \frac{1}{S_i}\iint_{S_i}\left\{\rho_0 c\bar{v}\sum_{m=0}^{\infty}\sum_{n=0}^{\infty}\frac{U(k_{z,mn},0)}{W(k_{z,mn})}\frac{k}{k_{z,mn,o}}\frac{\cos\left(\frac{m\pi x}{b_i}\right)\cos\left(\frac{n\pi y}{h}\right)\iint_{S_0}\cos\left(\frac{m\pi x}{b_i+l\tan\theta}\right)\cos\left(\frac{n\pi y}{h}\right)dxdy}{\iint_{S_0}\cos^2\left(\frac{m\pi x}{b_i+l\tan\theta}\right)\cos^2\left(\frac{n\pi y}{h}\right)dxdy}\right\}dxdy$$

$$= \rho_0 c\bar{v}\left\{\underbrace{\frac{U(k,z)}{W(k)}}_{E_{31}} + \underbrace{\sum_{m=1}^{\infty}\frac{U(k_{z,m0,i},0)}{W(k_{z,m0})}\frac{2k}{k_{z,m0,o}}\left(\frac{2}{m\pi}\right)^2\Psi_{lb2}\Psi_{b1}}_{E_{32}}\right.$$
$$\left. + \underbrace{\sum_{n=1}^{\infty}\frac{U(k_{z,0n,i},0)}{W(k_{z,0n})}\frac{2k}{k_{z,0n,o}}\left(\frac{2}{n\pi}\right)^2\Psi_{h2}\Psi_{h1}}_{E_{33}} \right.$$
$$\left. + \underbrace{\sum_{m=1}^{\infty}\sum_{n=1}^{\infty}\frac{U(k_{z,mn,i},0)}{W(k_{z,mn})}\frac{4k}{k_{z,mn,o}}\left(\frac{4}{mn\pi^2}\right)^2\Psi_{lb2}\Psi_{h2}\Psi_{b1}\Psi_{h1}}_{E_{34}}\right\} \tag{A27}$$

$$= \rho_0 c\bar{v}E_3$$

where

$$E_3 = E_{31} + E_{32} + E_{33} + E_{34}. \tag{A28}$$

The quantity E_{31} is obtained with $m = 0$ and $n = 0$. E_{32} is obtained with $m \geq 1$ and $n = 0$. E_{33} is obtained with $m = 0$ and $n \geq 1$. Additionally, E_{34} is obtained with $m \geq 1$ and $n \geq 1$.

Substituting Equation (A25) and $z = l$ into Equation (17) to obtain the average pressure at the outlet as follows:

$$p_o = \frac{1}{S_o}\iint_{S_o}\left\{\rho_0 c\bar{v}\sum_{m=0}^{\infty}\sum_{n=0}^{\infty}\frac{U(k_{z,mn},l)}{W(k_{z,mn})}\frac{k}{k_{z,mn,o}}\frac{\cos\left(\frac{m\pi x}{b_i+l\tan\theta}\right)\cos\left(\frac{n\pi y}{h}\right)\iint_{S_0}\cos\left(\frac{m\pi x}{b_i+l\tan\theta}\right)\cos\pi\left(\frac{n\pi y}{h}\right)dxdy}{\iint_{S_0}\cos^2\left(\frac{m\pi x}{b_i+l\tan\theta}\right)\cos^2\left(\frac{n\pi y}{h}\right)dxdy}\right\}dxdy$$

$$= \rho_0 c\bar{v}\left\{\underbrace{\frac{U(k,l)}{W(k)}}_{E_{41}} + \underbrace{\sum_{m=1}^{\infty}\frac{U(k_{z,m0,o},l)}{W(k_{z,m0})}\frac{2k}{k_{z,m0,o}}\left(\frac{2}{m\pi}\Psi_{lb2}\right)^2}_{E_{42}}\right.$$
$$\left. + \underbrace{\sum_{n=1}^{\infty}\frac{U(k_{z,0n,o},l)}{W(k_{z,0n})}\frac{2k}{k_{z,0n,o}}\left(\frac{2}{n\pi}\Psi_{h2}\right)^2}_{E_{43}}\right.$$
$$\left. + \underbrace{\sum_{m=1}^{\infty}\sum_{n=1}^{\infty}\frac{U(k_{z,mn,o},l)}{W(k_{z,mn})}\frac{4k}{k_{z,mn,o}}\left(\frac{4}{mn\pi^2}\Psi_{lb2}\Psi_{h2}\right)^2}_{E_{44}}\right\} \tag{A29}$$

$$= \rho_0 c\bar{v}E_4$$

where

$$E_4 = E_{41} + E_{42} + E_{43} + E_{44}. \tag{A30}$$

The quantity E_{41} is obtained with $m = 0$ and $n = 0$. E_{42} is obtained with $m \geq 1$ and $n = 0$. E_{43} is obtained with $m = 0$ and $n \geq 1$. Additionally, E_{44} is obtained with $m \geq 1$ and $n \geq 1$. The elements B and D can be calculated from Equations (19), (A18), (A19), (A27) and (A29) with the following equations

$$B = \frac{\rho_0 c\bar{v}E_3 - A\rho_0 c\bar{v}E_4}{v_o} = \frac{\rho_0 c\bar{v}E_3 - A\rho_0 c\bar{v}E_4}{\bar{v}} = \rho_0 c\left(E_3 - \frac{E_1 E_4}{E_2}\right) \tag{A31}$$

$$D = -\frac{C\rho_0 c\bar{v}E_4}{v_0} = -\frac{C\rho_0 c\bar{v}E_4}{\bar{v}} = \frac{E_4}{E_2}. \tag{A32}$$

Appendix B. FEM Methodology

Figure A1 shows the FEM model used the automatic matched layer (AML) method [33] in LMS Virtual.Lab software to calculate TL. Tetrahedral mesh (2 mm) was used to guarantee the calculation accuracy and the total number of grid cells of every REC was more than 400,000. The fluid material among the REC was defined as air, whose velocity was 340 m/s and density was 1.225 kg/m^3. Then, the outlet was AML property, which could simulate the nonreflecting boundary condition. The inlet acoustic boundary condition was defined as the plane wave with 1 W sound power.

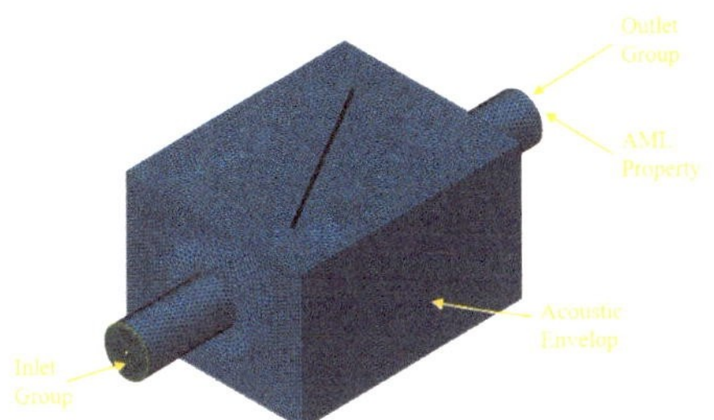

Figure A1. The FEM model of the REC.

References

1. Sun, Y.H.; Zhang, J.; Han, J.; Gao, Y.; Xiao, X.B. Acoustic transmission characteristics and optimum design of the wind ducts of high-speed train. *J. Mech. Eng. Chin. Ed.* **2018**, *54*, 129–137. (In Chinese) [CrossRef]
2. Sun, Y.H.; Zhang, J.; Han, J.; Gao, Y.; Xiao, X.B. Sound transmission characteristics of silencer in wind ducts of high-speed train. *J. Zhejiang Univ. Eng. Sci.* **2019**, *53*, 1389–1397. (In Chinese)
3. Gopalakrishnan, S.; Raut, M.S. Longitudinal wave propagation in one-dimensional waveguides with sinusoidally varying depth. *J. Sound Vib.* **2019**, *463*, 114945. [CrossRef]
4. Assis, G.F.C.A.; Beli, D.; Miranda, E.J.P., Jr.; Camino, J.F.; Dos Santos, J.M.C.; Arruda, J.R.F. Computing the complex wave and dynamic behavior of one-dimensional phononic systems using a state-space formulation. *Int. J. Mech. Sci.* **2019**, *163*, 105088. [CrossRef]
5. Liu, J.; Wang, T.; Chen, M. Analysis of sound absorption characteristics of acoustic ducts with periodic additional multi-local resonant cavities. *Symmetry* **2021**, *13*, 2233. [CrossRef]
6. Terashima, F.J.H.; de Lima, K.F.; Barbieri, N.; Barbieri, R.; Filho, N.L.M.L. A two-dimensional finite element approach to evaluate the sound transmission loss in perforated silencers. *Appl. Acoust.* **2022**, *192*, 108694. [CrossRef]
7. Junge, M.; Brunner, D.; Walz, N.-P.; Gaul, L. Simulative and experimental investigations on pressure-induced structural vibrations of a rear muffler. *J. Acoust. Soc. Am.* **2010**, *128*, 2782–2791. [CrossRef]
8. Liu, L.L.; Zheng, X.; Hao, Z.Y.; Qiu, Y. A computational fluid dynamics approach for full characterization of muffler without and with exhaust flow. *Phys. Fluids* **2020**, *32*, 066101.
9. Liu, L.L.; Hao, Z.Y.; Zheng, X.; Qiu, Y. A hybrid time-frequency domain method to predict insertion loss of intake system. *J. Acoust. Soc. Am.* **2020**, *148*, 2945. [CrossRef]
10. Liu, L.L.; Zheng, X.; Hao, Z.Y.; Qiu, Y. A time-domain simulation method to predict insertion loss of a dissipative muffler with exhaust flow. *Phys. Fluids* **2021**, *33*, 067114. [CrossRef]
11. Tolstoy, I. The WKB approximation, turning points, and the measurement of phase velocities. *J. Acoust. Soc. Am.* **1972**, *52*, 356–363. [CrossRef]
12. Subrahmanyam, P.B.; Sujith, R.I.; Lieuwen, T.C. A family of exact transient solutions for acoustic wave propagation in inhomogeneous, non-uniform area ducts. *J. Sound. Vib.* **2001**, *240*, 705–715. [CrossRef]
13. Li, J.; Morgans, A.S. The one-dimensional acoustic field in a duct with arbitrary mean axial temperature gradient and mean flow. *J. Sound. Vib.* **2017**, *400*, 248–269. [CrossRef]
14. Rani, V.K.; Rani, S.L. WKB solutions to the quasi 1-D acoustic wave equation in ducts with non-uniform cross-section and inhomogeneous mean flow properties—Acoustic field and combustion instability. *J. Sound. Vib.* **2018**, *436*, 183–219. [CrossRef]
15. Basu, S.; Rani, S.L. Generalized acoustic Helmholtz equation and its boundary conditions in a quasi 1-D duct with arbitrary mean properties and mean flow. *J. Sound. Vib.* **2021**, *512*, 116377. [CrossRef]
16. Eisner, E. Complete solutions of the "Webster" horn equation. *J. Acoust. Soc. Am.* **1967**, *41*, 1126–1146. [CrossRef]
17. Miles, J.H. Verification of a one-dimensional analysis of sound propagation in a variable area duct without flow. *J. Acoust. Soc. Am.* **1982**, *72*, 621–624. [CrossRef]
18. Lee, S.K.; Mace, B.R.; Brennan, M.J. Wave propagation, reflection and transmission in non-uniform one-dimensional waveguides. *J. Sound. Vib.* **2007**, *304*, 31–49. [CrossRef]
19. Martin, P.A. On Webster's horn equation and some generalizations. *J. Acoust. Soc. Am.* **2014**, *116*, 1381–1388. [CrossRef]
20. Donskoy, D.M. Directionality and gain of small acoustic velocity horns. *J. Acoust. Soc. Am.* **2017**, *142*, 3450–3458. [CrossRef]

21. Pagneux, V.; Amir, N.; Kergomard, J. A study of wave propagation in varying cross-section waveguides by modal decomposition. Part I. Theory and validation. *J. Acoust. Soc. Am.* **1996**, *100*, 2034–2048. [CrossRef]

22. Amir, N.; Pagneux, V.; Kergomard, J. A study of wave propagation in varying cross-section waveguides by modal decomposition. Part II. Results. *J. Acoust. Soc. Am.* **1997**, *101*, 2504–2517. [CrossRef]

23. Maurel, A.; Mercier, J.F.; Pagneux, V. Improved multimodal admittance method in varying cross section waveguides. *Proc. Math. Phys. Eng. Sci.* **2014**, *470*, 20130448. [CrossRef] [PubMed]

24. Pillai, M.A.; Ebenezer, D.D.; Deenadayalan, E. Transfer matrix analysis of a duct with gradually varying arbitrary cross-sectional area. *J. Acoust. Soc. Am.* **2019**, *146*, 4435–4445. [CrossRef] [PubMed]

25. Munjal, M.L. *Acoustics of Ducts and Mufflers*, 2nd ed.; John Wiley & Sons: New York, NY, USA, 2014.

26. Polyanin, A.D.; Zaitsev, V.F. *Handbook of Exact Solutions for Ordinary Differential Equations*; CRC Press: Boca Raton, FL, USA, 1995.

27. Fucci, G.; Kirsten, K. Expansion of infinite series containing modified Bessel functions of the second kind. *J. Phys. A Math. Theor.* **2015**, *48*, 435203. [CrossRef]

28. Torregrosa, A.J.; Broatch, A.; Payri, R. A study of the influence of mean flow on the acoustic performance of Herschel–Quincke tubes. *J. Acoust. Soc. Am.* **2000**, *107*, 1874–1879. [CrossRef]

29. Ih, J.G. The reactive attenuation of rectangular plenum chambers. *J. Sound. Vib.* **1992**, *157*, 93–122. [CrossRef]

30. Mechel, F.P. *Formulas of Acoustics*, 2nd ed.; Springer: Berlin/Heidelberg, Germany, 2008.

31. *ASTM E2611-09*; Standard Test Method for Measurement of Normal Incidence Sound Transmission of Acoustical Materials Based on the Transfer Matrix Method. ASTM International: West Conshohocken, PA, USA, 2009.

32. ANSYS Inc. *Ansys Fluent 19.1 Theory Guide*; ANSYS Inc.: Canonsburg, PA, USA, 2018.

33. Fu, J.; Chen, W.; Tang, Y.; Yuan, W.H.; Li, G.M.; Li, Y. Modification of exhaust muffler of a diesel engine based on finite element method acoustic analysis. *Adv. Mech. Eng.* **2015**, *7*, 1–2. [CrossRef]

MDPI AG

Grosspeteranlage 5

4052 Basel

Switzerland

Tel.: +41 61 683 77 34

Applied Sciences Editorial Office

E-mail: applsci@mdpi.com

www.mdpi.com/journal/applsci